Computational Fluid and Solid Mechanics

Series Editor:

Klaus-Jürgen Bathe
Massachusetts Institute of Technology
Cambridge, MA, USA

Advisors:

Franco Brezzi
University of Pavia
Pavia, Italy

Olivier Pironneau
Université Pierre et Marie Curie
Paris, France

Available Volumes

D. Chapelle, K.J. Bathe
The Finite Element Analysis of Shells – Fundamentals
2003

D. Drikakis, W. Rider
High-Resolution Methods for Incompressible and Low-Speed Flows
2005

M. Kojic, K.J. Bathe
Inelastic Analysis of Solids and Structures
2005

E.N. Dvorkin, M.B. Goldschmit
Nonlinear Continua
2005

B.Q. Li
Discontinuous Finite Elements in Fluid Dynamics and Heat Transfer
2006

Ben Q. Li

Discontinuous Finite Elements in Fluid Dynamics and Heat Transfer

With 167 Figures

 Springer

Author

Ben Q. Li, PhD
Professor of Mechanical Engineering
School of Mechanical and Materials Engineering
Washington State University
Pullman, Washington, USA

British Library Cataloguing in Publication Data
Li, Ben Q.
  Discontinuous finite elements in fluid dynamics and heat
  transfer. - (Computational fluid and solid mechanics)
  1. Fluid dynamics - Mathematical models 2. Heat -
  Transmission - Mathematical models 3. Finite element method
  4. Galerkin methods
  I. Title
  532'00151825
ISBN-10: 1852339888

Library of Congress Control Number: 2005935664

Computational Fluid and Solid Mechanics Series ISSN 1860-482X
ISBN-10: 1-85233-988-8   e-ISBN 1-84628-205-5   Printed on acid-free paper
ISBN-13: 978-1-85233-988-3

Printed in Germany

9 8 7 6 5 4 3 2 1

Springer Science+Business Media
springer.com

To my children: Thomas, Katherine and Lauren

Preface

Over the past several years, significant advances have been made in developing the discontinuous Galerkin finite element method for applications in fluid flow and heat transfer. Certain unique features of the method have made it attractive as an alternative for other popular methods such as finite volume and finite elements in thermal fluids engineering analyses.

This book is written as an introductory textbook on the discontinuous finite element method for senior undergraduate and graduate students in the area of thermal science and fluid dynamics. It also can be used as a reference book for researchers and engineers who intend to use the method for research in computational fluid dynamics and heat transfer. A good portion of this book has been used in a course for computational fluid dynamics and heat transfer for senior undergraduate and first year graduate students. It also has been used by some graduate students for self-study of the basics of discontinuous finite elements.

This monograph assumes that readers have a basic understanding of thermodynamics, fluid mechanics and heat transfer and some background in numerical analysis. Knowledge of continuous finite elements is not necessary but will be helpful. The book covers the application of the method for the simulation of both macroscopic and micro/nanoscale fluid flow and heat transfer phenomena. Background information on the subjects that are not covered in standard textbooks is also presented. Examples of different levels of difficulty are given, which help readers understand the concept and capability of the discontinuous finite element method and the computational procedures involved in the use of the method.

Chapter 1 of the book presents a brief review of fundamental laws and mathematical equations for thermal and fluid systems including both incompressible and compressible fluids and for generic boundary and initial conditions.

In Chapter 2, different approaches to formulate discontinuous finite element solutions for boundary and initial value problems are discussed. The numerical procedures for the discontinuous finite element formulation, elemental calculations and element-by-element solution are discussed in detail through simple, elementary and illustrative examples. The advantages and disadvantages of the discontinuous finite element formulations are also given.

Chapter 3 is concerned with the development of shape functions and elemental calculations for discontinuous finite elements. The Lagrangian basis functions, hierarchical shape functions, spectral elements and special elements are discussed. A majority of discontinuous finite element formulations use unstructured meshes made of triangular/tetrahedral elements. Construction of these elements from the Lagrangian interpolation functions or from coordinate transformations of the existing finite elements is presented. Numerical integration and elemental calculations are also given.

Starting with Chapter 4, we discuss the application of the discontinuous finite element methods for the solution of thermal and fluids problems. Chapter 4 deals with the heat conduction problems. Heat conduction is the first mode of heat transfer and the simple mathematical form of the governing equation serves as a good entry point for a numerical analysis of thermal problems. We present the detailed discontinuous formulations for both steady state and transient heat conduction problems. The stability analysis and selection of numerical fluxes for discontinuous finite element solution of diffusive systems are discussed.

Convection-dominant problems are discussed in Chapter 5, which covers pure convection, diffusion-convection, and inviscid and viscous nonlinear convection. The stability analysis and selection of various convection fluxes are also discussed. Both steady state and transient problems are considered. A good portion of the discussion is devoted to the numerical stability analysis and control of numerical oscillations.

Incompressible flow problems are discussed in Chapter 6, where the discontinuous finite element formulations for both isothermal and non-isothermal systems are given. The formulations are further used in the later chapters.

Chapter 7 is concerned about computational compressible flows using the discontinuous finite element method. The numerical procedure for both Euler and Navier–Stokes equations is presented. The use of various numerical fluxes, flux limiter and slope limiters in both 1-D and multidimensions is also discussed.

Chapter 8 discusses the discontinuous Galerkin boundary element method for the numerical solution of external radiation problems. Most numerical books on thermal and fluid flow analysis either have no or give very little coverage of the topic of radiation heat transfer. We present the discontinuous concept and its numerical implementation with the selection of kernel functions for external radiation calculations. Shadowing algorithms are discussed for detecting the internal blockages in 2-D, axisymmetric and 3-D enclosures.

Internal radiation occurs in many high temperature processes and is governed by the radiative transfer equation. The solution of the radiative transfer equation governing internal radiation is discussed in Chapter 9. This type of equation is difficult to solve using continuous finite elements but is almost ideal for discontinuous finite element computations. Detailed procedures for the numerical solution of the internal radiation problems are given. The analytical formulae for typical elements used for the discontinuous finite element formulation of 1-D, 2-D and 3-D simulations are given. Numerical examples include both simple pure internal radiation systems and complex thermal systems in which multiple heat transfer modes occur.

Chapter 10 discusses the use of the discontinuous finite element method for the solution of free and moving boundary problems. Both moving and fixed grid methods are discussed and the discontinuous finite element based algorithms for the solution of these problems are given. The concepts of the methods for moving boundary problems such as the volume of fluid method, the marker-and-cell method and the level set method are discussed. Incorporation of these fixed grid methods and moving grid methods into discontinuous finite element solvers using both structured and unstructured meshes are presented. The discontinuous finite element formulation of the phase field model, which has emerged as a powerful tool for modeling moving boundary problems at local scales, is also given, along with the 1-D, 2-D and 3-D examples of the evolution of very complex moving boundaries in phase change moving boundary problems.

The use of the discontinuous finite elements for the simulation of microscale and nanoscale heat transfer and fluid flow problems is discussed in Chapter 11. Some of the recently developed models describing the microscale heat transfer phenomena have mathematical forms that are particularly suited for the discontinuous finite element formulations. The numerical solution of non-Fourier heat transfer equations and the lattice Boltzmann equation is also given.

Chapter 12 deals with the discontinuous finite element solution of the thermal and fluid flow problems under the influence of applied electromagnetic fields. The basic theory of electromagnetism is presented. Numerical examples are given on the discontinuous finite element simulation of electroosmotive flows in microchannels, microwave heating and electrically-induced droplet deformation.

This book is printed in shades of grey. Color versions of some of the figures in this book can be downloaded in a pdf from springer.com. Computer codes used for some of the calculations may also be downloaded from the same web site.

Both the theory and applications of the discontinuous finite element method are still evolving and writing a book on this particular subject proved to be a major task. It was impossible to accomplish this task without assistance from various sources. I am most grateful to those whose contributions have made this monograph possible. I am indebted to Professors C.-W. Shu at Brown University and P. Castillo at University of Puerto Rico for helpful discussions on the mathematical theory of the discontinuous Galerkin finite element method. My appreciation also goes to Professor K. J. Bathe at Massachusetts Institute of Technology for sharing some of his latest work on the mixed finite element and the ALE methods and for stimulating comments. Professor H.-M. Yin of Department of Pure and Applied Mathematics, Washington State University, provided constructive comments on the basic theory of error analyses. I wish to thank my current and former graduate students, in particular, Drs. X. Ai, Y. Shu, B. Xu and X. Cui, who have helped in checking the examples and the exercises. I am also grateful to Ms. K. Faunce for her assistance in preparing the manuscript and formatting the final layout of the book, and to Messrs. A. Doyle and O. Jackson of Springer-Verlag and Ms. S. Moosdorf of LE-TeX for their continuous support.

Ben Q. Li
January, 2005

Contents

1

Introduction

In this chapter, the governing equations of fluid dynamics and heat transfer are described. The readers are assumed to have already acquired an adequate background in the area. Thus, a complete description of these governing equations is not included, but is well documented in the references at the end of the chapter. For convenient discussion in subsequent chapters, various forms of the governing equations used for discontinuous finite element formulations are also described. In general, fluid dynamics and heat transfer problems are classified as boundary value problems. Therefore, the standard boundary conditions and initial conditions required for the solution of these equations are given for various classes of problems.

1.1 Conservation Laws for a Continuum Medium

Conservation laws describe the physical principles governing the fluid motion and heat transfer in a continuum medium [1–26]. The continuum description is based on the basic continuum assumption that all macroscopic length and time scales are considerably larger than the largest molecular length and time scales. The mathematical formulations of these conservation laws are given below.

1.1.1 Conservation of Mass

Conservation of mass is a fundamental law governing the behavior of a continuum medium. It states that the total time rate of change of mass in a fixed region is identically zero. Physically, this means that the rate of change of the density of a fluid in motion is equal to the sum of the fluid convected into and out of the fixed region. Mathematically, the conservation of mass is described by the equation of continuity,

$$\frac{D\rho}{Dt} + \rho \nabla \cdot \mathbf{u} = 0 \qquad (1.1)$$

where ρ is the density of the fluid, $\mathbf{u}$ the velocity vector, and D/Dt the material derivative defined as follows:

$$\frac{D}{Dt} = \frac{\partial}{\partial t} + \mathbf{u} \cdot \nabla \tag{1.2}$$

Here ∇ is a vector differential operator,

$$\nabla = \hat{i}\frac{\partial}{\partial x} + \hat{j}\frac{\partial}{\partial y} + \hat{k}\frac{\partial}{\partial z} \tag{1.3}$$

with $\hat{i}$, $\hat{j}$, and $\hat{k}$ being the unit vectors. If the density is a constant or its material derivative is zero (that is, $D\rho/Dt = 0$), then the flow is incompressible and the equation of continuity is simplified to

$$\nabla \cdot \mathbf{u} = 0 \tag{1.4}$$

Mathematically, the above equation means that the velocity field of an incompressible flow is divergence free. This constraint is important in the numerical solution of the fluid dynamics equations.

1.1.2 Conservation of Momentum

The law of momentum conservation for a continuum medium is Newton's second law of motion. For a moving flow field, the law states that the total time rate of change of linear momentum or acceleration of a fluid element is equal to the sum of externally applied forces on a fixed region. Mathematically, Newton's second law is expressed as

$$\rho\frac{D\mathbf{u}}{Dt} = \nabla \cdot \mathbf{\sigma} + f_b \tag{1.5}$$

where $\mathbf{\sigma}$ is the Cauchy stress tensor and f_b is the body force per unit volume. The conservation of angular momentum leads to the symmetry condition on the stress tensor, that is, $\mathbf{\sigma} = \mathbf{\sigma}^T$.

1.1.3 Conservation of Energy

The law of conservation of energy is the first law of thermodynamics, which states that the time rate of change of the total energy is equal to the sum of work done by external forces and the change of heat content per unit time. For an incompressible fluid, the law of conservation of energy has the form of

$$\rho C_p \frac{DT}{Dt} = -\nabla \cdot \mathbf{q} + Q + \Phi \tag{1.6}$$

In the above equation, T is the temperature, $\mathbf{q}$ the flux, C_p the specific heat at constant pressure, and Φ the viscous dissipation,

$$\Phi = \boldsymbol{\tau} : \mathbf{D} \tag{1.7}$$

where $\boldsymbol{\tau}$ is the viscous part of the stress tensor $\boldsymbol{\sigma}$ and $\mathbf{D}$ the strain rate tensor. Physically, Φ represents the energy resulting from the friction between the fluid elements. The strain rate tensor is defined by the following relation:

$$\mathbf{D} = 0.5\left(\nabla \mathbf{u} + (\nabla \mathbf{u})^T\right) \tag{1.8}$$

In Equation 1.6, Q is the internal energy generation, which is a lumped sum of all source contributions,

$$Q = Q_s + Q_r + Q_R + Q_c \tag{1.9}$$

where Q_r is the heat source degenerated during chemical reactions; Q_R is the heat source resulting from internal radiation; $Q_c = -p\nabla \cdot \mathbf{u}$ is the energy source resulting from mechanical work, which is zero for incompressible fluids; and Q_s refers to all other applied heat sources.

1.1.4 Constitutive Relations

All materials are expected to satisfy the fundamental conservation principles of physics stated above. The dramatic differences in the behavior of different materials, such as solids, fluids, and viscoelastic materials, stem from the differences in the way that they resist deformation; or, more generally, in the way they respond when taken out of equilibrium. The mathematical specification of these "material response" laws is referred to as the set of constitutive relations for the material at hand. The mathematical expression of the constitutive relations is the statement of the dependence of the stress tensor and/or the heat flux $\mathbf{q}$ on the fields $\mathbf{D}(\mathbf{x},t)$, $T(\mathbf{x},t)$, and $\mathbf{u}(\mathbf{x},t)$. We note here that since the stress rate does not enter the constitutive relation for fluids, the frame independence principle is automatically satisfied [2].

In general, the Cauchy stress $\boldsymbol{\sigma}$ in a viscous fluid is decomposed into hydrostatic and viscous parts,

$$\boldsymbol{\sigma} = -p\mathbf{I} + \boldsymbol{\tau} \tag{1.10}$$

where p is the hydrostatic pressure (or the thermodynamic pressure) and $\mathbf{I}$ the unit tensor. The viscous stress tensor $\boldsymbol{\tau}$ is related to the strain rate tensor by the following constitutive relation:

$$\boldsymbol{\tau} = \Im : \mathbf{D} \tag{1.11}$$

where $\Im$ is the constitutive matrix and is a fourth order tensor.

For isotropic fluids, that is, fluids with properties independent of direction, the fourth order tensor is defined by two material constants, λ and μ. Consequently, the constitutive relation takes the Navier–Stokes relation, that is, the shear stress is linearly proportional to the strain rate,

$$\boldsymbol{\tau} = \lambda \, (\mathrm{tr}\mathbf{D})\mathbf{I} + 2\mu\mathbf{D} \tag{1.12}$$

Here the trace of the strain rate tensor, $\mathrm{tr}\mathbf{D}$, represents a volumetric deformation. For an incompressible fluid, $\mathrm{tr}\mathbf{D} = 0$.

For incompressible Newtonian fluids, the relation between the shear stress and the strain rate is given by Newton's hypothesis:

$$\boldsymbol{\tau} = 2\mu\mathbf{D} \tag{1.13}$$

where μ is the viscosity of the fluid. Note that for a solid the analogous linear constitutive relation is that of a Hookean solid.

The constitutive relation for heat conduction is the Fourier law, which states that the heat flux is proportional to the temperature gradient,

$$\mathbf{q} = -\boldsymbol{\kappa} \cdot \nabla T \tag{1.14}$$

where $\boldsymbol{\kappa}$ is the thermal conductivity tensor of order two. For an isotropic medium, $\boldsymbol{\kappa}$ is determined by a single constant,

$$\boldsymbol{\kappa} = \kappa \mathbf{I} \tag{1.15}$$

where κ is the thermal conductivity of the medium.

In general, the thermodynamic pressure of a fluid is a function of density and temperature, $p = p(\rho, T)$. For an ideal gas, the following relation exists:

$$p = \rho R T \tag{1.16}$$

where R is the gas constant.

For an incompressible fluid under non-isothermal conditions, the Boussinesq approximation is often used, which relates the density to the temperature in a linearized form,

$$\rho = \rho_0 (1 - \beta(T - T_r)) \tag{1.17}$$

where T_r is a reference temperature, ρ_0 is the density evaluated at T_r, and β denotes the thermal expansion coefficient,

$$\beta = \frac{1}{\rho_0} \frac{\partial \rho}{\partial T}\bigg|_{T_r} \tag{1.18}$$

In all the above equations, the fluid properties, such as C_p and μ, are a function of the fluid temperature and pressure, although the dependence on the latter is negligible for incompressible fluids. For non-Newtonian fluids, the viscosity and conductivity may also depend on the flow field, or strain rates, and temperature.

1.2 Governing Equations in Terms of Primitive Variables

For many fluid dynamics and heat transfer applications, the mathematical formulations described above are often written in either the vector form or in the component forms for various coordinate systems. These types of formulations are given below for convenience and will be used in subsequent chapters.

1.2.1 Vector Form

By directly applying the constitutive relations given in the last section to the conservation equations, one obtains the governing equations for fluid flow and heat transfer in primitive variables. For isotropic, Newtonian and incompressible fluids, the governing equations take the following familiar forms:

$$\nabla \cdot \mathbf{u} = 0 \tag{1.19}$$

$$\rho \frac{\partial \mathbf{u}}{\partial t} + \rho \mathbf{u} \cdot \nabla \mathbf{u} = -\nabla p + \nabla \cdot \mu \left(\nabla \mathbf{u} + (\nabla \mathbf{u})^T \right) + \boldsymbol{f} - \rho \mathbf{g} \beta (T - T_r) \tag{1.20}$$

$$\rho C_p \frac{\partial T}{\partial t} + \rho C_p \mathbf{u} \cdot \nabla T = \nabla \cdot \kappa \nabla T + Q + \Phi \tag{1.21}$$

where the Boussinesq approximation has been used and $\boldsymbol{f}$ is the body force excluding the gravitational force.

1.2.2 Component Form in Cartesian Coordinates

In the Cartesian coordinate system, the kinematic and constitutive relations are written in the indicial notation,

$$\sigma_{ij} = -p\delta_{ij} + \tau_{ij}; \qquad \tau_{ij} = 2\mu D_{ij} \tag{1.22}$$

$$D_{ij} = 0.5\left(\frac{\partial u_i}{\partial x_j} + \frac{\partial u_j}{\partial x_i} \right) \tag{1.23}$$

The conservation equations for mass, momentum, and energy now become

$$\frac{\partial u_i}{\partial x_i} = 0 \tag{1.24}$$

$$\rho \frac{\partial u_i}{\partial t} + \rho u_j \frac{\partial u_i}{\partial x_j} = -\frac{\partial p}{\partial x_i} + \frac{\partial}{\partial x_j}\left[\mu \left(\frac{\partial u_i}{\partial x_j} + \frac{\partial u_j}{\partial x_i} \right) \right] + f_i - \rho g_i \beta (T - T_r) \tag{1.25}$$

$$\rho C_p \frac{\partial T}{\partial t} + \rho C_p u_j \frac{\partial T}{\partial x_j} = \frac{\partial}{\partial x_j}\left(\kappa \frac{\partial T}{\partial x_j} \right) + Q + 2\mu D_{ij} D_{ij} \tag{1.26}$$

It is noted that the above equations are written in the Eulerian frame of reference, with indices $i, j = 1, 2, 3$. The Einstein convention on repeated indices is also applied.

1.2.3 Component Form in Cylindrical Coordinates

In a cylindrical coordinate system (r, θ, z), the vector differential operator and the material time derivative operator are defined by

$$\nabla = \hat{i}_r \frac{\partial}{\partial r} + \hat{i}_\theta \frac{1}{r}\frac{\partial}{\partial \theta} + \hat{i}_z \frac{\partial}{\partial z} \tag{1.27}$$

$$\frac{D}{Dt} = \frac{\partial}{\partial t} + u_r \frac{\partial}{\partial r} + \frac{u_\theta}{r}\frac{\partial}{\partial \theta} + u_z \frac{\partial}{\partial z} \tag{1.28}$$

where (u_r, u_θ, u_z) are the velocity components and $(\hat{i}_r, \hat{i}_\theta, \hat{i}_z)$ the unit vectors in the r, θ, and z directions, respectively. For curvilinear coordinates, the unit vectors are not necessarily constant. In fact, $\hat{i}_r$ and $\hat{i}_\theta$ are a function of angular coordinate θ,

$$\frac{\partial \hat{i}_r}{\partial \theta} = \hat{i}_\theta ; \quad \frac{\partial \hat{i}_\theta}{\partial \theta} = \hat{i}_r \tag{1.29}$$

The constitutive relations and the rate of strain tensors are

$$\sigma_{rr} = -p + 2\mu D_{rr} ; \quad \sigma_{\theta\theta} = -p + 2\mu D_{\theta\theta} ; \quad \sigma_{zz} = -p + 2\mu D_{zz} \tag{1.30}$$

$$\sigma_{r\theta} = 2\mu D_{r\theta} ; \quad \sigma_{\theta z} = 2\mu D_{\theta z} ; \quad \sigma_{zr} = 2\mu D_{zr} \tag{1.31}$$

$$D_{rr} = \frac{\partial u_r}{\partial r}; \quad D_{r\theta} = \frac{1}{2}\left(\frac{\partial u_\theta}{\partial r} - \frac{u_\theta}{r} + \frac{1}{r}\frac{\partial u_r}{\partial \theta}\right); \quad D_{\theta\theta} = \frac{1}{r}\frac{\partial u_\theta}{\partial \theta} + \frac{u_r}{r} \quad (1.32)$$

$$D_{zz} = \frac{\partial u_{zz}}{\partial z}; \quad D_{zr} = \frac{1}{2}\left(\frac{\partial u_r}{\partial z} + \frac{\partial u_z}{\partial r}\right); \quad D_{\theta z} = \frac{1}{2}\left(\frac{\partial u_\theta}{\partial z} + \frac{1}{r}\frac{\partial u_z}{\partial \theta}\right) \quad (1.33)$$

With the above notations, the governing equations for the cylindrical coordinate system can be written in the following form:

$$\frac{\partial u_r}{\partial r} + \frac{1}{r}\frac{\partial u_\theta}{\partial \theta} + \frac{\partial u_z}{\partial z} = 0 \quad (1.34)$$

$$\rho\left(\frac{Du_r}{Dt} - \frac{u_\theta^2}{r}\right) = f_r + \frac{1}{r}\left[\frac{\partial(r\sigma_{rr})}{\partial r} + \frac{\partial\sigma_{r\theta}}{\partial \theta} + \frac{\partial(r\sigma_{zr})}{\partial z}\right] - \frac{\sigma_{\theta\theta}}{r}$$
$$- \rho g_r \beta(T - T_r) \quad (1.35)$$

$$\rho\left(\frac{Du_\theta}{Dt} + \frac{u_r u_\theta}{r}\right) = f_\theta + \frac{1}{r}\left[\frac{\partial(r\sigma_{r\theta})}{\partial r} + \frac{\partial\sigma_{\theta\theta}}{\partial \theta} + \frac{\partial(r\theta_{\theta z})}{\partial z}\right] + \frac{\sigma_{r\theta}}{r}$$
$$- \rho g_\theta \beta(T - T_r) \quad (1.36)$$

$$\rho\left(\frac{Du_z}{Dt}\right) = f_z + \frac{1}{r}\left[\frac{\partial(r\sigma_{zr})}{\partial r} + \frac{\partial\sigma_{z\theta}}{\partial \theta} + \frac{\partial(r\sigma_{zz})}{\partial z}\right] - \rho g_z \beta(T - T_r) \quad (1.37)$$

$$\rho C_p\left(\frac{\partial T}{\partial t} + u_r\frac{\partial T}{\partial r} + \frac{u_\theta}{r}\frac{\partial T}{\partial \theta} + u_z\frac{\partial T}{\partial z}\right)$$
$$= \frac{1}{r}\frac{\partial}{\partial r}\left(rk_{rr}\frac{\partial T}{\partial r}\right) + \frac{1}{r^2}\frac{\partial}{\partial \theta}\left(k_{\theta\theta}\frac{\partial T}{\partial \theta}\right) + \frac{\partial}{\partial z}\left(k_{zz}\frac{\partial T}{\partial z}\right) + \Phi + Q \quad (1.38)$$

where the stress components are known in terms of the velocity components, and the viscous dissipation Φ may be explicitly written as

$$\Phi = 2\mu\left[\left(\frac{\partial u_r}{\partial r}\right)^2 + \left(\frac{1}{r}\frac{\partial u_\theta}{\partial \theta} + \frac{u_r}{r}\right)^2 + \left(\frac{\partial u_z}{\partial z}\right)^2\right] + \mu\left(\frac{\partial u_\theta}{\partial r} - \frac{u_\theta}{r} + \mu\frac{\partial u_r}{\partial \theta}\right)^2$$
$$+ \mu\left(\frac{1}{r}\frac{\partial u_z}{\partial \theta} + \frac{\partial u_\theta}{\partial z}\right)^2 + \mu\left(\frac{\partial u_r}{\partial z} + \frac{\partial u_z}{\partial r}\right)^2 \quad (1.39)$$

1.2.4 Summary

The fundamental equations presented above, in terms of primitive variables (velocity, pressure and temperature), are most commonly used in computational fluid dynamics and heat transfer. Other simplified forms of description for specific cases are also used. For example, equations utilizing the stream function are employed along with the energy equation to analyze convection problems in 2-D and axisymmetric geometries. The advantage of the stream function formulation is that the pressure does not need to be calculated directly during the solution phase. In discontinuous finite element literature, the fluid dynamics and heat transfer equations sometimes are written in conservation forms to facilitate computations. These conservation forms are particularly useful for the solution of compressible flow problems and will be discussed later in the chapter. For most of this book, the primitive variable formulations are used, as they are convenient for problems involving free surfaces, phase changes and regions with multiple connections.

1.3 Species Transport Equations

For flows in which transport processes for other species present in the system are important, scalar transport equations are needed. Because of convection, these equations are classified into the advection-diffusion category and are generic in the sense that they are not specially associated with a particular physical process. These equations are often used to simulate certain types of chemical reactions, and to predict the volume fraction of particle orientation for flows containing suspended particles or fibers. It is often convenient to write the transport equation for a scalar quantity C_i ($i = 1, 2, ..., n$) as follows:

$$\rho \left(\frac{\partial C_i}{\partial t} + u_j \frac{\partial C_i}{\partial x_j} \right) = \frac{\partial}{\partial x_j} \left(D_i \frac{\partial C_i}{\partial x_j} \right) + R_i \qquad (1.40)$$

where D_i is a diffusion coefficient and R_i is a volumetric source term for C_i, which may be attributed to chemical reactions or other mechanisms.

The presence of chemical species may change the density of the fluid and hence the flow field. To account for this effect in a non-isothermal multi-component flow system, a generic Boussinesq approximation may also be used to represent the buoyancy forces caused by variations in the auxiliary variables C_i,

$$\rho = \rho_0 [1 - \beta(T - T_r) - \beta_1(C_1 - C_{1,r}) \cdots - \beta_n(C_n - C_{n,r})] \qquad (1.41)$$

where $\beta_i = (1/\rho_0)(\partial \rho / \partial C)|_{C_{i,r}}$ is a scalar expansion coefficient and subscript r refers to a reference condition at which ρ_0 is evaluated. This variation in density is only permitted in the body force term and is used in place of the buoyancy term in Equation 1.20 for a multi-component system.

1.4 Governing Equations in Translating and Rotating Frames of Reference

For some applications, the external boundary to a fluid is also in motion. It is often more convenient to write the equation in a frame of reference in which the boundary is at rest. Cases like these happen when the moving frame of reference is translating with a given linear velocity $\mathbf{U}$, or rotating with a given angular velocity Ω. The equations for these cases can be easily reduced from the equations written for a frame of reference with arbitrary acceleration (or non-inertial frame). For momentum equations, the absolute acceleration of a fluid element in a fixed reference frame related to a reference frame with an arbitrary acceleration in the following fashion [7]:

$$\frac{D\mathbf{u}_f}{Dt} = \frac{D\mathbf{u}_{rf}}{Dt} + \frac{D\mathbf{u}_r}{Dt} + 2\Omega \times \mathbf{u}_r + \Omega \times (\Omega \times \mathbf{u}_r) + \frac{\partial \Omega_r}{\partial t} \times \mathbf{r} \qquad (1.42)$$

where subscripts f and r refer to fixed and arbitrary coordinates, and $\mathbf{u}_{rf}$ is the velocity with which the non-inertial frame moves relative to the fixed frame. Note that the derivative on the right refers to the arbitrary coordinate system, but not to the non-inertial frame. Thus, when the reference frame translates with a constant linear velocity $\mathbf{U}$, $\mathbf{u}_{rf} = \mathbf{U}$. For this case, the governing equations remain invariant, and only the boundary conditions are translated. In the case of a frame rotating with a constant, but without the translation $\mathbf{u}_{rf} = \mathbf{0}$, the acceleration of a fluid particle, with respect to this rotating frame of reference, becomes

$$\frac{D\mathbf{u}_f}{Dt} = \frac{\partial \mathbf{u}_r}{\partial t} + \mathbf{u}_r \cdot \nabla_r \mathbf{u}_r + 2\Omega \times \mathbf{u}_r + \Omega \times (\Omega \times \mathbf{u}_r) \qquad (1.43)$$

where ∇_r means the derivative with respect to the coordinates of the rotating frame. Thus, by substituting the above relation into the governing equations given in the above sections, one has the equations written for the rotating frame of reference in terms of the variable $\mathbf{u}_r$.

1.5 Boundary and Initial Conditions

The governing equations presented above will not have unique solutions unless appropriate boundary and initial conditions are specified. These conditions are, in essence, the physical constraints associated with specific thermal and fluids systems and are described below [1–26]. While specific forms may differ from application to application, the boundary conditions are derived based on the conservation principles: that is, the principles of mass, momentum and energy conservation, and the principle of thermodynamic equilibrium, applied across the boundary or interfaces. The jump conditions across a shock wave are discussed in Section 1.7, in the context of conservation forms of governing equations.

1.5.1 General Boundary Conditions

On each segment of the boundary of a computational domain Ω, it is necessary to prescribe appropriate boundary conditions. The boundary conditions relating to the momentum equation are either the specification of velocity components,

$$\mathbf{u} = \mathbf{u}(s,t) \tag{1.44}$$

or the specification of surface stresses,

$$\boldsymbol{\sigma}_n = \boldsymbol{\sigma} \cdot \mathbf{n}(s,t) = \boldsymbol{\sigma}_a(s,t) \tag{1.45}$$

where t is time, s is a parameter measuring position along the relevant boundary segment, and $\mathbf{n}(s,t)$ is the outward unit normal to the boundary. Note that $\boldsymbol{\sigma}_n$ and $\boldsymbol{\sigma}_a$ are vectors. Subscript a denotes a prescribed quantity. This use of the subscript is the same hereafter in this chapter until otherwise indicated.

For the solution of the temperature field, the thermal balance on the boundary demands that either the temperature profile is specified,

$$T = T_a(s,t) \tag{1.46}$$

or a heat flux is prescribed,

$$q = -\mathbf{n} \cdot k \nabla T = q_a(s,t) \tag{1.47}$$

where q_a includes contributions from both convective and/or radiative heat transfer along the boundary.

The boundary conditions relating to the chemical species are obtained based on the mass balance across the boundary, which leads to the specification of the concentration boundary conditions similar to those of temperatures,

$$C_i = C_{i,a}(s,t) \tag{1.48}$$

or to the prescription of the mass flux,

$$q_i = -\mathbf{n} \cdot D_i \nabla C_i = q_{i,a}(s,t,C_{i,a}) \tag{1.49}$$

For a mixture of M components, the boundary conditions take the form of

$$-\mathbf{n} \cdot D_i \nabla C_i + \alpha_T^{(i)} T \mathbf{n} \cdot \nabla T + C_i \mathbf{n} \cdot \mathbf{u} = r_i, \quad i = 1,..., M-1 \tag{1.50}$$

and

$$\rho \mathbf{u} \cdot \mathbf{n} = \sum_{n=1}^{M} r_n \tag{1.51}$$

where $\alpha_T^{(i)}$ is the thermal diffusion coefficient, and r_i is the rate of chemical reaction for species i.

Initial conditions for all the primitive variables take the following generic forms:

$$\mathbf{u}(\mathbf{x},0) = u^0(\mathbf{x}) \tag{1.52}$$

$$T(\mathbf{x},0) = T^0(\mathbf{x}) \tag{1.53}$$

$$C_i(\mathbf{x},0) = C_i^0(\mathbf{x}) \tag{1.54}$$

where superscript 0 stands for a prescribed initial quantity.

1.5.2 Free Boundary Conditions

A free boundary refers to the interface between a liquid and a gas phase, when the latter can be approximated as an effective vacuum. Situations like these occur where the gas phase has little or no motion. The boundary conditions along the interface can be derived from the kinematic continuity and mechanical equilibrium principles. The former requires that the interface remains a free surface, while the latter demands that the stress is continuous across the interface. Mathematically, these conditions are stated as follows:

$$S(\mathbf{x},t) = 0 \tag{1.55}$$

$$\frac{\partial S}{\partial t} + \mathbf{u} \cdot \nabla S = 0 \tag{1.56}$$

$$\boldsymbol{\sigma}_n = \boldsymbol{\sigma} \cdot \mathbf{n} = (2\gamma H_c - p_{amb})\mathbf{n} + \nabla_s \gamma \tag{1.57}$$

where S defines the free surface. Comparison with Equation 1.45 shows that the right hand terms replace $\boldsymbol{\sigma}_a$, and represent the gas phase effect. In Equation 1.57, p_{amb} is the ambient pressure in the gas, H_c is the mean curvature of the surface, and γ is the surface tension. Also, $\nabla_s = \nabla - \mathbf{n}(\mathbf{n} \cdot \nabla)$ is the surface vector differential operator [22, 27].

When contact lines (liquid–gas–solid interfaces) are present, an additional condition is needed to model the dynamics of the contact line. A commonly used approximation is the Navier slip condition on the liquid at the solid surface,

$$a \, \mathbf{t} \cdot \boldsymbol{\sigma} \cdot \mathbf{n} = (\mathbf{u} - \mathbf{u}_i^s) \cdot \mathbf{t} \tag{1.58}$$

Here $\mathbf{t}$ is the tangential vector to the surface, $\mathbf{u}_i^s$ is the velocity of the solid surface, and a is a constant of proportionality known as the slip coefficient.

1.5.3 Moving Interface Conditions

Across a moving interface between two fluids, the kinematic condition and the condition of stress continuity apply. In addition, a no-slip condition at the interface is required. With superscripts 1 and 2 denoting the two liquids across the interface, the interface boundary conditions are

$$S(\mathbf{x}, t) = 0 \tag{1.59}$$

$$\frac{\partial S}{\partial t} + \mathbf{u}^1 \cdot \nabla S = \frac{\partial S}{\partial t} + \mathbf{u}^2 \cdot \nabla S = 0 \tag{1.60}$$

$$\sigma_n^1 = \sigma_n^2 + 2\gamma H_c \mathbf{n} + \nabla_s \gamma \tag{1.61}$$

$$\mathbf{n} \times (\mathbf{u}^1 - \mathbf{u}^2) = 0 \tag{1.62}$$

where the normal $\mathbf{n}$ points from liquid 1 into liquid 2.

1.5.4 Phase Change Conditions

An important category of moving boundary problems in fluid dynamics and heat transfer is phase change problems, which describe the phase transition from liquid to solid, or liquid to gas, or *vice versa*. For these problems, the mass continuity, mechanical equilibrium and energy balance equations must also be satisfied across the interface. With superscripts L and S denoting the liquid and solid respectively, these conditions are

$$T^L = T^S = T^m - 2H_c \Gamma_G \tag{1.63}$$

$$k^L \mathbf{n} \cdot \nabla T^L - k^S \mathbf{n} \cdot \nabla T^S = \rho^S L \mathbf{n} \cdot (\mathbf{u}^S - \mathbf{u}^m) \tag{1.64}$$

$$\rho^L \mathbf{n} \cdot (\mathbf{u}^L - \mathbf{u}^m) = \rho^S \mathbf{n} \cdot (\mathbf{u}^S - \mathbf{u}^m) \tag{1.65}$$

$$\mathbf{n} \times (\mathbf{u}^L - \mathbf{u}^S) = 0 \tag{1.66}$$

Here T^m is the phase change temperature, $\mathbf{u}^m$ is the velocity of the interface; and $\Gamma_G = \gamma / \Delta s_f$ is the Gibbs–Thompson coefficient, with Δs_f being the entropy change between the liquid and solid per unit volume; and L is the latent heat per unit mass.

The normal **n** points from the liquid into the solid. If solidification involves solutal elements, their effects on the interface shape and the phase change temperature need to be included as well:

$$C^S \mathbf{n} \cdot (\mathbf{u}^S - \mathbf{u}^m) - (\rho_L / \rho_S) C^L \mathbf{n} \cdot (\mathbf{u}^L - \mathbf{u}^m)$$

$$= -(\rho_L / \rho_S) D^L \mathbf{n} \cdot \nabla C^L + D^S \mathbf{n} \cdot \nabla C^S \qquad (1.67)$$

$$C^S = \kappa C^L \qquad (1.68)$$

$$T^L = T^S = T^m + m C^L - 2 H_c \Gamma_G \qquad (1.69)$$

Thus, Equation 1.69 is used in place of Equation 1.63, and Equation 1.67 replaces Equation 1.65. In the above equations, $m = dT/dC^L$ is the rate of change of the melting temperature with concentration. It is important to note that Equations 1.68 and 1.69 are obtained from the thermodynamic equilibrium principle governing the behavior of materials at the liquid-solid interface.

In the case of liquid to vapor transition, some modifications are needed, though the basic principles are the same. The modification is to replace the above quantities for the solid phase (denoted by subscript S) in Equations 1.63–1.66 with those for the gas phase. Also, Equation 1.61 is used to satisfy mechanical equilibrium. In addition, the requirement of the equal chemical potential at the liquid–gas interface leads to the Clausius–Clapeyron expression for the relation between the pressure and the liquid–gas transition temperature:

$$\frac{dp}{dT^m} = \frac{C_m}{T^m} \qquad (1.70)$$

where $C_m = L/(v_g - v_l)$, and v_g and v_l are the specific volumes of gas and liquid, respectively.

1.6 Governing Equations for Flows Through Porous Media

The above equations are written for single-phase flows. Many engineering processes involve the movement of gases and liquids through porous media. Typical examples include distillation and absorption columns that are filled with various types of particles, fuel cells with porous electrodes and package beds and/or woods being dried. A complete discussion on the derivation of volume-averaged governing equations is given by Bear [16]. Here, the main ideas and mathematical formulations are summarized.

Let us suppose that within the domain of interest Ω, there is a region V containing a rigid porous material saturated with a viscous incompressible fluid, and a region Ω_f occupied entirely by fluid. The saturating fluid in V is the same as in Ω_f, if the two regions share a common permeable interface. Otherwise, the two fluids may be different. If V_f is the volume occupied by the fluid, and V_s is the volume occupied by the solid, then $V = V_s + V_f$; and the porosity of the porous medium is $\phi = V_f / V_s$. Assuming that the porous medium is homogeneous and isotropic and the fluid and solid are in local thermal equilibrium, the field variables can be averaged over V_f and V, with the former referring to the pore average and the latter, volume average. The averaging process gives rise to the following forms of the governing equations:

$$\frac{\partial \rho}{\partial t} + \nabla \cdot (\rho \mathbf{u}) = 0 \tag{1.71}$$

$$\frac{\rho}{\phi}\frac{\partial \mathbf{u}}{\partial t} + \left(\rho \hat{c} \|\mathbf{u}\|^m \mathbf{K}_{-1/2} + \mu \mathbf{K}_{-1} \right) \cdot \mathbf{u}$$
$$= -\nabla p + \nabla \cdot \left(\bar{\mu} (\nabla \mathbf{u} + (\nabla \mathbf{u})^T) \right) + \rho \mathbf{f} \tag{1.72}$$

$$(\rho c_p)_e \left(\frac{\partial T}{\partial t} + \mathbf{u} \cdot \nabla T \right) = \nabla \cdot (k_e \nabla T) + Q \tag{1.73}$$

$$\rho \left(\frac{\partial C}{\partial t} + \mathbf{u} \cdot \nabla C \right) = \nabla \cdot (\rho D)_e \nabla c + q_c + R \tag{1.74}$$

where the pore average and volume average quantities are denoted by an overbar and over-hat respectively,

$$\bar{a} = \frac{1}{V_f} \int_{V_f} a dV, \quad \hat{a} = \frac{1}{V} \int_V a dV \tag{1.75}$$

with a being any quantity (scalar, vector or tensor). Note here that in the above equations, the "^" on $\mathbf{u}$, p, and T were dropped for simplicity. Also, $\mathbf{K}$ is the permeability, a tensor of second order; it may take the following values: $K_{-1,ij} = (1/k_i)\delta_{ij}$, and $K_{-1/2,ij} = (1/\sqrt{k_i})_{ij}$; $\hat{c}$ is the inertia coefficient; $\bar{\mu}$ is an effective viscosity; and $\|u_i\|$ is the magnitude of the velocity. In the above equations, subscript e indicates an effective property, which is a function of porosity and the properties of fluids and solids and is calculated by the following expressions:

$$(\rho c_p)_e = \phi(\rho c_p)_f + (1-\phi)(\rho c_p)_s \tag{1.76}$$

$$k_e = \phi(k)_f + (1-\phi)k_s \tag{1.77}$$

where subscript s refer to solid matrix properties. Properties without subscripts are those of fluids.

The above equations represent a generalization of the standard Darcy equations for non-isothermal flows in a saturated porous medium. This system is sometimes referred to as the Forchheimer–Brinkman model for porous flows. With an appropriate selection of the coefficients, a number of standard flow models can be derived. For example, if $\hat{c} = 0$, a Brinkman model is obtained, while if $\bar{\mu} = \hat{c} = 0$, the standard Darcy formulation is recovered. The porous flow equations are very similar in form to the equations of a viscous fluid flow, except that the convection terms in the momentum equation are replaced by the Darcy–Forchheimer term.

1.7 Governing Equations in Conservation Form

While the governing equations given above are widely used in incompressible fluid flow applications, for compressible fluid flows, the governing equations often are written in the conservation form. By definition, a system of partial differential equations assumes a conservation form if it can be written as follows:

$$\frac{\partial \mathbf{u}}{\partial t} + \nabla \cdot \mathbf{F}(\mathbf{u}) = \mathbf{J}, \qquad \mathbf{u}(\mathbf{x},0) = \mathbf{u}_0(\mathbf{x}), \qquad \mathbf{x} \in \Omega, \ t > 0 \qquad (1.78)$$

where $\mathbf{F}$ is a tensor and $\mathbf{J}$ a vector. Systems not written in this form are non-conservative systems. By this definition, all the governing equations presented in the above sections, either in indicial, component, or vector notation, are in non-conservation form.

The conservation laws of mass, momentum and energy can be recast in the following conservation form:

$$\frac{\partial U}{\partial t} + \frac{\partial F}{\partial x} + \frac{\partial G}{\partial y} + \frac{\partial H}{\partial z} = J \qquad (1.79)$$

where U, F, G, H, and J are column vectors defined by

$$U = \left\{ \begin{array}{c} \rho \\ \rho u \\ \rho v \\ \rho w \\ \rho(e + (\mathbf{u} \cdot \mathbf{u})/2) \end{array} \right\} \qquad (1.80)$$

$$F = \left\{ \begin{array}{c} \rho u \\ \rho u^2 + p - \tau_{xx} \\ \rho v u - \tau_{xy} \\ \rho w u - \tau_{xz} \\ \rho(e + (\mathbf{u} \cdot \mathbf{u})/2)u + pu - k\, \partial T/\partial x - u\tau_{xx} - v\tau_{xy} - w\tau_{xz} \end{array} \right\} \qquad (1.81)$$

$$G = \left\{ \begin{array}{c} \rho v \\ \rho uv - \tau_{yx} \\ \rho v^2 + p - \tau_{yy} \\ \rho wv - \tau_{yz} \\ \rho(e + (\mathbf{u} \cdot \mathbf{u})/2)v + pv - k\,\partial T/\partial y - u\tau_{yx} - v\tau_{yy} - w\tau_{yz} \end{array} \right\} \qquad (1.82)$$

$$H = \left\{ \begin{array}{c} \rho w \\ \rho uw - \tau_{zx} \\ \rho vw - \tau_{zy} \\ \rho w^2 + p - \tau_{zz} \\ \rho(e + (\mathbf{u} \cdot \mathbf{u})/2)w + pw - k\,\partial T/\partial z - u\tau_{zx} - v\tau_{zy} - w\tau_{zz} \end{array} \right\} \qquad (1.83)$$

$$J = \left\{ \begin{array}{c} 0 \\ \rho f_x \\ \rho f_y \\ \rho f_z \\ \rho(u f_x + v f_y + w f_z) + \rho \dot{q} \end{array} \right\} \qquad (1.84)$$

In the above equations, the shear stresses τ_{ij} ($i, j = x, y, z$) are calculated by the following expressions:

$$\tau_{xx} = \frac{2}{3}\mu\left(2\frac{\partial u}{\partial x} - \frac{\partial v}{\partial y} - \frac{\partial w}{\partial z}\right); \qquad \tau_{yy} = \frac{2}{3}\mu\left(-\frac{\partial u}{\partial x} + 2\frac{\partial v}{\partial y} - \frac{\partial w}{\partial z}\right)$$

$$\tau_{zz} = \frac{2}{3}\mu\left(-\frac{\partial u}{\partial x} - \frac{\partial v}{\partial y} + 2\frac{\partial w}{\partial z}\right); \qquad \tau_{xy} = \mu\left(\frac{\partial v}{\partial x} + \frac{\partial u}{\partial y}\right) = \tau_{yx}$$

$$\tau_{xz} = \mu\left(\frac{\partial w}{\partial x} + \frac{\partial u}{\partial z}\right) = \tau_{zx}; \qquad \tau_{yz} = \mu\left(\frac{\partial w}{\partial y} + \frac{\partial v}{\partial z}\right) = \tau_{zy}$$

Here, some explanations of these terms may be helpful. The column vectors F, G and H are referred to as the flux vectors, and J represents a source term. The velocity vector $\mathbf{u}$ has three components: u, v, and w. The column vector U is the solution vector. In writing this generic equation in terms of column vectors, we note that the first elements of the U, F, G, H and J vectors, when added together via Equation 1.79, reproduce the continuity equation. The second elements of the U, F, G, H and J vectors, when added together via Equation 1.79, reproduce the x-momentum equation, and so forth.

When the viscous effects are neglected, $\tau_{ij} = 0$ $(i, j = x, y, z)$ and the above equations reduce to the Euler equations for inviscid fluids. The Euler equations are often used to predict the shock wave behavior in high speed flows. By definition, a shock wave is a surface of discontinuity (in the field variables), which is capable of propagating relative to the material. Common examples of shock waves include the crash of thunder, the crack of a whip, or the sound of a gunshot or firecracker. The jump condition across a shock wave can be obtained by integrating the balance equations in the conservation form, and for an inviscid fluid the jump condition is given by

$$[\rho \mathbf{V} \cdot \mathbf{n}] = 0 \tag{1.85}$$

$$[p + m\mathbf{V} \cdot \mathbf{n}] = 0 \tag{1.86}$$

$$[\mathbf{V} \times \mathbf{n}] = 0 \tag{1.87}$$

$$[h + 0.5(\mathbf{V} \cdot \mathbf{n})^2] = 0 \tag{1.88}$$

where $V = \mathbf{u} - \mathbf{v}$ is the velocity of the fluid relative to the shock surface, $\mathbf{v}$ is the velocity of the shock surface, $m = \rho_1 \, V \cdot \mathbf{n}_1 = \rho_2 \, V \cdot \mathbf{n}_2$ is the mass flux through the shock surface, and $h = e + p/\rho$ is the fluid enthalpy. The brackets denote the jump in the indicated quantity across the shock, $[A] = A_2 - A_1$, where A is any quantity and subscripts 1 and 2 denote conditions just before and just after the shock.

The conservation form of the governing equations provides a convenience in numerical computation in that the continuity, momentum and energy equations can all be expressed in the same generic equation. This is useful for developing numerical schemes and integral solutions to the equations. In fact, one can easily derive the integral control volume equations from the local (i.e., differential) equations once they have been placed in the conservation form. An important advantage of writing the governing equations in the conservation form is even more compelling when it comes to the compressible flow calculations. With the use of the conservation form, the shock capture scheme has a better numerical quality because the changes in the flux variables are zero across the shock waves.

Exercises

1. Set up a differential volume in a moving fluid and derive the mass conservation equation.
2. For the same differential volume, apply Newton's second law to derive the momentum balance equation.
3. For the same differential volume, apply the first law of thermodynamics to derive the energy balance equation.
4. Derive the differential form for axisymmetric flow field by simplifying the governing equations written in the cylindrical coordinate system.

5. Show that the conservation law forms of governing equations reduce to the Navier–Stokes equations, when the incompressibility condition is enforced.
6. Derive the boundary force balance equation for a curved liquid-liquid interface.
7. Starting with the Euler equations, derive the jump conditions across a shock wave.

References

[1]   Batchelor GK. An Introduction to Fluid Dynamics. London: Cambridge University Press, 1967.
[2]   Malvern LE. Introduction to the Mechanics of a Continuous Medium. Englewood Cliffs: Prentice Hall, 1969.
[3]   Aris R. Vectors, Tensors and the Basic Equations of Fluid Mechanics. New York: Dover Publications, Inc., 1962.
[4]   Carslaw HS, Jaeger JC. Conduction of Heat in Solids. Oxford University: Clarendon Press, 1959.
[5]   Bird RB, Stewart WE, Lightfoot EN. Transport Phenomena. New York: John Wiley & Sons, 1960.
[6]   Panton RL. Incompressible flow. New York: John Wiley, 1984.
[7]   Fox RW. McDonald AT. Introduction to Fluid Mechanics. New York: John Wiley and Sons, 1992.
[8]   Schlichting H. Boundary-Layer Theory. (translated by J. Kestin), 7th Ed, New York: McGraw-Hill, 1979.
[9]   White F. Viscous Fluid Flows. New York: McGraw Hill, 1974.
[10]   Kays WM, Crawford ME. Convective Heat and Mass Transfer. New York: McGraw Hill, 1980.
[11]   Gray DD, Giorgini A. The Validity of the Boussinesq Approximation for Liquids and Gases. Int. J. Heat Mass Trans. 1976; 19: 545–551.
[12]   Kaviany M. Principles of Heat Transfer in Porous Media. New York: Springer-Verlag, 1991.
[13]   Nield DA, Joseph DD. Effects of Quadratic Drag on Convection in a Saturated Porous Medium. Phys. Fluids 1985; 28: 995–997.
[14]   Nield DA. The Limitations of the Brinkman–Forchheimer Equation in Modeling Flow in a Saturated Porous Medium and at an Interface. Int. J. Heat Fluid Flow 1991; 12: 269–272.
[15]   Beavers GS, Joseph DD. Boundary Conditions at a Naturally Permeable Wall. J. Fluid Mech. 1967; 30: 197–207.
[16]   Bear J. Dynamics of Fluids in Porous Media. New York: American Elsevier, 1972.
[17]   Bonacini C, Comini G, Fasano A, Primicerio M. Numerical Solution of Phase-Change Problems. Int. J. Heat Mass Trans. 1973; 16: 1825–1832.
[18]   Comini G, DelGuidice S, Lewis R, Zienkiewicz OC. Finite Element Solution of Non-Linear Heat Conduction Problems with Special Reference to Phase Change. Int. J. Numer. Methods Eng. 1974; 8: 613–624.
[19]   Morgan K. A Numerical Analysis of Freezing and Melting with Convection. Comput. Meth. Appl. Mech. Eng. 1981; 28: 275–284.
[20]   Li BQ, Song SP. Finite Element Computation of Shapes and Marangoni Convection in Droplets Levitated by Electric Fields. Int. J. Comp. Fluid Dyn. 2001; 15: 293–308.
[21]   Li K, Li BQ. A 3-D Model for G-Jitter Driven Flows and Solidification in Magnetic Fields. J. Thermophys. Heat Transf. 2003; 17(4): 498–508.

[22] Huo Y, Li BQ. Three-Dimensional Marangoni Convection in Electrostatically Positioned Droplets under Microgravity. Int. J. Heat Mass Transf. 2004; 47: 3533–3547.

[23] Anderson JD. Modern Compressible Flows. New York: McGraw Hill, 1990.

[24] Chang CH, Raju MS, Sirignano WA. Numerical Analysis of Convecting, Vaporizing Fuel Droplet with Variable Properties. Int. J. Heat Mass Transf. 1992; 35(5): 1307–1324.

[25] Kurz W, Fisher DJ. Fundamentals of Solidification. Switzerland: Trans. Tech. Publications, 1989.

[26] Chattot JJ. Computational Aerodynamics and Fluid Dynamics: An Introduction. New York: Springer-Verlag, 2002.

[27] Weatherburn CE. Differential Geometry of Three Dimensions. London: Cambridge University Press, 1930.

2

Discontinuous Finite Element Procedures

The discontinuous finite element method makes use of the same function space as the continuous method, but with relaxed continuity at interelement boundaries. It was first introduced by Reed and Hill [1] for the solution of the neutron transport equation, and its history and recent development have been reviewed by Cockburn *et al.* [2, 3]. The essential idea of the method is derived from the fact that the shape functions can be chosen so that either the field variable, or its derivatives or generally both, are considered discontinuous across the element boundaries, while the computational domain continuity is maintained. From this point of view, the discontinuous finite element method includes, as its subsets, both the finite element method and the finite difference (or finite volume) method. Therefore, it has the advantages of both the finite difference and the finite element methods, in that it can be effectively used in convection-dominant applications, while maintaining geometric flexibility and higher local approximations through the use of higher order elements. This feature makes it uniquely useful for computational dynamics and heat transfer. Because of the local nature of a discontinuous formulation, no global matrix needs to be assembled; and thus, this reduces the demand on the in-core memory. The effects of the boundary conditions on the interior field distributions then gradually propagate through element-by-element connection. This is another important feature that makes this method useful for fluid flow calculations. Computational fluid dyanmics is an evolving subject, and very recent developments in the area are discussed in [4].

In the literature, the discontinuous finite element method is also called the discontinuous Galerkin method, or the discontinuous Galerkin finite element method, or the discontinuous method [1, 2, 3, 5, 6]. These terms will be used interchangeably throughout this book.

This chapter introduces the basic ideas of the discontinuous finite element method through simple and illustrative examples. The keyword here is *discontinuous*. Various views have been adapted to interpret the concept of discontinuity and three widely accepted ones are presented below [5, 7]. The discontinuous finite element formulation for boundary value problems, and overall procedures for numerical solutions are presented. The advantages and disadvantages of the various methods are also discussed, in comparison to the continuous finite element method. Examples used to illustrate the basic features and the solution procedures of the discontinuous finite element formulation are given.

2.1 The Concept of Discontinuous Finite Elements

To illustrate the basic ideas of the discontinuous finite element method, we consider a simple, one-dimensional, first order differential equation with u specified at one of the boundaries:

$$C(u)\frac{du}{dx}+f(u)=0; \qquad x \in [a,b] \tag{2.1}$$

$$u(x=a)=u_a \tag{2.2}$$

where, without loss of generality, the coefficient $C(u)$ is considered a function of the field variable u. By defining $dF = C(u)\,du$, the above differential equation may be further written as

$$\frac{dF}{dx}+f(u)=0 \tag{2.3}$$

The domain is discretized such that $\Omega_j = [x_j, x_{j+1}]$ with $j = 1, 2, ..., N$. Then, integrating the above equation over the element j, Ω_j, with respect to a weighting function $v(x)$,

$$\int_{x_j}^{x_{j+1}}\left[\frac{\partial F}{\partial x}+f(u)\right]v(x)\,dx = 0 \tag{2.4a}$$

and performing integration by parts on the differential operator, we have

$$F(u(x_{j+1}))v(x_{j+1})-F(u(x_j))v(x_j)$$

$$-\int_{x_j}^{x_{j+1}}\left[F\frac{\partial v(x)}{\partial x}-v(x)f(u)\right]dx = 0 \tag{2.4b}$$

On $\Omega_j = [x_j, x_{j+1}]$, u is approximated by $u_h \in H$, H being an appropriate function space of finite dimension, and v by v_h taken from the same function space as u_h, with $j = 1, 2, ..., N$. Upon substituting (u_h, v_h) for (u_h, v_h) in Equation 2.4b, we have the discontinuous Galerkin finite element formulation:

$$F(u_h(x_{j+1}))v_h(x_{j+1})-F(u_h(x_j))v_h(x_j)$$

$$-\int_{x_j}^{x_{j+1}} \left[F(u_h) \frac{\partial v_h(x)}{\partial x} - v_h(x) f(u_h) \right] dx = 0 \qquad (2.5)$$

In the continuous finite element approach, the field variable u_h is forced to be continuous across the boundary. As we know, this causes a problem of numerical instability, when $|c(u_h)|$ is large. The essential idea for the discontinuous method is that u_h is allowed to be discontinuous across the boundary. Therefore, across the element, the following two different values are defined at the two sides of the boundary:

$$u_j^+ = \lim_{x \downarrow x_j^+} u_h(x) \quad \text{and} \quad u_j^- = \lim_{x \uparrow x_j^-} u_h(x) \qquad (2.6)$$

Furthermore, we note that u_h is discontinuous only at the element boundaries. The solution u and $F(u)$ are smooth within (but excluding) the boundary. By this definition, the above equation contains the variables only within the integral limits Ω_j. There is no direct coupling with other intervals or other elements. The field values at a node, or the interface between two elements, are not unique. They are calculated using the two limiting values approaching the interface from the two adjacent elements. This feature is certainly desirable for problems with internal discontinuities, such as those pertaining to shock waves. We will discuss these specific problems in the chapters to follow.

The discontinuous formulation expressed in Equation 2.5 may be viewed from different perspectives, which all involve the cross-element treatments either by weakly imposing the continuity at the element interface, or by using numerical fluxes, or by boundary constraint minimization. These views are discussed below, so that the reader can fully appreciate the concept of discontinuity embedded in the formulation.

2.1.1 Weakly Imposed Cross-element Continuity

For the continuous finite element solution of boundary value problems, the consistency condition often requires that the field variable and its derivative be continuous in the computational domain, which implies the cross-element continuity requirement for these variables [8, 9]. In the continuous finite element formulation, the cross-element continuity is strongly enforced. The discontinuous formulation relaxes this continuity requirement, so that the cross-element continuity is weakly imposed. This is accomplished if $F(u)$, at the element boundaries, is chosen as follows [3, 5]:

$$+F(u_h(x_i)) = +F(u_i^+) \quad ; \quad -F(u_h(x_i)) = -F(u_i^-) \qquad (2.7)$$

so that the upstream value outside the element interval Ω_j is used, following the well known treatment for finite difference schemes. With Equation 2.7 substituted into Equation 2.5, the following integral equation is obtained:

$$F(u_h(x_{j+1}))v_h(x_{j+1}) - F(u_j^-)v_h(x_{j2})$$

$$-\int_{x_j}^{x_{j+1}} \left[F(u_h)\frac{\partial v_h(x)}{\partial x} - v_h(x)f(u_h) \right] dx = 0 \qquad (2.8)$$

This is one popular formulation for discontinuous finite element solutions. Equation 2.8 may be integrated once again with the result,

$$F(u_h(x_{j+1}))v_h(x_{j+1}) - F(u_j^-)v_h(x_j)$$

$$-(F(u_h(x_{j+1}))v_h(x_j) - F(u_j^+)v_h(x_j))$$

$$+\int_{x_j}^{x_{j+1}} \left[\frac{\partial F(u_h(x))}{\partial x} + f(u) \right] v(x)\,dx = 0 \qquad (2.9)$$

Here we stay with the upwinding rule at x_j, because only one boundary condition is available and it is applied at x_j. For this first order equation, $F(u_h^+) = F(u_h(x_{j+1}))$ at x_{j+1}. If one works with a second order equation, a similar rule may be applied at x_{j+1}. This point will be discussed further in Chapter 4. With these choices, the above equation is simplified as:

$$\left(F(u_j^+) - F(u_j^-)\right)v_h(x_j) + \int_{x_j}^{x_{j+1}} \left[\frac{\partial F(u_h(x))}{\partial x} + f(u_h) \right] v_h(x)\,dx = 0 \quad (2.10)$$

or more often, it is written in terms of a jump across the element boundary,

$$\int_{x_j}^{x_{j+1}} \left[\frac{\partial F(u_h(x))}{\partial x} + f(u_h) \right] v_h(x)\,dx + [F(u_j)]v_h(x_j) = 0 \qquad (2.11)$$

where the jump is defined by

$$[F(u_j)] = F(u_j^+) - F(u_j^-) \qquad (2.12)$$

In deriving the above equations, we have used the upwinding rule: $+F(u_h(x_j)) = + F(u_j^+)$. This procedure is graphically illustrated in Figure 2.1.

We now look at the implication of the above equation, i.e., Equation 2.12. Here, in essence, the continuity condition at x_j is satisfied *weakly* with respect to the weighting function $v(x)$. Note that x_j can be an internal boundary or external boundary. This is in contrast to the continuous finite element formulation, by which the continuity conditions are satisfied *strongly* across the element boundaries, $[F(u_j)] = 0$.

We note that since $v(x)$ is arbitrary, Equation 2.11 is equivalent to the following mathematical statement:

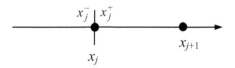

Figure 2.1. An illustration of the jump across x_j of element j: x_j and x_{j+1} mark the boundaries of the element

$$F(u_j^+) - F(u_j^-) = 0 \qquad \text{for} \quad x = x_j \qquad (2.13)$$

$$\frac{\partial F(u(x))}{\partial x} + f(u) = 0 \qquad \text{for} \quad x \in (x_j, x_{j+1}) \qquad (2.14)$$

Here, $F(u_j^+) - F(u_j^-) = 0$ also implies that $u_j^+ = u_j^-$ for monotone $F(u)$. Thus, Equation 2.11 is the weak form of Equations 2.13 and 2.14.

2.1.2 Numerical Boundary Fluxes for Discontinuity

Another treatment of the cross-element continuity is based on the use of a numerical flux to model $F(u)$. This is demonstrated by Cockburn *et al.* [2, 3]. Using this approach, $F(u)$ is replaced by the following flux expressions:

$$F(u_h(x_{j+1})) = h(u_{j+1}^-, u_{j+1}^+); \quad F(u_h(x_j)) = h(u_j^-, u_j^+) \qquad (2.15)$$

with an imposed consistency condition,

$$h(u, u) = F(u) \qquad (2.16)$$

Many different types of flux expressions have been used in the literature for this purpose, and have been reviewed in a recent paper by Arnold *et al.* [10]. To reproduce Equation 2.5, we may use the following definition for the numerical flux:

$$h(u_j^-, u_j^+) = F(u_j^-) \qquad (2.17)$$

which basically states that the flux at the element boundary is equal to the flux of the upstream element. With the numerical flux, the discontinuous finite element formulation for the 1-D problem is recast as

$$h(u_{j+1}^-, u_{j+1}^+) v_h(x_{j+1}) - h(u_j^-, u_j^+) v_h(x_j)$$

$$-\int_{x_j}^{x_{j+1}}\left[F(u_h)\frac{\partial v_h(x)}{\partial x}-v_h(x)f(u_h)\right]dx=0 \qquad (2.18)$$

It is apparent from the above discussion that construction of consistent numerical fluxes is important in discontinuous finite element calculations. These fluxes need to be chosen to satisfy numerical stability conditions and various forms of numerical fluxes and their stability conditions are given in [3, 10]. We note that different forms of numerical fluxes may be used to model various types of differential equations, and, as such, Equation 2.18 is more general. Selection of appropriate numerical fluxes for computational fluid dynamics applications is discussed in Chapters 4–7.

2.1.3 Boundary Constraint Minimization

The third view of the discontinuous treatment across the element boundaries is from the element boundary constraint minimization approach. To illustrate this view, we apply the Weight Residuals method to both the elements and their boundaries,

$$\sum_{j=1}^{N}\int_{x_j}^{x_{j+1}}\left[\frac{dF(u(x))}{dx}+f(u)\right]v(x)\,dx+\sum_{j=1}^{N}\int_{x_j^-}^{x_j^+}\frac{dF(u(x))}{dx}v(x)\,dx=0$$

$$(2.19)$$

Performing integrating by parts on Equation 2.19 and noting that the test function does not have to be continuous across the boundaries because of the intrinsic assumptions associated with a discontinuous finite element formulation, we have the following expression:

$$\sum_{j=1}^{N}\int_{x_j}^{x_{j+1}}\left[\frac{dF(u_h(x))}{dx}+f(u_h)\right]v_h(x)\,dx$$

$$+\sum_{j=1}^{N}\left(F(u_j^+)-F(u_j^-)\right)\;v_h(x)\;\;dx=0 \qquad (2.20)$$

where (u, v) are approximated by (u_h, v_h). Different forms of the weighting function may be used. One of the simple forms uses a linear combination of $v_h(x)$, defined on two adjacent elements as

$$v_h(x_j)=\alpha v_j^+ +(1-\alpha)v_j^- \qquad (2.21)$$

With the above equation substituted into Equation 2.20, one obtains the following formulation:

$$\sum_{j=1}^{N} \alpha \left(F(u_j^+) - F(u_j^-) \right) v_j^+ + (1-\alpha) \left(F(u_j^+) - F(u_j^-) \right) v_j^-$$

$$+ \sum_{j=1}^{N} \int_{x_j}^{x_{j+1}} \left[\frac{dF(u_h(x))}{dx} + f(u_h) \right] v_h(x)\, dx = 0 \qquad (2.22)$$

This expression of the formulation is also general. In fact, when $\alpha = 1$, one recovers the upwinding approach, as in Equation 2.11. On the other hand, if we carry out the integration once more and define the numerical flux as follows:

$$h(u_j^-, u_j^+) = \alpha F(u_j^-) + (1-\alpha) F(u_j^+) \qquad (2.23)$$

then Equation 2.22 reduces to Equation 2.18.

From the examples given above, a discontinuous element formulation can be constructed in three different ways: (1) by weakly imposed boundary conditions across element boundaries (Equation 2.11), (2) by the use of numerical flux expressions at the element boundaries (Equation 2.18), and (3) by the minimization of constraints across element boundaries (Equation 2.22). We note that while these three approaches treat cross-element discontinuities differently, they all fall into the general category of the Weighted Residuals method [6]. The first two involve the integration by parts, while the third one does not. If equations are written in non-conservative form, or if a conservative form does not exist, it is not straightforward to perform partial integration of the equations, because there is no "flux". In this case, the boundary minimization is more convenient for developing a discontinuous finite element formulation for these equations.

2.1.4 Treatment of Discontinuity for Non-conservative Systems

As stated in Section 1.7, a system of differential equations may be written in the "divergence" or "conservation law" form. By the definition given in Section 1.7, Equation 2.3 is in a conservative form, while Equation 2.1 is not.

In numerical analyses, the primitive variable is often solved instead of the flux function $F(u)$, and thus Equation 2.1 needs to be applied directly. In this case, from the definition, $dF(u) = C(u)\, du$, we may write,

$$F(u_i^+) - F(u_i^-) = \int_{u_i^-}^{u_i^+} C(u)\, du = [u]_i \int_{-1/2}^{+1/2} C\left([u]_i t + \tfrac{1}{2}(u_i^+ + u_i^-) \right) dt$$

$$(2.24)$$

where $[u]_i = u_i^+ - u_i^-$ is the jump across the element boundary. Since u is a smooth function, and $[u]_i$ is small, we may numerically approximate the integral by a mid-point rule,

$$\int_{-1/2}^{+1/2} C\big([u]_i t + \tfrac{1}{2}(u_i^+ + u_i^-)\big)dt = C\big(\tfrac{1}{2}(u_i^+ + u_i^-)\big) + O([u]_i^2) \qquad (2.25)$$

The relations given in the above two equations allow us to rewrite Equation 2.22 in the following form:

$$\int_{x_j}^{x_{j+1}} \left[C(u_h)\frac{du_h(x)}{dx} + f(u_h)\right] v_h(x)\,dx + \alpha\, C\big(\tfrac{1}{2}(u_j^+ + u_j^-)\big)[u]_j\, v_h(x_j)$$

$$+ (1-\alpha)C\big(\tfrac{1}{2}(u_j^+ + u_j^-)\big)[u]_j\, v_h(x_j) + O([u]_j^3) = 0 \qquad (2.26)$$

The last term, however, can be discarded without affecting the accuracy [3]. Thus, for the non-conservative equation stated in Equation 2.21, the discontinuous formulation is: find $u_h(x) \in P_l(\Omega_j)$ such that

$$\int_{x_j}^{x_{j+1}} \left[C(u_h)\frac{du_h(x)}{dx} + f(u_h)\right] v_h(x)\,dx$$

$$+ (1-\alpha)C\big(\tfrac{1}{2}(u_j^+ + u_j^-)\big)[u]_j\, v(x_j) + \alpha\, C\big(\tfrac{1}{2}(u_j^+ + u_j^-)\big)[u]_j v_h(x_j)$$

$$+ O([u_h]_j^3) = 0, \qquad \forall v_h(x) \in P_l(V_j) \qquad (2.27)$$

where $P_l(\Omega_j)$ is a piecewise polynomial of degree l defined over the interval $\Omega_j = [x_j, x_{j+1}]$. The boundary terms are set at $u_0^1 = u_0$ and $u_{N+1}^+ = u_{N+1}^-$.

2.1.5 Transient Problems

The discussion thus far has been limited to steady state problems. As with other methods, the treatment of the cross-element discontinuities can be readily extended to develop discontinuous finite element formulations for transient problems. Let us illustrate this point by considering a 1-D transient problem of hyperbolic type, sometimes referred to as convective wave equation, or convection equation, which is mathematically stated as

$$\frac{\partial u}{\partial t} + c\frac{\partial u}{\partial x} = 0, \qquad c > 0, x \in [a,b], \ t > 0 \qquad (2.28)$$

where c is a constant. Any of the above formulations can be applied to develop the needed integral formulation for a discontinuous finite element solution. Here we take the boundary constraint minimization approach and integrate the above partial differential equation with respect to a weighting function $v(x)$, whence we have the following result:

$$\sum_{j=1}^{N} \int_{x_j}^{x_{j+1}} \left[\frac{\partial u_h}{\partial t} + c \frac{\partial u_h}{\partial x} \right] v_h(x)\,dx + \sum_{j=1}^{N} \int_{x_j^-}^{x_j^+} \frac{\partial u_h(x)}{\partial x} v_h(x)\,dx = 0 \quad (2.29)$$

For a typical interval, $\Omega_j = [x_j, x_{j+1}], j = 1, \ldots, N$, the above equation reduces to the following form after integrating the second term:

$$\int_{x_j}^{x_{j+1}} \left[\frac{\partial u_h}{\partial t} + c \frac{\partial u_h}{\partial x} \right] v_h(x)\,dx + \alpha\, c[u]_j\, v_h(x_j) + (1-\alpha)c[u]_j\, v_h(x_j) = 0$$

$$(2.30)$$

Comparing this with Equation 2.22, and noticing that $F(u) = cu$ for this problem, we see that the transient term enters the integral description directly, as in the continuous finite element method.

2.2 Discontinuous Finite Element Formulation

We may extend the discussions on the 1-D examples to consider a more general class of problems and formally introduce the discontinuous finite element formulation for boundary value problems.

2.2.1 Integral Formulation

Let us consider a partial differential equation, written in the form of the conservation law for a scalar u,

$$\frac{\partial u}{\partial t} + \nabla \cdot \mathbf{F}(u) + b = 0 \; ; \;\; u(0,\mathbf{x}) = u_0(\mathbf{x})\,, \;\; \mathbf{x} \in \Omega, \;\; t > 0 \qquad (2.31)$$

To start, the computational domain is broken into a tessellation of finite elements $\Omega = \bigcup_{j=1}^{N} \Omega_j$. The field variable u is approximated by the interpolation function u_h, defined on each element Ω_j. Since the function u_h is allowed to be discontinuous across the element boundaries for discontinuous formulations, the finite element space, over which u_h is defined, is sometimes referred to as finite element broken space, to differentiate it from continuous finite element space [11]. The broken space is denoted by V_h^j and $V_h^j \subset L^2(\Omega)$, where $L^2(\Omega)$ is the Lebesgue space of square integrable functions, defined over Ω [12].

If the above equation is integrated over Ω_j with respect to a weighting function v, one has the weak form expression:

$$\int_{\Omega_j} v(\mathbf{x}) \left[\frac{\partial u}{\partial t} + \nabla \cdot \mathbf{F}(u) + b \right] dV = 0 \qquad (2.32)$$

We now perform integration by parts on the second term involving the divergence of flux and obtain the nornal fluxes along the boundary. This procedure yields the following result:

$$\int_{\Omega_j}\left[v_h\frac{\partial u_h}{\partial t}-\mathbf{F}(u_h)\cdot\nabla v_h+bv_h\right]dV+\int_{\partial\Omega_j}v_h\mathbf{F}(u_h)\cdot\mathbf{n}\,dS=0\,,$$

$$\forall v_h\in V_h^j \qquad (2.33)$$

where $\mathbf{n}$ is the local outnormal vector on the element boundary $\partial\Omega_j$. By substituting numerical fluxes along the element boundaries,

$$\mathbf{F}\cdot\mathbf{n}=F_n(u^-,\,u^+) \qquad (2.34)$$

Equation 2.33 can be integrated numerically. The construction of numerical fluxes is important, and there are many different fluxes for popular fluid flow and heat transfer problems. These fluxes will be discussed in subsequent chapters for specific applications.

The integration of Equation 2.33 with an appropriate choice of numerical fluxes will result in a set of ordinary differential equations,

$$\mathbf{M}\frac{d\mathbf{U}_{(j)}}{dt}+\mathbf{K}\mathbf{U}_{(j)}=\mathbf{F}_{(j)} \qquad (2.35)$$

where $\mathbf{U}_{(j)}$ is the vector of nodal values of variable u associated with element j, $\mathbf{K}$ the stiffness matrix, $\mathbf{M}$ the mass matrix, and $\mathbf{F}_{(j)}$ the force vector consisting of contributions from the source and boundary terms.

2.2.2 Time Integration

Time integration can proceed, in theory, by using the general approaches for the solution of initial value problems. Two important points, however, are noted when the time integration is carried out for Equation 2.35. First, since the discontinuous formulation is a local formulation, it often leads to standard explicit structures. Thus, the explicit methods for time integration are preferred with discontinuous finite element formulations, whenever possible. Of course, this does not mean that the implicit method is not possible. In practice, both explicit and implicit integrators can be applied, though the latter is much less frequently used with discontinuous formulations. Second, since the explicit methods are prone to numerical instability, appropriate stability analysis is needed for the time integration schemes [13–15]. Fortunately, stability criteria have been established for the most commonly used time integration methods for the fluid flow and heat transfer applications.

The following equations show some of the commonly used time integration schemes for the discontinuous finite element applications.

(1) First order Euler forward:

$$\mathbf{U}_j^{n+1} = \mathbf{U}_j^n + \Delta t\, H(\mathbf{U}_j^n, t^n) \tag{2.36}$$

(2) Second order scheme:

$$f_1 = \Delta t H(\mathbf{U}_j^n, t^n)\,; \qquad f_2 = \Delta t H(\mathbf{U}_j^n + f_1, t^n + \Delta t)\,;$$

$$\mathbf{U}_j^{n+1} = \mathbf{U}_j^n + 0.5(f_1 + f_2) \tag{2.37}$$

(3) Third order scheme:

$$f_1 = \Delta t H(\mathbf{U}_j^n, t^n)\,; \qquad f_2 = \Delta t H(\mathbf{U}_j^n + 0.5 f_1, t^n + 0.5\Delta t)\,;$$

$$f_3 = \Delta t H(\mathbf{U}_j^n - f_1 + 2 f_2, t^n + 0.5\Delta t)\,;$$

$$\mathbf{U}_j^{n+1} = \mathbf{U}_j^n + \tfrac{1}{6}(f_1 + 4 f_2 + f_3) \tag{2.38}$$

(4) Fourth order scheme:

$$f_1 = \Delta t H(\mathbf{U}_j^n, t^n)\,; \qquad f_2 = \Delta t H(\mathbf{U}_j^n + 0.5 f_1, t^n + 0.5\Delta t)\,;$$

$$f_3 = \Delta t H(\mathbf{U}_j^n 0.5 f_2, t^n + 0.5\Delta t)\,; \quad f_4 = \Delta t H(\mathbf{U}_j^n + f_3, t^n + \Delta t)\,;$$

$$\mathbf{U}_j^{n+1} = \mathbf{U}_j^n + \tfrac{1}{6}(f_1 + 2 f_2 + 2 f_3 + f_4) \tag{2.39}$$

In the above schemes, $H(\mathbf{U}_j^n, t^n) = \mathbf{M}^{-1}(\mathbf{F}_j(\mathbf{U}_j^n, t^n) - \mathbf{K}(\mathbf{U}_j^n, t^n)\mathbf{U}_j^n)$. Since these schemes are explicit, the time step has to satisfy the *CFL* (Courant–Friedrich–Lewy) condition for stability. While they represent some of the popular choices, other schemes are also possible. For example, a Total Variation Diminishing (*TVD*) scheme has been used for oscillation-free shock wave simulations [15]. It is noted that time integration can be most efficiently calculated if the mass matrix is diagonalized when an explicit scheme is used. For this purpose, the orthogonal hierarchical shape functions presented in Chapter 3 have been proven to be extremely useful. The use of these transient schemes will be discussed in subsequent chapters for the numerical solution of specific problems of fluid dynamics and heat transfer.

An implicit time integration scheme may also be used with a discontinuous finite element formulation. However, the use of an implicit scheme results in an even larger global matrix than a conventional finite element formulation, thereby eliminating the advantage of localized formulation associated with discontinuous finite elements. Consequently, almost all discontinuous finite element formulations presented thus far use the explicit time integration scheme for the solution of transient problems, for the purpose of facilitating the parallel computation associated with a local formulation.

2.3 Solution Procedures

We now consider the general computational procedure by which the discontinuous finite element method is used to obtain numerical solutions. From the above discontinuous formulations, it is clear that this method is local, in that the weakly imposed across-element boundary conditions permit the element-wise solutions. For each element, the elemental calculation is required, and is essentially the same as that used in the continuous finite element method. Also, as for the continuous counterpart, the discontinuous Galerkin formulation is obtained when the same interpolation functions are used for both unknowns and the trial functions.

One important implication of the above discontinuous formulation is that, because of a weakly imposed boundary condition across adjacent elements, a variety of elements or shape functions, including the discontinuous shape functions, can be chosen for computations. As a result, the discontinuous formulation embeds the continuous finite element and the finite volume/finite difference formulations. If, in particular, a constant element is chosen, then the formulation boils down to the traditional finite difference method. On the other hand, if the continuous function is chosen, and the cross-boundary continuity is enforced, one implements the continuous finite element method.

We note that the discontinuous finite element method falls also into the general category of the Weighted Residuals method for the solution of partial differential equations. Various familiar forms of domain- and boundary-based numerical methods can be derived from this general integral formulation, depending upon the choice of the weighting functions. For the Galerkin formulation, the weighting functions are chosen the same as the shape functions. The weighting functions, however, may be chosen differently from the shape functions. For example, if Green's functions are chosen as weighting functions, then the well known boundary element formulation of boundary value problems is obtained [16].

2.4 Advantages and Disadvantages of Discontinuous Finite Element Formulation

In comparison with the other numerical methods (finite difference and finite elements), the discontinuous finite element formulations have both advantages and disadvantages. It is important to understand these issues for developing specific applications.

2.4.1 Advantages

In discontinuous formulations, the interelement boundary continuity constraints are relaxed. Various upwinding schemes, proven successful for convection-dominant flows, can be easily incorporated through element boundary integrals that only involve the spatial derivative terms in the equations. Inside the elements, all terms are treated by the standard Galerkin method, leading to classical symmetric mass matrices and standard treatment of source terms.

Higher order approximations are obtained simply by increasing the order of the polynomials or other basis functions. The decoupling of the upwinding convection terms, and the other terms, yields a very attractive feature of the discontinuous method, especially in the case of convection-dominant problems. This method performs very well, and in fact it is often better than the SUPG method, for advection type problems. Even for linear elements, this method performs remarkably well [6].

The coupling between element variables is achieved through the boundary integrals only. This means that $\partial u/\partial t$ and the source terms are fully decoupled between elements. The mass matrices can be inverted at the element level, rendering the $\partial \mathbf{U}_{(j)}/\partial t$ explicit. With an appropriate choice of orthogonal shape functions, a diagonal mass matrix can be obtained, thereby resulting in a very efficient time marching algorithm.

The discontinuous finite element formulation is a local formulation and the action is focused on the element and its boundaries. Whatever the space dimension or the number of unknowns, the formulation remains basically the same and no special features need to be introduced.

Because of the local formulation, a discontinuous finite element algorithm will not result in an assembled global matrix and thus the in-core memory demand is not as strong. Also, the local formulation makes it very easy to parallelize the algorithm, taking advantage of either shared memory parallel computing or distributed parallel computing.

Also, because of the local formulation, both the h- and p-adaptive refinements are made easy and convenient. Compared with the continuous finite element method, the hp-adaptive algorithm based on the discontinuous formulation requires no additional cost associated with node renumbering.

2.4.2 Disadvantages

Like any other numerical methods, the discontinuous finite element method has its drawbacks. The blind use of this method would certainly result in a very inefficient algorithm. In comparison with finite elements using continuous basis functions, the number of variables is larger for an identical number of elements [7, 17]. This is obvious from the formulations given above, and is a natural consequence of relaxing the continuity requirements across the element boundaries.

Since the basis and test functions are discontinuous across element boundaries, second order spatial terms (diffusion) need to be handled by mixed methods, which enlarge the number of unknowns, or other special treatments. This is a serious drawback, when compared to the continuous methods where elliptic operators are handled relatively easily. Also, our experience with the heat conduction or diffusion problems indicates that if stabilization parameters are not used, the element matrix may become singular and thus pollute the numerical solution. The solution algorithm, based on the discontinuous formulation in general, is inferior to the continuous finite element method in its execution speed for pure conduction or diffusion problems, in particular steady state heat conduction and diffusion problems. Thus, for these problems, if memory is not a constraint in applications, the discontinuous formulation should be avoided.

In computer-aided thermal and fluids engineering design applications, complex numerical models are often required to represent a wide range of thermal and fluid flow phenomena. It is, therefore, unlikely that one single method would be best suited for modeling all the physical phenomena in a thermal/fluid system. Thus, a combination of methods, best suited for modeling certain types of phenomena, would be required, in order to develop the most efficient algorithms for specific applications. These issues are explored further in subsequent chapters of this book.

2.5 Examples

The examples in this chapter are selected for the purpose of illustrating the basic concepts of the discontinuous finite element formulation, and the general solution procedures for the numerical solution to boundary value problems. As a result, very simple problems are considered.

Example 2.1. Apply the discontinuous Galerkin finite element method to obtain the numerical solution of the following initial value problem:

$$\frac{du}{dx} = 1 \quad \text{with} \quad u(0) = 0; \qquad x \in [0,2] \qquad (2.1e)$$

and compare the numerical results with the analytical solution with the domain discretized by two linear elements.

Solution. The analytic solution to the problem is simple, $u = x$. Now, following the procedure in Section 2.1 leading to the element-wise formulation, we have the following integral equation with F replaced by u_h and v by ϕ_i:

$$\int_{x_j}^{x_{j+1}} \left[\frac{du_h(x)}{dx} + f(u_h) \right] \phi_i(x)\, dx + \left(u_j^+ - u_j^- \right) \phi_i(x_j) = 0 \qquad (2.2e)$$

where ϕ_i is the shape function. Now the domain is discretized into two elements, as shown in Figure 2.1e.

For simplicity, a linear interpolation is used for each of the elements. When an isoparametric shape function is used, we have the following relations:

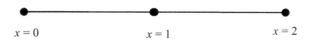

$$x = 0 \qquad\qquad x = 1 \qquad\qquad x = 2$$

Figure 2.1e. Discretization of the domain into two elements

$$u_h = \phi_1 u_1 + \phi_2 u_2 , \quad \phi_1(\xi) = 0.5(1-\xi) , \quad \phi_2(\xi) = 0.5(1+\xi)$$

$$x = \phi_1 x_1 + \phi_2 x_2 ; \quad dx = d\phi_1 x_1 + d\phi_2 x_2 = 0.5(x_2 - x_1) d\xi$$

$$\frac{du_h}{dx} = \frac{d\phi_1}{dx} u_1 + \frac{d\phi_2}{dx} u_2 ; \quad \frac{d\phi_1}{dx} = \frac{d\phi_1}{d\xi} \frac{d\xi}{dx} = -0.5 \times 2 = -1 ; \frac{d\phi_2}{dx} = 1$$

Applying Equation 2.11 to the first element $x \in [0, 1]$, and making use of the condition $u_0^- = u(0^-) = 0$, one has

$$\int_0^1 \left[\frac{du_h}{dx} - 1 \right] \phi_i(x) \, dx + u_0^+ \, \phi_i(0) = 0 \tag{2.3e}$$

Now with $u_0^+ = u_1$, $\phi_1(x = 0) = \phi_1(\xi = -1) = 1$, and $u(x^- = 1) = u_2$ and with the unknown variable replaced by its local approximation using interpolation functions in Equation 2.3e, the following expression is obtained for the first element:

$$\int_0^1 \binom{\phi_1}{\phi_2} \left(\frac{d\phi_1}{dx}, \frac{d\phi_2}{dx} \right) dx \binom{u_1}{u_2} + \begin{bmatrix} 1 & 0 \\ 0 & 0 \end{bmatrix} \binom{u_1}{u_2} = \int_0^1 \binom{\phi_1}{\phi_2} dx \tag{2.4e}$$

For this problem, the integration can be carried out analytically, whence we have the results,

$$\int_0^1 \binom{\phi_1}{\phi_2} \left(\frac{d\phi_1}{dx}, \frac{d\phi_2}{dx} \right) dx = \frac{1}{4} \int_{-1}^1 \binom{1-\xi}{1+\xi} (-1, \ 1) \, d\xi$$

$$= \frac{1}{4} \int_{-1}^1 \binom{-(1-\xi), \ (1-\xi)}{-(1+\xi), \ (1+\xi)} d\xi = \frac{1}{8} \binom{-4, \ 4}{-4, \ 4} = \frac{1}{2} \binom{-1, \ 1}{-1, \ 1} \tag{2.5e}$$

$$\int_0^1 \binom{\phi_1}{\phi_2} dx = \frac{1}{2} \int_{-1}^1 \binom{1-\xi}{1+\xi} \frac{dx}{d\xi} d\xi = \frac{1}{8} \binom{-(1-\xi)^2 \big|_{-1}^1}{(1+\xi)^2 \big|_{-1}^1} = \frac{1}{8} \binom{4}{4} = \frac{1}{2} \binom{1}{1}$$

$$\tag{2.6e}$$

Substituting Equations 2.5e–2.6e into Equation 2.4e yields the follwing matrix equation,

$$\frac{1}{2} \binom{-1, \ 1}{-1, \ 1} \binom{u_1}{u_2} + \begin{bmatrix} 1 & 0 \\ 0 & 0 \end{bmatrix} \binom{u_1}{u_2} = \frac{1}{2} \binom{1}{1} \tag{2.7e}$$

which can be solved for u_1 and u_2,

$$\begin{pmatrix} 1,1 \\ -1,1 \end{pmatrix} \begin{pmatrix} u_1 \\ u_2 \end{pmatrix} = \begin{pmatrix} 1 \\ 1 \end{pmatrix} \quad => \quad \begin{pmatrix} u_1 \\ u_2 \end{pmatrix} = \begin{pmatrix} 0 \\ 1 \end{pmatrix} \tag{2.8e}$$

where $u_1 = u(0^+)$ and $u_2 = u(1^-)$.

Now the same procedure is applied to the second element $x \in [1, 2]$ with the result,

$$\int_1^2 \left[\frac{du_h(x)}{dx} - 1 \right] \phi_i(x)\, dx + \left(u(1^+) - u(1^-) \right) \phi_i(x_j) = 0 \tag{2.9e}$$

At this point, $u(1^-) = u_2$ is known from Equation 2.8e. Furthermore, if the upwinding scheme is used, the following matrix equation is then obtained:

$$\int_1^2 \begin{pmatrix} \phi_1 \\ \phi_2 \end{pmatrix} \left(\frac{d\phi_1}{dx}, \ \frac{d\phi_2}{dx} \right) dx \begin{pmatrix} u_1 \\ u_2 \end{pmatrix} + \begin{bmatrix} 1 & 0 \\ 0 & 0 \end{bmatrix} \begin{pmatrix} u_1 \\ u_2 \end{pmatrix}$$

$$= \int_0^1 \begin{pmatrix} \phi_1 \\ \phi_2 \end{pmatrix} dx + \begin{bmatrix} 1 & 0 \\ 0 & 0 \end{bmatrix} \begin{pmatrix} u(1^-) \\ u_2 \end{pmatrix} \tag{2.10e}$$

The detailed integration is almost the same as for the first element,

$$\frac{1}{2} \begin{pmatrix} -1, 1 \\ -1, 1 \end{pmatrix} \begin{pmatrix} u_1 \\ u_2 \end{pmatrix} + \begin{bmatrix} 1 & 0 \\ 0 & 0 \end{bmatrix} \begin{pmatrix} u_1 \\ u_2 \end{pmatrix} = \frac{1}{2} \begin{pmatrix} 1 \\ 1 \end{pmatrix} + \begin{bmatrix} 1 & 0 \\ 0 & 0 \end{bmatrix} \begin{pmatrix} 1 \\ u_2 \end{pmatrix} \tag{2.11e}$$

Rearranging, we have the solution for the second element,

$$\begin{pmatrix} +1, 1 \\ -1, 1 \end{pmatrix} \begin{pmatrix} u_1 \\ u_2 \end{pmatrix} = \begin{pmatrix} 3 \\ 1 \end{pmatrix} \quad => \quad \begin{pmatrix} u_1 \\ u_2 \end{pmatrix} = \begin{pmatrix} 1 \\ 2 \end{pmatrix} \tag{2.12e}$$

where $u_1 = u(1^+)$ and $u_2 = u(2^-)$. The numerical results for this elementary example are: $u_0 = 0$, $u_1 = 0.5(u(1^+) + u(1^-)) = 1$ and $u_2 = 2$.

As discussed above, the discontinuous shape functions may be used because the field variable is considered discontinuous across the boundary. The use of discontinuous shape functions to obtain the same numerical results is illustrated in the following example.

Example 2.2. Re-solve the problem defined in Equation (2.1e) using geometrically discontinuous linear elements.

Solution. For the purpose of demonstration only, we consider the discontinuous shape function for the first element that is normalized at $x_1 = 0.2$ and $x_2 = 0.8$, which corresponds to $\xi=0$ and $\xi=1$ respectively, as shown in Figure 2.2e. Thus the following expressions are obtained:

$$u_h = \phi_1 u_1 + \phi_2 u_2 ; \quad \phi_1(\xi) = 0.5(1-\xi/0.8)$$

$$x = \phi_1 x_1 + \phi_2 x_2 ; \quad \phi_2(\xi) = 0.5(1+\xi/0.8)$$

$$dx = d\phi_1 x_1 + d\phi_2 x_2 = 0.5(x_2 - x_1) d\xi/0.8 = d\xi 0.5 \times 3/4$$

$$\frac{\partial u_h}{\partial x} = \frac{\partial \phi_1}{\partial x} u_1 + \frac{\partial \phi_2}{\partial x} u_2 ; \frac{\partial \phi_1}{\partial x} = \frac{\partial \phi_1}{\partial \xi} \frac{\partial \xi}{\partial x} = -\frac{1}{2 \times 0.8} \times \frac{2 \times 0.8}{(0.8-0.2)} = -\frac{1}{0.6}$$

$$\frac{\partial \phi_2}{\partial x} = \frac{\partial \phi_2}{\partial \xi} \frac{\partial \xi}{\partial x} = \frac{1}{2 \times 0.8} \times \frac{2 \times 0.8}{(0.8-0.2)} = \frac{1}{0.6}$$

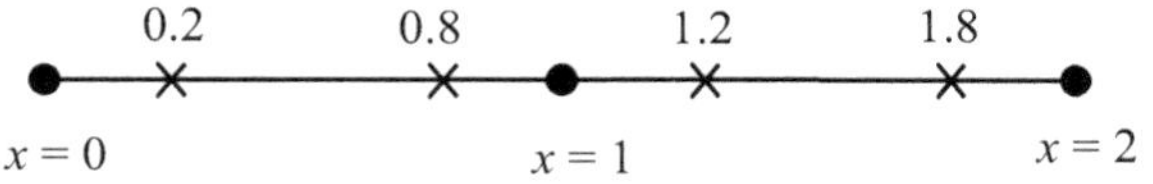

Figure 2.2e. An illustration of two linear discontinuous elements

To calculate the integration limits for the normalized coordinate ξ that correspond to $x = 0$ and $x = 1$, we make use of the isoparametric element to obtain the integration limits:

$$0 = x = \phi_1 x_1 + \phi_2 x_2 = 0.5(1-\xi/0.8)0.2 + 0.5(1+\xi/0.8)0.8 \implies \xi = -4/3$$

$$1 = x = \phi_1 x_1 + \phi_2 x_2 = 0.5(1-\xi/0.8)0.2 + 0.5(1+\xi/0.8)0.8 \implies \xi = 4/3$$

These expressions and integration limits are now substituted into Equation 2.11 for the first element and the resultant equation can be integrated analytically, whence we have

$$\int_0^1 \binom{\phi_1}{\phi_2}\left(\frac{d\phi_1}{dx}, \frac{d\phi_2}{dx}\right) dx \binom{u_1}{u_2}$$

$$+ \binom{\phi_1(-4/3)}{\phi_2(-4/3)}(\phi_1(-4/3), \phi_2(-4/3))\binom{u_1}{u_2} = \int_0^1 \binom{\phi_1}{\phi_2} dx \quad (2.13e)$$

$$\int_0^1 \binom{\phi_1}{\phi_2} dx = \frac{3}{16} \int_{-4/3}^{4/3} \binom{1-\xi/0.8}{1+\xi/0.8} d\xi = \frac{2.4}{32} \left[\begin{array}{c} -(1-\xi/0.8)^2 \\ (1+\xi/0.8)^2 \end{array} \right]_{-4/3}^{4/3}$$

$$= \frac{1}{2}\binom{1}{1} \tag{2.14e}$$

$$\int_0^1 \binom{\phi_1}{\phi_2}\left(\frac{d\phi_1}{dx}, \ \frac{d\phi_2}{dx}\right) dx = \frac{1}{3.2} \int_{-4/3}^{4/3} \binom{1-\xi/0.8}{1+\xi/0.8}(-1, \ 1) d\xi$$

$$= \frac{1}{3.2} \int_{-4/3}^{4/3} \binom{-(1-\xi/0.8), \ (1-\xi/0.8)}{-(1+\xi/0.8), \ (1+\xi/0.8)} d\xi$$

$$= \frac{1}{8}\left[\begin{array}{cc} (1-\xi/0.8)^2, & -(1-\xi/0.8)^2 \\ -(1+\xi/0.8)^2, & (1+\xi/0.8)^2 \end{array} \right]_{-4/3}^{4/3} = \frac{1}{1.2}\binom{-1, \ 1}{-1, \ 1} \tag{2.15e}$$

$$\binom{\phi_1(-4/3)}{\phi_2(-4/3)}(\phi_1(-4/3), \ \phi_2(-4/3)) = \binom{4/3}{-1/3}(4/3, \ -1/3)$$

$$= \frac{1}{9}\left[\begin{array}{cc} 16 & -4 \\ -4 & 1 \end{array} \right] \tag{2.16e}$$

The above results are combined to yield a matrix equation for the first element,

$$\frac{1}{1.2}\binom{-1, \ 1}{-1, \ 1} + \frac{1}{9}\left[\begin{array}{cc} 16 & -4 \\ -4 & 1 \end{array} \right] = \frac{1}{3.6}\binom{-3, \ 3}{-3, \ 3} + \frac{1}{3.6}\binom{6.4, \ -1.6}{-1.6, \ 0.4}$$

$$= \frac{1}{3.6}\binom{3.4, \ 1.4}{-4.6, \ 3.4} \tag{2.18e}$$

$$\frac{1}{3.6}\binom{3.4, \ 1.4}{-4.6, \ 3.4}\binom{u_1}{u_2} = \frac{1}{2}\binom{1}{1} \tag{2.17e}$$

Equation 2.17 is then solved to obtain the numerical solution,

$$\binom{3.4 \quad 1.4}{-4.6 \quad 3.4}\binom{u_1}{u_2} = \binom{1.8}{1.8} \Rightarrow \binom{u_1}{u_2} = \binom{0.2}{0.8} \tag{2.19e}$$

We see that $u_1 = u(x = 0.2) = 0.2$ and $u_2 = u(x = 0.8) = 0.8$, which match with the exact solutions. It is a simple exercise to show that the calculations for the second element yield the same results as in Example 2.1.

Example 2.3. Consider this internal radiation problem, defined by the following differential equation and boundary condition:

$$\frac{du}{dx}+u=1, \quad \text{with} \quad u(x=0)=0 \tag{2.20e}$$

Obtain the numerical solution using the discontinuous Galerkin finite element formulation with two linear discontinuous finite elements, and compare the results with the exact solution.

Solution. We first obtain the analytical solution to the problem above. The problem is solved by direct integration and the solution is $u(x) = 1 - e^{-x}$. Application of the discontinuous finite element formulation for the first element gives the result,

$$\int_0^1\left[\frac{du_h}{dx}+u_h-1\right]\phi_i(x)\,dx+u_0^+\,\phi_i(0)=0 \tag{2.21e}$$

where we have applied $u_0^- = u(0^-) = 0$. Now with $u_0^+ = u_1 = 0$, $\phi_1(x=0) = \phi_1(\xi = -1)$, $u(x^- = 1) = u_2$ substituted, one has

$$\int_0^1\binom{\phi_1}{\phi_2}\left(\frac{\partial\phi_1}{\partial x},\frac{\partial\phi_2}{\partial x}\right)dx\binom{u_1}{u_2}+\int_0^1\binom{\phi_1}{\phi_2}(\phi_1,\phi_2)\,dx\binom{u_1}{u_2}+\begin{bmatrix}1&0\\0&0\end{bmatrix}\binom{u_1}{u_2}$$

$$=\int_0^1\binom{\phi_1}{\phi_2}dx \tag{2.22e}$$

The detailed calculations are the same as before, whence we have the following expressions:

$$\int_0^1\binom{\phi_1}{\phi_2}dx=\frac{1}{2}\binom{1}{1};\quad \int_0^1\binom{\phi_1}{\phi_2}\left(\frac{\partial\phi_1}{\partial x},\frac{\partial\phi_2}{\partial x}\right)dx=\frac{1}{2}\binom{-1,\ 1}{-1,\ 1} \tag{2.23e}$$

The additional term comes from the treatment of $u(x)$,

$$\int_{x_j}^{x_{j+1}}\binom{\phi_1}{\phi_2}(\phi_1,\phi_2)\,dx=\frac{1}{8}\int_{-1}^1\binom{1-\xi}{1+\xi}(1-\xi,\ 1+\xi)\,d\xi$$

$$=\frac{1}{8}\int_{-1}^1\binom{(1-\xi)^2,\ (1-\xi^2)}{(1-\xi^2),\ (1+\xi)^2}d\xi=\frac{1}{8}\binom{-\frac13(1-\xi)^3,\ (\xi-\frac13\xi^3)}{(\xi-\frac13\xi^3),\ \frac13(1+\xi)^3}\Big|_{-1}^1$$

$$=\frac{1}{8}\binom{\frac83,\ \frac43}{\frac43,\ \frac83}=\frac{1}{6}\binom{2,1}{1,2} \tag{2.24e}$$

Assembling these expressions into Equation 2.33, one has the matrix equation,

$$\frac{1}{2}\begin{bmatrix} -1, 1 \\ -1, 1 \end{bmatrix}\begin{pmatrix} u_1 \\ u_2 \end{pmatrix} + \frac{1}{6}\begin{bmatrix} 2, 1 \\ 1, 2 \end{bmatrix}\begin{pmatrix} u_1 \\ u_2 \end{pmatrix} + \begin{bmatrix} 1 & 0 \\ 0 & 0 \end{bmatrix}\begin{pmatrix} u_1 \\ u_2 \end{pmatrix} = \frac{1}{2}\begin{pmatrix} 1 \\ 1 \end{pmatrix} \qquad (2.25e)$$

Simplifying, we have the following numerical results for the first element:

$$\begin{pmatrix} 5, 4 \\ -2, 5 \end{pmatrix}\begin{pmatrix} u_1 \\ u_2 \end{pmatrix} = \begin{pmatrix} 3 \\ 3 \end{pmatrix} \ => \ \begin{pmatrix} u_1 \\ u_2 \end{pmatrix} = \begin{pmatrix} u(0^+) \\ u(1^-) \end{pmatrix} = \begin{pmatrix} 0.091 \\ 0.636 \end{pmatrix} \qquad (2.26e)$$

where $u_1 = u(0^+)$ and $u_2 = u(1^-)$. This compares with the analytical solution: $u(0) = 0$ and $u(1) = 0.632$.

For the second element, the same procedure is applied with the result,

$$\frac{1}{2}\begin{pmatrix} -1, 1 \\ -1, 1 \end{pmatrix}\begin{pmatrix} u_1 \\ u_2 \end{pmatrix} + \frac{1}{6}\begin{bmatrix} 2, 1 \\ 1, 2 \end{bmatrix}\begin{pmatrix} u_1 \\ u_2 \end{pmatrix} + \begin{bmatrix} 1 & 0 \\ 0 & 0 \end{bmatrix}\begin{pmatrix} u_1 \\ u_2 \end{pmatrix} = \frac{1}{2}\begin{pmatrix} 1 \\ 1 \end{pmatrix} + \begin{bmatrix} 1 & 0 \\ 0 & 0 \end{bmatrix}\begin{pmatrix} u_1^- \\ u_2 \end{pmatrix}$$

$$(2.27e)$$

Rearranging and setting $u_1^- = u(1^-)$, we have the numerical values for the second element,

$$\begin{pmatrix} 5, 4 \\ -2, 5 \end{pmatrix}\begin{pmatrix} u_1 \\ u_2 \end{pmatrix} = \begin{pmatrix} 75/11 \\ 3 \end{pmatrix} \ => \ \begin{pmatrix} u_1 \\ u_2 \end{pmatrix} = \begin{pmatrix} u(1^+) \\ u(2^-) \end{pmatrix} = \begin{pmatrix} 0.669 \\ 0.868 \end{pmatrix} \qquad (2.28e)$$

which compares with the analytical solution: $u(2) = 0.865$.

For this simple example, the solutions can be readily obtained using the continuous finite element method or the finite difference method. For comparison, the numerical results using different methods are listed in Table 2.1e, and compared with those calculated using the analytic solution.

Table 2.1e. Comparison of numeric results with analytical solution

x	0	1	2
DFEM $u(x)$	0.04545	0.653	0.868
Analytic $(1-e^{-x})$	0	0.632	0.865
DFEM $u(x^-)$	0	0.636	0.868
FEM	0	0.643	0.857
FD	0	0.500	0.750

In Table 2.1e, the values of DFEM $u(x)$ are obtained using the averaged quantities across the element boundary: that is, $u(x) = 0.5(u(x^-) + u(x^+))$. The solution is better approximated if we take $u(x) = u(x^-)$, as shown by those given in the row associated with DFEM $u(x^-)$. The standard continuous finite element solution (FEM) is reasonably good, although not as good as the discontinuous finite element solution (DFEM $u(x^-)$). The standard finite difference approximation (FD), with upwinding, seems to be least accurate for this problem.

Example 2.4. Consider a two-dimensional convection problem defined by the following differential equation and boundary condition:

$$\frac{\partial u}{\partial t}+\frac{\partial u}{\partial x}+\frac{\partial u}{\partial y}=0 \quad x\in[-\pi,\pi]\times[-\pi,\pi]\times(0,T] \tag{2.29e}$$

with periodic boundary conditions and initial data

$$u(x,y,t=0)=\sin(\pi x)\sin(\pi y) \tag{2.30e}$$

Obtain the numerical solutions using the discontinuous finite element method and discuss parallel computing performance.

Solution. This problem was solved by Biswas *et al.* [18], and is used here as an example to demonstrate the parallel performance of the discontinuous finite element method. Their algorithm employed a discontinuous Galerkin finite element discretization, with a basis of piecewise Legendre polynomials. Temporal discretization employes a Runge–Kutta method. Dissipative fluxes and projection limiting prevent oscillations near the solution discontinuities. Parallel computing used from 1 to 256 processors. The computed results are given in Table 2.2e. It is seen from the results that, as the number of processors increases while keeping the work per processor constant, the discontinuous finite element method achieves a very impressive parallel computing performance.

Table 2.2e. Scaled parallel efficiency: solution times (without I/O) and total execution times, measured on the nCUEE/2

Number of processors	Work (W)	Solution time (s)	Solution parallel efficiency	Total time (s)	Total parallel efficiency
1	18432	926.92	–	927.16	–
2	36864	927.06	99.98%	927.31	99.98%
4	73728	927.13	99.97%	927.45	99.96%
8	147456	927.17	99.97%	927.58	99.95%
16	294912	927.38	99.95%	928.13	99.89%
32	589824	927.89	99.89%	929.90	99.70%
64	1179648	928.63	99.81%	931.28	99.55%
128	2359296	930.14	99.65%	937.67	98.88%
256	4718592	933.97	99.24%	950.25	97.57%

Exercises

1. Show that when a delta function is chosen as the weighting function, the Weighted Residuals formulation gives the finite volume scheme.
2. Solve the problem defined by Equation 2.1e using five linear elements and compare with the analytic solution.
3. Solve Example 2.1e using five linear continuous finite elements and five finite volume cells. Compare the results with the results in Exercise 1 and the analytical solution.
4. Complete the calculations in Example 2.2e for the second element and compare with the analytic solution.
5. Apply a discontinuous finite element formulation to solve the problem defined by Equation 2.20e when the domain is discretized into six linear elements.
6. Solve Equation 2.20e using six linear continuous finite elements and six finite volume cells respectively and compare the results with those obtained in Exercise 5.
7. Use discontinuous finite element formulation and three quadratic elements to solve Equation 2.20e and compare with the results obtained in Exercise 5.
8. Develop a computer code for a discontinuous finite element solution to Equation 2.20e, and compare the results obtained from the code with those calculated in Exercises 6 and 7.

References

[1]   Reed WH, Hill TR. Triangular Mesh Methods for the Neutron Transport Equation. Los Alamos Scientific Laboratory Report LA-UR-73-479, Los Alamos, NM, 1973.
[2]   Cockburn B, Karniadakis G, Shu CW. The Development of Discontinuous Galerkin Methods. In: Cockburn B, Karniadakis G, Shu CW, editors. Discontinuous Galerkin Methods: Theory, Computation and Applications. Lecture Notes in Computational Science and Engineering, 11. New York: Springer-Verlag, 2000; 3–50.
[3]   Cockburn B. Devising Discontinuous Galerkin Methods for Non-Linear Hyperbolic Conservation Laws. J. Comput. Appl. Math. 2001; 128: 187–204.
[4]   Bathe KJ. (Editor) Computational Fluid and Solid Mechanics 2005, the 3rd MIT Conference on Computational Fluid and Solid Mechanics. Amsterdam: Elsevier, 2005.
[5]   Hulsen MA. The Discontinuous Galerkin Method with Explicit Runge–Kutta Time Integration Hyperbolic and Parabolic Systems with Source Terms. The Netherlands: Deft University, 1991.
[6]   Johnson C. Finite Element Solution of Partial Differential Equations. London: Cambridge University, 1986.
[7]   Zienkiewicz OC, Taylor RL, Sherwin SJ, Peiro I. On Discontinuous Galerkin Methods. Int. J. Numer. Meth. Eng. 2003; 58: 1119–1148.
[8]   Bathe KJ. Finite Element Procedures. Englewood Cliffs: Prentice Hall, 1996.
[9]   Zienkiewicz OC, Taylor RL. The Finite Element Method, 4th Ed. New York: McGraw-Hill, 1989.
[10] Arnold DN, Brezzi F, Cockburn B, Marni D. Unified Analysis of Discontinuous Galerkin Methods for Elliptic Problems. SIAM J. Numer. Anal. 2002; 39:1749–1779.

[11] Oden, JT, Babuska I, Baumann CE. A Discontinuous *hp*-Finite Element Method for Diffusion Problems. J. Comput. Phys.1998; 146: 491–519.

[12] Brenner SC, Scott LR. The Mathematical Theory of Finite Element Methods. New York: Springer-Verlag, 2002.

[13] Johnson C, Pitkaranta. An Analysis of the Discontinuous Galerkin Method for Scalar Hyperbolic Equation. SIAM J. Sci. Stat. Comput. 1986; 46: 1–26.

[14] Ai X and Li BQ. Solution of Hyperbolic Heat Conduction Problems by Discontinuous Finite Element Method. J. Eng. Comput. 2004; 21(16): 577–597.

[15] Shu CW. Total-Variation-Diminishing Time Discretization. SIAM J. Sci. Stat. Comput. 1988; 48(9): 1073–1085.

[16] Brebbia CA, Telles JF, Wrobel LC. Boundary Element Techniques. New York: Springer-Verlag, 1984.

[17] Hughes TJR, Engel G, Mazzei L, Larson MG. A Comparison of Discontinuous and Continuous Galerkin Methods Based on Error Estimates, Conservation, Robustness and Efficiency. In: Cockburn B, Karniadakis G, Shu CW, editor. Discontinuous Galerkin Methods: Theory, Computation and Applications. Lecture Notes in Computational Science and Engineering, 11. New York: Springer-Verlag, Feb. 2000; 135–146.

[18] Biswas R, Devine KD, Flaherty JE. Parallel Adaptive Finite Element Methods for Conservation Laws. Appl. Numer. Math. 1994; 14(1): 255–283.

3

Shape Functions and Elemental Calculations

Like its continuous counterpart, the discontinuous finite element method employs shape functions for local approximations. The use of 1-D linear shape functions was demonstrated in the last chapter for the discontinuous Galerkin solution of boundary and initial value problems. There, it was also shown that a discontinuous finite element formulation hinges critically on local interpolation functions and involves computations at the element level. The calculations for higher dimensions require interpolation functions or shape functions defined in multi-dimensions. For many engineering applications, elements based on linear and quadratic interpolations may be sufficient. Because the elements for the discontinuous finite element formulations are embedded in a finite element broken space, and because the formulations are localized, higher order polynomial interpolation functions can be easily incorporated in the solution procedure. This flexibility in local approximation has motivated researchers to develop and apply more accurate, local spectral basis functions for higher order analysis.

In this chapter, the shape functions, numerical integration, and elemental calculations used for discontinuous finite element calculations are discussed. The chapter starts with the procedure and basic criteria for the construction of the simple 1-D finite elements. The idea is then extended to establish a general framework for developing standard finite elements in multi-dimensions. A brief discussion is given on the construction of spectral elements, which are known for their higher order accuracy of approximations. Hierarchical elements are widely used in p-type adaptive analysis; and the construction of a variety of hierarchical shape functions and the selection of hierarchical polynomials for certain applications are also discussed. The techniques for constructing special elements and transition elements through coordinate transformation are also presented. Interpolation error estimates for uniform and non-uniform meshes are presented by introducing various norms for error measures. These error estimates are of crucial importance in developing adaptive algorithms for discontinuous finite element computations. Numerical integration is discussed and the Gaussian quadrature is given for various classes of elements. Elemental calculations are an important part of a discontinuous formulation, and are discussed for 2-D and 3-D volume and surface elements.

3.1 Shape Functions

Shape functions are local functions restricted to an element, and are of vital importance to discontinuous finite element approximations. Here, we discuss the basic ideas for constructing shape functions developed from the Lagrangian interpolation theory. The Lagrangian interpolation functions make a natural candidate for local finite element approximations, because of their simplicity [1–3]. The 1-D shape functions are discussed first and the concept is extended to the higher dimensions. The elements discussed in this section are generally classified as standard elements.

3.1.1 1-D Shape Functions

Application of the discontinuous finite element formulation starts with breaking the computational domain into a tessellation of elements. The shape function is then constructed over each of the elements. Consider the 1-D discretization as shown in Figure 3.1, where the domain is discretized into $N = 5$ elements.

Element j

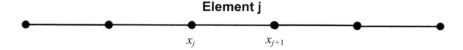

$$x_j \qquad x_{j+1}$$

Figure 3.1. Discretization of a 1-D domain into 5 linear finite elements. Dots denote the nodal points

For element j, $\Omega_j = [x_j, x_{j+1}]$, and by the Lagrangian interpolation theory, the unknown function u is approximated by a function $u_h(x)$ local to the element,

$$u_h(x) = ax + b \qquad (3.1)$$

At two end points of Ω_j, we require that the approximate function $u_h(x)$ assumes the values of the unknown function,

$$u_h(x_j) = u_j ; \quad u_h(x_{j+1}) = u_{j+1} \qquad (3.2)$$

With these relations substituted into Equation 3.1, the two constants, a and b, can be determined. The linear functions are re-written in terms of the nodal values of u,

$$u_h(x) = u_j \phi_j(x) + u_{j+1} \phi_{j+1}(x) \qquad (3.3)$$

In the above equation, $\phi_i(x)$ $(i = j, j+1)$ is the shape function defined over Ω_j,

$$\phi_j(x) = \frac{x_{j+1} - x}{x_{j+1} - x_j} \text{ and } \phi_{j+1}(x) = \frac{x - x_j}{x_{j+1} - x_j}, \quad \forall x \in V_j \tag{3.4}$$

Here $x \in \Omega_j$ applies to every x in Ω_j. While simple, the shape functions in Equation 3.4 display two important properties: (i) $\phi_j(x)$ is unit at the node j or x_j, and (ii) it is non-zero on Ω_j and is zero everywhere else. The first property ensures that a local interpolation function satisfies the conditions stated in Equation 3.3, while the second property is the statement of element locality. The shape functions above are called linear shape functions because they vary linearly over Ω_j.

For practical applications, the isoparametric approximation is often used, by which the shape functions are constructed over a normalized coordinate system. The element defined in a normalized coordinate system is referred to as a canonical element. To recast the shape function in normalized coordinates, a normalized parameter ($\xi \in [-1,+1]$) is introduced, and is related to x in the following way:

$$\xi = \frac{2x - x_{j+1} - x_j}{x_{j+1} - x_j} \tag{3.5}$$

This is equivalent to transforming Ω_j from the x coordinate to the ξ coordinate, or natural coordinate, as shown in Figure 3.2.

Figure 3.2. Mapping of an element, defined on $x \in [x_j, x_{j+1}]$, to a canonical element defined on $\xi \in [-1, +1]$. The known functions u_j and u_{j+1} are mapped as u_1 and u_2 in the ξ coordinate system

Using the transformation rule given in Equation 3.5, the linear shape functions $\phi_i(x)$ ($i = 1, 2$) may now be written in terms of the ξ coordinate system,

$$\phi_1(\xi) = \tfrac{1}{2}(1-\xi) \text{ and } \phi_2(\xi) = \tfrac{1}{2}(1+\xi) \tag{3.6}$$

The shape functions, as shown in Figure 3.3, vary linearly over the normalized range, $[-1, +1]$. With the shape functions defined in this way, the unknown function is transformed as a function of ξ,

$$u(\xi) = u(\xi) = u_1 \phi_1(\xi) + u_2 \phi_2(\xi) \tag{3.7}$$

As shown above, a shape function can be constructed either in the original coordinate system or, more conveniently, in a normalized coordinate system. While the above discussion illustrates the construction of a linear shape function, the same procedure can be applied to develop higher order shape functions. As seen above, a linear element has two nodes at its ends. For higher order elements, more nodes are required and at each node the basic condition is satisfied: $u_h(x_j) = u_j$, with j being the node number. For example, a quadratic element has 3 nodes and a cubic element has 4 nodes. These higher order shape functions can also be expressed in terms of the natural coordinate $\xi \in [-1,1]$. For a quadratic element, the above procedure leads to a set of three shape functions, all of quadratic order:

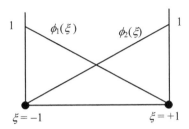

Figure 3.3. Linear element shape functions

$$\phi_1(\xi) = -\tfrac{1}{2}\xi(1-\xi); \quad \phi_2(\xi) = (1+\xi)(1-\xi); \quad \phi_3(\xi) = \tfrac{1}{2}\xi(1+\xi) \qquad (3.8)$$

These shape functions and node arrangement for a quadratic element defined over $[-1,1]$ are graphically displayed in Figure 3.4.

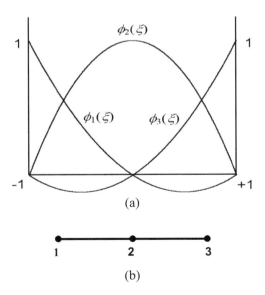

Figure 3.4. Quadratic element shape functions for a 1-D canonical element: (a) shape functions and (b) node arrangement

Similarly, cubic shape functions can be constructed, and when written for a canonical element $\{\xi \in [-1,1]\}$, they assume the form of

$$\phi_1(\xi) = -\tfrac{9}{16}(\tfrac{1}{3}+\xi)(\tfrac{1}{3}-\xi)(1-\xi); \quad \phi_2(\xi) = \tfrac{27}{16}(1+\xi)(\tfrac{1}{3}-\xi)(1-\xi)$$

$$\phi_3(\xi) = \tfrac{27}{16}(1+\xi)(\tfrac{1}{3}+\xi)(1-\xi); \quad \phi_4(\xi) = -\tfrac{9}{16}(1+\xi)(\tfrac{1}{3}+\xi)(\tfrac{1}{3}-\xi) \qquad (3.9)$$

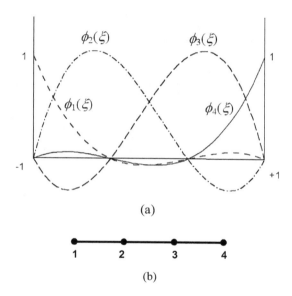

(a)

(b)

Figure 3.5. Cubic element shape functions for a 1-D canonical element: (a) shape functions and (b) node arrangement

The shape functions of even higher orders may now be constructed readily by following the same procedure. These functions are in the general class of the Lagrangian interpolation polynomials of the nth order, $\psi_k^n(x)$, which are defined by

$$\psi_k^n(x) = \sum_{j=0,j\ne k}^{n} \frac{(x-x_j)}{(x_k-x_j)}$$

$$= \frac{(x-x_0)(x-x_1)\cdots(x-x_{k-1})(x-x_{k+1})\cdots(x-x_n)}{(x_k-x_0)(x_k-x_1)\cdots(x_k-x_{k-1})(x_k-x_{k+1})\cdots(x_k-x_n)} \qquad (3.10)$$

where the subscript k is the kth interpolation point, and the superscript $n+1$ the total number of interpolation points. Clearly, the Lagrangian interpolation basis given above maintains the two important properties for shape functions of any order,

$$\psi_k^n(x) = \begin{cases} \delta_{kj} & x = x_j \in [x_0, x_n] \\ 0 & x \notin [x_0, x_n] \end{cases} \tag{3.11}$$

where δ_{kj} is the delta function: $\delta_{kj} = 1$ if $k = j$ and $\delta_{kj} = 0$ if $k \neq j$.

As we shall show below, the Lagrangian polynomials are also useful for constructing elements in multi-dimensions. Before we do so, let us look at one simple example illustrating the use of Equation 3.10.

Example 3.1. Use the Lagrangian interpolation formulae to construct the quadratic shape functions over a canonical element $\{\xi \in [-1,1]\}$.

Solution. Let $x = \xi$. For a quadratic approximation, three interpolation points are taken: $x_0 = \xi_0 = -1$, $x_1 = \xi_1 = 0.0$, and $x_2 = \xi_2 = 1$. From Equation 3.10, we have

$$\phi_1(\xi) = \psi_{k=0}^2(x) = \frac{(x - x_1)(x - x_2)}{(x_0 - x_1)(x_0 - x_2)} = \frac{(\xi - 0.0)(\xi - 1)}{(-1 - 0.0)(-1 - 1)} = -\frac{1}{2}\xi(1 - \xi)$$
$$\tag{3.1e}$$

Similarly, the other two shape functions, i.e. $\phi_2(\xi)$ and $\phi_3(\xi)$, are obtained by setting $k = 1$ and $k = 2$ in Equation 3.10.

3.1.2 2-D Shape Functions

For two dimensional analyses, shape functions defined over 2-D geometries are needed. The triangular and quadrilateral elements are perhaps the most frequently used elements in thermal and fluids engineering analysis.

3.1.2.1 Triangular Elements

As in the 1-D case, linear elements are considered first. For a linear triangular element, the unknown function u within each element is approximated as

$$u_h(x, y) = a + bx + cy \tag{3.12}$$

where a, b, and c are constant coefficients to be determined. A linear triangular element has three nodes located at the vertices of the triangle (see Figure 3.6a). The nodes are numbered counterclockwise by numerals 1, 2, and 3, with the corresponding values of u denoted by u_1, u_2 and u_3, respectively. Enforcing Equation 3.2 at the three nodes, one has

$$u_1 = u_h(x_1, y_1) = a + bx_1 + cy_1; \quad u_2 = u_h(x_2, y_2) = a + bx_2 + cy_2;$$

$$u_3 = u_h(x_3, y_3) = a + bx_3 + cy_3$$

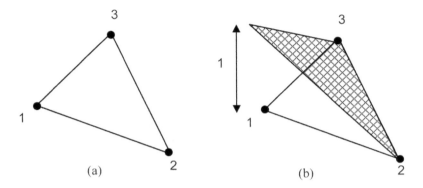

Figure 3.6. A 3–node triangular element and its shape function: (a) numbering sequence and (b) distribution of function $\phi_1(x,y)$

Solving for the constant coefficients a, b, and c in terms of u_j, and substituting them back into Equation 3.12, one has

$$u_h(x,y) = \sum_{j=1}^{3} \phi_j(x,y)u_j \tag{3.13}$$

where $\phi_j(x, y)$ is the interpolation or shape function given by

$$\phi_j(x,y) = \frac{1}{2\Delta^e}(a_j^e + b_j^e x + c_j^e y), \quad j = 1, 2, 3 \tag{3.14}$$

in which

$$a_1^e = x_2^e y_3^e - y_2^e x_3^e; \quad b_1^e = y_2^e - y_3^e; \quad c_1^e = x_3^e - x_2^e$$

$$a_2^e = x_3^e y_1^e - y_3^e x_1^e; \quad b_2^e = y_3^e - y_1^e; \quad c_2^e = x_1^e - x_3^e$$

$$a_3^e = x_1^e y_2^e - y_1^e x_2^e; \quad b_3^e = y_1^e - y_2^e; \quad c_3^e = x_2^e - x_1^e$$

and

$$\Delta^e = \frac{1}{2}\begin{vmatrix} 1 & x_1^e & y_1^e \\ 1 & x_2^e & y_2^e \\ 1 & x_3^e & y_3^e \end{vmatrix} = \tfrac{1}{2}(b_1^e c_2^e - b_2^e c_1^e) = \text{area of the } e\text{th element}$$

In the above equations, x_j^e and y_j^e (j = 1,2,3) denote the coordinate values of the jth node in the eth element.

From Equation 3.13, we see that the interpolation functions have the following important property:

$$\phi_i(x_j^e, y_j^e) = \delta_{ij} = \begin{cases} 1 & i = j \\ 0 & i \neq j \end{cases} \tag{3.15}$$

Thus, at node i, $u(x_i, y_i)$ in Equation 3.13 reduces to its nodal value u_i. Another important feature of $\phi_i(x,y)$ is that it vanishes when the observation point (x, y) is on the element side opposite to the jth node. Therefore, the value of u at an element side is not related to the value of u at the opposite node; rather it is determined by the values at the two endpoints of its associated side. This important feature guarantees the continuity of the solution across the element sides. These features are shown in Figure 3.6a, which displays the interpolation functions $\phi_j(x,y)$ for a triangular element.

Shape functions for triangular elements can also be expressed in normalized area coordinates. If we join any point P in the triangle to the vertices of the triangle, we have three sub-triangles with areas, A_1, A_2, and A_3, as shown in Figure 3.7. Then the shape functions can be written in terms of normalized area coordinates η_j,

$$\phi_j(x, y) = \phi_j(\eta_1, \eta_2, \eta_3) = \eta_j = A_j/\Delta^e, \, j = 1, 2, 3 \tag{3.16}$$

with η_j satisfying the normalization relation,

$$\eta_1 + \eta_2 + \eta_3 = 1 \tag{3.17}$$

The higher order shape functions can be constructed similarly.

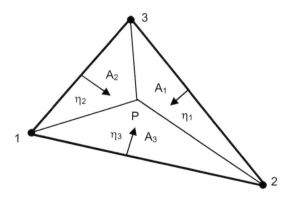

Figure 3.7. Area coordinates for a triangular element

Alternatively, the shape functions can be constructed directly, in terms of the normalized area coordinates. One simple recursive relation is derived for shape functions for triangular elements [1],

$$\phi_j(\eta_1,\eta_2,\eta_3) = \phi_{(I,J,K)} = \psi_I^I(\eta_1)\psi_J^J(\eta_2)\psi_K^K(\eta_3) \tag{3.18}$$

where $\psi_k^n(\eta_l)$ ($l=1, 2, 3$) is given by the Lagrangian polynomials, and $I+J+K=M$ is the order of the polynomial. This relation can be used to generate shape functions of any order.

Example 3.2. Develop shape functions from the recursive relation for 3– and 6– node triangular elements.

Solution. For a linear triangle, the highest order of a polynomial is $M = 1$. From Equation 3.14, we have in reference to Figure 3.1e (a),

$$\phi_1(\eta_1,\eta_2,\eta_3) = \phi_{(1,0,0)} = \psi_1^1(\eta_1) = \frac{\eta_1 - (\eta_1)_0}{(\eta_1)_1 - (\eta_1)_0} = \frac{\eta_1 - 0}{1 - 0} = \eta_1 \tag{3.2e}$$

Similarly,

$$\phi_2(\eta_1,\eta_2,\eta_3) = \phi_{(0,1,0)} = \psi_1^1(\eta_2) = \eta_2; \; \phi_3(\eta_1,\eta_2,\eta_3) = \phi_{(0,0,1)} = \psi_1^1(\eta_3) = \eta_3$$

This is a different approach to the same problem.

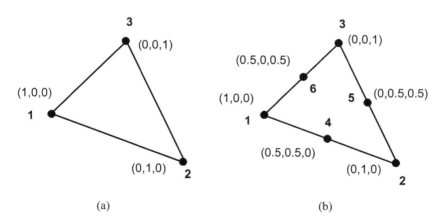

(a)   (b)

Figure 3.1e. Triangular elements and their shape functions: (a) linear element and (b) quadratic element

For a quadratic triangle or 6–node triangle, the polynomial has an order of $M = 2$. Let us consider node 1,

$$\phi_1(\eta_1,\eta_2,\eta_3) = \phi_{(2,0,0)} = \psi_2^2(\eta_1) = \frac{[\eta_1 - (\eta_1)_0][\eta_1 - (\eta_1)_1]}{[(\eta_1)_2 - (\eta_1)_0][(\eta_1)_2 - (\eta_1)_1]}$$

$$= \frac{(\eta_1 - 0)(\eta_1 - 0.5)}{(1-0)(1-0.5)} = \eta_1(2\eta_1 - 1) \tag{3.3e}$$

By the same token, the corner nodes have a similar form, that is, $\phi_j(\eta_1, \eta_2, \eta_3) = \eta_j(2\eta_j - 1)$. For the mid-node, say, node 4, we have the following result in reference to Figure 3.1e(b):

$$\phi_4(\eta_1,\eta_2,\eta_3) = \phi_{(1,1,0)} = \psi_1^1(\eta_1)\psi_1^1(\eta_2) = \frac{[\eta_1 - (\eta_1)_0]}{[(\eta_1)_1 - (\eta_1)_0]}\frac{[\eta_2 - (\eta_2)_0]}{[(\eta_2)_1 - (\eta_2)_0]}$$

$$= \frac{(\eta_1 - 0)}{(1-0.5)}\frac{(\eta_2 - 0)}{(1-0.5)} = 4\eta_1\eta_2 \tag{3.4e}$$

Other mid-node terms can be obtained similarly [1].

3.1.2.2 Quadrilateral Elements
There are four nodes in a rectangular element. This allows us to construct an interpolation function in the following form, which contains four coefficients a, b, c, and d,

$$u_h(x,y,z) = a + bx + cy + dxy \tag{3.19}$$

Now, we may follow the same procedure given above to obtain shape functions ϕ_j, $j = 1, 2, 3, 4$, which may then be written in normalized coordinates through coordinate transformation.

Alternatively, the shape functions are formed by taking products of one-dimensional Lagrangian polynomials. Multi-dimensional polynomials formed in this manner are called "tensor-product" approximations. Let us consider the 2×2 square $\{(\xi,\eta)|-1\leq\xi,\eta\leq1\}$ shown in Figure 3.8. For simplicity, the vertices of the element are indexed with double subscripts as (0,0), (1,0), (0,1), and (1,1), which correspond to node numbers 1, 2, 4 and 3, respectively. With nodes at each vertex, we now construct a bilinear Lagrangian polynomial by tensor-product [1],

$$\phi_{(k,l)}(\xi,\eta) = \psi_k^m(\xi)\psi_l^n(\eta), \quad k, l = 0, 1 \tag{3.20}$$

where the Lagrangian polynomials ψ_i^j are used and m and n are orders of interpolation functions. Written with the node number as the subscript, the shape functions for the 4–node element take the following form:

$$\phi_1(\xi,\eta) = \psi_0^1(\xi)\psi_0^1(\eta) = \tfrac{1}{4}(1-\xi)(1-\eta)$$

$$\phi_2(\xi,\eta) = \psi_1^1(\xi)\psi_0^1(\eta) = \tfrac{1}{4}(1+\xi)(1-\eta)$$

$$\phi_3(\xi,\eta) = \psi_1^1(\xi)\psi_1^1(\eta) = \tfrac{1}{4}(1+\xi)(1+\eta)$$

$$\phi_4(\xi,\eta) = \psi_0^1(\xi)\psi_1^1(\eta) = \tfrac{1}{4}(1-\xi)(1+\eta)$$

where the Lagrangian interpolation functions in Equation 3.10 have been used. The tensor product is general, and can be easily extended to the shape functions of higher order. This is illustrated through an example below.

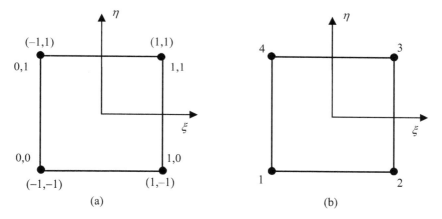

Figure 3.8. A canonical square element with 4 nodes: (a) double indexed and (b) node arrangement

Example 3.3. Construct the shape functions for a 9–node element defined over a square $\{(\xi,\eta)|\ -1 \le \xi, \eta \le 1\}$ and selectively display the function over the 2-D domain.

Solution. For a 9–node element, the tensor product formula given in Equation 3.14 is extended to the quadratic order in both ξ and η,

$$\phi_{(k,l)}(\xi,\eta) = \psi_k^m(\xi)\psi_l^n(\eta), \quad k,l = 0,1,2;\ m,n = 2 \tag{3.5e}$$

Using Equation 3.5e and noting the correspondence between the node number and subscript index in 1-D shape functions, as shown in Figure 3.2e, one has the shape functions for a 9–node element,

$$\phi_1(\xi,\eta) = \tfrac{1}{4}\xi\eta(1-\xi)(1-\eta); \qquad \phi_2(\xi,\eta) = \tfrac{1}{4}\xi\eta(1+\xi)(1-\eta)$$

$$\phi_3(\xi,\eta) = \tfrac{1}{4}\xi\eta(1+\xi)(1+\eta); \qquad \phi_4(\xi,\eta) = \tfrac{1}{4}\xi\eta(1-\xi)(1+\eta)$$

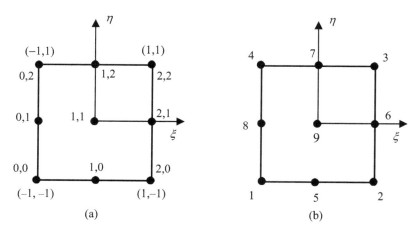

Figure 3.2e. A canonical square element with 9 nodes: (a) coordinates and double indexed node arrangement, and (b) single indexed node arrangement

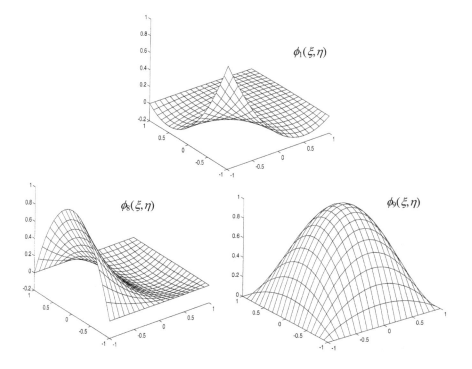

Figure 3.3e. The distribution of selected shape functions for a 9–node quadratic element

$$\phi_5(\xi,\eta) = -\tfrac{1}{2}\eta(1-\xi^2)(1-\eta); \quad \phi_6(\xi,\eta) = \tfrac{1}{2}\xi(1+\xi)(1-\eta^2)$$

$$\phi_7(\xi,\eta) = \tfrac{1}{2}\eta(1-\xi^2)(1+\eta); \quad \phi_8(\xi,\eta) = -\tfrac{1}{2}\xi(1-\xi)(1-\eta^2)$$

$$\phi_9(\xi,\eta) = (1-\xi^2)(1-\eta^2) \tag{3.6e}$$

Three of these shape functions are selectively plotted in Figure 3.3e.

For engineering analyses, 8–node elements are used frequently. The shape functions for this type of element may be constructed by starting with the general interpolation function, solving for the coefficient constants, and then transforming them into normalized coordinates. This procedure can be very tedious. Alternatively, they can be constructed by a combination of 1-D and 2-D shape functions of mixed orders, as shown in the example below.

Example 3.4. Develop shape functions for an 8–node element from shape functions given for a 9–node element.

Solution. A shape function at a node i needs to satisfy two conditions: (1) $\phi_i = 1$ at node i, and (2) $\phi_i = 0$ at other nodes. We start with the mid-point, say, node 5, as shown in Figure 3.4e. The conditions are satisfied if a Lagrangian interpolation of a quadratic × linear type is used,

$$\phi_5(\xi,\eta) = \tfrac{1}{2}(1-\xi^2)(1-\eta) \tag{3.7e}$$

The other mid-nodes have a similar form; e.g., the shape function for node 8 is

$$\phi_8(\xi,\eta) = \tfrac{1}{2}(1-\xi)(1-\eta^2) \tag{3.8e}$$

We next consider corner node 1. If we start with the a linear interpolation,

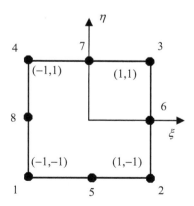

Figure 3.4e. A canonical square element with 8 nodes

$$\phi_{1,trial}(\xi,\eta) = \tfrac{1}{4}(1-\xi)(1-\eta) \tag{3.9e}$$

It is clear that $\phi_{1,trial} = 1$ at node 1, $\phi_{1,trial} = 0$ at nodes 2, 3, 4, 6, 7, but $\phi_{1,trial} = 0.5$ at nodes 5 and 8. The shape function is zero at nodes 5 and 8 if the following correction is made,

$$\phi_1(\xi,\eta) = \phi_{1,trial}(\xi,\eta) - 0.5\phi_5(\xi,\eta) - 0.5\phi_8(\xi,\eta)$$
$$= \tfrac{1}{4}(1-\xi)(1-\eta) - \tfrac{1}{4}(1-\xi^2)(1-\eta) - \tfrac{1}{4}(1-\xi)(1-\eta^2)$$
$$= \tfrac{1}{4}(1-\xi)(1-\eta)(-1-\xi-\eta) \tag{3.10e}$$

The shape functions for other corner nodes can be obtained similarly. These shape functions are given in Table 3.1.

3.1.3 3-D Shape Functions

Three-dimensional element shape functions can be constructed very similarly, following the procedures illustrated for 2-D elements.

3.1.3.1 Tetrahedral Elements

These elements are the simplest geometric units to approximate 3-D geometries. In a similar fashion to that discussed for 1-D and 2-D elements, an unknown function u can be interpolated in a tetrahedron [1, 2].

$$u_h(x,y) = a + bx + cy + dz \tag{3.21}$$

The coefficients a, b, c, and d can be obtained in terms of u_i at the nodal points, and then substituted back to yield the final form of the interpolation,

$$u_h(x,y) = \phi_1 u_1 + \phi_2 u_2 + \phi_3 u_3 + \phi_4 u_4 \tag{3.22}$$

where ϕ_i is defined as

$$\phi_i(x,y,z) = \frac{1}{6V^e}\left(a_i^e + b_i^e x + c_i^e y + d_i^e z\right) \tag{3.23}$$

Here the above coefficients are determined from the following relations:

$$a = \frac{1}{6V^e}\begin{vmatrix} u_1 & u_2 & u_3 & u_4 \\ x_1^e & x_2^e & x_3^e & x_4^e \\ y_1^e & y_2^e & y_3^e & y_4^e \\ z_1^e & z_2^e & z_3^e & z_4^e \end{vmatrix} = \frac{1}{6V^e}\left(a_1^e u_1 + a_2^e u_2 + a_3^e u_3 + a_4^e u_4\right)$$

$$b = \frac{1}{6V^e} \begin{vmatrix} 1 & 1 & 1 & 1 \\ u_1 & u_2 & u_3 & u_4 \\ y_1^e & y_2^e & y_3^e & y_4^e \\ z_1^e & z_2^e & z_3^e & z_4^e \end{vmatrix} = \frac{1}{6V^e} \left(b_1^e u_1 + b_2^e u_2 + b_3^e u_3 + b_4^e u_4 \right)$$

$$c = \frac{1}{6V^e} \begin{vmatrix} 1 & 1 & 1 & 1 \\ x_1^e & x_2^e & x_3^e & x_4^e \\ u_1 & u_2 & u_3 & u_4 \\ z_1^e & z_2^e & z_3^e & z_4^e \end{vmatrix} = \frac{1}{6V^e} \left(c_1^e u_1 + c_2^e u_2 + c_3^e u_3 + c_4^e u_4 \right)$$

$$d = \frac{1}{6V^e} \begin{vmatrix} 1 & 1 & 1 & 1 \\ x_1^e & x_2^e & x_3^e & x_4^e \\ y_1^e & y_2^e & y_3^e & y_4^e \\ u_1 & u_2 & u_3 & u_4 \end{vmatrix} = \frac{1}{6V^e} \left(d_1^e u_1 + d_2^e u_2 + d_3^e u_3 + d_4^e u_4 \right)$$

$$V^e = \frac{1}{6} \begin{vmatrix} 1 & 1 & 1 & 1 \\ x_1^e & x_2^e & x_3^e & x_4^e \\ y_1^e & y_2^e & y_3^e & y_4^e \\ z_1^e & z_2^e & z_3^e & z_4^e \end{vmatrix} = \text{Volume of element}$$

where x_i^e, y_i^e, z_i^e, u_i^e are x, y, z coordinates, and u_i values at node i. From the above equations, the coefficients a_i^e, b_i^e, c_i^e, d_i^e are readily obtained by expansion. In fact, they are the determinants of the cofactors of the relevant matrices.

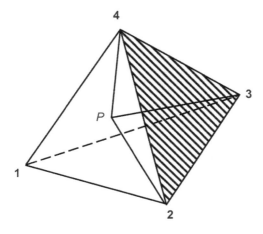

Figure 3.9. Tetrahedral edge element with edge and node numbering. P is an internal point

Similar to the 2-D triangular case, we may also use the volume coordinates to express the shape functions. This is constructed in the same manner as the area coordinates. As shown in Figure 3.9, the volume coordinate is defined by the ratio of the subvolume, constructed from point P inside the tetrahedron, connected to three vertices of a face triangle,

$$\eta_1 = V_{P234}/V^e \; ; \; \eta_2 = V_{1P34}/V^e \; ; \; \eta_3 = V_{12P4}/V^e \; ; \; \eta_4 = V_{123P}/V^e \quad (3.24)$$

Thus, the shape function takes the following form for a linear element:

$$\phi_i = \eta_i \, , \, i = 1, 2, 3, 4 \quad (3.25)$$

The following simple recursive relation may also be used to express the shape functions in terms of the volume coordinates [1],

$$\phi_j(\eta_1, \eta_2, \eta_3, \eta_4) = \phi_{(I,J,K,L)} = \psi_I^I(\eta_1)\psi_J^J(\eta_2)\psi_K^K(\eta_3)\psi_L^L(\eta_4) \quad (3.26)$$

where $I + J + K + L = M$ is the higher order of the polynomial. This relation can be used to generate shape functions of any order for 3-D tetrahedral elements.

3.1.3.2 Hexahedral Elements

Following the above procedure, an interpolation function is constructed for an 8–node brick element, which contains 8 coefficients $a, b, c, d, e, f, g,$ and h:

$$u_h(x, y) = a + bx + cy + dz + exy + fyz + gxz + hxyz \quad (3.27)$$

The coefficients are now determined to obtain shape functions $\phi_j, j=1,\ldots,8$. A typical 8–node brick element is shown in Figure 3.10. These shape functions, written in (x, y, z) coorindates may then be written in normalized coordinates through coordinate transformation.

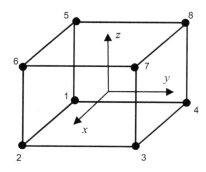

Figure 3.10. A brick element with 8 nodes

An easier way is to construct the shape functions by taking products of one-dimensional Lagrangian interpolation functions. Thus, for the $2 \times 2 \times 2$ cube, $\{(\xi, \eta, \zeta) | -1 \le \xi, \eta, \zeta \le 1\}$, shown in Figure 3.11, the vertices of the element may be indexed with triple subscripts as $(0,0,0)$, $(1,0,0)$, $(1,1,0)$, $(0,1,0)$, $(0,0,1)$, $(1,0,1)$, $(1,1,1)$, and $(0,1,1)$, which correspond to single number indexed shape functions 1, 2, 3, 4, 5, 6, 7, and 8. With a node at a vertex, we construct a bilinear Lagrangian polynomial by tensor product,

$$\phi_{(k,l)}(\xi, \eta, \zeta) = \psi_k^I(\xi)\psi_l^J(\eta)\psi_m^K(\zeta), \; k, l, m = 0, 1 \qquad (3.28)$$

A hexahedron can be transformed into a canonical element, as in the 1-D and 2-D cases. With the isoparametric transformation, this is easily achieved, as shown in Figure 3.11. Often the isoparametric transformation is made using the following transformation rules:

$$x(\xi, \eta, \zeta) = \sum_{i=1}^{N_e} \phi_i(\xi, \eta, \zeta)x_i, \quad y(\xi, \eta, \zeta) = \sum_{i=1}^{N_e} \phi_i(\xi, \eta, \zeta)y_i,$$

$$z(\xi, \eta, \zeta) = \sum_{i=1}^{N_e} \phi_i(\xi, \eta, \zeta)z_i \qquad (3.29)$$

where N_e is the number of nodes of the element or number of nodes used for parametric transformation.

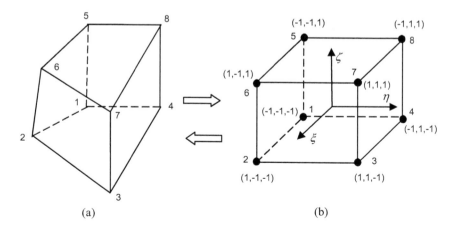

(a)   (b)

Figure 3.11. A hexagon is transformed into a canonical cube in 3-D: (a) the hexagonal element and (b) the canonical cube

Both linear and higher order shape functions can be obtained using this relation. The procedure is identical to the 2-D case and thus is omitted here. Also, shape functions for 20–node elements can be described in a precisely analogous way to that given for 2-D 8–node elements.

The commonly used shape functions for 1-D, 2-D and 3-D elements are summarized in Table 3.1.

Table 3.1. Elements and shape functions

Elements	Shape functions
	$\phi_1(\xi) = \frac{1}{2}(1-\xi), \quad \phi_2(\xi) = \frac{1}{2}(1+\xi)$
	$\phi_1(\xi) = -\frac{1}{2}\xi(1-\xi), \quad \phi_2(\xi) = (1+\xi)(1-\xi)$ $\phi_3(\xi) = \frac{1}{2}\xi(1+\xi)$
	$\phi_1(\xi) = -\frac{9}{16}(\frac{1}{3}+\xi)(\frac{1}{3}-\xi)(1-\xi)$ $\phi_2(\xi) = \frac{27}{16}(1+\xi)(\frac{1}{3}-\xi)(1-\xi)$ $\phi_3(\xi) = \frac{27}{16}(1+\xi)(\frac{1}{3}+\xi)(1-\xi)$ $\phi_4(\xi) = -\frac{9}{16}(1+\xi)(\frac{1}{3}+\xi)(\frac{1}{3}-\xi)$
	$\phi_1(\xi_1,\xi_2,\xi_3) = \xi_1$ $\phi_2(\xi_1,\xi_2,\xi_3) = \xi_2$ $\phi_3(\xi_1,\xi_2,\xi_3) = \xi_3$
	$\phi_i(\xi_1,\xi_2,\xi_3) = \xi_i(2\xi_i - 1)$ $\phi_{i+3}(\xi_1,\xi_2,\xi_3) = 4\xi_i\xi_j$ $i = 1,2,3$ $j = i+1$ if $i=1,2$ $j = 1$ if $i=3$
	$\phi_i(\xi_1,\xi_2,\xi_3) = \frac{1}{2}\xi_i(3\xi_i - 1)(3\xi_i - 2)$ $\phi_{2i+2}(\xi_1,\xi_2,\xi_3) = \frac{9}{2}\xi_i\xi_j(3\xi_i - 1)$ $\phi_{2i+3}(\xi_1,\xi_2,\xi_3) = \frac{9}{2}\xi_i\xi_j(3\xi_j - 1)$ $\phi_{10}(\xi_1,\xi_2,\xi_3) = 27\xi_1\xi_2\xi_3$ $i = 1,2,3$ $j = i+1$ if $i=1,2$ $j = 1$ if $i=3$

Table 3.1. Continued

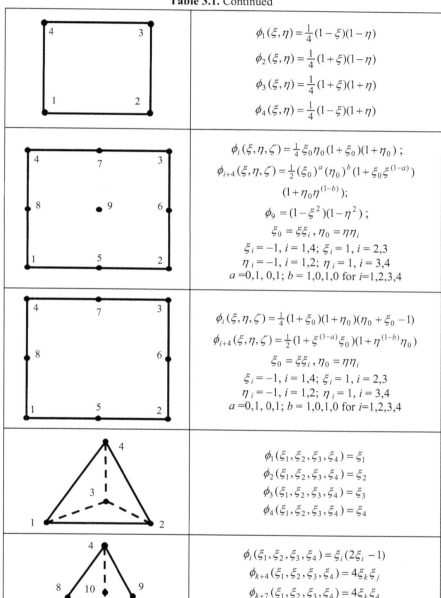

$$\phi_1(\xi,\eta) = \tfrac{1}{4}(1-\xi)(1-\eta)$$

$$\phi_2(\xi,\eta) = \tfrac{1}{4}(1+\xi)(1-\eta)$$

$$\phi_3(\xi,\eta) = \tfrac{1}{4}(1+\xi)(1+\eta)$$

$$\phi_4(\xi,\eta) = \tfrac{1}{4}(1-\xi)(1+\eta)$$

$$\phi_i(\xi,\eta,\zeta) = \tfrac{1}{4}\xi_0\eta_0(1+\xi_0)(1+\eta_0)\ ;$$

$$\phi_{i+4}(\xi,\eta,\zeta) = \tfrac{1}{2}(\xi_0)^a(\eta_0)^b(1+\xi_0\xi^{(1-a)})$$

$$(1+\eta_0\eta^{(1-b)});$$

$$\phi_9 = (1-\xi^2)(1-\eta^2)\ ;$$

$$\xi_0 = \xi\xi_i,\ \eta_0 = \eta\eta_i$$

$$\xi_i = -1, i = 1,4;\ \xi_i = 1, i = 2,3$$

$$\eta_i = -1, i = 1,2;\ \eta_i = 1, i = 3,4$$

$$a = 0,1, 0,1;\ b = 1,0,1,0 \text{ for } i=1,2,3,4$$

$$\phi_i(\xi,\eta,\zeta) = \tfrac{1}{4}(1+\xi_0)(1+\eta_0)(\eta_0+\xi_0-1)$$

$$\phi_{i+4}(\xi,\eta,\zeta) = \tfrac{1}{2}(1+\xi^{(1-a)}\xi_0)(1+\eta^{(1-b)}\eta_0)$$

$$\xi_0 = \xi\xi_i,\ \eta_0 = \eta\eta_i$$

$$\xi_i = -1, i = 1,4;\ \xi_i = 1, i = 2,3$$

$$\eta_i = -1, i = 1,2;\ \eta_i = 1, i = 3,4$$

$$a = 0,1, 0,1;\ b = 1,0,1,0 \text{ for } i=1,2,3,4$$

$$\phi_1(\xi_1,\xi_2,\xi_3,\xi_4) = \xi_1$$

$$\phi_2(\xi_1,\xi_2,\xi_3,\xi_4) = \xi_2$$

$$\phi_3(\xi_1,\xi_2,\xi_3,\xi_4) = \xi_3$$

$$\phi_4(\xi_1,\xi_2,\xi_3,\xi_4) = \xi_4$$

$$\phi_i(\xi_1,\xi_2,\xi_3,\xi_4) = \xi_i(2\xi_i-1)$$

$$\phi_{k+4}(\xi_1,\xi_2,\xi_3,\xi_4) = 4\xi_k\xi_j$$

$$\phi_{k+7}(\xi_1,\xi_2,\xi_3,\xi_4) = 4\xi_k\xi_4$$

$$i = 1,2,3,4\ ;\ k = 1,2,3$$

$$j = k+1 \text{ if } k = 1,2$$

$$j = 1 \text{ if } k = 3$$

Table 3.1. Continued

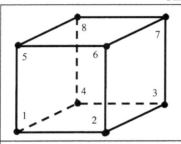

$$\phi_i(\xi,\eta,\zeta) = \tfrac{1}{8}(1+\xi_0)(1+\eta_0)(1+\zeta_0)$$

$$\xi_0 = \xi\xi_i,\, \eta_0 = \eta\eta_i,\, \zeta_0 = \zeta\zeta_i$$

$$\xi_i = -1,\, i = 1,4,5,8$$
$$\xi_i = 1,\, i = 2,3,6,7$$
$$\eta_i = -1,\, i = 1,2,5,6$$
$$\eta_i = 1,\, i = 3,4,7,8$$
$$\zeta_i = -1,\, i = 1,2,3,4$$
$$\zeta_i = 1,\, i = 5,6,7,8$$

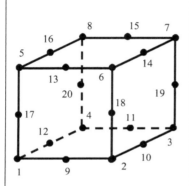

$$\phi_i(\xi,\eta,\zeta) = \tfrac{1}{8}(1+\xi_0)(1+\eta_0)(1+\zeta_0)$$
$$\times (\xi_0 + \eta_0 + \zeta_0 - 2);$$
$$\phi_{i+8}(\xi,\eta,\zeta) = \tfrac{1}{4}(1+\xi_0\xi^{(1-a)})$$
$$\times (1+\eta_0\eta^{(1-b)})(1+\zeta_0\zeta^{(1-c)});$$
$$\phi_{i+16}(\xi,\eta,\zeta) = \tfrac{1}{4}(1+\xi_0\xi^{(1-d)})(1+\eta_0\eta^{(1-e)})$$
$$\times (1+\zeta_0\zeta^{(1-f)}),\, i = 1,2,3,4;$$
$$\xi_0 = \xi\xi_i,\, \eta_0 = \eta\eta_i,\, \zeta_0 = \zeta\zeta_i$$
$$\xi_i = -1,\, i = 1,4,5,8;\; \xi_i = 1,\, i = 2,3,6,7$$
$$\eta_i = -1,\, i = 1,2,5,6;\; \eta_i = 1,\, i = 3,4,7,8$$
$$\zeta_i = -1,\, i = 1,2,3,4;\; \zeta_i = 1,\, i = 5,6,7,8$$
$$a = 0,\, b = c = 1 \text{ for } i = 1,3,5,7$$
$$a = c = 1,\, b = 0 \text{ for } i = 2,4,6,8$$
$$d = e = 1,\, f = 0,\, \text{for } i = 1,2,3,4$$

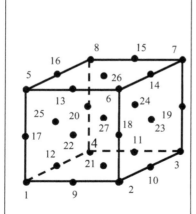

$$\phi_i(\xi,\eta,\zeta) = \tfrac{1}{8}\xi_0\eta_0\zeta_0(1+\xi_0)(1+\eta_0)(1+\zeta_0)$$
$$\phi_{i+8}(\xi,\eta,\zeta) = \tfrac{1}{4}\xi_0^a\eta_0^b\zeta_0^c(1+\xi_0\xi^{(1-a)})$$
$$\times (1+\eta_0\eta^{(1-b)})(1+\zeta_0\zeta^{(1-c)});$$
$$\phi_{i+16}(\xi,\eta,\zeta) = \tfrac{1}{4}\xi_0^d\eta_0^e\zeta_0^f(1+\xi_0\xi^{(1-d)})$$
$$\times (1+\eta_0\eta^{(1-e)})(1+\zeta_0\zeta^{(1-f)}),\, i = 1,2,3,4;$$
$$\phi_{i+20}(\xi,\eta,\zeta) = \tfrac{1}{2}\xi_0^l\eta_0^m\zeta_0^n(1+\xi_0\xi^{(1-l)})$$
$$\times (1+\eta_0\eta^{(1-m)})(1+\zeta_0\zeta^{(1-n)}),\, i = 1,...,6;$$
$$\phi_{27} = (1-\xi^2)(1-\eta^2)(1-\zeta^2);$$
$$\xi_0 = \xi\xi_i,\, \eta_0 = \eta\eta_i,\, \zeta_0 = \zeta\zeta_i$$
$$\xi_i = -1,\, i = 1,4,5,8;\; \xi_i = 1,\, i = 2,3,6,7$$
$$\eta_i = -1,\, i = 1,2,5,6;\; \eta_i = 1,\, i = 3,4,7,8$$
$$\zeta_i = -1,\, i = 1,2,3,4;\; \zeta_i = 1,\, i = 5,6,7,8$$
$$a = 0,\, b = c = 1 \text{ for } i = 1,3,5,7$$
$$a = c = 1,\, b = 0 \text{ for } i = 2,4,6,8$$
$$d = e = 1,\, f = 0,\, \text{for } i = 1,2,3,4$$
$$l = m = 0,\, n = 1 \text{ for } i = 1,6$$
$$l = n = 0,\, m = 1 \text{ for } i = 2,4$$
$$m = n = 0,\, l = 1 \text{ for } i = 3,5$$

3.2 Construction of Special Elements

For many applications, elements of mixed order are needed for analysis. When this happens, transition elements are often used. These elements are not listed in standard element libraries. Below, we consider a technique to construct these elements from those in the standard libraries.

3.2.1 Non-standard Elements

We consider a case illustrated in Figure 3.12, where the transition elements are needed to patch the domain approximation from 4–node elements to 9–node elements. The transition element has 5 nodes with three nodes on the side shared with the 9–node element. To construct the shape functions for the 5–node element, we first consider node 5 for which the shape function has a quadratic × linear type,

$$\phi_5(\xi,\eta) = \tfrac{1}{2}(1+\xi)(1-\eta^2) \tag{3.30}$$

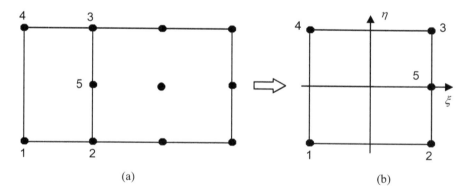

Figure 3.12. A 5–node rectangular element patches meshes of 4–node elements to meshes of 9–node elements: (a) mesh and (b) a 5–node transition element

The addition of node 5 affects only nodes 2 and 3, for which shape functions need to be corrected. We consider the correction for node 2 first. The shape function for a linear interpolation has the form,

$$\phi_{2,trial}(\xi,\eta) = \tfrac{1}{4}(1+\xi)(1-\eta) \tag{3.31}$$

which is equal to 0.5 at node 5. This leads us to the following correction to ensure that the final shape function is zero at node 5 as well:

$$\phi_2(\xi,\eta) = \phi_{2,trial}(\xi,\eta) - 0.5\phi_5(\xi,\eta)$$

$$= \tfrac{1}{4}(1+\xi)(1-\eta) - \tfrac{1}{4}(1+\xi)(1-\eta^2)$$

$$= \tfrac{1}{4}(1+\xi)(1-\eta)(-\eta) \tag{3.32}$$

Similarly, the shape function for node 3 needs to be corrected,

$$\phi_3(\xi,\eta) = \tfrac{1}{4}(1+\xi)(1+\eta) - \tfrac{1}{4}(1+\xi)(1-\eta^2) = \tfrac{1}{4}\eta(1+\xi)(1+\eta) \tag{3.33}$$

Techniques for constructing variable-number-nodes elements are also given in [4].

3.2.2 Construction of Element Shape Functions by Node Collapsing

There are occasions where linear triangle elements need to be expressed in terms of normalized elements in (ξ,η). While one can follow the procedure illustrated in Section 3.1 to obtain the shape functions, an easier way is to collapse a 4–node element into a 3–node element. As shown in Figure 3.13, one intends to merge nodes 3 and 4 of a quadrilateral element into node 3 of a new triangular element. The shape functions for nodes 1 and 2 are unchanged as they are not affected by the action along the edge defined by nodes 3 and 4. For the new node 3, the shape function is constructed by adding the shape functions for the old nodes 3 and 4,

$$\phi_3(\xi,\eta) = \phi_{3,old}(\xi,\eta) + \phi_{4,old}(\xi,\eta)$$

$$= \tfrac{1}{4}(1-\xi)(1+\eta) + \tfrac{1}{4}(1+\xi)(1+\eta) = \tfrac{1}{2}(1+\eta) \tag{3.34}$$

This technique also can be applied to other elements in both 2– and 3-dimensions.

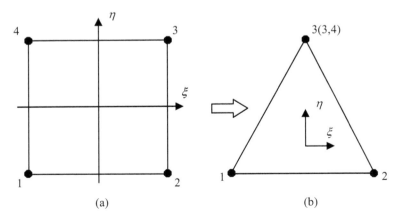

(a)   (b)

Figure 3.13. Nodes 3 and 4 are collapsed to transform a rectangular element (a) to a triangular element (b)

3.2.3 Spectral Elements

For some applications, spectral elements are required to provide needed high order spatial resolution. The spectral elements are constructed using the orthogonal functions, which can be useful to develop diagonalized element matrices and thus speed up calculations. We consider a 1-D case here. The extension of the 1-D shape functions to multidimensional elements can be obtained using the tensor product approach discussed in Section 3.1.2.2.

For a 1-D element, the following interpolation function is constructed using the Chebychev polynomial $T_i(\xi)$ [5]:

$$u(\xi) = \sum_{i=0}^{N_s} a_i T_i(\xi) \tag{3.35}$$

where N_s is the number of Chebychev polynomial terms used in the approximation. The expansion coefficient a_i is calculated using the orthogonality condition for the Chebychev polynomials:

$$\int_{-1}^{+1} T_i(\xi) T_j(\xi) w(\xi) d\xi = 0 \qquad i \ne j \tag{3.36}$$

where $w(\xi) = 1/\sqrt{1-\xi^2}$ is the weighting function. Thus, we have

$$\int_{-1}^{+1} T_i(\xi) w(\xi) u(\xi) d\xi = a_i \int_{-1}^{+1} T_i(\xi) T_i(\xi) w(\xi) d\xi \tag{3.37}$$

In the spectral approximation, the Gauss–Lobatto rules are used to evaluate the above integrals on both the left and right sides, although the latter can be evaluated analytically [6]. Applying this rule, we have

$$\int_{-1}^{+1} T_i(\xi) T_i(\xi) w(\xi) d\xi = \bar{C}_i \frac{\pi}{2} , \qquad \bar{C}_i = \begin{cases} 2, & i = 0, N_s \\ 1, & 1 \le i \le N_s - 1 \end{cases} \tag{3.38}$$

and

$$\int_{-1}^{+1} u(\xi) T_i(\xi) w(\xi) d\xi = \sum_{j=0}^{N_s} T_i(\xi_j) u(\xi_j) w_j \tag{3.39}$$

where

$$\xi_j = \cos \frac{j\pi}{N_s} ; \quad w_j = \begin{cases} 0.5 \, \pi/N_s , & j = 0, \ N_s \\ \pi/N_s , & 1 \le j \le N_s - 1 \end{cases} \tag{3.40}$$

Substituting back into Equation 3.37, we have the coefficients for the Chebychev polynomials,

$$a_i = \frac{2}{N_s \overline{C}_i} \sum_{j=0}^{N_s} \frac{1}{\overline{C}_j} T_i(\xi_j) u(\xi_j)$$

(3.41)

Since $T_i(\xi)$ is defined as

$$T_i(\xi) = \cos(i \cos^{-1}\xi), \qquad i = 0,1,2, ...$$

(3.42)

the use of the Gauss–Lobatto [6] rules leads to the following expression:

$$T_i(\xi_j) = \cos \frac{ij\pi}{N_s}$$

(3.43)

which may be substituted back into Equation 3.41 with the result,

$$a_i = \frac{2}{N_s \overline{C}_i} \sum_{j=0}^{N_s} \frac{r(\xi_j)}{\overline{C}_j} \cos \frac{ij\pi}{N_s}$$

(3.44)

Clearly, this is nothing but a cosine transformation, and the fast Fourier Transformation (FFT) can be readily applied to expedite the calculations. With Equation 3.44 substituted back into the original approximation for u, one has the following expression:

$$u(\xi) = \sum_{j=0}^{N_s} \phi_j(\xi) u(\xi_j)$$

(3.45)

where $\phi_j(\xi)$ is the spectral interpolation shape function, and is calculated by

$$\phi_j(\xi) = \frac{2}{N_s \overline{C}_j} \sum_{i=0}^{N_s} \frac{T_i(\xi_j) T_i(\xi)}{\overline{C}_i}$$

(3.46)

From the above two equations, it is obvious that the shape function has the desired property,

$$\phi_j(\xi_i) = \delta_{ij}$$

(3.47)

While the spectral elements here are constructed using the Chebychev polynomial, the procedure applies to any other orthogonal functions.

3.3 Hierarchical Shape Functions

Shape functions are the functions restricted to an element, and all the above elements satisfy two important local constraints: (1) $\phi_j(x)$ is unity at node j and vanishes at all other nodes and (2) $\phi_j(x)$ is only non-zero on those elements containing node j. The elements satisfying these constraints are called standard shape functions [1]. Hierarchical shape functions, however, maintain only the second property. They are constructed by adding, hierarchically, higher-degree corrections to the lower-degree shape functions. Specifically, a hierarchical basis of degree $p + 1$ is constructed as a correction to the degree p basis, and the entire basis need not be reconstructed when the polynomial degree is increased. In this section, we consider the construction of hierarchical shape functions, which have been shown to be particularly useful for the discontinuous finite element solution of certain types of problems [7].

3.3.1 1-D Hierarchical Correction

Let us consider what happens when we add the quadratic correction to the linear shape function over an element. This means that we construct a piecewise quadratic hierarchical shape function. The restriction of this function to a canonical element $\xi \in [-1, +1]$ has the hierarchical form,

$$u(\xi) = u_0 \phi_0(\xi) + u_1 \phi_1(\xi) + a_2 \phi_2(\xi)$$

(3.48)

with the hierarchical shape functions,

$$\phi_0(\xi) = \tfrac{1}{2}(1 - \xi) \; ; \; \phi_1(\xi) = \tfrac{1}{2}(1 + \xi) \; ; \; \phi_2(\xi) = (1 + \xi)(1 - \xi)$$

(3.49)

Here a_2 is a constant, but not equal to u_2. In fact, $du(\xi)/d\xi = -2a_2$. This interpretation gives a general meaning, but is not necessary. The coefficient a_2 needs to be obtained as part of the numerical solution. We note, however, that $\phi_0(-1) = \phi_1(+1) = 1$ for the hierarchical functions. The above hierarchical shape functions are plotted in Figure 3.14. It transpires that only one shape function is quadratic, and the other two are actually linear, unchanged from the linear approximations. This is different from the standard quadratic shape functions, which are all quadratic (see Equation 3.8). Also for the standard element shape functions, all the coefficients of ϕ_j were equal to the variables u_j.

Hierarchical polynomials of higher order can be constructed similarly by simply adding the higher order corrections. For example, for a cubic hierarchical approximation, we may add to Equation 3.48 the following term:

$$\phi_3(\xi) = \xi(\xi^2 - 1)$$

(3.50)

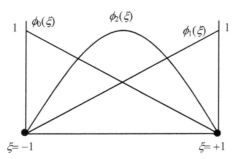

Figure 3.14. A hierarchical element of quadratic order

with a_3 being the corresponding coefficient. In general, we can conveniently construct the hierarchical approximations by adding the hierarchical shape functions in increased orders,

$$\phi_k(\xi) = \begin{cases} (\xi^k - 1)/k! & k = \text{even} \\ \xi(\xi^{k-1} - 1)/k! & k = \text{odd} \end{cases} \tag{3.51}$$

where $k \geq 2$ is the order of the polynomial and $k!$ is included for convenience, but not necessary.

For discontinuous formulation with explicit time integration for transient problems, there is an advantage of using orthogonal functions to diagonalize the mass matrix [7]. One of these hierarchical shape functions is based on the Legendre polynomials [2], which are defined by the following differential equation [6]:

$$(1 - \xi^2)\frac{d^2 P_i(\xi)}{d\xi^2} - 2\xi\frac{dP_i(\xi)}{d\xi} + i(i+1)P_i(\xi) = 0,$$

$$\text{for } \xi \in [-1, +1], \quad i \geq 0 \tag{3.52a}$$

or by the Rodrigues formula [8]:

$$P_i(\xi) = \frac{1}{2^i \, i!} \frac{d^i}{d\xi^i} [(\xi^2 - 1)^i], \, i \geq 0 \tag{3.52b}$$

The Legendre polynomials have the following useful properties:

$$\int_{-1}^{+1} P_i(\xi)P_j(\xi)d\xi = \frac{2\delta_{ij}}{2i+1}; \quad P_i(1) = 1; \quad P_i(-\xi) = (-1)^i P_i(\xi), \, i \geq 0 \tag{3.53}$$

$$\frac{dP_{i+1}(\xi)}{dx} = (2i+1)P_i(\xi) + \frac{dP_{i-1}(\xi)}{dx} \quad , i \geq 1 \tag{3.54}$$

$$(i+1)P_{i+1}(\xi) = (2i+1)\xi P_i(\xi) - iP_{i-1}(\xi) \quad , i \geq 1 \tag{3.55}$$

The hierarchical shape functions using the Legendre polynomials are defined by

$$\phi_{0,p}(\xi) = \tfrac{1}{2}(1-\xi), \ \phi_{1,p}(\xi) = \tfrac{1}{2}(1+\xi) \tag{3.56}$$

$$\phi_{i,p}(\xi) = \sqrt{\frac{2i-1}{2}} \int_{-1}^{\xi} P_{i-1}(\mu)d\mu = \frac{P_i(\xi) - P_{i-2}(\xi)}{\sqrt{2i-1}},$$

$$\text{for } \phi_{i,p}(\pm 1) = 0, \ \ i \geq 2 \tag{3.57}$$

where the subscript p on ϕ denotes the basis of the Legendre polynomials for $p \geq 2$. A useful property of the above functions is that the bilinear form of the first order derivatives is orthogonal,

$$\int_{-1}^{+1} \frac{d\phi_{j,p}(\xi)}{d\xi} \frac{d\phi_{i,p}(\xi)}{d\xi} d\xi = \frac{2i-1}{2} \int_{-1}^{+1} P_{j-1}(\mu)P_{i-1}(\mu)d\mu = \delta_{i,j},$$

$$i,j \geq 2 \tag{3.58}$$

where the first property of Equation 3.53 has been used. This procedure will diagonalize the submatrix of the diffusion stiffness matrix, starting from ($i = 2, j = 2$).

3.3.2 Canonical Square and Cubic Elements

As for the standard shape functions, the hierarchical shape functions for the rectangular and brick elements can be constructed using the tensor product of the 1-D shape functions. Thus, making use of Equation 3.51, we have the hierarchical shape functions

$$\phi_{kl}(\xi,\eta) = \phi_k(\xi)\phi_l(\eta) , \ k, l > 1 \tag{3.59}$$

for a canonical square element $\{(\xi, \eta) \mid \xi, \eta \in [-1, +1]\}$, and

$$\phi_{klm}(\xi,\eta,\zeta) = \phi_k(\xi)\phi_l(\eta)\phi_m(\zeta) , \ k, l, m > 1 \tag{3.60}$$

for a canonical cubic element $\{(\xi, \eta, \zeta) \mid \xi, \eta, \zeta \in [-1, +1]\}$. In the above equation, the basis function $\phi_j(\xi)$ is given by $\phi_0(\xi) = 0.5(1 - \xi)$ and $\phi_1(\xi) = 0.5(1 + \xi)$, and for $j \geq 2$, $\phi_j(\xi)$ takes either the regular polynomials (Equation 3.51) or the Legendre polynomials (Equation 3.57) [2, 3, 9].

The field variable can be interpolated using the hierarchical shape functions over the canonical element,

$$u(\xi,\eta,\zeta) = \sum_{k=0}^{p}\sum_{l=0}^{p}\sum_{m=0}^{p} a_{(k,l,m)}\phi_{klm}(\xi,\eta,\zeta)$$

$$= \sum_{k=0}^{p}\sum_{l=0}^{p}\sum_{m=0}^{p} a_{(k,l,m)}\phi_k(\xi)\phi_l(\eta)\phi_m(\zeta) \qquad (3.61)$$

where p is the order of approximation, and $a_{(k,l,m)}$ ($k, l, m = 0, 1$) corresponds to the u value at each corner of the element, and $a_{(k,l,m)}$ ($k, l, m \geq 2$) is constant.

Example 3.5. Find the hierarchical shape functions for an 8–node square element using the 1-D interpolation function given by Equations 3.49 and 3.51, and write the explicit form of the interpolation function for the field variable u.

Solution. We use the tensor product rule to obtain the needed shape functions for vertices and sides of the 8–node element shown in Figure 3.5e.
  For the four vertices, we have the linear functions,

$$\phi_{0,0}(\xi,\eta) = \tfrac{1}{4}(1-\xi)(1-\eta); \quad \phi_{1,0}(\xi,\eta) = \tfrac{1}{4}(1+\xi)(1-\eta)$$

$$\phi_{1,1}(\xi,\eta) = \tfrac{1}{4}(1+\xi)(1+\eta); \quad \phi_{0,1}(\xi,\eta) = \tfrac{1}{4}(1-\xi)(1+\eta) \qquad (3.11e)$$

For the four sides, we have the quadratic functions,

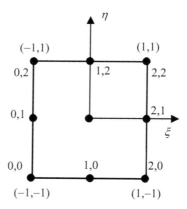

Figure 3.5e. An 8–node element showing the number convention used for the hierarchical shape functions. The numbers in the brackets are coordinates, and the number pairs without brackets are used to denote the nodes

$$\phi_{2,0}(\xi) = (1+\xi)(1-\xi)(1+\eta); \quad \phi_{1,2}(\xi) = (1+\xi)(1-\eta)(1+\eta)$$

$$\phi_{2,1}(\xi) = (1+\xi)(1-\xi)(1-\eta); \quad \phi_{0,2}(\xi) = (1-\xi)(1-\eta)(1+\eta) \quad (3.12e)$$

With the above hierarchical shape functions, we obtain the interpolation for $u(\xi,\eta)$ over the element,

$$u(\xi,\eta) = \sum_{k=0}^{p}\sum_{l=0}^{p} a_{(k,l)}\phi_k(\xi)\phi_l(\eta)$$

$$= u_{0,0}\phi_0(\xi)\phi_0(\eta) + u_{1,0}\phi_1(\xi)\phi_0(\eta) + u_{1,1}\phi_1(\xi)\phi_1(\eta)$$

$$+ u_{0,1}\phi_0(\xi)\phi_1(\eta) + a_{2,0}\phi_2(\xi)\phi_0(\eta) + a_{2,1}\phi_2(\xi)\phi_1(\eta)$$

$$+ a_{1,2}\phi_1(\xi)\phi_2(\eta) + a_{0,2}\phi_0(\xi)\phi_2(\eta) \quad (3.13e)$$

3.3.3 Triangular and Tetrahedral Elements

For triangular and tetrahedral elements, the area (volume) coordinates can be used. Again, for linear functions, we have

$$\phi_1 = \eta_1; \quad \phi_2 = \eta_2; \quad \phi_3 = \eta_3 \quad (3.62)$$

Consider the triangle shown in Figure 3.15. Along side 1–2, $\eta_3 = 0$ and thus,

$$\eta_1 + \eta_2 = 1 \quad (3.63)$$

The higher order function has the form given by Equation 3.51. Since $-1 \le \xi \le +1$ along the side 1–2, a simple coordinate transformation yields along the side,

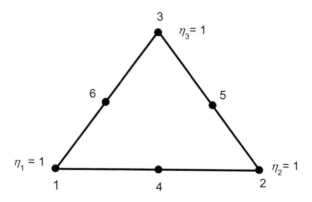

Figure 3.15. A quadratic triangular element with 6 nodes

$$\xi = 2\eta_2 - 1 = \eta_2 - \eta_1 \tag{3.64}$$

This then allows us to write the higher order polynomials given by Equation 3.51 in terms of $\eta_2 - \eta_1$,

$$\phi_{k,1-2}(\eta_1,\eta_2) = \begin{cases} [(\eta_2 - \eta_1)^k - (\eta_1 + \eta_2)^k]/k! & k = \text{even} \\ (\eta_2 - \eta_1)[(\eta_2 - \eta_2)^{k-1} - (\eta_1 + \eta_2)^{k-1}]/k! & k = \text{odd} \end{cases} \tag{3.65}$$

where $k \geq 2$ is the order of the polynomial. Similarly, we can write the higher order functions ($k \geq 2$) for the other two sides,

$$\phi_{k,2-3}(\eta_2,\eta_3) = \begin{cases} [(\eta_3 - \eta_2)^k - (\eta_3 + \eta_2)^k]/k! & k = \text{even} \\ (\eta_3 - \eta_2)[(\eta_3 - \eta_2)^{k-1} - (\eta_3 + \eta_2)^{k-1}]/k! & k = \text{odd} \end{cases} \tag{3.66}$$

$$\phi_{k,3-1}(\eta_2,\eta_3) = \begin{cases} [(\eta_1 - \eta_3)^k - (\eta_3 + \eta_1)^k]/k! & k = \text{even} \\ (\eta_1 - \eta_3)[(\eta_1 - \eta_3)^{k-1} - (\eta_3 + \eta_1)^{k-1}]/k! & k = \text{odd} \end{cases} \tag{3.67}$$

The above functions define the polynomials along the three edges. Internal hierarchical functions are also needed for cases where $k \geq 3$. These internal functions are called bubble functions, and satisfy the condition that they become zero along all three edges of the triangular element,

$$\phi_{k,i}(\eta_1,\eta_2,\eta_3) = \eta_1^I \eta_2^J \eta_3^K, \quad I + J + K = k - 3 \geq 0 \tag{3.68}$$

where k is the highest order, and the subscript i denotes the interior interpolation. In constructing the hierarchical shape functions for the orders of $p \geq 2$, we have the following expressions by the tensor product rule:

$$\phi_m(\eta_1,\eta_2,\eta_3) = \eta_i \eta_j \phi_{m,i-j}, \text{ for mid-nodes on the edge } i-j \tag{3.69}$$

$$\phi_n(\eta_1,\eta_2,\eta_3) = \eta_1 \eta_2 \eta_3 \phi_{n,i}, \text{ for nodes inside triangle} \tag{3.70}$$

The construction of the hierarchical polynomials for tetrahedral elements follows the same procedure as described above for each face, and then adds the additional higher functions for the interior nodes. The numerical details involved in construction are given in [10]. Thus, in addition to the above functions for each face of a tetrahedron, the hierarchical shape functions are needed for nodes inside the tetrahedron,

$$\phi_q(\eta_1,\eta_2,\eta_3)=\eta_1\eta_2\eta_3\eta_4\phi_{q,v} \tag{3.71}$$

where ϕ_q vanishes on all faces as desired, and $\phi_{q,v}$ is the polynomial correction,

$$\phi_{q,v}(\eta_1,\eta_2,\eta_3)=\eta_1^I\eta_2^J\eta_3^K\eta_4^L\,,\,I+J+K+L=q-4\geq0 \tag{3.72}$$

with v denoting the nodes inside the polynomial.

The above procedure is general. If the Legendre polynomials are used for higher approximations where $k\geq2$, then one can show, by exactly the same arguments given above, that the polynomial corrections along the edges are

$$\phi_{k,p,i-j}(\eta_i,\eta_j)=\frac{4\phi_{k,p}(\eta_j-\eta_i)}{1-(\eta_j-\eta_i)^2}\ \text{ for edge } i-j \tag{3.73}$$

where $\phi_{k,p}(\eta_j-\eta_i)$ is defined by Equation 3.57. Szabo and Babuska [2] suggest that the correction to the interior points in the triangle takes the following form:

$$\phi_{k,p,\lambda\mu}(\eta_1,\eta_2,\eta_3)=P_\lambda(\eta_2-\eta_1)P_\mu(2\eta_3-1)\,,\ \lambda+\mu=k-3\,,\,k\geq3 \tag{3.74}$$

With these corrections, the hierarchical shape functions of quadratic order or higher can be obtained by the simple multiplication process,

$$\phi_m(\eta_1,\eta_2,\eta_3)=\eta_i\eta_j\phi_{m,p,i-j}\,,\text{ for mid-nodes on the edge } i-j \tag{3.75}$$

$$\phi_n(\eta_1,\eta_2,\eta_3)=\eta_1\eta_2\eta_3\phi_{n,p,\lambda\mu}(\eta_1,\eta_2,\eta_3)\,,\text{ for inside nodes} \tag{3.76}$$

Similarly, the expression for a tetrahedron is obtained by repeating the above process for each face. For the nodes inside the tetrahedron, the hierarchical shape functions have the form [3],

$$\phi_k(\eta_1,\eta_2,\eta_3)=\eta_1\eta_2\eta_3\eta_4P_\lambda(\eta_2-\eta_1)P_\mu(2\eta_3-1)P_\theta(2\eta_4-1)\,,$$

$$\lambda+\mu+\theta=k-4\,,\ k\geq4 \tag{3.77}$$

Example 3.6. For a 7–node triangular element, derive the hierarchical order term for the mid-edge nodes and the interior node using the Legendre polynomials as correction terms. Develop hierarchical shape functions for the element.

Solution. We first draw a 7–node triangular element as shown in Figure 3.6e. Let us then consider node 4, the mid-node for edge 1–2. Along this edge, $\eta_3=0$ and η_1

$+\ \eta_2 = 1$. Establishing the coordinate ξ along the edge such that $-1\leq\xi\leq+1$, we have the following coordinate transform:

$$\eta_1 = (1-\xi)/2 ; \ \eta_2 = (1+\xi)/2 ; \ \eta_3 = 0 , \ \xi = \eta_2 - \eta_1 \qquad (3.14e)$$

Also, from Equation 3.57, the edge shape function is described by

$$\phi_{2,p}(\xi) = \sqrt{3}^{-1}(P_2(\xi)-1) ; \ P_2(\xi) = 0.5(3\xi^2 - 1) \qquad (3.15e)$$

Since $\eta_2 = 0$ along the edge 1–3, and $\eta_1 = 0$ along edge 2–3, the quadratic hierarchical shape function for node 4 in general should have the form of

$$\phi_2^4(\eta_1,\eta_2,\eta_3) = \eta_1\eta_2\phi_{2,p,1-2}(\xi) \qquad (3.16e)$$

where the superscript denotes the node number. Along the edge 1–2, the above two equations must be equal, and thus we have

$$\phi_2^4(\eta_1,\eta_2,0) = \eta_1\eta_2\phi_{2,p,1-2}(\xi) = 0.25(1+\xi)(1-\xi)\phi_{2,p,1-2}(\xi) = \phi_{2,p}(\xi) \qquad (3.17e)$$

that is,

$$\phi_{2,p,1-2}(\xi) = 4\phi_{2,p}(\xi)/(1-\xi^2) = 2(\xi^2 -1)\sqrt{3}/(1-\xi^2) = -\sqrt{6} \qquad (3.18e)$$

The hierarchical shape function for node 4 then becomes

$$\phi_2^4(\eta_1,\eta_2,\eta_3) = \eta_1\eta_2\phi_{2,p,1-2}(\eta_2 - \eta_1) = -\sqrt{6}\,\eta_1\eta_2 \qquad (3.19e)$$

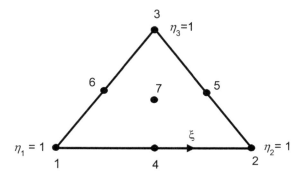

Figure 3.6e. A 7–node triangular element

The shape functions for nodes 5 and 6 can be obtained similarly. For node 7 inside the triangle, $\lambda = \mu = 0$, which leads to

$$\phi_3^7 (\eta_1,\eta_2,\eta_3) = \eta_1\eta_2\eta_3\phi_{3,p,00} (\eta_1,\eta_2,\eta_3) = \eta_1\eta_2\eta_3 \qquad (3.20e)$$

In summary, the hierarchical shape functions for the 7–node element are given by

$$\phi^1 (\eta_1,\eta_2,\eta_3) = \eta_1; \quad \phi^2 (\eta_1,\eta_2,\eta_3) = \eta_2; \quad \phi^3 (\eta_1,\eta_2,\eta_3) = \eta_3$$

$$\phi^4 (\eta_1,\eta_2,\eta_3) = -\sqrt{6}\,\eta_1\eta_2; \quad \phi^5 (\eta_1,\eta_2,\eta_3) = -\sqrt{6}\,\eta_2\eta_3$$

$$\phi^6 (\eta_1,\eta_2,\eta_3) = -\sqrt{6}\,\eta_3\eta_1; \quad \phi_3^7 (\eta_1,\eta_2,\eta_3) = \eta_1\eta_2\eta_3 \qquad (3.21e)$$

3.3.4 Obtaining Hierarchical Elements Through Coordinate Transformations

Hierarchical shape functions for various element shapes can also be obtained through coordinate transformations between other shapes and a 2-D or 3-D canonical element [7, 9, 11].

We illustrate this point by considering a transformation between a triangle and a square as shown in Figure 3.16. Here the triangular element defined in (s_1, s_2) coordinates is mapped to a square element defined in (t_1, t_2). Note that the vertical lines (t_1=constant) in the square domain become the lines radiating from the point $(-1, 1)$ in the triangular domain. The ray (t_1 = constant) is multi-valued at ($s_1 = -1$, $s_2 = 1$). We then easily develop a hierarchical shape function for the triangular elements by the substitution of (s_1, s_2) for (t_1, t_2),

$$\phi_{kl} (s_1,s_2) = \phi_k (t_1)\phi_l (t_2) = \phi_k (t_1(s_1,s_2))\phi_l (t_2(s_1,s_2)) \qquad (3.78)$$

The above procedure may also be applied to develop 3-D transformations between various shapes, such as hexahedrons, prisms, pyramids and tetrahedrons, and canonical cubic elements $\{(t_1, t_2, t_3) \mid t_1, t_2, t_3 \in [-1, +1]\}$ [7, 11]. For a tetrahedron defined in coordinates (s_1, s_2, s_3), the hierarchical shape functions are

$$\phi_{klm} (s_1,s_2,s_3) = \phi_k (t_1(s_1,s_2,s_3))\phi_l (t_2(s_1,s_2,s_3))\phi_m (t_3(s_1,s_2,s_3)) (3.79)$$

where the (t_1, t_2, t_3) to (s_1, s_2, s_3) transformation is given by

$$s_1 = \frac{2(1+t_1)}{-t_2 - t_3} - 1; \quad s_2 = \frac{2(1+t_2)}{1-t_3} - 1; \quad s_3 = t_3 \qquad (3.80)$$

The transformation rule given by Equation 3.79 maps the four vertices of the tetrahedron in (s_1, s_2, s_3); which are defined at $(-1,-1,-1)$, $(1,-1,-1)$, $(1,-1,-1)$ and $(-1,-1,1)$, to the eight vertices of a cubic element as shown in Figure 3.17.

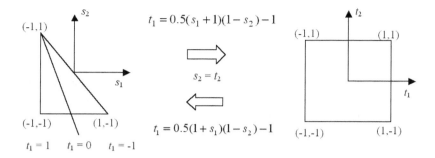

Figure 3.16. Transformation between a triangular element and a rectangular element

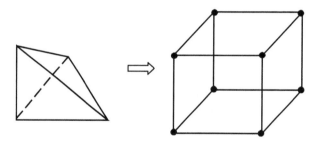

Figure 3.17. Transformation from a tetrahedron to a canonical cube element

3.3.5 Orthogonal Mass Matrix Construction

For explicit time integration, a diagonalized mass matrix is beneficial for stabilizing the time marching process with a larger time step. For continuous finite element approximations, a common procedure to obtain a diagonalized mass matrix is to use either the Newton–Cotes integration rules; or more simply, to sum the off-diagonal terms of a row, and then add to the diagonal term of the row [1, 12]. While this procedure can still be used in the discontinuous algorithm, the localized nature of the discontinuous formulation makes it very convenient to develop basis functions for diagonalizing the mass matrix. In this regard, there are many candidates from the class of special functions [6].

Let us illustrate this point by considering a 1-D case. The idea is similar to Equation 3.58, except that we seek the following term for diagonalizing

$$\int_{-1}^{+1} \phi_{j,p}(\xi)\phi_{i,p}(\xi)d\xi = \frac{2i+1}{2}\int_{-1}^{+1} P_j(\mu)P_i(\mu)d\mu = \delta_{i,j} \quad \text{for } i,j \geq 2$$

$$(3.81)$$

Clearly, this property is easily satisfied when the shape function is defined by

$$\phi_{i,p}(\xi) = \sqrt{\tfrac{2i+1}{2}}P_i(\xi) \tag{3.82}$$

Thus if the approximate solution is expressed as

$$u(\xi) = \sum_{i=1}^{k} a_i \phi_{i,p}(\xi) = \sum_{i=1}^{k} a_i \sqrt{\tfrac{2i+1}{2}}P_i(\xi) \tag{3.83}$$

then the mass matrix will be diagonalized with the ith element calculated by

$$\int_{-1}^{1} \phi_{j,p}(\xi)u(\xi)d\xi = \sum_{i=1}^{k} a_i \int_{-1}^{1} \phi_{i,p}(\xi)\phi_{j,p}(\xi)d\xi = \delta_{ij}a_i \tag{3.84}$$

In this approximation, however, the coefficient a_i is not equal to the value of u at node i. This is unimportant to the discontinuous formulation since the field values are updated during element-by-element calculations. This approach is equivalent to treating each element as a separate domain, over which the spectral method is applied [5]. Of course, when needed, shape functions with this orthogonal property can be constructed using the procedure given in Section 3.2.3.

This idea can be readily extended to multidimensional problems, and can also be extended to other element shapes. One of these approaches was reported by Lomtev *et al.* [7], who proposed to use the Jacobi polynomial as the principal function for hierarchical shape functions that preserve the orthogonality property for mass matrices. These principal functions assume the following form:

$$\phi_i^a(\xi) = P_i^{0,0}(\xi); \quad \phi_{ij}^b(\xi) = \left(\tfrac{1-\xi}{2}\right)^i P_j^{2i+1,0}(\xi);$$

$$\phi_{ijk}^c(\xi) = \left(\tfrac{1-\xi}{2}\right)^{i+j} P_k^{2i+2j+2,0}(\xi) \tag{3.85}$$

where $P_i^{\alpha,\beta}(\xi)$ is the Jacobi polynomial with the following properties,

$$P_n^{\alpha,\beta}(x) = \frac{(1-x)^{-\alpha}(1+x)^{-\beta}}{(-2)^n n!}\frac{d^n}{dx^n}\left[(1-x)^{\alpha}(1+x)^{\beta}(1-x^2)^n\right] \tag{3.86}$$

$$\int_{-1}^{1}(1-x)^{\alpha}(1+x)^{\beta}P_n^{\alpha,\beta}(x)P_m^{\alpha,\beta}(x)dx = \delta_{nm} \tag{3.87}$$

With these definitions, the hierarchical shape functions for various multidimensional elements can be constructed, which possess the property of mass matrix diagonalization [7].

3.4 Interpolation Error Analysis

Error estimates are important for numerical analysis. In finite element analyses, both *post priori* and *a priori* estimates are computed using various error measures. Here we consider interpolation errors associated with shape functions.

3.4.1 Hilbert Space and Various Error Measures

Different measures are used for the purpose of error estimates in finite element calculations. We discuss below the interpolation error analysis, which is perhaps the most important error analysis.

Useful concepts associated with the mathematical analysis of the finite element methods, which also apply to local analysis, are discussed in detail in [13, 14]. Some concepts and definitions from linear analysis, which are essential for the measurement of basic interpolation errors, are presented here. Additional information may be found in references [13–15]. First, if V is a real linear space, then the mapping $a(.,.) : V \times V \to R$ is called an inner product if (*i*) $a(.,.)$ is symmetric, i.e., $a(v, w) = a(w, v)$, v, $w \in V$, (*ii*) $a(v, v)$ is a scalar product, i.e., $a(v, v) \geq 0$, $\forall v \in V$, (*iii*) $a(v, v) = 0$ if and only if $v = 0$, and (*iv*) $a(v, w)$ is linear with respect to v, i.e., $a(\alpha v_1 + \beta v_2, w) = \alpha a(v_1, w) + \beta a(v_2, w)$ for any scalars α, $\beta \in R$ and $v_1, v_2, w \in V$. Second, if $a(.,.)$ is an inner product, then for $v \in V$, the associated norm $\|\bullet\|$ is defined by

$$\| v \| = [a(v, v)]^{1/2}, \qquad \forall v \in V \tag{3.88}$$

If $a(.,.)$ is an inner product with the corresponding norm $\|\bullet\|$, then we have the Cauchy–Schwarz inequality,

$$| a(v, w) | \leq \| v \| \| w \|, \qquad \forall v, w \in V \tag{3.89}$$

We now define the Hilbert space: V is a Hilbert space if V is complete (that is, every Cauchy sequence is convergent with respect to $\|\bullet\|$) and V is a linear space with an inner product with the corresponding norm $\|\bullet\|$.

A sequence of $v_1, v_2, v_3, \ldots$, of elements v_i in the linear space V is a Cauchy sequence if for all $\varepsilon > 0$ there is always a natural number $N > 0$ such that $\| v_i - v_j \| < \varepsilon$ if $i, j > N$. Further v_i converges to v if $\| v_i - v \| \to 0$ as $i \to \infty$.

An L^2-function is often used in the literature and it refers to a class of functions that are square integrable. An L^2-function is a member of the L^2 space,

$$L^2(\Omega) = \{f \mid f \text{ is defined on } \Omega \text{ and } \int_\Omega f^2 dx < \infty\} \tag{3.90}$$

with the corresponding L^2-norm defined as

$$\| f \|_{L^2} = \left(\int_{\Omega} f^2 dx \right)^{1/2} \tag{3.91}$$

where $f = f(x)$ is a real-valued function defined on Ω. The L^2-norm is a widely used measure for error analysis in finite element calculations. We note that the L^2 space is a special case of the Lebesgue space [14]. A Lebesgue space is defined by the following expression:

$$L^p(\Omega) = \{ f : f \text{ is measuable on } \Omega \text{ and } \int_{\Omega} | f(x) |^p dx < \infty \} \tag{3.92}$$

with the corresponding norm given by

$$\| f \|_{L^p} := \left(\int_{\Omega} | f(x) |^p dx \right)^{1/p} \quad \text{for } 1 \le p < \infty \tag{3.93a}$$

$$\| f \|_{L^\infty(\Omega)} = \lim_{m \to \infty} \left(\int_{\Omega} | f(x) |^m dx \right)^{1/m} = \text{ess} \sup_{x \in \Omega} | f(x) | \quad \text{for } p = \infty \tag{3.93b}$$

In finite element literature, the class $H^m(\Omega)$ of functions is also used to quantify the smoothness and regularity of functions. To define the class $H^m(\Omega)$, we first define g as the αth weak partial derivative, written as [14]

$$g = D_w^\alpha f = \frac{\partial^{|\alpha|} f}{\partial x_1^{\alpha_1} \cdots \partial x_n^{\alpha_n}} = \frac{\partial^{\alpha_1} f}{\partial x_1^{\alpha_1}} \cdots \frac{\partial^{\alpha_n} f}{\partial x_n^{\alpha_n}} \tag{3.94a}$$

if $f, g \in L_{loc}^1(\Omega)$ and if there exists

$$\int_{\Omega} f D_w^\alpha \phi \, dx = (-1)^{|\alpha|} \int_{\Omega} g \phi dx \tag{3.94b}$$

for all testing functions $\phi \in C_c^\infty(\Omega)$, with $C_c^\infty(\Omega)$ being the space of infinitely differentiable functions with compact support. In the above equations, $\alpha_1, \alpha_2, \cdots, \alpha_n$ are non-negative integers, and $| \alpha | = \alpha_1 + \alpha_2 + \cdots + \alpha_n$.

Now we define the class $H^m(\Omega)$ by

$$H^m(\Omega) = \{ f \mid f \in L^2(\Omega), D_w^\alpha f \in L^2(\Omega), | \alpha | \le m \} \tag{3.95a}$$

where $D_w^\alpha f$ denotes the weak derivatives of order $|\alpha|$ [14] with the following inner product,

$$a(f_1, f_2) = \sum_{|\alpha| \leq m} \int_\Omega (D_w^\alpha f_1 \cdot D_w^\alpha f_2) dx, \text{ for } f_1, f_2 \in H^m(\Omega) \qquad (3.95b)$$

It is noted that $H^m(\Omega)$ is a special case of the more general Sobolev space, which is defined by $W_p^k(\Omega)$,

$$W_p^k(\Omega) = \left\{ f \in L_{loc}^1(\Omega) \mid \| f \|_{W_p^k(\Omega)} < \infty \right\} \qquad (3.96)$$

where $L_{loc}^1(\Omega)$ is the locally integrable Lebesgue function and the Sobolev norm is defined by [14]

$$\| f \|_{W_p^k(\Omega)} = \left(\sum_{|\alpha| \leq k} \| D_w^\alpha f \|_{L^p(\Omega)}^p \right)^{1/p} \quad \text{for } 1 \leq p < \infty \qquad (3.97a)$$

$$\| f \|_{W_\infty^k(\Omega)} = \max_{|\alpha| \leq k} \| D_w^\alpha f \|_{L^\infty(\Omega)} \qquad \text{for } p = \infty \qquad (3.97b)$$

In finite element analysis, seminorms are also used. The Sobolev semi-norm is defined by [14]

$$| f |_{W_p^k(\Omega)} = \left(\sum_{|\alpha| = k} \| D_w^\alpha f \|_{L^p(\Omega)}^p \right)^{1/p} \quad \text{for } 1 \leq p < \infty \qquad (3.97c)$$

$$| f |_{W_\infty^k(\Omega)} = \max_{|\alpha| = k} \| D_w^\alpha f \|_{L^\infty(\Omega)} \qquad \text{for } p = \infty \qquad (3.97d)$$

The Sobolev space has important properties, and some of these useful for error analysis are summarized here. First, we have $W_p^m(\Omega) \subset W_p^k(\Omega)$ for $0 < k \leq m$ and $1 \leq p \leq \infty$, and $W_q^k(\Omega) \subset W_p^k(\Omega)$ for $k < 0$ and $1 \leq p \leq q \leq \infty$. By definition, we have $| f |_{W_p^k(\Omega)} \leq \| f \|_{W_p^k(\Omega)}$. Also, for an n-dimensional domain Ω with Lipschitz boundary, we have the following important estimate:

$$\| f \|_{W_\infty^m(\Omega)} \leq C \| f \|_{W_p^k(\Omega)}, \quad \forall f \in W_p^k(\Omega) \qquad (3.97e)$$

for $k > m > 0$, $1 \leq p \leq \infty$, $k - m \geq n$ when $p = 1$, and $k - m > n/p$ when $p > 1$. A particularly useful relation is also obtained when $m = 0$,

$$\| f \|_{L^{\infty}(\Omega)} \leq C \| f \|_{W_p^k(\Omega)} \, , \, \forall f \in W_p^k(\Omega) \tag{3.97f}$$

The boundary estimates are also important. For Ω with boundary $\partial\Omega$ being of Lipschitz type, we have the following relation [14]:

$$\| f \|_{L^p(\partial\Omega)} \leq C \| f \|_{L^p(\Omega)}^{1-1/p} \| f \|_{W_p^1(\Omega)}^{1/p} \, , \, \forall f \in W_p^1(\Omega) \tag{3.97g}$$

for $1 \leq p \leq \infty$.

We note that, from the above definitions, we establish the following useful relations: $H^k(\Omega) = W_2^k(\Omega)$ and $H^0(\Omega) = L^2(\Omega) = W_2^0(\Omega)$.

We may further define two measures for error analysis. The first is the H^m-norm, which is defined by

$$\| f \|_m^2 = \| f \|_{H^m(\Omega)}^2 = \sum_{|\alpha| \leq m} \int_{\Omega} \left(D_w^\alpha f \right)^2 d\Omega \tag{3.98a}$$

The second is the *seminorm*, which is also used in finite element calculations, and is defined in this book as

$$| f |_m = | f |_{H^m(\Omega)} = \left(\sum_{|\alpha| = m} \int_{\Omega} \left| D_w^\alpha f \right|^2 dx \right)^{1/2} \tag{3.98b}$$

Here $| f |_{H^m(\Omega)}$ measures the L^2-norm of the partial derivatives of f of order exactly equal to m. Obviously, these two measures are special cases of the Solobev norm and seminorm, respectively.

With the above definitions, the following norms can be used as error measures. For illustrative purposes, 1-D problems are applied to define the measure. Similar forms can be constructed for multidimensional problems.

The L^2-norm error measure is given by

$$\| e(x) \|_0 = \| u(x) - u_h(x) \|_0 = \left(\int_a^b | u(x) - u_h(x) |^2 dx \right)^{1/2} \tag{3.99}$$

The maximum norm measure is defined as

$$\| e(x) \|_{\infty} = \| u(x) - u_h(x) \|_{\infty} = \max_{a \leq x \leq b} | u(x) - u_h(x) | \tag{3.100}$$

The H^1-norm measure is given by

$$\| e(x) \|_1 = \left(\int_a^b (| u(x) - u_h(x)|^2 + | u'(x) - u_h'(x)|^2) dx \right)^{1/2} \tag{3.101}$$

Here to simplify the notation, we have used the following defintions: $\|\bullet\|_1 = \|\bullet\|_{H^1(\Omega)}$ and $\|\bullet\|_0 = \|\bullet\|_{L^2(\Omega)}$.

3.4.2 Interpolation Error Analysis for 1-D Elements

An important part of error analysis in finite element solutions is related to the errors associated with interpolation functions. Consider a function $u(\xi)$ over an interval $\xi \in [-1, +1]$ and its Lagrangian interpolation,

$$u(\xi) = \sum_{k=0}^{p} u_k \phi_k(\xi) + e(\xi) \tag{3.102}$$

where $e(\xi)$ is the error of the interpolation and is calculated by the following expression [6]:

$$e(\xi) = \frac{u^{(p+1)}(\zeta(\xi))}{(p+1)!} \prod_{i=0}^{p} (\xi - \xi_i) = C u^{(p+1)}(\zeta) \prod_{i=0}^{p} (\xi - \xi_i),$$

$$\zeta(\xi), \ \xi \in (-1, +1) \tag{3.103}$$

Here C is a constant independent of ξ. The interpolation is exact up to order p.

Like the continuous finite element analysis, the discontinuous solutions have a variety of error estimates. We consider below the L^2 and H^1 error estimates, which are most frequently used in the analysis. Let us consider a linear $(p = 1)$ approximation for a 1-D problem, $\xi \in [-1, +1]$. Here, the error for the approximation is

$$e(\xi) = C u''(\zeta)(1 + \xi)(1 - \xi), \ \xi \in (-1, +1) \tag{3.104}$$

where $C_0 = 1/2! = \frac{1}{2}$. Its maximum possible error is obviously bound by

$$|e(\xi)| = C |u''(\zeta)||1 - \xi^2| \le C \max_{-1 \le \xi \le 1} |u''(\xi)| \max_{-1 \le \xi \le 1} |1 - \xi^2| = C \max_{-1 \le \xi \le 1} |u''(\xi)| \tag{3.105}$$

An element, for example, element j, $x \in [x_j, x_{j+1}]$ is transformed into a canonical element $\xi \in [-1, +1]$ by $\xi = (2x - x_{j+1} - x_j)/(x_{j+1} - x_j) = (2x - x_{j+1} - x_j)/h_j$ so that one has the result below:

$$u''(\xi) = \frac{d^2 u(\xi)}{d\xi^2} = \frac{d^2 u(x)}{dx^2} \left(\frac{dx}{d\xi} \right)^2 = \frac{h_j^2}{4} \frac{d^2 u(x)}{dx^2} \tag{3.106}$$

Furthermore, the local maximum norm for a function $f(x)$ defined on $[x_j, x_{j+1}]$ is given by

$$\| f(\bullet) \|_{\infty, j} = \max_{x_j \le x \le x_{j+1}} |f(x)| \tag{3.107}$$

where ($\bullet$) means that the actual normal is independent of $x \in [x_j, x_{j+1}]$. We can write the above error estimate (Equation 3.105) as

$$|e(\bullet)| \le C \max_{-1 \le \xi \le 1} |u''(\xi)| = C \tfrac{1}{4} h_j^2 \| u''(\bullet) \|_{\infty, j} \le C h^2 \| u''(\bullet) \|_{\infty} \tag{3.108}$$

where $h = \max_{1 \le j \le N} h_j$ and $\| u''(\bullet) \|_{\infty} = \max_{1 \le j \le N} \| u''(\bullet) \|_{\infty, j}$, with N being the number of elements.

The L^2-norm in local definition is given by

$$\| e(\bullet) \|_{0, j} = \left(\int_{x_j}^{x_{j+1}} e^2(x) dx \right)^{1/2} \tag{3.109}$$

where the subscript j means that the estimate is local to element j. For $x \in [x_j, x_{j+1}]$,

$$\| e(\bullet) \|_{0, j}^2 = \int_{x_j}^{x_{j+1}} e^2(x) dx$$

$$= \frac{h_j}{2} \int_{-1}^{1} C[u''(\zeta(\xi))(\xi^2 - 1)]^2 d\xi \le C h_j \int_{-1}^{1} [u''(\zeta(\xi))]^2 d\xi \tag{3.110}$$

where we have used $(\xi^2 - 1)^2 \le 1$. We assume that ζ varies smoothly with ξ, and expand u'' in the Taylor series,

$$u''(\zeta(\xi)) = u''(\xi) + u^{(3)}(\mu)(\zeta - \xi),$$

$$= u''(\xi) + O(\zeta - \xi) \le C_1 u''(\xi) \quad \text{where} \quad \mu \in (\zeta, \xi) \tag{3.111}$$

with C_1 being a constant. Here, we have also assumed that for a smooth function, the term $O(\zeta - \xi)$ is proportional to $u''(\xi)$ [14], and the proportional constant is absorbed into the constant C_1. Thus, the error estimate is determined by the following expression:

$$\| e(\bullet) \|_{0,j}^2 \le C_1 \tfrac{1}{8} h_j \int_{-1}^{1} [u''(\zeta(\xi))]^2 d\xi \le C h_j^4 \int_{x_j}^{x_{j+1}} [u''(x)]^2 dx$$

$$= C h_j^4 \| u''(\bullet) \|_{0,j}^2 \tag{3.112}$$

where C is a constant and use has been made of Equation 3.106. The L^2-norm error estimate is now calculated by summing the contributions from all the elements,

$$\| e(\bullet) \|_0^2 \le C h^4 \| u''(\bullet) \|_0^2 \tag{3.113}$$

where

$$\| f(\bullet) \|_0^2 = \sum_{j=1}^{N} \int_{x_j}^{x_{j+1}} f^2(x) dx = \sum_{j=1}^{N} \| f(\bullet) \|_{0,j}^2 \tag{3.114}$$

We now consider the error estimate measured by the H^1-norm. By definition, the H^1-norm for the error takes the following form:

$$\| e(\bullet) \|_1^2 = \int_{x_j}^{x_{j+1}} [e^2(x) + e'^2(x)] dx \tag{3.115}$$

We need to calculate the term involving $e'(x)$. Differentiating the error term (i.e. Equation 3.104), and assuming that $d\zeta/d\xi$ is bounded, we obtain

$$e'(\xi) = u''(\zeta(\xi))\xi + \frac{u^{(3)}(\zeta(\xi))}{2} \frac{d\zeta}{d\xi}(\xi^2 - 1) \tag{3.116}$$

and

$$\int_{x_j}^{x_{j+1}} e'^2(x) dx = \frac{2}{h_j} \int_{-1}^{+1} \left[u''(\zeta(\xi))\xi + \frac{u^{(3)}(\zeta(\xi))}{2} \frac{d\zeta}{d\xi}(\xi^2 - 1) \right]^2 d\xi \tag{3.117}$$

We then notice that from Equation 3.106

$$[u''(\zeta(\xi))\xi]^2 \le [u''(\zeta(\xi))]^2 \le C[u''(\xi)]^2 = C u''^2(x) h_j^4 \tag{3.118}$$

$$[u^{(3)}(\zeta(\xi))(d\zeta/d\xi)(\xi^2 - 1)]^2 \le [C_2 u''(\zeta(\xi))(d\zeta/d\xi) h]^2$$

$$\le [C_3 u''(\zeta(\xi)) h]^2 \tag{3.119}$$

where C_1, C_2 and C are arbitrary constants. Using the same argument, we obtain

$$\| e'(\bullet) \|_{0,j}^2 \leq Ch_j^2 \| u''(\bullet) \|_{0,j}^2 \qquad (3.120)$$

Summing over the N elements, we have

$$\| e'(\bullet) \|_0^2 \leq Ch^2 \| u''(\bullet) \|_0^2 \qquad (3.121)$$

With this we have the final error estimate as $h \to 0$,

$$\| e(\bullet) \|_1^2 = \| e(\bullet) \|_0^2 + \| e'(\bullet) \|_0^2 \leq Ch^2 (1 + ch^2) \| u''(\bullet) \|_0^2 \leq Ch^2 \| u''(\bullet) \|_0^2 \qquad (3.122)$$

where c and C are two constants.

The above analysis is for a linear element. The procedure can also be extended to higher order elements. Thus, we have the general expression,

$$\| u - U \|_0 \leq C_p h^{p+1} \| u^{(p+1)} \|_0, \qquad (3.123a)$$

$$\| u - U \|_1 \leq C_p h^p \| u^{(p+1)} \|_0 \qquad (3.123b)$$

for the polynomial interpolant U of order p and $u \in H^{p+1}$. Here C_p depends on p. It is further shown [13] that

$$\| u - U \|_m \leq Ch^{n-m} \| u \|_n \qquad (3.124)$$

for $0 \leq m \leq n$ and C is a constant. Additional informion on interpolation error estimates in terms of the Sobolev norms can be found in [14].

3.4.3 Interpolation Error Analysis for 2-D/3-D Elements

The interpolation errors for multidimensional elements are obviously more difficult to assess than the 1-D elements as shown above. The basic procedure, however, remains the same. The error estimates for a non-uniform mesh are given by [13]

$$\| u - U \|_s \leq \frac{Ch^{p+1-s}}{(\sin\alpha)^s} | u |_{p+1} \quad \text{for } u \in H^{p+1}(\Omega) \text{ and } s = 0,1 \qquad (3.125)$$

for triangular and tetrahedral elements, and

$$\| u - U \|_s \leq \frac{C_p}{\beta^s} h^{p+1-s} | u |_{p+1} \quad \text{for } u \in H^{p+1}(\Omega) \text{ and } s = 0,1 \qquad (3.126)$$

for rectangles and bricks. Here h is the largest edge of the elements, α is the smallest angle of the triangle or tetrahedron, and β is the smallest aspect ratio of the rectangle or brick elements.

A mesh is considered uniform if all angles of all elements are bounded away from 0 and p and all aspect ratios bounded from zero as the element mesh size $h \to 0$. For a uniform mesh, it can be shown that the error estimate is given by [13, 14]

$$\| u - U \|_s \le C' h^{p+1-s} \, |u|_{p+1} \le C h^{p+1-s} \, \| u \|_{p+1}$$

$$\text{for } u \in H^{p+1}(\Omega) \text{ and } s = 0, 1, \cdots, p+1 \quad (3.127)$$

where the polynomial interpolant U is of order p and $u \in H^{p+1}$ and C and C' are two constants.

3.5 Numerical Integration

Numerical integration for discontinuous finite element calculations often is carried out using the numerical quadrature formulae. For some simple cases, analytical integration is possible and formulae for these cases are included. Since the Gaussian integration quadrature formulae present the best accuracy for a given number of points, emphasis will be given to this type of numerical integration procedure.

3.5.1 1-D Numerical Integration

One-dimensional numerical integration primarily consists of finding the area under the curve defined by a function $f(s)$. By the Gaussian integration rule, an integral can be computed using the following formula:

$$\int_{-1}^{+1} f(s)\,d\xi = \sum_{i=1}^{n} w_i \, f(s_i) \qquad (3.128)$$

where s_i denotes the integration points, w_i are the associated weighting parameters, and n is the total number of integration points. These points are given in Table 3.2. The error associated with the above quadrature is $O(d^{2n}f(s)/d^{2n}s)$. This means that an exact integration can be obtained for a polynomial $f(s)$ of up to $(2n-1)$ degree, if n integration points are used.

Example 3.7. Evaluate the integral $I = \int_{-1}^{+1} [x^2 + \cos(x/2)]\,dx$ using three-point Gaussian quadrature.

Table 3.2. Abscissae and weight coefficients of the Gaussian quadrature formula

$$\int_{-1}^{+1} f(x)\,dx = \sum_{j=1}^{n} w_j f(a_j)$$

$\pm a$	w
$n = 1$	
0.000 000 000 000 000	2.000 000 000 000 000
$n = 2$	
0.577 350 269 189 626	1.000 000 000 000 000
$n = 3$	
0.774 596 669 241 483	0.555 555 555 555 555
0.000 000 000 000 000	0.888 888 888 888 888
$n = 4$	
0.861 136 311 594 953	0.347 854 845 137 454
0.339 981 043 584 856	0.652 145 154 862 546
$n = 5$	
0.906 179 845 938 664	0.236 926 885 056 189
0.538 469 310 105 683	0.478 628 670 499 366
0.000 000 000 000 000	0.568 888 888 888 889
$n = 6$	
0.932 469 514 203 152	0.171 324 492 379 170
0.661 209 386 466 265	0.360 761 573 048 139
0.238 619 186 083 197	0.467 913 934 572 691
$n = 7$	
0.946 107 912 342 759	0.129 484 966 168 870
0.741 531 185 599 394	0.279 705 391 489 277
0.405 845 151 377 397	0.381 830 050 505 119
0.000 000 000 000 000	0.417 959 183 673 469
$n = 8$	
0.960 289 856 497 536	0.101 228 536 290 376
0.796 666 477 413 627	0.222 381 034 453 374
0.525 532 409 916 329	0.313 706 645 877 887
0.183 434 642 495 650	0.362 683 783 378 362
$n = 9$	
0.968 160 239 507 626	0.081 274 388 361 574
0.836 031 107 326 636	0.180 648 160 694 857
0.613 371 432 700 590	0.260 610 696 402 935
0.324 253 423 403 809	0.312 347 077 040 003
0.000 000 000 000 000	0.330 239 355 001 260
$n = 10$	
0.973 906 528 517 172	0.066 671 344 308 688
0.865 063 366 688 985	0.149 451 349 150 581
0.679 409 568 299 024	0.219 086 362 515 982
0.433 395 394 129 247	0.269 266 719 309 996
0.148 874 338 981 631	0.295 524 224 714 753

Solution. From Table 3.2 for the three Gauss points and weights, we have $x_1 = x_3 = \pm 0.77459...$, $x_2 = 0.000...$, $w_1 = w_3 = 5/9$, and $w_2 = 8/9$. Then we have

$$I = \left((-0.77459)^2 + \cos(-0.5 \times 0.77459) \right) \times \tfrac{5}{9} + \left[0^2 + \cos 0 \right] \times \tfrac{8}{9}$$

$$+ \left((0.77459)^2 + \cos(0.5 \times 0.77459) \right) \times \tfrac{5}{9} = 2.585 \qquad (3.22e)$$

The integral can be calculated analytically with the result,

$$I = \int_{-1}^{1} [x^2 + \cos(x/2)]\, dx = \left(\frac{1}{3}x^3 - 2\sin\frac{x}{2} \right)\Big|_{-1}^{1} = 2.585 \qquad (3.23e)$$

In this example, the three-point Gaussian quadrature yields the exact answer to four significant figures.

3.5.2 2-D and 3-D Numerical Integration

The above 1-D integration quadrature formulae can be easily extended to multi-dimensions,

$$\int_{-1}^{+1} \int_{-1}^{+1} f(s,t)\, ds\, dt = \sum_{i=1}^{n} \sum_{j=1}^{n} w_i\, w_j\, f(s_i, t_j)\, ds\, dt \qquad (3.129)$$

and

$$\int_{-1}^{+1} \int_{-1}^{+1} \int_{-1}^{+1} f(s,t,u)\, ds\, dt\, du = \sum_{i=1}^{n} \sum_{j=1}^{n} \sum_{k=1}^{n} w_i\, w_j\, w_k\, f(s_i, t_j, u_k)\, ds\, dt\, du \qquad (3.130)$$

The same integration points listed in Table 3.2 can be used to carry out the numerical integration in each of the dimensions.

Example 3.8. Evaluate the integral $\int_{-1}^{+1} \int_{-1}^{+1} r^4 s^2\, dr\, ds$ using two-point Gauss quadrature and compare it with the analytic solution.

Solution. The weights and abscissas are given in Table 3.2. Choosing points $n = 3$ for the r direction and $n = 2$ for the s direction, we have

$$\int_{-1}^{+1} \int_{-1}^{+1} r^4 s^2\, dr\, ds = \int_{-1}^{+1} \left[\left(\frac{5}{9} \right)\left(-\frac{\sqrt{3}}{\sqrt{5}} \right)^4 + \left(\frac{5}{9} \right)\left(\frac{\sqrt{3}}{\sqrt{5}} \right)^4 \right] s^2\, ds$$

$$= \frac{2}{5} \int_{-1}^{+1} s^2 ds = \frac{2}{5} \left[(1) \left(-\frac{1}{\sqrt{3}} \right)^2 + (1) \left(\frac{1}{\sqrt{3}} \right)^2 \right] = \frac{4}{15}$$

The analytical solution for the problem can be easily calculated as well,

$$\int_{-1}^{+1} \int_{-1}^{+1} r^4 s^2 drds = \frac{1}{15} \left(r^5 \right)_{-1}^{+1} \left(s^3 \right)_{-1}^{+1} = \frac{4}{15}$$

We can see that the selected integration points are enough to exactly evaluate the integral.

3.5.3 Integration for Triangular and Tetrahedral Elements

Numerical integration over a triangle can be performed more conveniently using the area coordinates, ξ_1, ξ_2 and ξ_3,

$$\int_0^1 \int_0^{1-\xi_2} f(\xi_1, \xi_2, \xi_3) d\xi_1 d\xi_2 = \sum_{i=1}^m f(\xi_{1,i}, \xi_{2,i}, \xi_{3,i}) w_i \qquad (3.131)$$

This formula is credited to Hammar et al. [15]. The triangular integration points and the associated weighting factors are listed in Table 3.3, where m indicates the integration order. If $m = 1$, an exaction integration can be obtained for a polynomial of degree $p = 1$. For $m = 3$, integration is exact for $p \le 2$; and for $m = 7$, $p \le 4$. Similar weights also exist for tetrahedral elements, which are listed in Table 3.4.

Besides the numerical quadrature rule above, for certain cases the following analytical integration formulae may also be used for triangular elements to expedite the calculations:

$$\int_0^1 \int_0^{1-\xi_2} \xi_1^l \xi_2^m \xi_3^n d\xi_1 d\xi_2 = \frac{l! m! n!}{(l+m+n+2)!} \qquad (3.132)$$

The above numerical and analytical formulae may also be readily extended to 3-D integration over a tetrahedral element,

$$\int_0^1 \int_0^{1-\xi_2} \int_0^{1-\xi_2-\xi_3} f(\xi_1^l, \xi_2^m, \xi_3^n, \xi_4^q) d\xi_1 d\xi_2 d\xi_3$$

$$= \sum_{i=1}^m f(\xi_{1,i}, \xi_{2,i}, \xi_{3,i}, \xi_{4,i}) w_i \qquad (3.133)$$

for the use of numerical quadrature, and

$$\int_0^1 \int_0^{1-\xi_2} \int_0^{1-\xi_2-\xi_3} \xi_1{}^l \xi_2{}^m \xi_3{}^n \xi_4{}^q d\xi_1 d\xi_2 d\xi_3 = \frac{l!\,m!\,n!\,q!}{(l+m+n+q+3)!} \quad (3.134)$$

for analytical integration.

Table 3.3. Numerical intergration formulae for triangles

Order (m)	Figure	Error	Points	Triangular coordinates	Weights
Linear ($m=1$)		$R=O(h^2)$	1	$\frac{1}{3},\frac{1}{3},\frac{1}{3}$	1
Quadratic ($m=2$)		$R=O(h^3)$	3 3 3	$\frac{1}{2},\frac{1}{2},\frac{1}{2}$ $0,\frac{1}{2},\frac{1}{2}$ $\frac{1}{2},0,\frac{1}{2}$	$\frac{1}{3}$ $\frac{1}{3}$ $\frac{1}{3}$
Cubic ($m=3$)		$R=O(h^4)$	1 2 3 4	$\frac{1}{3},\frac{1}{3},\frac{1}{3}$ $\begin{bmatrix}0.6,0.2,0.2\\0.2,0.6,0.2\\0.2,0.2,0.6\end{bmatrix}$	$-\frac{27}{48}$ $\frac{25}{48}$
Quintic ($m=7$)		$R=O(h^6)$	1 2 3 4 5 6 7	$\frac{1}{3},\frac{1}{3},\frac{1}{3}$ $\begin{bmatrix}\alpha_1,\beta_1,\beta_1\\\beta_1,\alpha_1,\beta_1\\\beta_1,\beta_1,\alpha_1\end{bmatrix}$ $\begin{bmatrix}\alpha_2,\beta_2,\beta_2\\\beta_2,\alpha_2,\beta_2\\\beta_2,\beta_2,\alpha_2\end{bmatrix}$	0.225 000 0000 0.132 394 1527 0.125 939 1805

$$\alpha_1 = 0.059\ 715\ 8717, \quad \beta_1 = 0.470\ 142\ 0641$$
$$\alpha_2 = 0.797\ 426\ 9853, \quad \beta_2 = 0.101\ 286\ 5073$$

Table 3.4. Numerical integration formulae for tetrahedra

Order (m)	Figure	Error	Points	Triangular coordinates	Weights
Linear ($m=1$)		$R=O(h^2)$	1	$\dfrac{1}{4},\dfrac{1}{4},\dfrac{1}{4},\dfrac{1}{4}$	1
Quadratic ($m=2$)		$R=O(h^3)$	1 2 3 4	α,β,β,β β,α,β,β β,β,α,β β,β,β,α	0.25 0.25 0.25 0.25
				$\alpha = 0.585\ 410\ 20,\quad \beta = 0.138\ 196\ 60$	
Cubic ($m=3$)		$R=O(h^4)$	1 2 3 4 5	$\dfrac{1}{4},\dfrac{1}{4},\dfrac{1}{4},\dfrac{1}{4}$ $\dfrac{1}{2},\dfrac{1}{6},\dfrac{1}{6},\dfrac{1}{6}$ $\dfrac{1}{6},\dfrac{1}{2},\dfrac{1}{6},\dfrac{1}{6}$ $\dfrac{1}{6},\dfrac{1}{6},\dfrac{1}{2},\dfrac{1}{6}$ $\dfrac{1}{6},\dfrac{1}{6},\dfrac{1}{6},\dfrac{1}{2}$	$-\dfrac{4}{5}$ $\dfrac{9}{20}$ $\dfrac{9}{20}$ $\dfrac{9}{20}$ $\dfrac{9}{20}$

3.6 Elemental Calculations

The first step in applying the discontinuous finite element method to a given boundary value problem is the discretization of a computational domain into a collection of elements. The calculations are then performed over an element. In almost all cases, the isoparametric elements are used in calculations. We present below the use of isoparametric elements for 2-D and 3-D domain calculations and boundary calculations.

3.6.1 Domain Calculations

For an isoparametric formulation, the points in the elements are mapped to a unit (or canonical) element, and all ensuing calculations are performed in the mapped element, which is called the master element. This is shown in Figure 3.18, where a curvilinear element is mapped to its corresponding master element in the normalized coordinate system. The transformation between the finite element and the master element of a simple shape is invertible; or simply put, the Jacobian of the transformation matrix is positive. For simplicity, the master element is chosen to be a square, where local coordinates ξ and η are normalized, $\{(\xi,\eta) \mid \xi, \eta \in [-1, +1]\}$.

As discussed above, any variables defined on the element can be approximated using the shape functions in the form,

$$u_h(\xi,\eta) = \sum_{i=1}^{N_e} u_i \phi_i(\xi,\eta) \tag{3.135}$$

where N_e is the number of nodes of the element under consideration.

If we now treat the coordinate variables x and y themselves as functions on Ω, then the shape functions may be used to construct the mapping,

$$x(\xi,\eta) = \sum_{i=1}^{N_e} x_i \phi_i(\xi,\eta) \; ; \; y(\xi,\eta) = \sum_{i=1}^{N_e} y_i \phi_i(\xi,\eta) \tag{3.136}$$

Here (x_i, y_i) are the (x,y) coordinates of local nodal point i in element Ω_e. Note that by this transformation, every element in the discretized mesh can be mapped onto the master element. This will make the program phase convenient.

Several important properties of this isoparametric mapping need to be discussed before we consider the detailed computational procedures. First, we note that the functions x and y are differentiable with respect to the local coordinates ξ and η,

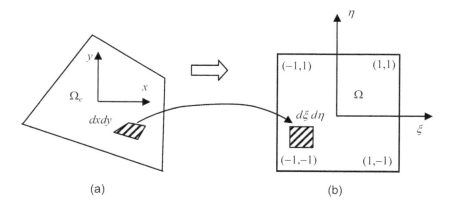

(a)   (b)

Figure 3.18. Transformation of an element in (x,y) (a) into a canonical element in (ξ,η) (b)

$$dy(\xi,\eta) = \frac{\partial y}{\partial \xi}d\xi + \frac{\partial y}{\partial \eta}d\eta = \sum_{i=1}^{N_e} y_i \frac{\partial \phi_i(\xi,\eta)}{\partial \xi}d\xi + \sum_{i=1}^{N_e} y_i \frac{\partial \phi_i(\xi,\eta)}{\partial \eta}d\eta$$

(3.137)

$$dx(\xi,\eta) = \frac{\partial x}{\partial \xi}d\xi + \frac{\partial x}{\partial \eta}d\eta = \sum_{i=1}^{N_e} x_i \frac{\partial \phi_i(\xi,\eta)}{\partial \xi}d\xi + \sum_{i=1}^{N_e} x_i \frac{\partial \phi_i(\xi,\eta)}{\partial \eta}d\eta$$

(3.138)

In matrix notation,

$$\begin{bmatrix} dx \\ dy \end{bmatrix} = \begin{bmatrix} \dfrac{\partial x}{\partial \xi} & \dfrac{\partial x}{\partial \eta} \\ \dfrac{\partial y}{\partial \xi} & \dfrac{\partial y}{\partial \eta} \end{bmatrix} \begin{bmatrix} d\xi \\ d\eta \end{bmatrix} = \mathbf{J} \begin{bmatrix} d\xi \\ d\eta \end{bmatrix}$$

(3.139)

where the 2×2 matrix is the Jacobian matrix of the transformation. The invertibility condition requires that the determinant of the Jacobian be non-zero at $\xi,\eta \in \Omega$. Thus we have

$$\begin{bmatrix} d\xi \\ d\eta \end{bmatrix} = \mathbf{J}^{-1} \begin{bmatrix} dx \\ dy \end{bmatrix}$$

(3.140)

where the inverse of the Jacobian matrix is calculated by

$$\mathbf{J}^{-1} = \frac{1}{|\mathbf{J}|} \begin{bmatrix} \dfrac{\partial \xi}{\partial x} & \dfrac{\partial \xi}{\partial y} \\ \dfrac{\partial \eta}{\partial x} & \dfrac{\partial \eta}{\partial y} \end{bmatrix}$$

(3.141)

For an affine transformation, that is, for a one-to-one mapping from Ω_e to the master element Ω, it is necessary that

$$|\mathbf{J}(\xi,\eta)| > 0, \quad \forall \xi,\eta \in \Omega$$

(3.142)

Let us now take a closer look at the physical meaning of the Jacobian $\mathbf{J}$. For this purpose, we construct a differential area in the master element, $da = d\xi d\eta$. When inverted back, dA has its image in the x–y plane as $dA = dxdy = |\mathbf{J}|d\xi d\eta$. Thus, $|\mathbf{J}|$ is just the ratio of areas of elements at points (x,y) and (ξ,η), i.e., $|\mathbf{J}| = dA/da$.

The above procedure can be directly extended to 3-D calculations. In 3-D domains, the master element is often chosen to be a cube defined by $\{(\xi,\eta,\varsigma) \mid \xi,\eta,$

$\zeta \in [-1, +1]$}. The transformation between a finite element and the master element can be constructed via

$$x = \sum_{i=1}^{N_e} x_i \phi_i(\xi,\eta,\varsigma) \; ; \; y = \sum_{i=1}^{N_e} y_i \phi_i(\xi,\eta,\varsigma) \; ; \; z = \sum_{i=1}^{N_e} z_i \phi_i(\xi,\eta,\varsigma) \quad (3.143)$$

where N_e is the number of nodes of the element under consideration. This will allow element Ω_e to be completely determined by specifying the (x, y, z) coordinates of all the nodal points of Ω_e. With this, the following transformation property is established between (x,y,z) and (ξ,η,ς):

$$\begin{pmatrix} dx \\ dy \\ dz \end{pmatrix} = \begin{bmatrix} \dfrac{\partial x}{\partial \xi} & \dfrac{\partial x}{\partial \eta} & \dfrac{\partial x}{\partial \varsigma} \\ \dfrac{\partial y}{\partial \xi} & \dfrac{\partial y}{\partial \eta} & \dfrac{\partial y}{\partial \varsigma} \\ \dfrac{\partial z}{\partial \xi} & \dfrac{\partial z}{\partial \eta} & \dfrac{\partial z}{\partial \varsigma} \end{bmatrix} \begin{pmatrix} d\xi \\ d\eta \\ d\varsigma \end{pmatrix} \; ; \quad \begin{pmatrix} d\xi \\ d\eta \\ d\varsigma \end{pmatrix} = \begin{bmatrix} \dfrac{\partial \xi}{\partial x} & \dfrac{\partial \xi}{\partial y} & \dfrac{\partial \xi}{\partial z} \\ \dfrac{\partial \eta}{\partial x} & \dfrac{\partial \eta}{\partial y} & \dfrac{\partial \eta}{\partial z} \\ \dfrac{\partial \varsigma}{\partial x} & \dfrac{\partial \varsigma}{\partial y} & \dfrac{\partial \varsigma}{\partial z} \end{bmatrix} \begin{pmatrix} dx \\ dy \\ dz \end{pmatrix} \quad (3.144)$$

from which the Jacobian of the transformation and its inverse are calculated by

$$\mathbf{J} = \begin{bmatrix} \dfrac{\partial x}{\partial \xi} & \dfrac{\partial x}{\partial \eta} & \dfrac{\partial x}{\partial \varsigma} \\ \dfrac{\partial y}{\partial \xi} & \dfrac{\partial y}{\partial \eta} & \dfrac{\partial y}{\partial \varsigma} \\ \dfrac{\partial z}{\partial \xi} & \dfrac{\partial z}{\partial \eta} & \dfrac{\partial z}{\partial \varsigma} \end{bmatrix} \; ; \quad \mathbf{J}^{-1} = \begin{bmatrix} \dfrac{\partial \xi}{\partial x} & \dfrac{\partial \xi}{\partial y} & \dfrac{\partial \xi}{\partial z} \\ \dfrac{\partial \eta}{\partial x} & \dfrac{\partial \eta}{\partial y} & \dfrac{\partial \eta}{\partial z} \\ \dfrac{\partial \varsigma}{\partial x} & \dfrac{\partial \varsigma}{\partial y} & \dfrac{\partial \varsigma}{\partial z} \end{bmatrix} \quad (3.145)$$

We may apply the above mapping to calculate the quantities needed for the element matrices for each element in the mesh. With the function values treated as boundary conditions at the cross element boundaries, the element matrix is then inverted and unknowns are obtained at the nodal points.

Example 3.9. Consider a rectangular element as shown in Figure 3.7e(a) below. The temperature is $(100, 105, 120, 89)$ at the four corner points. Evaluate $\mathbf{J}$ and the derivatives of the temperature field at $\xi = 0$ and $\eta = 0$.

Solution. For a 4–node element, $N_e = 4$ and the transformation Jacobian is calculated explicitly by substituting the transformation rule given in Equation 3.136,

$$J = \frac{1}{4} \begin{bmatrix} \begin{matrix} -(1-\eta)x_1 + (1-\eta)x_2 \\ +(1+\eta)x_3 - (1+\eta)x_4 \end{matrix} & \begin{matrix} -(1-\eta)y_1 + (1-\eta)y_2 \\ +(1+\eta)y_3 - (1+\eta)y_4 \end{matrix} \\ \begin{matrix} -(1+\xi)x_1 - (1+\xi)x_2 \\ +(1+\xi)x_3 + (1+\xi)x_4 \end{matrix} & \begin{matrix} -(1+\xi)y_1 - (1+\xi)y_2 \\ +(1+\xi)y_3 + (1+\xi)y_4 \end{matrix} \end{bmatrix} \equiv \begin{bmatrix} J_{11} & J_{12} \\ J_{21} & J_{22} \end{bmatrix} \qquad (3.24e)$$

Substituting the coordinates of (x,y) at the four nodal points, we have the Jacobian evaluated at $(\xi = 0, \eta = 0)$,

$$J = \frac{1}{4} \begin{bmatrix} 2(1-\eta) + 2(1+\eta) & (1+\eta) - (1+\eta) \\ -2(1+\xi) + 2(1+\xi) & (1+\xi) + (1-\xi) \end{bmatrix} = \begin{bmatrix} 1 & 0 \\ 0 & \frac{1}{2} \end{bmatrix} \qquad (3.25e)$$

or

$$\det J = |J| = 1/2 \qquad (3.26e)$$

The inverse of matrix J at $(\xi = 0, \eta = 0)$ becomes

$$J^{-1} = \frac{1}{\det J} \begin{bmatrix} J_{22} & -J_{12} \\ -J_{21} & J_{11} \end{bmatrix} = \begin{bmatrix} 1 & 0 \\ 0 & 2 \end{bmatrix} \qquad (3.27e)$$

$$dxdy = \det J d\xi\, d\eta = 0.5 d\xi\, d\eta \qquad (3.28e)$$

To calculate the temperature derivatives at $(\xi = 0, \eta = 0)$, we use the following formulae:

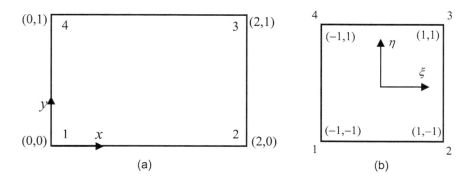

Figure 3.7e. Transformation of a rectangular element (a) into a canonical element (b)

$$\begin{Bmatrix} \dfrac{\partial T}{\partial x} \\[2mm] \dfrac{\partial T}{\partial y} \end{Bmatrix} = \mathbf{J}^{-1} \begin{Bmatrix} \dfrac{\partial T}{\partial \xi} \\[2mm] \dfrac{\partial T}{\partial \eta} \end{Bmatrix} = \dfrac{1}{\det \mathbf{J}} \begin{bmatrix} J_{22} & -J_{12} \\ -J_{21} & J_{11} \end{bmatrix} \begin{Bmatrix} \dfrac{\partial T}{\partial \xi} \\[2mm] \dfrac{\partial T}{\partial \eta} \end{Bmatrix} \tag{3.29e}$$

$$\begin{Bmatrix} \dfrac{\partial T}{\partial \xi} \\[2mm] \dfrac{\partial T}{\partial \eta} \end{Bmatrix} = \begin{Bmatrix} \displaystyle\sum_{i=1}^{4} T_i \dfrac{\partial \phi_i(\xi)}{\partial \xi} \\[4mm] \displaystyle\sum_{i=1}^{4} T_i \dfrac{\partial \phi_i(\eta)}{\partial \eta} \end{Bmatrix} = \begin{Bmatrix} 36 \\ 7 \end{Bmatrix} \tag{3.30e}$$

With these quantities, we have the needed temperature derivatives with respect to the x and y coordinates,

$$\begin{Bmatrix} \dfrac{\partial T}{\partial x} \\[2mm] \dfrac{\partial T}{\partial y} \end{Bmatrix} = \begin{bmatrix} 1 & 0 \\ 0 & 2 \end{bmatrix} \begin{Bmatrix} 36 \\ 7 \end{Bmatrix} = \begin{Bmatrix} 36 \\ 14 \end{Bmatrix} \tag{3.31e}$$

These expressions will be used in the derivation of the element stiffness matrix for the element.

3.6.2 Boundary Calculations

The boundary for a 2-D domain is basically a curve. For a curved element boundary, the same isoparametric principles presented above for domain calculations are applicable. In this case, the curve is mapped onto a canonical 1-D element by the following transformation:

$$x(\xi) = \sum_{i=1}^{n_e} x_i \phi_i(\xi); \quad y(\xi) = \sum_{i=1}^{n_e} y_i \phi_i(\xi) \tag{3.146}$$

where n_e is the number of nodes per boundary element. This is illustrated in Figure 3.19.

Thus, a differential arc-length ds in the (x,y) plane can be written in terms of the normalized coordinate ξ,

$$ds(\xi) = |\mathbf{J}(\xi)| \, d\xi = \left[\left(\dfrac{dx}{d\xi} \right)^2 + \left(\dfrac{dy}{d\xi} \right)^2 \right]^{1/2} \tag{3.147}$$

with the Jacobian of the transformation calculated by

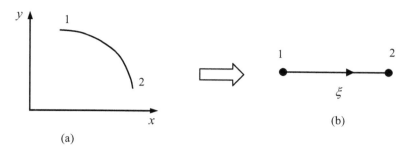

Figure 3.19. Mapping of a 2-D curve into a 1-D unit element: (a) 2-D boundary element and (b) 1-D canonical element

$$| \mathbf{J}(\xi) | = \left[\left(\frac{dx}{d\xi} \right)^2 + \left(\frac{dy}{d\xi} \right)^2 \right]^{1/2}$$

$$= \left[\left(\sum_{i=1}^{n_e} x_i \frac{d\phi_i(\xi)}{d\xi} \right)^2 + \left(\sum_{i=1}^{n_e} y_i \frac{d\phi_i(\xi)}{d\xi} \right)^2 \right]^{1/2} \tag{3.148}$$

The isoparametric treatment of a 3-D surface element is not as simple as that of the 2-D curve boundary. The calculations can be complex if the local normal and tangential components of the velocity field need to be specified along a curvilinear surface. This requires a tedious geometric treatment that involves differential geometry operations, and the rotation of the matrix in local coordinates at the surface, for the purpose of appropriately imposing velocity and surface stress boundary conditions [16, 17]. One approach is presented here, which makes use of local surface coordinates and of sharp edges with a specified local coordinate system, as well as consistent surface normals. With reference to Figure 3.20, a local coordinate system (η, ζ, n) is defined at a point on the surface. Note that during calculations, this local system may be (although it is not required to be) chosen conveniently so as to be coincident with the normalized coordinate system for the isoparametric calculations at the element level. The xyz and $\eta\zeta n$ coordinate systems are related by the following coordinate transformation,

$$\begin{pmatrix} \dfrac{\partial}{\partial \eta} \\ \dfrac{\partial}{\partial \zeta} \\ \dfrac{\partial}{\partial n} \end{pmatrix} = \mathbf{J} \begin{pmatrix} \dfrac{\partial}{\partial x} \\ \dfrac{\partial}{\partial y} \\ \dfrac{\partial}{\partial z} \end{pmatrix} = \begin{bmatrix} x_{,\eta} & y_{,\eta} & z_{,\eta} \\ x_{,\zeta} & y_{,\zeta} & z_{,\zeta} \\ x_{,n} & y_{,n} & z_{,n} \end{bmatrix} \begin{pmatrix} \dfrac{\partial}{\partial x} \\ \dfrac{\partial}{\partial y} \\ \dfrac{\partial}{\partial z} \end{pmatrix} \tag{3.149}$$

where the subscript , refers to the derivative, e.g., $x_{,n} = \partial x / \partial n$.

In constructing the Jacobian matrix, we used the following differential geometry relations:

$$\mathbf{r}_1 = x_{,\eta}\,\hat{i} + y_{,\eta}\,\hat{j} + z_{,\eta}\,\hat{k}\;;\; \mathbf{r}_2 = x_{,\zeta}\,\hat{i} + y_{,\zeta}\,\hat{j} + z_{,\zeta}\,\hat{k} \qquad (3.150)$$

$$\mathbf{r}_n = \mathbf{r}_1 \times \mathbf{r}_2 = x_{,n}\,\hat{i} + y_{,n}\,\hat{j} + z_{,n}\,\hat{k}$$

$$= \left(y_{,\eta}\,z_{,\zeta} - z_{,\eta}\,y_{,\zeta}\right)\hat{i} - \left(x_{,\eta}\,z_{,\zeta} - z_{,\eta}\,x_{,\zeta}\right)\hat{j} + \left(x_{,\eta}\,y_{,\zeta} - y_{,\eta}\,x_{,\zeta}\right)\hat{k} \quad (3.151)$$

The Jacobian matrix may be inverted analytically with the following result:

$$[\mathbf{J}]^{-1} = \frac{1}{|\mathbf{J}|}\begin{bmatrix} y_{,\zeta}\,z_{,n} - z_{,\zeta}\,y_{,n} & -(y_{,\eta}\,z_{,n} - z_{,\eta}\,y_{,n}) & y_{,\eta}\,z_{,\zeta} - z_{,\eta}\,y_{,\zeta} \\ -(x_{,\zeta}\,z_{,n} - z_{,\zeta}\,x_{,n}) & x_{,\eta}\,z_{,n} - z_{,\eta}\,x_{,n} & x_{,\eta}\,z_{,\zeta} - z_{,\eta}\,x_{,\zeta} \\ x_{,\zeta}\,y_{,n} - y_{,\zeta}\,x_{,n} & -(x_{,\eta}\,y_{,n} - y_{,\eta}\,x_{,n}) & x_{,\eta}\,y_{,\zeta} - y_{,\eta}\,x_{,\zeta} \end{bmatrix}$$

$$(3.152)$$

Furthermore, a shape (or any) function $f(\eta, \zeta)$ defined over the surface is a function of (η,ζ) only, and hence $\partial f(\eta,\zeta)/\partial n$. With these relations, one may then relate the volume differential operator to the surface operator,

$$\begin{pmatrix} \dfrac{\partial}{\partial x} \\ \dfrac{\partial}{\partial y} \\ \dfrac{\partial}{\partial z} \end{pmatrix} = [\mathbf{J}]^{-1}\begin{pmatrix} \dfrac{\partial}{\partial \eta} \\ \dfrac{\partial}{\partial \zeta} \\ \dfrac{\partial}{\partial n} \end{pmatrix} \qquad (3.153)$$

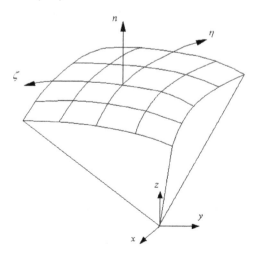

Figure 3.20. Transformation between local curvilinear and global Cartesian coordinate systems

which may be written in the terminology of differential geometry [18],

$$\nabla = \frac{\partial}{\partial x}\hat{i} + \frac{\partial}{\partial y}\hat{j} + \frac{\partial}{\partial y}\hat{k} = \frac{1}{H^2}\mathbf{r}_1\left(G\frac{\partial}{\partial \eta} - F\frac{\partial}{\partial \zeta}\right) + \frac{1}{H^2}\mathbf{r}_2\left(E\frac{\partial}{\partial \zeta} - F\frac{\partial}{\partial \eta}\right)$$

(3.154)

with $E = \mathbf{r}_1^2 = x_{,\eta}^2 + y_{,\eta}^2 + z_{,\eta}^2$, $F = \mathbf{r}_1 \cdot \mathbf{r}_2 = x_{,\eta} x_{,\zeta} + y_{,\eta} y_{,\zeta} + z_{,\eta} z_{,\zeta}$, $H^2 = EG - F^2$, $G = \mathbf{r}_2^2 = x_{,\zeta}^2 + y_{,\zeta}^2 + z_{,\zeta}^2$. It is stressed that in Equation 3.154, (η, ζ) is not necessarily orthogonal so long as they are not collinear. This is important in that irregular quadrilateral surface elements can be readily handled in 3-D finite element calculations presented here.

For flow calculations, the consistent normal of the surface at node i is required, which must satisfy the continuity equation [19],

$$n_x^i = \frac{1}{n_i}\int_\Omega \frac{\partial \phi_i}{\partial x}dV; \ n_y^i = \frac{1}{n_i}\int_\Omega \frac{\partial \phi_i}{\partial y}dV; \ n_z^i = \frac{1}{n_i}\int_\Omega \frac{\partial \phi_i}{\partial z}dV$$

(3.155)

where

$$n_i = \left[\left(\int_\Omega \frac{\partial \phi_i}{\partial x}dV\right)^2 + \left(\int_\Omega \frac{\partial \phi_i}{\partial y}dV\right)^2 + \left(\int_\Omega \frac{\partial \phi_i}{\partial z}\right)^2 dV\right]^{1/2}$$

(3.156)

Note that the integration is carried over all the elements sharing node i. Once the normal is known, the two tangential directions $\mathbf{t}_1 = (t_x^1, t_y^1, t_z^1)$ and $\mathbf{t}_2 = (t_x^2, t_y^2, t_z^2)$ can be easily calculated using the cross product relations, $\mathbf{t}_1 = \mathbf{b} \times \mathbf{n}$, in which $\mathbf{b}$ is an arbitrary space vector such that $\mathbf{b} \times \mathbf{n} \neq 0$, and $\mathbf{t}_2 = \mathbf{n} \times \mathbf{b}$. This ensures that the local coordinate system defined by $\mathbf{t}_1 \times \mathbf{t}_2 \times \mathbf{n}$ forms an orthogonal triplet at any node (e.g., node i). This is a strict condition, different from that imposed on (η, ζ, n). The velocities (or any other vector quantities) defined in the $\mathbf{t}_1 \times \mathbf{t}_2 \times \mathbf{n}$ system are now related to those in the xyz system through the following transformation:

$$\begin{pmatrix} U_{t2} \\ U_{t1} \\ U_n \end{pmatrix} = \begin{bmatrix} t_x^1 & t_y^1 & t_z^1 \\ t_x^2 & t_y^2 & t_z^2 \\ n_x & n_y & n_z \end{bmatrix}\begin{pmatrix} U_x \\ U_y \\ U_z \end{pmatrix}$$

(3.157)

Note that the above transformation also applies to force vectors.

The above procedure can be particularly useful for surface tension driven flows in which the surface stress tensor is a function of the local surface gradient of temperature or concentration or both. One such calculation is presented for a 3-D surface driven flow [17]. There, to calculate the surface tension contributions, the integration of the ∇ term is calculated using the relation given by Equation 3.154.

The velocities defined in the (x,y,z) coordinates will be converted to those in the $t_1 \times t_2 \times n$ system using Equation 3.157. For 3-D flow calculations, sharp edges are formed by the intersection of two surfaces where the $t_1 \times t_2 \times n$ system is not uniquely defined by the above computational procedure. This causes difficulty when appropriate velocity and stress boundary conditions are specified along the edge. To overcome the problem, the normal of the edge needs to be associated with one of the two joining surfaces, and an additional constraint needs to be imposed on the selection of a tangential direction. The detailed treatment and examples of the above surface calculations can be found in [17].

Exercises

1. Starting with the general interpolation function $u_h(x) = a + bx + cx^2$, determine the coefficients a, b, c, and construct the shape functions on an element. Show that these shape functions are the same as given by Equation 3.8 when defined on $\xi \in [-1, 1]$.

2. Obtain the coefficients $a_i^e, b_i^e, c_i^e, d_i^e$ by expanding the tetrahedral equations and prove that these coefficients are determents of a relevant cofactor matrix.

3. Derive the shape functions for a linear tetrahedral element using the recursive relation in terms of volume coordinates.

4. Obtain shape functions for an 8–node element using tensor product.

5. Obtain shape functions for a 20–node element using the analogous approach to that given for the 8–node element in Example 3.3.

6. Using a 2×2 integration rule, evaluate the integral numerically by Gaussian integration,

$$\int_0^1 \int_1^3 (x^2 y + xy^3)\, dx dy$$

   Integrate the above expression analytically and compare it with the numerical integration.

7. Develop a computer program to perform the element calculations for triangular and quadrilateral elements.

8. Consider a quadrilateral element with four corners defined at $(1,1)$, $(4,1)$, $(6,6)$ and $(1,5)$, which correspond to nodes 1, 2, 3, and 4. Using the 2×2 integration rule and Gaussian integration, apply the computer code to calculate the inverse of the Jacobian matrix at every integration point.

9. Prove the relation given in Equation 3.154. Note that the surface vector differential operator is related to the volume operator: $\nabla_s = \nabla - \mathbf{n}(\mathbf{n} \cdot \nabla)$, where ∇_s is the surface vector differential operator.

10. Construct a 2-D spectral shape function using tensor product.

11. Derive shape functions for a 5–node pyramid element by collapsing the nodes 5, 6, 7, 8 into one node as shown in the figure below.

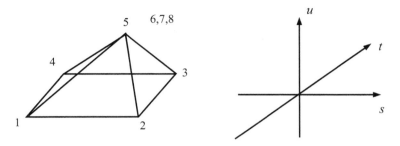

References

[1]   Zienkiewicz OC, Taylor RL. The Finite Element Method, 4th Ed. New York: McGrawHill, 1989.

[2]   Szabo B, Babuska I. Finite Element Analysis. New York: John Wiley and Sons, 1991.

[3]   Kardestuncer H, Norrie DH. Finite Element Handbook. New York: McGrawHill, 1987.

[4]   Bathe KJ. Finite Element Procedures. Englewood Cliffs: Prentice Hall, 1996.

[5]   Gottlieb D, Orszag SA. Numerical Analysis of Spectral Methods. Philadelphia: Soc. for Indust. & Appl. Math. 1977.

[6]   Abromowitz M, Stegun IA. Handbook of Mathematical Functions, Volume 55 of Applied Mathematics Series. Gathersburg: National Bureau of Standards, 1964.

[7]   Lomtev I, Kirby RM, Karniadakis GE. A Discontinuous Galerkin ALE Method for Compressible Viscous Flows in Moving Domains. J. Comput. Phys. 1999; 155: 128–159.

[8]   Botkov E. Mathematical Physics. Reading MA: Addison-Wesley Publishing Company, 1968.

[9]   Shephard MS, Dey S, Flaherty JE. A Straightforward Structure to Construct Shape Functions for Variable P-order Meshes. Comput. Methods Appl. Mech. Eng. 1997; 147: 209–233.

[10]   Peano AG. Hierarchics of Conforming Finite Elements for Elasticity and Plate Bending. Comp. Math. Appl. 1976; 2–4.

[11]   Lomtev I. Discontinuous Spectral/hp Element Methods for High Speed Flows. Ph.D. Thesis. Providence RI: Applied Mathematics, Brown University, 1999.

[12]   Bathe KJ. Finite Element Procedures. Englewood Cliffs: Prentice Hall, 1996.

[13]   Chapelle D, Bathe KJ. The Finite Element Analysis of Shells – Fundamentals. New York: Springer, 2003.

[14]   Brenner SC, Scott LR. The Mathematical Theory of Finite Element Methods. New York: Springer-Verlag, 1994.

[15]   Hammer PC, Marlowe OJ, Stroud AH. Numerical Integration over Simplexes and Cones. Math. Tables Other Aids Comput. 1956; 10: 130–139.

[16]   Song SP, Li BQ. A Hybrid Boundary/Finite Element Method for Simulating Viscous Flows and Shapes of Droplets in Electric Fields. Int. J. Comp. Fluid Dyn. 2001; 15: 293–308.

[17]   Huo Y, Li BQ. Three-dimensional Marangoni Convection in Electrostatically Positioned Droplets under Microgravity. Int. J. Heat Mass Trans. 2004; 47: 3533–3547.

[18]   Weatherburn CE. Differential Geometry of Three Dimensions. London: Cambridge University Press, 1930.

[19]   Engleman MS, Sani RL, Gresho PM, Bercovier M. Consistent vs. Reduced Quadrature Penalty Methods for Incompressible Media Using Several Old and New Elements. Int. J. Numer. Meth. Fluid 1984; 2: 25–51.

4

Conduction Heat Transfer and Potential Flows

In this chapter, the discontinuous finite element formulation for heat conduction and potential flow problems is discussed. Heat conduction is perhaps the simplest heat transfer mode, but plays an important role in thermal science and engineering analyses. For introductory purposes, a simple, steady state 1-D heat conduction problem in a solid slab is considered first. Through this example, we will illustrate the similarities and dissimilarities between the discontinuous and continuous methods for the solution of this class of problems. This is followed by a discussion on steady state heat conduction problems in multidimensional geometries. The same discussion style is then followed to present the discontinuous finite element solution of the transient heat conduction problems. The potential flow problems and flows in porous media have a similar mathematical structure, and thus can be analyzed similarly.

One of the important issues in a discontinuous finite element formulation is the appropriate selection of numerical fluxes across the discontinuous inter-element boundaries. The theory underlying the selection of numerical fluxes is discussed. By the general classification of partial differential equations, the steady state conduction equations are elliptic, and the transient heat conduction equations are parabolic. The former have no real characteristics, while the latter have one family of characteristic curves. A simple difference between these two types of equations is that for a steady state heat conduction problem, a solution is required over an entire domain, whereas for the transient heat conduction problems, it is possible to obtain a solution in a small time interval. This difference also determines the numerical schemes to be applied, and the stability results for the selection of numerical fluxes. Various consistent and stable numerical fluxes are given for the discontinuous formulation of both steady and transient heat conduction problems.

For time dependent problems, an appropriate time step is important for numerical solutions. This remains true with the discontinuous finite element method; in fact it is even more so because most discontinuous formulations use the explicit time scheme for marching in time. Higher order accuracy schemes, such as Runge–Kutta time integrators, are also discussed. The matrix method and the Nuemann method are presented for analyzing the numerical stability of a time integration scheme and for selection of a critical time step for transient problems.

4.1 1-D Steady State Heat Conduction

From the standpoint of heat transfer, this is perhaps the simplest type of problem; and it offers an entry point into thermal analysis. Many numerical methods such as the finite element and finite difference/finite volume methods treat this type of problem as an introductory example. Here it is also used as the first heat transfer problem for the discontinuous Galerkin finite element solution. The 1-D steady state heat conduction problem is given by the following equations:

$$\frac{d^2T}{dx^2} = 0 \ , \ T(x=0) = 0 \ \text{and} \ T(x=1) = 1 \ \text{on} \ x \in [0,1] \tag{4.1}$$

for which the analytical solution is of a simple form: $T(x) = x$. Here T is the temperature.

In finite element literature, Equation 4.1 is referred to as the irreducible form [1]. The equation can also be written in a mixed form, which splits the second order equation into two first order differential equations,

$$-\frac{dq}{dx} = 0 \ ; \ q - \frac{dT}{dx} = 0 \ \ x \in [0,1] \tag{4.2}$$

The boundary condition remains the same. For this system of equations, the first equation in Equation 4.2 has q as its variable and the temperature is not involved. Consequently, there is a lack of direct coupling of the temperature T with the heat flux q. This lack of $T-q$ coupling has an implication in numerical stability, which will be discussed later.

We now apply the discontinuous solution procedure to solve for the temperature distribution. To do that, the domain is first discretized into N elements with $(N + 1)$ nodes. For an element defined on $\Omega_j = [x_j, x_{j+1}]$, we follow the procedures given in Chapter 2 to integrate the two equations with respect to testing functions w and v, and subsequently perform integration by parts, with the following result:

$$\int_{x_j}^{x_{j+1}} q_h \frac{dw}{dx} dx - \hat{q}_{j+1} w_{j+1}^- + \hat{q}_j w_j^+ = 0 \tag{4.3a}$$

$$\int_{x_j}^{x_{j+1}} q_h v dx + \int_{x_j}^{x_{j+1}} T_h \frac{dv}{dx} dx - \hat{T}_{j+1} v_{j+1}^- + \hat{T}_j v_j^+ = 0 \tag{4.3b}$$

where the temperature T and the heat flux q at the element boundaries are replaced by the generic numerical fluxes, $\hat{T}$ and $\hat{q}$. Other quantities used in the formulation are given in Figure 4.1.

Selecting the appropriate flux expressions is by no means trivial. A good choice of the numerical fluxes should satisfy the conditions for the existence and uniqueness of the solution, and should make the numerical scheme stable. Fortunately, for this type of problem, a variety of numerical fluxes that meet these conditions have been proposed, and their numerical properties have been studied comparatively in two recent papers [2, 3]. We will return to this subject in the next section when we discuss the multidimensional problems.

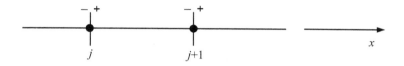

Figure 4.1. Illustration of boundary interfacial quantities for element j

At this moment, let us say we have found and constructed suitable numerical fluxes, which take the following forms:

$$\hat{T}_j = \begin{cases} T_j^-, & j = \\ 0.5(T_j^- + T_j^+) + C_{12}(T_j^- - T_j^+), & j = 2, \cdots N \\ T_j^+, & j = N+ \end{cases} \tag{4.4a}$$

$$\hat{q}_j = \begin{cases} q_j^+ - C_{11}(T_j^- - T_j^+), & j = 1 \\ 0.5(q_j^+ + q_j^-) - C_{11}(T_j^- - T_j^+) - C_{12}(q_j^- - q_j^+), & j = 2, \cdots, N \\ q_j^- - C_{11}(T_j^- - T_j^+), & j = N+1 \end{cases} \tag{4.4b}$$

where C_{11} and C_{12} are two constants, the selection of which is discussed later, and where we have incorporated the boundary conditions at nodes $j = 1$ and $j = N + 1$ into the numerical flux expressions.

With these numerical fluxes, the element matrix can be formed from the integral equations. For simplicity, we select the linear elements for the calculations. Then for the jth element, $\Omega_j = [x_j, x_{j+1}]$, the shape functions are given by

$$\phi_1(\xi) = \tfrac{1}{2}(1 - \xi) ; \ \phi_2(\xi) = \tfrac{1}{2}(1 + \xi) \tag{4.5}$$

The temperature T and flux q are interpolated over an isoparametric linear element,

$$T_h(\xi) = \phi_1(\xi)T_j^+ + \phi_2(\xi)T_{j+1}^- ; \quad q_h(\xi) = \phi_1(\xi)q_j^+ + \phi_2(\xi)q_{j+1}^-$$

$$x(\xi) = \phi_1(\xi)x_j + \phi_2(\xi)x_{j+1} ; \quad h_{(j)} = x_{j+1} - x_j$$

$$\frac{\partial \phi_1}{\partial x} = -\frac{1}{h_{(j)}} \; ; \; \frac{\partial \phi_2}{\partial x} = \frac{1}{h_{(j)}} \tag{4.6}$$

By the Galerkin method, the weighting functions are taken the same as the shape functions,

$$v = \{\phi_1(\xi), \phi_2(\xi)\}^T \; ; \; w = \{\phi_1(\xi), \phi_2(\xi)\}^T \tag{4.7}$$

Substituting Equations 4.4–4.7 into the integral formulation (Equation 4.3) yields the following expressions for element j:

$$\int_{x_j}^{x_{j+1}} \begin{pmatrix} d\phi_1/dx \\ d\phi_2/dx \end{pmatrix} (\phi_1, \phi_2) \, dx \begin{pmatrix} q_j^+ \\ q_{j+1}^- \end{pmatrix} + \begin{pmatrix} (C_{12}+0.5)q_j^+ \\ (C_{12}-0.5)q_{j+1}^- \end{pmatrix} + \begin{pmatrix} C_{11}T_j^+ \\ C_{11}T_{j+1}^- \end{pmatrix}$$
$$= \begin{pmatrix} (C_{12}-0.5)q_j^- \\ (C_{12}+0.5)q_{j+1}^+ \end{pmatrix} + \begin{pmatrix} C_{11}T_j^- \\ C_{11}T_{j+1}^+ \end{pmatrix} \tag{4.8a}$$

$$\int_{x_j}^{x_{j+1}} \begin{pmatrix} \phi_1 \\ \phi_2 \end{pmatrix} (\phi_1, \phi_2) \, dx \begin{pmatrix} q_j^+ \\ q_{j+1}^- \end{pmatrix} + \int_{x_j}^{x_{j+1}} \begin{pmatrix} d\phi_1/dx \\ d\phi_2/dx \end{pmatrix} (\phi_1, \phi_2) \, dx \begin{pmatrix} T_j^+ \\ T_{j+1}^- \end{pmatrix}$$
$$+ \begin{pmatrix} (0.5-C_{12})T_j^+ \\ -(C_{12}+0.5)T_{j+1}^- \end{pmatrix} = \begin{pmatrix} -(C_{12}+0.5)T_j^- \\ (0.5-C_{12})T_{j+1}^+ \end{pmatrix} \tag{4.8b}$$

These matrix elements are readily calculated using the computational procedures detailed in Chapter 3. After some algebra, the following results are obtained:

$$\int_{x_j}^{x_{j+1}} \begin{pmatrix} \phi_1 \\ \phi_2 \end{pmatrix} (\phi_1, \phi_2) \, dx = \frac{h_{(j)}}{6} \begin{bmatrix} 2 & 1 \\ 1 & 2 \end{bmatrix} \tag{4.9a}$$

$$\int_{x_j}^{x_{j+1}} \begin{pmatrix} d\phi_1/dx \\ d\phi_2/dx \end{pmatrix} (\phi_1, \phi_2) \, dx = \frac{1}{2} \begin{bmatrix} -1 & -1 \\ 1 & 1 \end{bmatrix} \tag{4.9b}$$

With Equations 4.8 and 4.9 substituted into Equations 4.8a and 4.8b and combining relevant terms, one has the matrix equations for the discretized temperatures and heat fluxes,

$$\begin{bmatrix} C_{12} & -1/2 \\ 1/2 & C_{12} \end{bmatrix} \begin{pmatrix} q_j^+ \\ q_{j+1}^- \end{pmatrix} + \begin{bmatrix} C_{11} & 0 \\ 0 & C_{11} \end{bmatrix} \begin{pmatrix} T_j^+ \\ T_{j+1}^- \end{pmatrix} = \begin{pmatrix} -(1/2-C_{12})q_j^- + C_{11}T_j^- \\ (1/2+C_{12})q_{j+1}^+ + C_{11}T_{j+1}^+ \end{pmatrix} \tag{4.10a}$$

$$\frac{h_{(j)}}{6}\begin{bmatrix} 2 & 1 \\ 1 & 2 \end{bmatrix}\begin{pmatrix} q_j^+ \\ q_{j+1}^- \end{pmatrix}+\begin{bmatrix} -C_{12} & -1/2 \\ 1/2 & -C_{12} \end{bmatrix}\begin{pmatrix} T_j^+ \\ T_{j+1}^- \end{pmatrix}=\begin{pmatrix} -(1/2+C_{12})T_j^- \\ (1/2-C_{12})T_{j+1}^+ \end{pmatrix} \quad (4.10b)$$

Incorporating the boundary conditions and combining the above two equations, one has the following numerical implementation:

(1) Element $j = 1$

$$\begin{bmatrix} h_{(j)}/3 & h_{(j)}/6 & -1/2 & -1/2 \\ h_{(j)}/6 & h_{(j)}/3 & 1/2 & -C_{12} \\ 1/2 & -1/2 & C_{11} & 0 \\ 1/2 & C_{12} & 0 & C_{11} \end{bmatrix}\begin{pmatrix} q_1^+ \\ q_2^- \\ T_1^+ \\ T_2^- \end{pmatrix}=\begin{pmatrix} -T_1^- \\ (1/2-C_{12})T_2^+ \\ C_{11}T_1^- \\ (1/2+C_{12})q_2^- + C_{11}T_2^+ \end{pmatrix}$$

(4.11a)

(2) Element $j = 2, \cdots, N-1$

$$\begin{bmatrix} h_{(j)}/3 & h_{(j)}/6 & -C_{12} & -1/2 \\ h_{(j)}/6 & h_{(j)}/3 & 1/2 & -C_{12} \\ C_{12} & -1/2 & C_{11} & 0 \\ 1/2 & C_{12} & 0 & C_{11} \end{bmatrix}\begin{pmatrix} q_j^+ \\ q_{j+1}^- \\ T_j^+ \\ T_{j+1}^- \end{pmatrix}=\begin{pmatrix} -(1/2+C_{12})T_j^- \\ (1/2-C_{12})T_{j+1}^+ \\ -(1/2-C_{12})q_j^- + C_{11}T_j^- \\ (1/2+C_{12})q_{j+1}^- + C_{11}T_{j+1}^+ \end{pmatrix}$$

(4.11a)

(3) Element $j = N$

$$\begin{bmatrix} h_{(j)}/3 & h_{(j)}/6 & -C_{12} & -1/2 \\ h_{(j)}/6 & h_{(j)}/3 & 1/2 & 1/2 \\ C_{12} & -1/2 & C_{11} & 0 \\ 1/2 & -1/2 & 0 & C_{11} \end{bmatrix}\begin{pmatrix} q_N^+ \\ q_{N+1}^- \\ T_N^+ \\ T_{N+1}^- \end{pmatrix}=\begin{pmatrix} -(1/2+C_{12})T_N^- \\ T_{N+1}^+ \\ -(1/2-C_{12})q_N^- + C_{11}T_N^- \\ C_{11}T_{N+1}^+ \end{pmatrix}$$

(4.11c)

Here the boundary conditions are applied such that $T_1^- = T(x = 0)$ and $T_{N+1}^+ = T(x = 1)$. The above equations may be written in matrix form,

$$\mathbf{KU} = \mathbf{F}, \mathbf{U}_{(j)} = \{q_j^+, q_{j+1}^-, T_j^+, T_{j+1}^-\}^T, j = 1, 2, \ldots, N \quad (4.12)$$

which can be easily inverted to obtain solution $\mathbf{U}$,

$$\mathbf{U} = \mathbf{K}^{-1}\mathbf{F} \quad (4.13)$$

The numerical solution to the above equations can be solved iteratively, using the successive substitution method. A typical computational procedure is as follows. To start, all the variables are initialized to zero. Equation 4.11a is solved to obtain the data for the first element, where boundary conditions are specified. Some of these are then used as input data into Equation 4.11b until the last element is reached, and then Equation 4.11c is completed for solution. Then one can move backward to start from Equation 4.11a with updated data, and so on, until the convergence is achieved. Like the continuous finite element method, the L^2-norm error for all unknowns is used to determine the convergence for a discontinuous finite element solution,

$$\left(\frac{\sum_{j=1}^{N}(T_{j,k+1}^{-} - T_{j,k}^{-})^2 + (T_{j,k+1}^{+} - T_{j,k}^{+})^2 + (q_{j,k+1}^{-} - q_{j,k}^{-})^2 + (q_{j,k+1}^{+} - q_{j,k}^{+})^2}{\sum_{j=1}^{N}{T_{j,k+1}^{-}}^2 + {T_{j,k+1}^{+}}^2 + {q_{j,k+1}^{-}}^2 + {q_{j,k}^{+}}^2} \right)^{1/2} \le \varepsilon$$

(4.14)

where ε is the preset tolerance and the subecript k refers to the kth iteration.

For this problem, one may also form a large global matrix and then invert the matrix to obtain the solution once, as is done using the continuous finite element method. This would involve solving the matrix approximately of dimension $4N \times 4N$. This approach would generally defeat the advantage of locality associated with the discontinuous Galerkin finite element method, since the matrix is considerably larger than when the standard finite element method is applied with the irreducible form.

Another way of solving the above equations could be to eliminate the variable q and solve for the temperature T only. This would almost certainly speed up the calculations, but also increase the bandwidth of the element matrix. We will show this through an example in Section 4.6, where the stability analysis is carried out. Further comparison shows that the saving in CPU time for solving T alone is less significant than the q–T iterative solution, in particular, for 3-D problems. Let us look at two specific examples.

Example 4.1. Discretize the domain $[0, 1]$ into three linear elements and use the successive substitution method to solve element-by-element for the steady state temperature distribution stated by Equation 4.2. Compare the results obtained, using different combinations of two stabilization constants C_{12} and C_{11}.

Solution. Let us illustrate the solution procedure by setting $C_{11} = 4$, $C_{12} = 1/2$ and $h = 1/3$. Also $T_1^{-} = 0$ and $T_4^{+} = 1$. The coefficient matrix $\mathbf{K}$ for the three elements can be analytically calculated and inverted.

(1) Element $j = 1$

$$\mathbf{K}_1 = \begin{bmatrix} 1/9 & 1/18 & -1/2 & -1/2 \\ 1/18 & 1/9 & 1/2 & -1/2 \\ 1/2 & -1/2 & 4 & 0 \\ 1/2 & 1/2 & 0 & 4 \end{bmatrix}; \begin{pmatrix} q_1^+ \\ q_2^- \\ T_1^+ \\ T_2^- \end{pmatrix} = (\mathbf{K}_1)^{-1} \begin{pmatrix} 0 \\ 0 \\ 0 \\ q_2^+ + 4T_2^+ \end{pmatrix}$$ (4.1e)

(2) Element $j = 2$

$$\mathbf{K}_2 = \begin{bmatrix} 1/9 & 1/18 & -1/2 & -1/2 \\ 1/18 & 1/9 & 1/2 & -1/2 \\ 1/2 & -1/2 & 4 & 0 \\ 1/2 & 1/2 & 0 & 4 \end{bmatrix}; \begin{pmatrix} q_2^+ \\ q_3^- \\ T_2^+ \\ T_3^- \end{pmatrix} = (\mathbf{K}_2)^{-1} \begin{pmatrix} -T_2^- \\ 0 \\ 4T_2^- \\ q_3^+ + 4T_3^+ \end{pmatrix}$$ (4.2e)

(3) Element $j = 3$

$$\mathbf{K}_3 = \begin{bmatrix} 1/9 & 1/18 & -1/2 & -1/2 \\ 1/18 & 1/9 & 1/2 & 1/2 \\ 1/2 & -1/2 & 4 & 0 \\ 1/2 & -1/2 & 0 & 4 \end{bmatrix}; \begin{pmatrix} q_3^+ \\ q_4^- \\ T_3^+ \\ T_4^- \end{pmatrix} = (\mathbf{K}_3)^{-1} \begin{pmatrix} -T_3^- \\ 1 \\ 4T_3^- \\ 4 \end{pmatrix}$$ (4.3e)

The calculation starts with all the variables initialized to zero except the boundary values. The successive substitution method is used to carry out the iteration process. No under- or over-relaxation parameters are used. The iteration sweep is from elements $j = 1$ through $j = 3$. The results are shown in Table 4.1e, where the nodal values of q and T are averaged values. As is seen from the table, after the first iteration, the effect of node 4 is felt at node 3 and the other two nodes are practically zero, because of the boundary condition at node 1. The results converge to the analytical solution ($T(x) = x$, and $q = 1$) in about 30 iteration sweeps.

Table 4.1e. Convergence of the 1-D solution ($C_{11} = 4$)

Number of Iterations	Node 1		Node 2		Node 3		Node 4	
	T	q	T	q	T	q	T	q
1	0.0000	0.0000	0.0000	0.0000	0.1378	1.4841	0.9523	2.2049
10	0.0002	1.0198	0.3393	1.0052	0.6729	0.9931	1.0039	0.9783
20	0.0000	0.9998	0.3333	0.9999	0.6666	1.0001	1.0000	1.0002
30	0.0000	1.0000	0.3333	1.0000	0.6667	1.0000	1.0000	1.0000

The procedure above is applied to the same problem, but with different combinations of C_{12} and C_{11}, to test the sensitivity of these parameters. The results are given in Table 4.2e, where all the results are terminated at the 20th iteration for the purpose of comparison. All these results eventually converge to the exact solution. From the table, we see that a combination of $C_{11} = 4$ and $C_{12} = 0$ gives better accuracy than other combinations. This is not totally unexpected, because for this problem, there is a lack of conviction as to which direction to upwind; and thus, an averaged value ($C_{12}=0$) seems to be a reasonable choice [4]. Stability theory states that other C_{12} values are also possible [2].

Table 4.2e. Effect of stabilization constants on convergence (terminated at the 20th iteration)

		Node 1		Node 2		Node 3		Node 4	
		T	q	T	q	T	q	T	q
$C_{11}=4$	$C_{12}=-1/2$	0.0000	0.9996	0.3332	0.9997	0.6666	1.0001	1.0000	1.0003
	$C_{12}=0$	0.0000	0.9998	0.3333	0.9999	0.6666	1.0001	1.0000	1.0002
	$C_{12}=1/2$	0.0000	0.9996	0.3332	0.9999	0.6665	1.0004	1.0000	1.0007
$C_{12}=0$	$C_{11}=1$	0.0007	1.0121	0.3363	0.9995	0.6694	0.9961	1.0011	0.9819
	$C_{11}=4$	0.0000	0.9998	0.3333	0.9999	0.6666	1.0001	1.0000	1.0002
	$C_{11}=10$	-0.0003	0.9807	0.3270	0.9908	0.6604	1.0092	0.9997	1.0193

Example 4.2. Study the effects of mesh on the accuracy of the solution for the temperature distribution for 1-D steady state heat conduction, with a prescribed heating source $Q = 1$,

$$\frac{d^2T}{dx^2} + Q = 0 \ , \ T(x=0) = T(x=1) = 0 \ \text{ on } \ x \in [0,1] \tag{4.4e}$$

Solution. The discretized equations are the same as those given in Equation 4.1e, except that the body source is added to the force term. The results obtained using different meshes, with the parameters $C_{12} = 0$, and $C_{11} = 4$, are given in Figure 4.1e. As expected, the denser mesh yields a closer match between the numerical and analytical solutions.

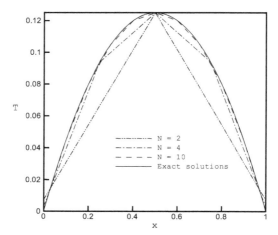

Figure 4.1e. Effect of the mesh size on the accuracy of numerical solution

4.2 Steady State Heat Conduction in Multidimensions

The above example has demonstrated the basic procedure for the discontinuous finite element solution of 1-D steady state heat conduction problems. We may now continue with the ideas and solve the heat conduction problems in multidimensions. Again, convection and other modes are not considered. We have seen that the numerical fluxes are important. In fact, since the discontinuous formulation of the heat conduction problem is very similar to the mixed finite element formulation of second order differential equations, certain conditions must be satisfied in order to guarantee the stability of the numerical method. We will discuss this issue more specifically in Section 4.6.

In what follows, we present the formulation and the computational procedure for the discontinuous finite element solution of the heat conduction problems over a multidimensional domain. For these purposes, we consider the heat conduction problem schematically shown in Figure 4.2,

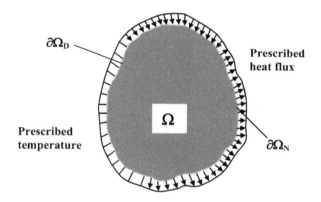

$\partial\Omega_D$

Prescribed heat flux

Prescribed temperature

Ω

$\partial\Omega_N$

Figure 4.2. Definition of 2-D steady state heat conduction problem

$$\nabla \cdot \kappa \nabla T + Q = 0 \qquad\qquad \in \Omega \qquad\qquad (4.15\text{a})$$

$$T = T_D \qquad\qquad \in \partial\Omega_D \qquad\qquad (4.15\text{b})$$

$$-\mathbf{n} \cdot \kappa \nabla T = -\mathbf{n} \cdot \mathbf{q}_N = h(T - T_\infty) \qquad\qquad \in \partial\Omega_N \qquad\qquad (4.15\text{c})$$

where subscripts D and N refer to the Dirichlet and Neumann boundary conditions, respectively, T is the temperature, κ the thermal conductivity, Q the heat source, h the heat transfer coefficient and T_∞ the temperature of the environment.

To develop a discontinuous Galerkin finite element formulation, we first re-write the problem as a system of first order differential equations, as has been done for the 1-D problem illustrated in the last section. In doing so, we have the following set of differential equations:

$$\mathbf{q} = \kappa\nabla T \; ; \; \nabla\cdot\mathbf{q} = -Q \qquad\qquad \in\Omega \qquad\qquad (4.16)$$

The domain is now discretized with finite elements. Let us consider the formulation over an element, say, the jth element, as shown in Figure 4.3, which schematically shows the geometric arrangement of element j and its neighbors in a typical 2-D mesh. The local numbers of the elements are also marked to show the relation between element j and its neighbors. Furthermore, the elements are separated in order to show the inter-element boundary quantities. In the real mesh, however, the elements are not geometrically separated. Multiplying the first and second equations in Equation 4.16 by test functions $\mathbf{w}$ and v and integrating over the element, one has

$$\int_{\Omega_j} \mathbf{q}\cdot\mathbf{w}\, dV = \int_{\Omega_j} \kappa\nabla T\cdot\mathbf{w} dV \qquad\qquad (4.17a)$$

$$\int_{\Omega_j} v\nabla\cdot\mathbf{q}\, dV = -\int_{\Omega_j} vQdV \qquad\qquad (4.17b)$$

Integration-by-parts once yields the following weak formulation for the discontinuous finite element solution:

$$\int_{\Omega_j} \mathbf{q}\cdot\mathbf{w}\, dV = -\int_{\Omega_j} T\nabla\cdot(\kappa\mathbf{w})dV + \int_{\partial\Omega_j} \kappa T\mathbf{n}_j\cdot\mathbf{w} dS \qquad\qquad (4.18a)$$

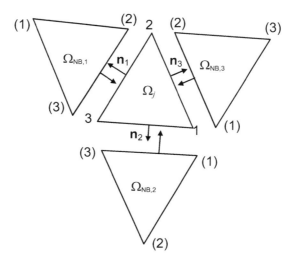

Figure 4.3. Geometric relation between element j and its neighbors. The elements are separated to better illustrate inter-element quantities

$$\int_{\Omega_j} \mathbf{q} \cdot \nabla v \, dV = \int_{\Omega_j} vQ \, dV + \int_{\partial\Omega_j} \mathbf{q} \cdot \mathbf{n}_j v \, dS \qquad (4.18b)$$

where $\mathbf{n}_j$ is the outward normal unit vector to $\partial\Omega_j$, the boundary of the element.

We now seek to approximate the exact solution $(\mathbf{q}, T)$ with functions $(\mathbf{q}_h, T_h)$ in the finite element broken space, whence we have the following results:

$$\int_{\Omega_j} \mathbf{q}_h \cdot \mathbf{w} \, dV = -\int_{\Omega_j} T_h \nabla \cdot (\kappa \mathbf{w}) dV + \int_{\partial\Omega_j} \kappa \hat{T}_h \mathbf{n}_j \cdot \mathbf{w} dS \qquad (4.19a)$$

$$\int_{\Omega_j} \mathbf{q}_h \cdot \nabla v \, dV = \int_{\Omega_j} vQ dV + \int_{\partial\Omega_j} \hat{\mathbf{q}}_h \cdot \mathbf{n}_j v dS \qquad (4.19b)$$

where the numerical fluxes ($\hat{\mathbf{q}}_h$, $\hat{T}_h$) are the approximations to $(\mathbf{q}, T)$ on the boundary of element j. To complete the discontinuous finite element solution, these numerical fluxes must be specified in terms of $\mathbf{q}_h$ and T_h, and in terms of the boundary conditions. These numerical fluxes must be suitably selected to render the discontinuous formulation stable.

Many numerical fluxes have been reported in the literature [2, 3, 5]. Recently, Arnold et al. [3] have presented a critical review of these fluxes and analyzed their suitability for the numerical solution of steady state heat conduction problems. Castillo et al. [2] also performed error analysis of these numerical fluxes for elliptical problems. Table 4.1 lists the numerical fluxes that are considered consistent and stable for the solution of the steady state heat conduction problems.

Table 4.1. Numerical fluxes for steady state heat diffusion calculations

Method	$\hat{\mathbf{q}}_h$	$\hat{T}_h$
LDG [6]	$\{\mathbf{q}_h\} - C_{11}[T_h] - \mathbf{C}_{12}[\mathbf{q}_h]$	$\{T_h\} + \mathbf{C}_{12}[T_h]$
DG [2]	$\{\mathbf{q}_h\} - C_{11}[T_h] - \mathbf{C}_{12}[\mathbf{q}_h]$	$\{T_h\} + \mathbf{C}_{12}[T_h] - C_{22}[\mathbf{q}_h]$
Brezzi et al. [7]	$\{\mathbf{q}_h\} - \alpha^r([T_h])$	$\{T_h\}$
IP [8]	$\{\nabla T_h\} - C_{11}[T_h]$	$\{T_h\}$
Bassi-Rebay [9]	$\{\nabla T_h\} - \alpha^r([T_h])$	$\{T_h\}$
NIPG [10]	$\{\nabla T_h\} - C_{11}[T_h]$	$\{T_h\} + \mathbf{n}_j \cdot [T_h]$

Note: C_{11}, C_{22}, C_{12} and C_{21} are constant matrices, and $\mathbf{n}_j$ is the outward normal of the boundary of element j.

In Table 4.1, the operators that calculate the averages and jumps between the local element and its neighbors are defined for the local element. Referring to Figure 4.4, the unit vectors $\mathbf{n}^+$ and $\mathbf{n}^-$ are the boundary outnormal vectors to elements j^+ and its neighbor j^-, respectively. The average and jump operators in the table are defined as follows:

$$\{\mathbf{q}\} = 0.5(\mathbf{q}^+ + \mathbf{q}^-) ; \quad [\mathbf{q}] = \mathbf{q}^+ \cdot \mathbf{n}^+ + \mathbf{q}^- \cdot \mathbf{n}^- \tag{4.20a}$$

$$\{T\} = 0.5(T^+ + T^-) ; \quad [T] = T^+\mathbf{n}^+ + T^-\mathbf{n}^- \tag{4.20b}$$

where we have used the underscored curly and square brackets to denote these special averages. It is noted that this use of underlined brackets will be implied throughout this book, unless indicated otherwise.

By these definitions, the jump $[\mathbf{q}]$ is a scalar function of $\mathbf{q}$, which involves the normal components only; and the jump $[T]$ is a vector function of T. The advantage of these definitions is that they do not depend on assigning an ordering to the elements.

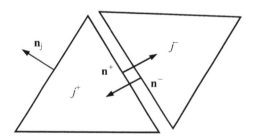

Figure 4.4 Element j (j^+) and its neighbor j^-

With the numerical fluxes taken from [2], and substituted into Equation 4.19, one has the final integral representation:

$$\int_{\Omega_j} \mathbf{q}_h \cdot \mathbf{w} \, dV + \int_{\Omega_j} T_h \nabla \cdot (\kappa \mathbf{w}) dV$$

$$- \int_{\partial\Omega_j} (\{T_h\} + \mathbf{C}_{12} \cdot [T_h] - C_{22}[\mathbf{q}_h])\mathbf{n}_j \cdot \kappa \mathbf{w} dS = 0 \tag{4.21a}$$

$$\int_{\Omega_j} \mathbf{q}_h \cdot \nabla v \, dV - \int_{\partial\Omega_j} (\{\mathbf{q}_h\} - C_{11}[T_h] - \mathbf{C}_{12}[\mathbf{q}_h]) \cdot \mathbf{n}_j v dS = \int_{\Omega_j} v Q dV \tag{4.21b}$$

The functions T and $\mathbf{q}$ are assumed to vary over the element, according to the space shape functions such that

$$T_h = \sum_{j=1}^{N_e} \varphi_j(\mathbf{r})T_j = \Phi^T(\mathbf{r})\underline{\mathbf{T}} \; ; \; q_{h,i} = \sum_{j=1}^{N_e} \varphi_j(\mathbf{r})q_{i,j} = \Phi^T(\mathbf{r})\underline{\mathbf{q}}_i \, , i = x, y, z$$

$$\tag{4.22}$$

where $\Phi = (\phi_1, \phi_2, \phi_3, \ldots, \phi_{N_e})^T$ and N_e is the number of nodes per element.

Substituting the above local approximation into Equation 4.21, we have the following expressions:

$$\left(\int_{\Omega_j} \Phi \Phi^T \, dV \right) \underline{\mathbf{q}} + \left(\int_{\Omega_j} [\nabla(\kappa\Phi)] \Phi^T \, dV \right) \underline{\mathbf{T}} + \left(\int_{\partial\Omega_j} \kappa C_{22} \Phi \Phi^T \mathbf{nn} \, dS \right) \cdot \underline{\mathbf{q}}$$

$$- \left(\int_{\partial\Omega_j} \kappa(1/2 + C_{12}) \Phi \Phi^T \mathbf{n} dS \right) \underline{\mathbf{T}} - \left(\int_{\partial\Omega_j} \kappa(1/2 - C_{12}) \Phi \Phi^T \mathbf{n} dS \right) \underline{\mathbf{T}}_{(NB)}$$

$$- \left(\int_{\partial\Omega_j} \kappa C_{22} \Phi \Phi^T \mathbf{n} \otimes \mathbf{n} \, dS \right) \cdot \underline{\mathbf{q}}_{(NB)} = 0 \qquad (4.23a)$$

$$\left(\int_{\Omega_j} (\nabla\Phi) \Phi^T \, dV \right) \cdot \underline{\mathbf{q}} - \left(\int_{\partial\Omega_j} (1/2 - C_{12}) \Phi \Phi^T \mathbf{n} dS \right) \cdot \underline{\mathbf{q}}$$

$$- \left(\int_{\partial\Omega_j} (1/2 + C_{12}) \Phi \Phi^T \mathbf{n} dS \right) \cdot \underline{\mathbf{q}}_{(NB)} + \left(\int_{\partial\Omega_j} C_{11} \Phi \Phi^T dS \right) \underline{\mathbf{T}}$$

$$- \left(\int_{\partial\Omega_j} C_{11} \Phi \Phi^T dS \right) \underline{\mathbf{T}}_{(NB)} = \int_{\Omega_j} \Phi Q dV \qquad (4.23b)$$

where $\mathbf{n} = \mathbf{n}_j$ and $\mathbf{n} = (n_x, n_y, n_z)$ to simplify notation; $\otimes$ is the dyadic operator, $(\mathbf{u} \otimes \mathbf{v}) \cdot \mathbf{w} = (\mathbf{w} \cdot \mathbf{v})\mathbf{u}$ or $\mathbf{u} \otimes \mathbf{v} = \mathbf{uv}$; subscript (NB) refers to the quantities belonging to the neighboring elements; and $\mathbf{q} = (\underline{\mathbf{q}}_x, \underline{\mathbf{q}}_y, \underline{\mathbf{q}}_z)$. Also, in deriving the above equation, the following relations have been used (see Figures 4.3 and 4.4 for relevant geometric relations):

$$q^+ = \mathbf{q}^+ \cdot \mathbf{n}^+, \; \mathbf{n}^+ = -\mathbf{n}^-; \; q^- = \mathbf{q}^- \cdot \mathbf{n}^+ = -\mathbf{q}^- \cdot \mathbf{n}^-$$

$$\mathbf{C}_{12} \cdot \mathbf{n}^+ = C_{12} = -\mathbf{C}_{12} \cdot \mathbf{n}^-, \; \mathbf{n}^+ = \mathbf{n}_j,$$

By the Galerkin procedure, the trial functions (v and w) are approximated in the same way as the unknown variables (T and $\mathbf{q}$). With these functions substituted into Equations 4.23a and 4.23b, the following matrix equation is obtained for the element j:

$$\begin{bmatrix} \mathbf{E} & \mathbf{0} & \mathbf{0} & \mathbf{H}_x \\ \mathbf{0} & \mathbf{E} & \mathbf{0} & \mathbf{H}_y \\ \mathbf{0} & \mathbf{0} & \mathbf{E} & \mathbf{H}_z \\ \mathbf{J}_x & \mathbf{J}_y & \mathbf{J}_z & \mathbf{0} \end{bmatrix} \begin{pmatrix} \underline{\mathbf{q}}_x \\ \underline{\mathbf{q}}_y \\ \underline{\mathbf{q}}_z \\ \underline{\mathbf{T}} \end{pmatrix} + \sum_{i=1}^{NS} \begin{bmatrix} \mathbf{E}_{xx,i} & \mathbf{E}_{xy,i} & \mathbf{E}_{xz,i} & \mathbf{H}_{x,i} \\ \mathbf{E}_{yx,i} & \mathbf{E}_{yy,i} & \mathbf{E}_{yz,i} & \mathbf{H}_{y,i} \\ \mathbf{E}_{zx,i} & \mathbf{E}_{zy,i} & \mathbf{E}_{zz,i} & \mathbf{H}_{z,i} \\ \mathbf{J}_{x,i} & \mathbf{J}_{y,i} & \mathbf{J}_{z,i} & \mathbf{G}_{T,i} \end{bmatrix} \begin{pmatrix} \underline{\mathbf{q}}_x \\ \underline{\mathbf{q}}_y \\ \underline{\mathbf{q}}_z \\ \underline{\mathbf{T}} \end{pmatrix}$$

$$\sum_{i=1}^{NS} \begin{bmatrix} \mathbf{E}_{xx,B,i} & \mathbf{E}_{xy,B,i} & \mathbf{E}_{xy,B,i} & \mathbf{H}_{x,B,i} \\ \mathbf{E}_{yx,B,i} & \mathbf{E}_{yy,B,i} & \mathbf{E}_{yz,B,i} & \mathbf{H}_{y,B,i} \\ \mathbf{E}_{zx,B,i} & \mathbf{E}_{zy,B,i} & \mathbf{E}_{zz,B,i} & \mathbf{H}_{z,B,i} \\ \mathbf{J}_{x,B,i} & \mathbf{J}_{y,B,i} & \mathbf{J}_{z,B,i} & \mathbf{G}_{T,B,i} \end{bmatrix} \begin{pmatrix} \mathbf{q}_x \\ \mathbf{q}_y \\ \mathbf{q}_z \\ \mathbf{T} \end{pmatrix}_{(NB,i)} = \begin{pmatrix} \mathbf{0} \\ \mathbf{0} \\ \mathbf{0} \\ \mathbf{S}_T \end{pmatrix} \tag{4.24}$$

where NS is the number of sides of the element and the matrices are calculated as follows:

$$H_{l,km} = \int_{\Omega_j} \frac{d(\kappa\phi_k)}{dl} \phi_m dV , \quad l = x, y, z$$

$$J_{l,km} = \int_{\partial\Omega_j} \frac{d\phi_k}{dl} \phi_m dV , \quad l = x, y, z$$

$$E_{km} = \int_{\Omega_j} \phi_k \phi_m dV ; \quad G_{T,i,km} = -G_{T,B,i,km} = C_{11} \int_{\partial\Omega_{j,i}} \phi_k \phi_m dS$$

$$E_{lr,i,km} = -E_{lr,B,i,km} = C_{22} \int_{\partial\Omega_{j,i}} \kappa\phi_k\phi_m n_l n_r dS , \quad l, r = x, y, z$$

$$H_{l,i,km} = -(\tfrac{1}{2} + C_{12}) \int_{\partial\Omega_{j,i}} \kappa\phi_k\phi_m n_l dS , \quad l = x, y, z$$

$$H_{l,B,i,km} = -(\tfrac{1}{2} - C_{12}) \int_{\partial\Omega_{j,i}} \kappa\phi_k\phi_m n_l dS , \quad l = x, y, z$$

$$J_{l,i,km} = -(\tfrac{1}{2} - C_{12}) \int_{\partial\Omega_{j,i}} \phi_k\phi_m n_l dS , \quad l = x, y, z$$

$$J_{l,B,i,km} = -(\tfrac{1}{2} + C_{12}) \int_{\partial\Omega_{j,i}} \phi_k\phi_m n_l dS , \quad l = x, y, z$$

$$S_{T,k} = \int_{\Omega_j} \phi_k Q dV$$

with $\partial\Omega_{j,i}$ denoting the interface between the element j and its neighbor element i: $\partial\Omega_j = \bigcup_{i=1}^{NS} \partial\Omega_{j,i}$.

The above matrices can be further combined to yield the following resultant matrix:

$$KU = F \qquad (4.25)$$

where $\mathbf{K}$ is the stiffness matrix, $\mathbf{U} = [\mathbf{q}_x{}^T, \mathbf{q}_y{}^T, \mathbf{q}_z{}^T, \mathbf{T}^T]^T = [q_{x,1}, q_{x,2}, \ldots, q_{x,Ne}, q_{y,1}, q_{y,2}, \ldots, q_{y,Ne}, q_{x,1}, q_{x,2}, \ldots, q_{x,Ne}, T_1, T_2, \ldots, T_{Ne}]^T.$, and $\mathbf{F}$ is the source vector.

The computational procedure is similar to the 1-D case. Once again, three different approaches can be implemented to solve the problem. It is recommended that the element-by-element sweeping method, with successive substitution, be used for a large-scale problem, for which the discontinuous finite element method has the distinct advantage. If the element-by-element iterative solution scheme is employed, then information on the neighboring elements that is available during computation can be treated as the source term, and moved to the right hand side. Let us now consider an example below.

Example 4.3. Consider a two-dimensional steady heat conduction problem defined over a square,

$$\frac{\partial^2 T}{\partial x^2} + \frac{\partial^2 T}{\partial y^2} + Q = 0 \quad \text{on } x \times y \in [-1,1] \times [-1,1] \qquad (4.5e)$$

where the heating source Q is given by $Q(x,y) = \exp[-(x^2 + y^2)]$, and the boundary condition is $T = 0$ at all the edges of the square.

Solution. We consider the use of linear elements for the calculations. The domain is discretized using the linear triangular elements, which are shown in Figure 4.2e. For an element, the discontinuous finite element formulation can be developed as described above. For this problem, we set $C_{22} = 0$.

The definition of the numerical fluxes on the boundary is given as follows:

$$\mathbf{q}^- = \mathbf{q}^+, \ T^- = g_D, \ \mathbf{C}_{12} \cdot \mathbf{n}^- = 1/2 \ \text{and} \ \mathbf{C}_{12} \cdot \mathbf{n}^+ = -1/2 \ \text{on} \ \partial\Omega_D \qquad (4.6e)$$

$$\mathbf{q}^- = g_N, \ T^- = T^+, \ \mathbf{C}_{12} \cdot \mathbf{n}^- = -1/2 \ \text{and} \ \mathbf{C}_{12} \cdot \mathbf{n}^+ = 1/2 \ \text{on} \ \partial\Omega_N \qquad (4.7e)$$

For a linear triangular element, the matrices given in Equation 4.24 can be calculated analytically. Consider a canonical element shown in Figure 4.3e. The shape function for this element takes a simple form,

$$\Phi(x,y) = \begin{bmatrix} \phi_1(x,y) \\ \phi_2(x,y) \\ \phi_3(x,y) \end{bmatrix} = \frac{1}{2S} \begin{bmatrix} x_2 y_3 - x_3 y_2 & y_{23} & x_{32} \\ x_3 y_1 - x_1 y_3 & y_{31} & x_{13} \\ x_1 y_2 - x_2 y_1 & y_{12} & x_{21} \end{bmatrix} \begin{bmatrix} 1 \\ x \\ y \end{bmatrix} \qquad (4.7e)$$

where S is the area of the element. For element Ω_j, S is calculated by the following expression:

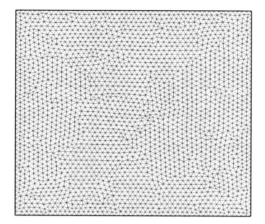

Figure 4.2e. Unstructured mesh used for computation of the 2-D steady heat conduction problem

$$S = \frac{1}{2}\begin{vmatrix} x_{13} & x_{23} \\ y_{13} & y_{23} \end{vmatrix} = \frac{1}{2}[(x_1 - x_3)(y_2 - y_3) - (x_2 - x_3)(y_1 - y_3)] \qquad (4.8e)$$

and the single-indexed subscripts refer to the node number local to the element. Also, $x_{ij} = x_i - x_j$ and $y_{ij} = y_i - y_j$.

Taking the derivative of the shape functions, we have

$$\nabla\Phi(x, y) = \begin{bmatrix} \nabla\phi_1(x, y) \\ \nabla\phi_2(x, y) \\ \nabla\phi_3(x, y) \end{bmatrix} = \frac{1}{2S}\begin{bmatrix} y_{23} \\ y_{31} \\ y_{12} \end{bmatrix}\hat{x} + \frac{1}{2S}\begin{bmatrix} x_{32} \\ x_{13} \\ x_{21} \end{bmatrix}\hat{y} \qquad (4.9e)$$

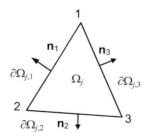

Figure 4.3e. Geometric relation in a single element used for elemental calculations

With Equations 4.7e and 4.8e, the element stiffness matrices are calculated analytically with the results,

$$\mathbf{E} = \int_{\Omega_j} \Phi\Phi^T \, dV = \frac{S}{12} \begin{bmatrix} 2 & 1 & 1 \\ 1 & 2 & 1 \\ 1 & 1 & 2 \end{bmatrix} \tag{4.10e}$$

$$\mathbf{H}_x = \mathbf{J}_x = \int_{\Omega_j} \left(\frac{\partial \Phi}{\partial x} \right) \Phi^T \, dV = \frac{1}{6} \begin{bmatrix} y_{23} & y_{23} & y_{23} \\ y_{31} & y_{31} & y_{31} \\ y_{12} & y_{12} & y_{12} \end{bmatrix} \tag{4.11e}$$

$$\mathbf{H}_y = \mathbf{J}_y = \int_{\Omega_j} \left(\frac{\partial \Phi}{\partial y} \right) \Phi^T \, dV = \frac{1}{6} \begin{bmatrix} x_{32} & x_{32} & x_{32} \\ x_{13} & x_{13} & x_{13} \\ x_{21} & x_{21} & x_{21} \end{bmatrix} \tag{4.12e}$$

$$\mathbf{E}_{lr,i} = -\mathbf{E}_{lr,B,i} = 0 \tag{4.13e}$$

$$\mathbf{H}_{l,i} = \mathbf{J}_{l,B,i} = -(\tfrac{1}{2} + C_{12}) \int_{\partial\Omega_{j,i}} \Phi\Phi^T n_l dS = -(\tfrac{1}{2} + C_{12}) n_l \mathbf{C}_i \tag{4.14e}$$

$$\mathbf{H}_{l,B,i} = \mathbf{J}_{l,i} = -(\tfrac{1}{2} - C_{12}) \int_{\partial\Omega_{j,i}} \Phi\Phi^T n_l dS = -(\tfrac{1}{2} - C_{12}) n_l \mathbf{C}_i \tag{4.15e}$$

$$\mathbf{G}_{T,i} = -\mathbf{G}_{T,B,i} = C_{11} \mathbf{C}_i \tag{4.16e}$$

where $l,r = x, y; i = 1, 2, 3$, and

$$\mathbf{C}_1 = \frac{L_1}{6} \begin{bmatrix} 0 & 0 & 0 \\ 0 & 2 & 1 \\ 0 & 1 & 2 \end{bmatrix}; \quad \mathbf{C}_2 = \frac{L_2}{6} \begin{bmatrix} 2 & 0 & 1 \\ 0 & 0 & 0 \\ 1 & 0 & 2 \end{bmatrix}; \quad \mathbf{C}_3 = \frac{L_3}{6} \begin{bmatrix} 2 & 1 & 0 \\ 1 & 2 & 0 \\ 0 & 0 & 0 \end{bmatrix}$$

Note that numerical integration may also be applied to obtain the same results.

With Equations (4.10e–16e) substituted into Equation 4.24, we have the following matrix equation for element j:

$$\begin{bmatrix} \mathbf{E} & 0 & \mathbf{H}_x \\ 0 & \mathbf{E} & \mathbf{H}_y \\ \mathbf{J}_x & \mathbf{J}_y & 0 \end{bmatrix} \begin{Bmatrix} \mathbf{q}_x \\ \mathbf{q}_y \\ \mathbf{T} \end{Bmatrix} + \sum_{i=1}^{3} \begin{bmatrix} 0 & 0 & \mathbf{H}_{x,i} \\ 0 & 0 & \mathbf{H}_{y,i} \\ \mathbf{J}_{x,i} & \mathbf{J}_{y,i} & \mathbf{G}_{T,i} \end{bmatrix} \begin{Bmatrix} \mathbf{q}_x \\ \mathbf{q}_y \\ \mathbf{T} \end{Bmatrix}$$

$$+\sum_{i=1}^{3}\begin{bmatrix} 0 & 0 & \mathbf{H}_{x,B,i} \\ 0 & 0 & \mathbf{H}_{y,B,i} \\ \mathbf{J}_{x,B,i} & \mathbf{J}_{y,B,i} & \mathbf{G}_{T,B,i} \end{bmatrix}\begin{pmatrix} \mathbf{q}_x \\ \mathbf{q}_y \\ \mathbf{T} \end{pmatrix}_{(NB,i)} = 0 \tag{4.17e}$$

The above matrices can be combined with the following resultant matrix:

$$\mathbf{KU} = \mathbf{F} \tag{4.18e}$$

where $\mathbf{U}$ is the unknown vector, and $\mathbf{K}$ and $\mathbf{F}$ are defined as follows:

$$\mathbf{K} = \begin{bmatrix} \mathbf{E} & 0 & \mathbf{H}_x \\ 0 & \mathbf{E} & \mathbf{H}_y \\ \mathbf{J}_x & \mathbf{J}_y & 0 \end{bmatrix} + \sum_{i=1}^{3}\begin{bmatrix} 0 & 0 & \mathbf{H}_{x,i} \\ 0 & 0 & \mathbf{H}_{y,i} \\ \mathbf{J}_{x,i} & \mathbf{J}_{y,i} & \mathbf{G}_{T,i} \end{bmatrix}$$

$$\mathbf{F} = -\sum_{i=1}^{3}\begin{bmatrix} 0 & 0 & \mathbf{H}_{x,B,i} \\ 0 & 0 & \mathbf{H}_{y,B,i} \\ \mathbf{J}_{x,B,i} & \mathbf{J}_{y,B,i} & \mathbf{G}_{T,B,i} \end{bmatrix}\begin{pmatrix} \mathbf{q}_x \\ \mathbf{q}_y \\ \mathbf{T} \end{pmatrix}_{(NB,i)}$$

The calculations start with an element located at the boundary, and progressively sweep into the domain element-by-element. The successive substitution method may be used for the iterative solution. The calculations require 3000 iterations to converge within a tolerance of $\varepsilon = 10^{-6}$. Considerably fewer iterations are needed if the tolerance is not set so stringent. Stabilization constants chosen are $C_{12} = 0$ and $C_{11} = 50$. The computed temperature contours and a 3-D view of temperature distribution are given in Figure 4.4e. It is seen that the highest temperature occurs at the center of the square where the heating source is at its maximum, which is consistent with the principle of heat conduction. It is noted that this problem has a four-fold symmetry and, as a result, only a quadrant is needed to speed up the computations.

4.3 1-D Transient Heat Conduction

In the above two sections, we discussed the discontinuous finite element formulation for steady state heat conduction problems. Following the same development style, we now consider the solution of transient heat conduction problems using the discontinuous method. Again, to illustrate the basic ideas and solution procedures, we start with a simple transient 1-D heat conduction problem,

$$\frac{\partial T}{\partial t} = \frac{\partial^2 T}{\partial x^2}, \quad x \in [a,b] \tag{4.26}$$

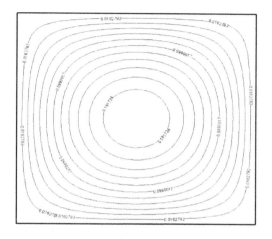

(a) 2-D contours

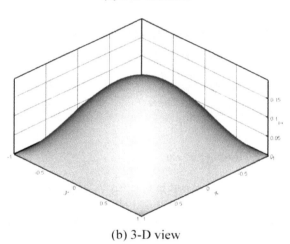

(b) 3-D view

Figure 4.4e. Discontinuous finite element solution for the 2-D steady heat conduction problem: (a) 2-D isothermal contour plot and (b) 3-D view of the temperature distribution where $\mathbf{U} = [\mathbf{q}_x^T, \mathbf{q}_y^T, \mathbf{T}^T]^T = [q_{x,1}, q_{x,2}, \dots, q_{x,Ne}, q_{y,1}, q_{y,2}, \dots, q_{y,Ne}, T_1, \quad T_2, \dots, T_{Ne}]^T$, and K and F are calculated as follows:

with periodic boundary conditions, and the initial condition $T(x,0) = \sin(x)$. For simplicity, all properties assumed to be unity.

To apply the local discontinuous Galerkin method, the heat conduction problem is first split into a system of two first order equations for variables $q(x,t)$ and $T(x,t)$,

$$\frac{\partial T}{\partial t} - \frac{\partial q}{\partial x} = 0 \; ; \; \frac{\partial T}{\partial x} - q = 0 \tag{4.27}$$

To obtain a discontinuous finite element formulation, the domain is discretized into N number of elements. Integration of the above equation over element $j \in [x_j, x_{j+1}]$, with respect to weighting functions v and w, yields

$$\int_{x_j}^{x_{j+1}} \left(\frac{\partial T}{\partial t} - \frac{\partial q}{\partial x} \right) v(x)\, dx = 0 \; ; \quad \int_{x_j}^{x_{j+1}} \left(-q + \frac{\partial T}{\partial x} \right) w(x)\, dx = 0 \qquad (4.28)$$

We follow the same procedure as discussed in the last section and integrate by parts once on the spatial derivatives to obtain

$$\int_{x_j}^{x_{j+1}} \left(v\frac{\partial T}{\partial t} + \frac{\partial v}{\partial x} q \right) dx - q(x_{j+1}^-)v(x_{j+1}^-) + q(x_j^+)v(x_j^+) = 0 \qquad (4.29a)$$

$$-\int_{x_j}^{x_{j+1}} \left(qw + T\frac{\partial w}{\partial x} \right) dx + T(x_{j+1}^-)w(x_{j+1}^-) - T(x_j^+)w(x_j^+) = 0 \qquad (4.29b)$$

We now face the choice of selecting the quantities at the element boundaries, $q(x_{j+1}^+)$, $q(x_{j+1}^-)$, $T(x_{j+1}^+)$ and $T(x_{j+1}^-)$. Before generalizing the choice of numerical fluxes, we consider below two other simpler options for the choices. It is noted that for the transient problems, the time dependent term makes a coupling between the temperature field and the heat fluxes, and this linkage provides a natural stabilization factor for the discontinuous numerical scheme.

4.3.1 Alternating Upwinding Scheme

One choice is intuitive, as suggested in Chapter 2, which is to take the upwinding value. By this choice, the following values may be used for the interface quantities in reference to Figure 4.1,

$$q(x_{j+1}^-) = q(x_{j+1}^+) = q_{j+1}^+ \; ; \; T(x_{j+1}^+) = T(x_{j+1}^-) = T_{j+1}^- \qquad (4.30a)$$

$$q(x_j^-) = q(x_j^+) = q_j^+ \; ; \; T(x_j^+) = T(x_j^-) = T_j^- \qquad (4.30b)$$

These selections may be interpreted as meaning that the available values at the neighboring boundary are considered known, and are applied as the boundary conditions for element j, whenever they become available during the iteration. This approach was first proposed by Cockburn and Shu [6] and has an order of accuracy of $k+1$ for an interpolation of order k [4,11]. With these selections, the weak form equations can be written as

$$\int_{x_j}^{x_{j+1}} \left(v\frac{\partial T}{\partial t} + \frac{\partial v}{\partial x} q \right) dx + q_j^+ = q_{j+1}^+ \qquad (4.31a)$$

$$\int_{x_j}^{x_{j+1}} \left(qw + T\frac{\partial w}{\partial x} \right) dx - T_{j+1}^- = T_j^- \qquad (4.31b)$$

where $v(x_j^+) = v(x_j^-) = w(x_{j+1}^+) = w(x_{j+1}^-) = 1$ has been used.

4.3.2 Central Fluxes

Bassi and Rebay were the first to apply the discontinuous finite element method for the solution of diffusion-type problems. In their original approach [12], they proposed the simplest central flux expression, and applied the scheme for compressible flow calculations, obtaining quite satisfactory results. Their central scheme basically uses the average of the two values across the boundary,

$$q(x_{j+1}^-) = q(x_{j+1}^+) = 0.5(q_{j+1}^- + q_{j+1}^+) \qquad (4.32a)$$

$$T(x_{j+1}^-) = T(x_{j+1}^+) = 0.5(T_{j+1}^- + T_{j+1}^+) \qquad (4.32b)$$

Here, we could choose q and T so that they satisfy the *LBB* condition for the mixed finite element formulation. This means that different functional spaces may need to be used for q and T. However, Bassi and Rebay [12] argue that the fluxes can be chosen from the same function space. Cockburn and Shu [6] later studied this method and proved that the method converges at a rate of one order lower from optimal; that is, where the error estimate is of order k for piecewise polynomials of degree k [11].

With Equation 4.32 substituted into Equation 4.29, the following weak form solution for q and T is obtained,

$$\int_{x_j}^{x_{j+1}} \left(v\frac{\partial T}{\partial t} + \frac{\partial v}{\partial x}q \right) dx - 0.5q_{j+1}^- + 0.5q_j^+ = 0.5q_{j+1}^+ - 0.5q_j^- \qquad (4.33a)$$

$$\int_{x_j}^{x_{j+1}} \left[qw + T\frac{\partial w}{\partial x} \right] dx - 0.5T_{j+1}^- + 0.5T_j^+ = 0.5T_{j+1}^+ - 0.5T_j^- \qquad (4.33b)$$

4.3.3 Unified Representation

The above discontinuous finite element formulations can be expressed in a unified fashion as follows:

$$\int_{x_j}^{x_{j+1}} v_h \left(\frac{\partial T_h}{\partial x} + q_h \right) dx - \hat{q}_{j+1} + \hat{q}_j = 0 \qquad (4.34a)$$

$$\int_{x_j}^{x_{j+1}} v_h \left(\frac{\partial T_h}{\partial t} + \frac{\partial q_h}{\partial x} \right) dx + \hat{T}_j - \hat{T}_{j+1} = 0 \qquad (4.34b)$$

where T_h is substituted for T, q_h for q, and v_h for v and w, $j = 1, 2, \cdots, N$ and the numerical fluxes are defined by

$$\hat{q} = 0.5(q^+ + q^-) - C_{12}(q^+ - q^-) ; \quad \hat{T} = 0.5(T^+ + T^-) + C_{12}(T^+ - T^-) \quad (4.35)$$

with $C_{12} = 0$ for the central scheme, and $C_{12} = 0.5$ for the alternating upwinding scheme.

4.3.4 Numerical Implementation

We further introduce the element shape functions $\phi_k(x)$ as follows:

$$T_h(x) = \sum_{k=1}^{N_e} \phi_k(x)T^{(k)} = \Phi^T \mathbf{T} ; \quad q_h(x) = \sum_{k=1}^{N_e} \phi_k(x)q^{(k)} = \Phi^T \mathbf{q} \quad (4.36)$$

where N_e is the number of nodes of an element. Also, subscript (k) on T and q refers to the kth node local to the element. Substituting these equations into Equation 4.34, followed by numerical integration, one obtains the results below:

$$\sum_{m=1}^{N_e} S_{(j)}^{(km)} q_{(j)}^{(m)} + \sum_{m=1}^{N_e} L_{(j)}^{(km)} T_{(j)}^{(m)} + \hat{T}_j \phi_k(x_j) - \hat{T}_{j+1} \phi_k(x_{j+1}) = 0 \qquad (4.37a)$$

$$\sum_{m=1}^{N_e} S_{(j)}^{(km)} \dot{T}_{(j)}^{(m)} + \sum_{m=1}^{N_e} L_{(j)}^{(km)} q_{(j)}^{(m)} + \hat{q}_j \phi_k(x_j) - \hat{q}_{j+1} \phi_k(x_{j+1}) = 0 \qquad (4.37b)$$

for $k = 1, 2, \ldots, N_e$ (k being the kth node local to element j), with matrices $S_{(j)}^{(km)}$ and $L_{(j)}^{(km)}$ calculated by

$$S_{(j)}^{(km)} = \int_{x_j}^{x_{j+1}} \phi_k \phi_m dx ; \quad L_{(j)}^{(km)} = \int_{x_j}^{x_{j+1}} \frac{d\phi_k}{dx} \phi_m dx \qquad (4.38)$$

where subscript $(j) = 1, 2, \ldots, N$ (N being the number of elements) refers to the jth element. Note that subscripts j and $j+1$ without a round bracket refer to the two corner nodes of the jth element numbered globally, $\Omega_j = [x_j, x_{j+1}]$. For a piecewise constant approximation, (j) and j refer to the same, in which case the the node and the element are numbered the same way.

The above equation can be rewritten in the matrix form of

$$\mathbf{S}_{(j)}\underline{\mathbf{q}}_{(j)} + \mathbf{L}_{(j)}\underline{\mathbf{T}}_{(j)} + \hat{T}_j \begin{pmatrix} \phi_1(x_j) \\ \phi_2(x_j) \\ \vdots \\ \phi_{Ne}(x_j) \end{pmatrix} - \hat{T}_{j+1} \begin{pmatrix} \phi_1(x_{j+1}) \\ \phi_2(x_{j+1}) \\ \vdots \\ \phi_{Ne}(x_{j+1}) \end{pmatrix} = 0 \qquad (4.39)$$

$$\mathbf{S}_{(j)}\underline{\dot{\mathbf{T}}}_{(j)} + \mathbf{L}_{(j)}\underline{\mathbf{q}}_{(j)} + \hat{q}_j \begin{pmatrix} \phi_1(x_j) \\ \phi_2(x_j) \\ \vdots \\ \phi_{Ne}(x_j) \end{pmatrix} - \hat{q}_{j+1} \begin{pmatrix} \phi_1(x_{j+1}) \\ \phi_2(x_{j+1}) \\ \vdots \\ \phi_{Ne}(x_{j+1}) \end{pmatrix} = 0 \qquad (4.40)$$

The mass matrices can be inverted on an element level and we find

$$\underline{\mathbf{q}}_{(j)} + \mathbf{K}_{(j)}\underline{\mathbf{T}}_{(j)} + \hat{T}_j\boldsymbol{\theta}_j - \hat{T}_{j+1}\boldsymbol{\theta}_{j+1} = 0 \qquad (4.41a)$$

$$\underline{\dot{\mathbf{T}}}_{(j)} + \mathbf{K}_{(j)}\underline{\mathbf{q}}_{(j)} + \hat{q}_j\boldsymbol{\theta}_j - \hat{q}_{j+1}\boldsymbol{\theta}_{j+1} = 0 \qquad (4.41b)$$

where the following defintions have been used:

$$\boldsymbol{\theta}_j = (\mathbf{S}_{(j)})^{-1} \begin{pmatrix} \phi_1(x_j) \\ \phi_2(x_j) \\ \vdots \\ \phi_{Ne}(x_j) \end{pmatrix} \quad ; \quad \boldsymbol{\theta}_{j+1} = (\mathbf{S}_{(j)})^{-1} \begin{pmatrix} \phi_1(x_{j+1}) \\ \phi_2(x_{j+1}) \\ \vdots \\ \phi_{Ne}(x_{j+1}) \end{pmatrix}$$

$$\mathbf{K}_{(j)} = (\mathbf{S}_{(j)})^{-1}\mathbf{L}_{(j)} \qquad (4.42)$$

For convenience, the matrices for the interpolations of up to quadratic order are given below:

(1) Piecewise continuous, $N_e = 1$:

$$q_j + \frac{1}{h_j}(a_1 T_j - a_2 T_{j-1}) - \frac{1}{h_j}(a_1 T_{j+1} - a_2 T_j) = 0 \qquad (4.43a)$$

$$\dot{T}_j + \frac{1}{h_j}(a_3 q_j - a_4 q_{j-1}) + \frac{1}{h_j}(a_3 q_{j+1} - a_4 q_j) = 0 \qquad (4.43b)$$

(2) Linear interpolation, $N_e = 2$:

$$\begin{pmatrix} q_{(j)}^{(1)} \\ q_{(j)}^{(2)} \end{pmatrix} + \frac{1}{h_{(j)}} \begin{pmatrix} -1 & 1 \\ -1 & 1 \end{pmatrix} \begin{pmatrix} T_{(j)}^{(1)} \\ T_{(j)}^{(2)} \end{pmatrix} + \frac{1}{h_{(j)}} \begin{pmatrix} 4 \\ -2 \end{pmatrix} (a_1 T_{(j)}^{(1)} - a_2 T_{(j-1)}^{(2)})$$

$$+ \frac{1}{h_{(j)}} \begin{pmatrix} -2 \\ 4 \end{pmatrix} (T_{(j+1)}^{(1)} - T_{(j)}^{(2)}) = 0 \qquad (4.44a)$$

$$\begin{pmatrix} \dot{T}_{(j)}^{(1)} \\ \dot{T}_{(j)}^{(2)} \end{pmatrix} + \frac{1}{h_{(j)}} \begin{pmatrix} -1 & 1 \\ -1 & 1 \end{pmatrix} \begin{pmatrix} q_{(j)}^{(1)} \\ q_{(j)}^{(2)} \end{pmatrix} + \frac{1}{h_{(j)}} \begin{pmatrix} 4 \\ -2 \end{pmatrix} (a_3 q_{(j)}^{(1)} - a_4 q_{(j-1)}^{(2)})$$

$$+ \frac{1}{h_{(j)}} \begin{pmatrix} -2 \\ 4 \end{pmatrix} (q_{(j+1)}^{(1)} - q_{(j)}^{(2)}) = 0 \qquad (4.44b)$$

(3) Quadratic interpolation, $N_e = 3$:

$$\begin{pmatrix} q_{(j)}^{(1)} \\ q_{(j)}^{(2)} \\ q_{(j)}^{(3)} \end{pmatrix} + \frac{1}{h_{(j)}} \begin{pmatrix} -3 & 4 & -1 \\ -1 & 0 & 1 \\ 1 & -4 & 3 \end{pmatrix} \begin{pmatrix} T_{(j)}^{(1)} \\ T_{(j)}^{(2)} \\ T_{(j)}^{(3)} \end{pmatrix} + \frac{1}{h_{(j)}} \begin{pmatrix} 9 \\ -\frac{3}{2} \\ 3 \end{pmatrix} (a_1 T_{(j)}^{(1)} - a_2 T_{(j-1)}^{(3)})$$

$$+ \frac{1}{h_{(j)}} \begin{pmatrix} 3 \\ -\frac{3}{2} \\ 9 \end{pmatrix} (T_{(j+1)}^{(1)} - T_{(j)}^{(3)}) = 0 \qquad (4.45a)$$

$$\begin{pmatrix} \dot{T}_{(j)}^{(1)} \\ \dot{T}_{(j)}^{(2)} \\ \dot{T}_{(j)}^{(3)} \end{pmatrix} + \frac{1}{h_{(j)}} \begin{pmatrix} -3 & 4 & -1 \\ -1 & 0 & 1 \\ 1 & -4 & 3 \end{pmatrix} \begin{pmatrix} q_{(j)}^{(1)} \\ q_{(j)}^{(2)} \\ q_{(j)}^{(3)} \end{pmatrix} + \frac{1}{h_{(j)}} \begin{pmatrix} 9 \\ -\frac{3}{2} \\ 3 \end{pmatrix} (a_3 q_{(j)}^{(1)} - a_4 q_{(j-1)}^{(3)})$$

$$+ \frac{1}{h_{(j)}} \begin{pmatrix} 3 \\ -\frac{3}{2} \\ 9 \end{pmatrix} (q_{(j+1)}^{(1)} - q_{(j)}^{(3)}) = 0 \qquad (4.45b)$$

In the above equations, the overdot stands for time derivative and also we take $a_1 = a_3 = -a_2 = -a_4 = 0.5$ for the central scheme and $a_1 = -a_4 = 1$ and $a_3 = a_2 = 0$ for the alternating upwinding scheme.

4.3.5 The Runge–Kutta Time Integration

Equation 4.41b requires a time integrator for a numerical solution. A commonly used time marching algorithm for discontinuous calculations is the explicit time integration. The various orders of the Runge–Kutta (RK) schemes are given below.

(1) The first order RK method (the Euler forward method):

$$\mathbf{T}_{(j)}^{k+1} = \mathbf{T}_{(j)}^{k} + (\Delta t)\dot{\mathbf{T}}_{(j)}^{k} \tag{4.46}$$

(2) The second order RK method:

$$\mathbf{k}_1 = (\Delta t)\dot{\mathbf{T}}_{(j)}(\mathbf{T}_{(j)}^{k}) \tag{4.47a}$$

$$\mathbf{k}_2 = (\Delta t)\dot{\mathbf{T}}_{(j)}(\mathbf{T}_{(j)}^{k} + \Delta t \mathbf{k}_1) \tag{4.47b}$$

$$\mathbf{T}_{(j)}^{k+1} = \mathbf{T}_{(j)}^{k} + \frac{1}{2}(\mathbf{k}_1 + \mathbf{k}_2) \tag{4.47c}$$

(3) The third order RK method:

$$\mathbf{k}_1 = (\Delta t)\dot{\mathbf{T}}_{(j)}(\mathbf{T}_{(j)}^{k}) \tag{4.48a}$$

$$\mathbf{k}_2 = (\Delta t)\dot{\mathbf{T}}_{(j)}(\mathbf{T}_{(j)}^{k} + \frac{1}{2}\Delta t \mathbf{k}_1) \tag{4.48b}$$

$$\mathbf{k}_3 = (\Delta t)\dot{\mathbf{T}}_{(j)}(\mathbf{T}_{(j)}^{k} - \Delta t \mathbf{k}_1 + 2\Delta t \mathbf{k}_2) \tag{4.48c}$$

$$\mathbf{T}_{(j)}^{k+1} = \mathbf{T}_{(j)}^{k} + \frac{1}{6}(\mathbf{k}_1 + 4\mathbf{k}_2 + \mathbf{k}_3) \tag{4.48d}$$

where the superscript k refers to the kth time step.

It is important to realize that the above explicit time schemes require time steps that are smaller than a critical value. Normally the *CFL* criterion for the time step applies. In Section 4.6.3, we show through stability analysis how to determine an adequate time step for numerical integration in time.

It should be noted here that the implicit time scheme is also possible, but it can be cumbersome to use with the discontinuous finite element method and a global matrix of size larger than the continuous finite element method needs to be solved (see also discussion following Equation 4.14).

4.3.6 Computational Procedures

One way to solve the above system is to use the iterative procedure as has been done for the steady state case. By this approach, q is solved first on the given values of T, and then time integration is applied to obtain T at $t + \Delta t$. This procedure continues from the boundary and sweeps through the entire domain, element by element, for every time step.

Another approach is to eliminate the equation for q, and solve the combined equation for T. The element calculations, and time marching remain the same as above. This approach is possible because q is, in essence, an intermediate variable and can be eliminated. This is point is further discussed at the end of Example 4.4 below.

Yet another approach is to combine all the variables together to form a global matrix, which is then inverted by LU decomposition to obtain the solution and march in time. This approach is rarely used, but can be useful for implicit time marching schemes.

Example 4.4. Calculate the evolution of the temperature distribution in a 1-D slab with a heating source $Q = 1$ using linear elements and the alternating upwinding scheme with the first order time integration. The boundary and initial conditions are: $T(x=0,t)=T(x=1,t)=0$, $T(x,0)=0$. Take all the properties to be one and derive a matrix equation for the temperature values **T** by eliminating the variable **q**.

Solution. The numerical implementation for the linear interpolation is given in the previous section. Therefore, we have

$$\begin{pmatrix} q_{(j)}^{(1)} \\ q_{(j)}^{(2)} \end{pmatrix} = -\frac{1}{h_{(j)}}\begin{pmatrix} -1 & 1 \\ -1 & 1 \end{pmatrix}\begin{pmatrix} T_{(j)}^{(1)} \\ T_{(j)}^{(2)} \end{pmatrix} - \frac{1}{h_{(j)}}\begin{pmatrix} 4 \\ -2 \end{pmatrix}(a_1 T_{(j)}^{(1)} - a_2 T_{(j-1)}^{(2)})$$

$$-\frac{1}{h_{(j)}}\begin{pmatrix} -2 \\ 4 \end{pmatrix}(T_{(j+1)}^{(1)} - T_{(j)}^{(2)}) = 0 \tag{4.19e}$$

$$\begin{pmatrix} T_{(j)}^{(1)} \\ T_{(j)}^{(2)} \end{pmatrix}^{k+1} = \begin{pmatrix} T_{(j)}^{(0)} \\ T_{(j)}^{(1)} \end{pmatrix}^{k} - \frac{\Delta t}{h_{(j)}}\begin{pmatrix} -1 & 1 \\ -1 & 1 \end{pmatrix}\begin{pmatrix} q_{(j)}^{(1)} \\ q_{(j)}^{(2)} \end{pmatrix}^{k} - \frac{\Delta t}{h_{(j)}}\begin{pmatrix} 4 \\ -2 \end{pmatrix}(a_3 q_{(j)}^{(1)} - a_4 q_{(j-1)}^{(2)})^{k}$$

$$-\frac{\Delta t}{h_{(j)}}\begin{pmatrix} -2 \\ 4 \end{pmatrix}(q_{(j+1)}^{(1)} - q_{(j)}^{(3)})^{k} + \begin{pmatrix} 1 \\ 1 \end{pmatrix}\Delta t \tag{4.20e}$$

where superscript k denotes the kth time step and the last term results from the contribution of the body source.

Choosing the time step $\Delta t = 10^{-4}$ and $h_{(j)} = 0.01$, we can proceed to do the calculations. The computational procedure is as follows. First, all the variables are initially set to the initial temperature and then q^k is solved using the first equation. Then the time integration is performed to calculate T^{k+1} using the second equation, and the result is then substituted back into the first equation to calculate q^{k+1}. This process continues until the time reaches the specified total time. The calculated results are shown in Figure 4.5e.

For illustration, we have chosen to solve the variables q and T simultaneously. For this problem, q is in fact an auxiliary variable, and thus can be eliminated at the element level to obtain a form for T as follows:

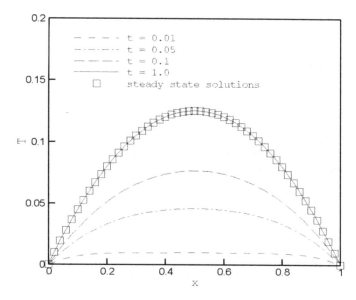

Figure 4.5e. Temperature distributions for the 1-D transient heat conduction problem

$$\underline{\mathbf{T}}_{(j)}^{k+1} = \underline{\mathbf{T}}_{(j)}^{k} - \Delta t (\mathbf{A}\underline{\mathbf{T}}_{(j-2)} + \mathbf{B}\underline{\mathbf{T}}_{(j-1)} + \mathbf{C}\underline{\mathbf{T}}_{(j)} + \mathbf{D}\underline{\mathbf{T}}_{(j+1)} + \mathbf{E}\underline{\mathbf{T}}_{(j+2)} + \mathbf{Q}_{(j)})^{k}$$

(4.21e)

which spreads over five nodes. In the above equation, **A**, **B**, **C**, **D**, and **E** are coefficient matrices and **Q** is the source vector. This would speed up the calculations. The values of q^k, if needed, can then be obtained by processing the data from $\{T\}^k$. We note also that, for the alternating upwinding scheme, the spread is only over 3 points, and thus should be faster than the central scheme.

4.4 Transient Heat Conduction in Multidimensions

Let us now consider the discontinuous formulation for multidimensional problems. The mathematical equation for heat conduction in a multidimensional domain is given by

$$\rho C_p \frac{\partial T}{\partial t} = \nabla \cdot k \nabla T + Q \qquad \in \Omega \qquad (4.49a)$$

$$T = T_0 \qquad \in \partial\Omega_D \qquad (4.49b)$$

$$-\mathbf{n} \cdot k \nabla T = h(T - T_\infty) \qquad \in \partial\Omega_N \qquad (4.49c)$$

$$T(\mathbf{r},t=0) = T^0(\mathbf{r}) \qquad\qquad\qquad \in \Omega \qquad\qquad (4.49d)$$

To develop a discontinuous finite element formulation, Equation 4.49(a) is first split into two first order equations,

$$\mathbf{q} = k\nabla T \; ; \; \rho C_p \frac{\partial T}{\partial t} = \nabla \cdot \mathbf{q} + Q \qquad\qquad (4.50)$$

After the domain is discretized, we apply the Galerkin procedure to develop a discontinuous finite element formulation for an element. This involves integrating the above equations with respect to the weighting functions $(\mathbf{w}, v)$ over element j ($\mathbf{r} \in \Omega_j$),

$$\int_{\Omega_j} \mathbf{q} \cdot \mathbf{w} dV = \int_{\Omega_j} \mathbf{w} \cdot k\nabla T dV \qquad\qquad (4.51a)$$

$$\int_{\Omega_j} v\rho C_p \frac{\partial T}{\partial t} dV = \int_{\Omega_j} v(\nabla \cdot \mathbf{q} + Q) dV \qquad\qquad (4.51b)$$

The relevant geometric arrangement of the element j and its neighbors is given in Figure 4.3. We now integrate by parts to obtain

$$\int_{\Omega_j} \mathbf{q} \cdot \mathbf{w} \, dV = -\int_{\Omega_j} T\nabla \cdot (k\mathbf{w}) dV + \int_{\partial\Omega_j} kT \mathbf{n}_j \cdot \mathbf{w} dS \qquad\qquad (4.52a)$$

$$\int_{\Omega_j} v\rho C_p \frac{\partial T}{\partial t} dV + \int_{\Omega_j} \mathbf{q} \cdot \nabla v \, dV = \int_{\Omega_j} vQ dV + \int_{\partial\Omega_j} \mathbf{q} \cdot \mathbf{n}_j v dS \qquad (4.52b)$$

where $\mathbf{n}_j$ is the outward normal unit vector to $\partial\Omega_j$, the boundary of the element.

We now seek to approximate the exact solution $(\mathbf{q}, T)$ with functions $(\mathbf{q}_h, T_h)$ in the finite element broken space, whence we have the following results:

$$\int_{\Omega_j} \mathbf{q}_h \cdot \mathbf{w} \, dV = -\int_{\Omega_j} T_h\nabla \cdot (k\mathbf{w}) dV + \int_{\partial\Omega_j} k\hat{T}_h \mathbf{n}_j \cdot \mathbf{w} dS \qquad\qquad (4.53a)$$

$$\int_{\Omega_j} v\rho C_p \frac{\partial T_h}{\partial t} dV + \int_{\Omega_j} \mathbf{q}_h \cdot \nabla v \, dV = \int_{\Omega_j} vQ dV + \int_{\partial\Omega_j} \hat{\mathbf{q}}_h \cdot \mathbf{n}_j v dS \qquad\qquad (4.53b)$$

where the numerical fluxes $(\hat{\mathbf{q}}_h, \hat{T}_h)$ are approximations to $(\mathbf{q},T)$ on the boundary of element j. To complete the discontinuous finite element solution, these numerical fluxes must be specified in terms of $\mathbf{q}_h$ and T_h, and in terms of the boundary conditions. These numerical fluxes must be suitably selected to render the discontinuous formulation stable. Besides those shown in Table 4.1, additional fluxes, simpler in form, also satisfy the stability criteria for transient calculations. For convenience, the widely used numerical fluxes tested for transient heat conduction calculations are listed in Table 4.2. Selection of the numerical flux in discontinuous continuous finite element formulations for diffusion problems is recently discussed in [13].

Table 4.2. Numerical fluxes for transient heat diffusion calculations

Method	$\hat{\mathbf{q}}_h$	$\hat{T}_h$
LDG [6]	$\{\mathbf{q}_h\} - C_{11}[T_h] - \mathbf{C}_{12}[\mathbf{q}_h]$	$\{T_h\} + \mathbf{C}_{12}[T_h]$
DG [2]	$\{\mathbf{q}_h\} - C_{11}[T_h] - \mathbf{C}_{12}[\mathbf{q}_h]$	$\{T_h\} + \mathbf{C}_{12}[T_h] - \mathbf{C}_{22}[\mathbf{q}_h]$
Brezzi *et al.* [7]	$\{\mathbf{q}_h\} - \alpha^r([T_h])$	$\{T_h\}$
IP [8]	$\{\nabla T_h\} - C_{11}[T_h]$	$\{T_h\}$
Bassi *et al.* [9]	$\{\nabla T_h\} - \alpha^r([T_h])$	$\{T_h\}$
NIPG [10]	$\{\nabla u_h\} - C_{11}[u_h]$	$\{T_h\} + \mathbf{n}_j \cdot [T_h]$
Bubaska–Zlamal [14]	$-\alpha_j[T_h]$	$T_h\mid_{\Omega_j}$
Brezzi *et al.* [15]	$-\alpha_r[T_h]$	$T_h\mid_{\Omega_j}$
Bassi –Rebay [12]	$\{\mathbf{q}_h\}$	$\{T_h\}$
Baumann–Oden [16]	$\{\nabla T_h\}$	$\{T_h\} + \mathbf{n}_j \cdot [T_h]$

Note: C_{11}, C_{22}, $\mathbf{C}_{12}$ and $\mathbf{C}_{21}$ are constant matrices, and $\mathbf{n}_j$ is the outward normal of element j.

The operators are the same as defined in Section 4.2. With the numerical fluxes defined by Equation 4.20 substituted into Equation 4.52a, b, one has the final integral formulation,

$$\int_{\Omega_j} \mathbf{q}_h \cdot \mathbf{w}\, dV + \int_{\Omega_j} T_h \nabla \cdot (k\mathbf{w}) dV$$

$$- \int_{\partial\Omega_j} (\{T\} + \mathbf{C}_{12} \cdot [T] - C_{22}[q])\mathbf{n}_j \cdot \mathbf{w} dS = 0 \qquad (4.53\text{a})$$

$$\int_{\Omega_j} v\rho C_p \frac{\partial T_h}{\partial t} dV + \int_{\Omega_j} \mathbf{q}_h \cdot \nabla v\, dV$$

$$- \int_{\partial\Omega_j} (\{\mathbf{q}\} - C_{11}[T] - \mathbf{C}_{12}[q]) \cdot \mathbf{n}_j v dS = \int_{\Omega_j} vQ dV \qquad (4.53\text{b})$$

This is very similar to Equation 4.21, except for the transient term.

Now with the unknowns approximated using a polynomial basis function, followed by tedious algebra, one has the final matrix equation for the discontinuous finite element formulation for the transient heat conduction problem,

$$
\begin{bmatrix} 0 & 0 & 0 & 0 \\ 0 & 0 & 0 & 0 \\ 0 & 0 & 0 & 0 \\ 0 & 0 & 0 & M_T \end{bmatrix}
\begin{pmatrix} \dot{\underline{q}}_x \\ \dot{\underline{q}}_y \\ \dot{\underline{q}}_z \\ \dot{\underline{T}} \end{pmatrix}
+
\begin{bmatrix} E & 0 & 0 & H_x \\ 0 & E & 0 & H_y \\ 0 & 0 & E & H_z \\ J_x & J_y & J_y & 0 \end{bmatrix}
\begin{pmatrix} \underline{q}_x \\ \underline{q}_y \\ \underline{q}_z \\ \underline{T} \end{pmatrix}
$$

$$
+ \sum_{i=i}^{NS}
\begin{bmatrix} \mathbf{E}_{xx,i} & \mathbf{E}_{xy,i} & \mathbf{E}_{xz,i} & \mathbf{H}_{x,i} \\ \mathbf{E}_{yx,i} & \mathbf{E}_{yy,i} & \mathbf{E}_{yz,i} & \mathbf{H}_{y,i} \\ \mathbf{E}_{zx,i} & \mathbf{E}_{zy,i} & \mathbf{E}_{zz,i} & \mathbf{H}_{z,i} \\ \mathbf{J}_{x,i} & \mathbf{J}_{y,i} & \mathbf{J}_{z,i} & \mathbf{G}_{T,i} \end{bmatrix}
\begin{pmatrix} \underline{q}_x \\ \underline{q}_y \\ \underline{q}_z \\ \underline{T} \end{pmatrix}
$$

$$
+ \sum_{i=i}^{NS}
\begin{bmatrix} \mathbf{E}_{xx,B,i} & \mathbf{E}_{xy,B,i} & \mathbf{E}_{xy,B,i} & \mathbf{H}_{x,B,i} \\ \mathbf{E}_{yx,B,i} & \mathbf{E}_{yy,B,i} & \mathbf{E}_{yz,B,i} & \mathbf{H}_{y,B,i} \\ \mathbf{E}_{zx,B,i} & \mathbf{E}_{zy,B,i} & \mathbf{E}_{zz,B,i} & \mathbf{H}_{z,B,i} \\ \mathbf{J}_{x,B,i} & \mathbf{J}_{y,B,i} & \mathbf{J}_{z,B,i} & \mathbf{G}_{T,B,i} \end{bmatrix}
\begin{pmatrix} \underline{q}_x \\ \underline{q}_y \\ \underline{q}_z \\ \underline{T} \end{pmatrix}_{(NB,i)}
=
\begin{pmatrix} 0 \\ 0 \\ 0 \\ S_T \end{pmatrix}
\tag{4.54a}
$$

where the overdot denotes time derivative, e.g., $\dot{\underline{T}} = \partial \underline{T}/\partial t$, the mass matrix is calculated by

$$
M_{T,km} = \int_{\Omega_j} \rho C_p \phi_k \phi_m \, dV
\tag{4.54b}
$$

and other matrices are calculated using the expressions given in Section 4.2.

The above matrix form is useful to illustrate the effects of each term and their interactions. It is, however, inconvenient for computations, because the mass matrix is singular when an explicit scheme is used. To facilitate computations, the above equation is re-written in two separate matrx equations,

$$
\begin{bmatrix} E & 0 & 0 \\ 0 & E & 0 \\ 0 & 0 & E \end{bmatrix}
\begin{pmatrix} \underline{q}_x \\ \underline{q}_y \\ \underline{q}_z \end{pmatrix}
+
\begin{pmatrix} H_x \\ H_y \\ H_z \end{pmatrix} \underline{T}
+ \sum_{i=1}^{NS} \Bigg\{
\begin{bmatrix} \mathbf{E}_{xx,i} & \mathbf{E}_{xy,i} & \mathbf{E}_{xz,i} \\ \mathbf{E}_{yx,i} & \mathbf{E}_{yy,i} & \mathbf{E}_{yz,i} \\ \mathbf{E}_{zx,i} & \mathbf{E}_{zy,i} & \mathbf{E}_{zz,i} \end{bmatrix}
\begin{pmatrix} \underline{q}_x \\ \underline{q}_y \\ \underline{q}_z \end{pmatrix}
+
\begin{pmatrix} \mathbf{H}_{x,i} \\ \mathbf{H}_{y,i} \\ \mathbf{H}_{z,i} \end{pmatrix} \underline{T} \Bigg\}
$$

$$
+ \sum_{i=1}^{NS} \Bigg\{
\begin{bmatrix} \mathbf{E}_{xx,B,i} & \mathbf{E}_{xy,B,i} & \mathbf{E}_{xz,B,i} \\ \mathbf{E}_{yx,B,i} & \mathbf{E}_{yy,B,i} & \mathbf{E}_{yz,B,i} \\ \mathbf{E}_{zx,B,i} & \mathbf{E}_{zy,B,i} & \mathbf{E}_{zz,B,i} \end{bmatrix}
\begin{pmatrix} \underline{q}_x \\ \underline{q}_y \\ \underline{q}_z \end{pmatrix}_{(NB,i)}
+
\begin{pmatrix} \mathbf{H}_{x,B,i} \\ \mathbf{H}_{y,B,i} \\ \mathbf{H}_{z,B,i} \end{pmatrix} \mathbf{T}_{(NB,i)} \Bigg\}
=
\begin{pmatrix} 0 \\ 0 \\ 0 \end{pmatrix}
$$

$$
\tag{4.55a}
$$

$$\mathbf{M}_T \frac{d\mathbf{T}}{dt} + \sum_{l=x,y,z} \mathbf{J}_l \underline{\mathbf{q}}_l + \sum_{i=1}^{NS} \mathbf{G}_{T,i} \underline{\mathbf{T}} + \sum_{i=1}^{NS} \mathbf{G}_{T,B,i} \underline{\mathbf{T}}_{(NB,i)}$$

$$+ \sum_{i=1}^{NS} \sum_{l=x,y,z} \mathbf{J}_{l,B,i} (\underline{\mathbf{q}}_l)_{(NB,i)} = \mathbf{S}_T \qquad (4.55b)$$

With the above equations, the iterative solution can be obtained using the computational procedure as discussed in Section 4.3.6.

Example 4.5. Show that the LDG numerical flux scheme in Table 4.2 is the same as the alternating upwinding scheme for the 1-D transient heat conduction problem.

Solution. Let us consider element $j \in [x_j, x_{j+1}]$, as shown in Figure 4.6e. As usual, the second order differential equation is split into two first order differential equations, and this is followed by integration with respect to the weighting function pair (v, w),

$$\int_{x_j}^{x_{j+1}} \left[\frac{\partial T}{\partial t} - \nabla_x \cdot \mathbf{q} \right] v(x)\, dx = 0 \text{ and } \int_{x_j}^{x_{j+1}} [-\mathbf{q} + \nabla_x T]\, w(x)\, dx = 0$$

$$(4.22e)$$

$$\int_{x_j}^{x_{j+1}} \left[v \frac{\partial T}{\partial t} + \nabla_x v \cdot \mathbf{q} \right] dx$$

$$- \left(\mathbf{n}^- \cdot \mathbf{q}(x_{j+1}^-) v(x_{j+1}^-) + \mathbf{n}^+ \cdot \mathbf{q}(x_j^+) v(x_j^+) \right) = 0 \qquad (4.23e)$$

$$\int_{x_j}^{x_{j+1}} [\mathbf{q} \cdot \mathbf{w} + T \nabla_x \cdot \mathbf{w}]\, dx$$

$$- \left(T(x_{j+1}^-) \mathbf{n}^- \cdot \mathbf{w}(x_{j+1}^-) + T(x_j^+) \mathbf{n}^+ \cdot \mathbf{w}(x_j^+) \right) = 0 \qquad (4.24e)$$

where $\mathbf{q}(x) = q(x)\hat{i}$, $\nabla_x = \hat{i}\, \partial/\partial x$, and $\mathbf{w}(x) = w(x)\hat{i}$.

To demonstrate that the LDG method listed in Table 4.2 leads to the alternating upwinding scheme, we start from the general form of the numerical fluxes,

$$\hat{\mathbf{q}}_h = \{\mathbf{q}_h\} - C_{11}[T_h] - \mathbf{C}_{12}[\mathbf{q}_h]; \quad \hat{T}_h = \{T_h\} + \mathbf{C}_{12} \cdot [T_h] \qquad (4.25e)$$

At point j, we have $\mathbf{q}^+ = q^+ \hat{i}$, and $\mathbf{q}^- = q^- \hat{i}$, $C_{12} = \mathbf{n}^- \cdot \mathbf{C}_{12}$

$$\hat{\mathbf{q}} \cdot \mathbf{n}^- = \mathbf{n}^- \cdot \{\mathbf{q}\} - C_{11}\mathbf{n}^- \cdot [T] - \mathbf{n}^- \cdot \mathbf{C}_{12}[\mathbf{q}] = 0.5(\mathbf{q}^+ \cdot \mathbf{n}^- + \mathbf{q}^- \cdot \mathbf{n}^-)$$

$$- C_{11}\mathbf{n}^- \cdot [T^+\mathbf{n}^+ + T^-\mathbf{n}^-] - \mathbf{n}^- \cdot \mathbf{C}_{12}[\mathbf{q}^+ \cdot \mathbf{n}^+ + \mathbf{q}^- \cdot \mathbf{n}^-]$$

$$= 0.5(q^+ + q^-) - C_{11}(-T^+ + T^-) - C_{12}[-q^+ + q^-] \qquad (4.26e)$$

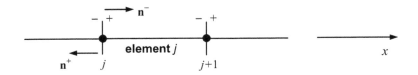

Figure 4.6e. Illustration of various quantities for element j

$$\hat{T} = \{T\} + \mathbf{C}_{12} \cdot [T] = 0.5(T^+ + T^-) + \mathbf{C}_{12} \cdot [T^+\mathbf{n}^+ + T^-\mathbf{n}^-]$$

$$= 0.5(T^+ + T^-) + C_{12}[-T^+ + T^-] \qquad (4.27e)$$

If we take $C_{11} = 0$, $C_{12} = 0.5$, then we have

$$\hat{\mathbf{q}} \cdot \mathbf{n}^- = 0.5(q^+ + q^-) - C_{12}[-q^+ + q^-] = q^+ ;$$

$$\hat{T} = 0.5(T^+ + T^-) + C_{12}[-T^+ + T^-] = T^- \qquad (4.28e)$$

At point $j+1$, $\mathbf{q}^+ = q^+\hat{i}$, $\mathbf{q}^- = q^-\hat{i}$ and $C_{12} = -\mathbf{n}^+ \cdot \mathbf{C}_{12}$, whence one has

$$\hat{\mathbf{q}} \cdot \mathbf{n}^+ = \mathbf{n}^+ \cdot \{\mathbf{q}\} - C_{11}\mathbf{n}^+ \cdot [T] - \mathbf{n}^+ \cdot \mathbf{C}_{12}[\mathbf{q}] = 0.5(\mathbf{q}^+ \cdot \mathbf{n}^+ + \mathbf{q}^- \cdot \mathbf{n}^+)$$

$$- C_{11}\mathbf{n}^+ \cdot [T^+\mathbf{n}^+ + T^-\mathbf{n}^-] - \mathbf{n}^+ \cdot \mathbf{C}_{12}[\mathbf{q}^+ \cdot \mathbf{n}^+ + \mathbf{q}^- \cdot \mathbf{n}^-]$$

$$= 0.5(-q^+ - q^-) - C_{11}(T^+ - T^-) + C_{12}[-q^+ + q^-] \qquad (4.29e)$$

$$\hat{T} = \{T\} + \mathbf{C}_{12} \cdot [T] = 0.5(T^+ + T^-) + \mathbf{C}_{12} \cdot [T^+\mathbf{n}^+ + T^-\mathbf{n}^-]$$

$$= 0.5(T^+ + T^-) + C_{12}[-T^+ + T^-] \qquad (4.30e)$$

If we take $C_{11} = 0$, $C_{12} = 0.5$, then the above two equations are simplified,

$$\hat{\mathbf{q}} \cdot \mathbf{n}^+ = 0.5(-q^+ - q^-) + C_{12}[-q^+ + q^-] = -q^+ ,$$

$$\hat{T} = 0.5(T^+ + T^-) + C_{12}[-T^+ + T^-] = T^- \qquad (4.31e)$$

With these relations, we have for the boundary terms of element j,

$$-\left(\mathbf{n}^- \cdot \mathbf{q}(x_{j+1}^-)v(x_{j+1}^-) + \mathbf{n}^+ \cdot \mathbf{q}(x_j^+)v(x_j^+)\right)$$

$$= -q_{j+1}^+ v(x_{j+1}^-) + q_j^+ v(x_j^+) = -q_{j+1}^+ + q_j^+ \qquad (4.32e)$$

$$-\left(T(x_{j+1}^{-})\,\mathbf{n}^{-}\cdot\mathbf{w}\,(x_{j+1}^{-})+T(x_{j}^{+})\,\mathbf{n}^{+}\cdot\mathbf{w}(x_{j}^{+})\right)=-T_{j+1}^{-}+T_{j}^{-} \qquad (4.33e)$$

Substituting the above two equations into Equations 4.22e and 4.24e, and writing the results in terms of scalar quantities, we have the following expressions:

$$\int_{x_j}^{x_{j+1}}\left[v\frac{\partial T}{\partial t}+\frac{\partial w}{\partial x}q\right]dx-q_{j+1}^{+}+q_{j}^{+}=0 \qquad (4.34e)$$

$$\int_{x_j}^{x_{j+1}}\left[q\cdot w+u\frac{\partial w}{\partial x}\right]dx-T_{j+1}^{-}+T_{j}^{-}=0 \qquad (4.35e)$$

which are the same as Equation 4.53.

The discontinuous finite element algorithm developed above may be applied to solve the transient heat conduction problems defined in a multidimensional domain. One of these calculations is given here. For this case, we consider the transient temperature distribution in a unit circle with a fixed temperature at its edge. The circle is heated by a Gaussian heating source. A problem of this type often occurs in laser heating processes.

The mathematical statement of the problem is given by

$$\frac{\partial T}{\partial t}=\frac{\partial^{2}T}{\partial x^{2}}+\frac{\partial^{2}T}{\partial y^{2}}+Q\ ,\ r^{2}=x^{2}+y^{2}\leq 1 \qquad (4.56)$$

where the heating source Q is in the form of $Q=Q_0\exp[-(x^2+y^2)/a^2]$, and the boundary condition is $T=0$ at $r=1$. Set $Q_0=1$ and $a=1$ for the calculations. The properties of the material are also set to unity.

For this problem, the calculations used an unstructured triangular mesh with linear elements for both temperature and heat flux unknowns. The central flux approximation is used to approximate the numerical flux. The mesh distribution and computed temperature results are given in Figure 4.5. The temperature field calculated assumes perfect rotational symmetry, despite the unstructured mesh used, indicating the accuracy of the method. It is noted that for this simple problem the condition of rotational symmetry would allow us to use a 1-D model rather than a 2-D geometry.

4.5 Potential Flows and Flows in Porous Media

Both potential flows and flows in porous media have the same mathematical structure as the heat conduction problems discussed above (see Chapter 1). They are all classified as diffusion problems. The discontinuous formulation and solution procedures are basically identical. For these cases, the flow field is derived from the gradient of a potential, which satisfies a Poisson equation. Some examples of the discontinuous finite element solution of porous flow problems are given in references [10, 17, 18].

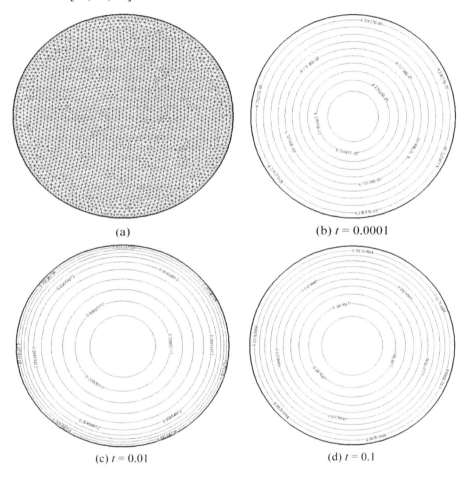

Figure 4.5. Computed temperature evolution in a unit circle heated by a Gaussian heating source: (a) unstructured mesh with linear triangular elements, and (b)–(d) temperature contour at different time steps

4.6 Selection of Numerical Fluxes

This section discusses the selection of numerical fluxes for diffusion problems. The selection of these numerical fluxes is based on the stability of a discontinuous finite element formulation. By the stability of a numerical formulation, one generally means that the consistency and convergence criteria of the formulation are met in a discretized form. Mathematically, the transient and steady state heat conduction problems are classified into different types of differential equations and thus the stability criteria are not necessarily the same, though they may be closely related. For discontinuous calculations, numerical fluxes need to be selected so as to ensure that the formulation satisfies the consistency and stability conditions.

4.6.1 Stability for Steady State Problems

For numerical schemes, convergence of a numerical solution to an exact solution is required. For this convergence to occur, the requirements of consistency and stability must be met. The consistency requirement means that a numerical scheme represents the differential equations, and their boundary conditions, as the size of the elements approaches zero. The numerical stability requirement states that the solution of the numerical scheme for solving a well-conditioned problem changes only a small amount if the input data change a little. This means that any error committed in the early stages of iteration will not grow in an uncontrolled manner. A well-conditioned mathematical problem is one whose solution changes by only a small amount if the problem data are changed by a small amount. The convergence rate, or order of accuracy, is measured by the power of the element size h. In this context, a stable solution of a discretized system will not contain spurious modes that may pollute the solution, regardless of the size of the elements used. If the discretized system is represented in a matrix format, $\mathbf{KU} = \mathbf{F}$, as it often is, then the stability condition requires that the stiffness matrix $\mathbf{K}$ is non-singular with a uniformly bounded condition number [19]. The selection of the numerical fluxes thus needs to satisfy these basic requirements.

4.6.1.1 Stability and Numerical Fluxes
In this section, some basic steps are discussed regarding the stability and consistency studies, for the purpose of developing or selecting appropriate numerical fluxes for the discontinuous finite element simulation for steady state heat conduction problems. In the literature, these problems are classified as elliptic problems. There are several approaches to this type of problem. Here we follow the procedures outlined in a recent paper by Cockburn [5], using the theory of partial differential equations detailed by Evans [20].

We consider below a heat conduction problem with $T = 0$ along the boundary,

$$\nabla^2 T = -f \ \in \Omega ; \qquad T = 0 \ \in \partial\Omega \qquad (4.57)$$

To check the consistency requirement, an integral form of the differential equation is derived, following essentially the same steps given in [16]. Integrating the above equation with respect to the exact solution T, and carrying out the integration by parts, we have the integral representation for Equation 4.57,

$$\int_{\Omega} |\nabla T|^2 \, dV = \int_{\Omega} |\mathbf{q}|^2 \, dV = \int_{\Omega} fT dV \qquad (4.58)$$

where $\mathbf{q} = \nabla T$, and use has been made of the condition $T = 0$ on the boundary. Equation 4.58 simply means that if the exact solution is used, this integral condition should be satisfied. We note that unlike the analysis of the finite difference method, the analysis of a discontinuous finite element method starts with the integral form of the differential equation. The consistency and stability requirements will then be analyzed for the integral form.

If the consistency requirement is to be met, then a choice of discretized $\mathbf{q}_h$ will satisfy the above integral equation as the size h goes to zero. The discontinuous finite element formulation for the above problem is given by two integral equations for element j,

$$\int_{\Omega_j} \mathbf{q}_h \cdot \mathbf{w} \, dV = - \int_{\Omega_j} T_h \nabla \cdot \mathbf{w} dV + \int_{\partial\Omega_j} \hat{T}_h \mathbf{n}_j \cdot \mathbf{w} dS \qquad (4.59a)$$

$$\int_{\Omega_j} \mathbf{q}_h \cdot \nabla v dV = \int_{\Omega_j} v f dV + \int_{\partial\Omega_j} \hat{\mathbf{q}}_h \cdot \mathbf{n}_j v dS \qquad (4.59b)$$

Letting $\mathbf{w} = \mathbf{q}_h$ in the first equation and $v = T_h$ in the second equation, adding on the elements, one obtains

$$\int_{\Omega} |\mathbf{q}_h|^2 \, dV - \sum_j \left(- \int_{\Omega_j} T_h \nabla \cdot \mathbf{q}_h dV + \int_{\partial\Omega_j} \hat{T}_h \mathbf{q}_h \cdot \mathbf{n}_j dS \right) = 0 \qquad (4.60a)$$

$$\sum_j \left(\int_{\Omega_j} \mathbf{q}_h \cdot \nabla T_h dV - \int_{\partial\Omega_j} T_h \hat{\mathbf{q}}_h \cdot \mathbf{n}_j dS \right) = \int_{\Omega} f T_h dV \qquad (4.60b)$$

where $\Omega = \cup \Omega_j$. Adding the above two equations together yields the final expression for the discretized solution,

$$\int_{\Omega} |\mathbf{q}_h|^2 dV + \Theta_h = \int_{\Omega} f T_h dV \qquad (4.61)$$

where the extra terms are collected as the sum of flux terms across the element boundaries,

$$\Theta_h = - \sum_j \left(- \int_{\Omega_j} \nabla \cdot (T_h \mathbf{q}_h) dV + \int_{\partial\Omega_j} (\hat{T}_h \mathbf{q}_h \cdot \mathbf{n}_j + T_h \hat{\mathbf{q}}_h \cdot \mathbf{n}_j) dS \right) \quad (4.62)$$

Comparison of Equations 4.58 and 4.61 suggests that if a numerical scheme is stable and iterates to a converged solution, then Θ_h must be non-negative and $\Theta_h \to 0$ as $h \to 0$. Note that if $\Theta_h < 0$, $\mathbf{q}_h$ may be unbounded, which will cause Equation 4.61 to differ from Equation 4.58 and the solution becomes unstable.

Thus, consistent numerical fluxes $\hat{T}_h$ and $\hat{q}_h$ must be chosen such that they render Θ_h non-negative. Let us examine Equation 4.62 again:

$$
\begin{aligned}
\Theta_h &= \sum_{j \in N} \int_{\partial \Omega_j} (T_h \mathbf{q}_h \cdot \mathbf{n}_j - \hat{T}_h \mathbf{q}_h \cdot \mathbf{n}_j - T_h \hat{\mathbf{q}}_h \cdot \mathbf{n}_j)dS \\
&= \sum_{j \in N} \int_{\partial \underline{\Omega}_j} \left[T_h \mathbf{q}_h - \hat{T}_h \mathbf{q}_h - T_h \hat{\mathbf{q}}_h \right] dS \\
&= \sum_{j \in N} \int_{\partial \underline{\Omega}_{j,i}} ([T_h \mathbf{q}_h] - \hat{T}_h [\mathbf{q}_h] - [T_h] \cdot \hat{\mathbf{q}}_h)dS \\
&\quad + \int_{\partial \Omega} (T_h \mathbf{q}_h - \hat{T}_h \mathbf{q}_h \cdot \mathbf{n} - T_h \hat{\mathbf{q}}_h \cdot \mathbf{n})dS \\
&= \sum_{j \in N} \int_{\partial \underline{\Omega}_{j,i}} ((\{T_h\} - T_h)[\mathbf{q}_h] + [T_h] \cdot (\{\mathbf{q}_h\} - \hat{\mathbf{q}}_h)) \, dS \\
&\quad + \int_{\partial \Omega} (T_h (\mathbf{q}_h - \hat{\mathbf{q}}_h) \cdot \mathbf{n} - \hat{T}_h \mathbf{q}_h \cdot \mathbf{n}) \, dS
\end{aligned}
\tag{4.63}
$$

where N is the total number of elements, and $\partial \Omega_{j,i}$ means integration once over a shared internal element boundary, $\partial \Omega_{j,i} \in \partial V_m^+ \cap \partial V_m^-$.

If we take the following expressions for the fluxes at the internal element boundaries inside the domain Ω,

$$
\hat{\mathbf{q}}_h = \{\mathbf{q}_h\} - C_{11}[T_h] - C_{12}[\mathbf{q}_h]; \quad \hat{T}_h = \{T_h\} + C_{12} \cdot [T_h] - C_{22}[\mathbf{q}_h]
\tag{4.64}
$$

and for those at the element boundaries that share with the exterior boundary of the domain $\partial \Omega$,

$$
\hat{\mathbf{q}}_h = \mathbf{q}_h - C_{11} T_h \mathbf{n}; \qquad T_h = 0
\tag{4.65}
$$

then Θ_h is calculated by

$$
\Theta_h = \sum_{e \in E_{ih}} \int_e (C_{22}[\mathbf{q}_h]^2 + C_{11}[T_h]^2)dS + \int_{\partial \Omega} C_{11} T_h^2 dS \geq 0
\tag{4.66}
$$

provided C_{11} and C_{22} are non-negative. Note that the boundary conditions are imposed weakly through the definition of the numerical traces. It is apparent from the proof above that C_{12}, which can be either negative or positive or zero, is selected to set boundary conditions [2]. The above proof shows that the parameters C_{11} and C_{22} play an important role in ensuring the stability and the accuracy of the discontinuous formulations.

Castillo *et al.* [2] show that to guarantee the existence and uniqueness of the approximate solution of the DG methods, the parameter C_{11} has to be greater than zero, and the local finite element spaces $U(\Omega_j)$ and $Q(\Omega_j)$ must satisfy the following *compatibility* condition:

$$T_h \in U(\Omega_j): \int_{\Omega_j} \nabla T_h v dV = 0, \forall v \in Q(\Omega_j) \quad \text{then } \nabla T_h = 0 \qquad (4.67)$$

To have a well-posed problem, the approximate solution to Equation 4.57 with $f = 0$ should be the trivial solution [20]. For this case, Equation 4.61 becomes

$$\int_\Omega |\mathbf{q}_h|^2 dV + \sum_{e \in E_{ih}} \int_e \left(C_{22}[\mathbf{q}_h]^2 + C_{11}[\underline{T_h}]^2 \right) dS + \int_{\partial\Omega} C_{11} T_h^2 dS = 0 \,(4.68)$$

which implies that $\mathbf{q}_h = 0$, $[\underline{T_h}] = 0$ on $\partial\Omega$, provided that $C_{11} > 0$. We can now rewrite the first equation defining the discontinuous method as follows:

$$\int_{\Omega_j} \nabla T_h v dV = 0, \qquad \forall v \in Q_h \qquad (4.69)$$

which, by the compatibility condition, implies that $\nabla T_h = 0$. Hence $T_h = 0$, i.e., the trivial solution.

Cockburn and Shu [6] further show that when all the local spaces contain the polynomials of degree k, the orders of convergence of the L^2-norms of the errors in $\mathbf{q}$ and T are k and $k+1$, respectively, when C_{11} is of order $O(h^{-1})$.

The discontinuous method presented above is locally conservative. To see that, we re-write the two integral equations as follows:

$$\int_{\Omega_j} \mathbf{q}_h \cdot \mathbf{w} dV = -\int_{\Omega_j} T_h \cdot \mathbf{w} dV + \int_{\partial\Omega_j} \hat{T}_h \mathbf{w} \cdot \mathbf{n}_j dS \qquad (4.70a)$$

$$\int_{\Omega_j} \mathbf{q}_h \cdot \nabla v dV = \int_{\Omega_j} f v dV + \int_{\partial\Omega_j} v \hat{\mathbf{q}}_h \cdot \mathbf{n}_j dS \qquad (4.70b)$$

for all $\mathbf{w}, v \in Q(\Omega_j) \times U(\Omega_j)$.

4.6.1.2 Discontinuous and Mixed Finite Element Formulations
Zienkiewicz *et al.* [21] argue that a discontinuous formulation is a mixed finite element method,

$$a(\mathbf{q}_h, v) + b(T_h, v) = 0 \,; \; -b(w, \mathbf{q}_w) + c(T_h, w) = F(w) \,, \; \forall \mathbf{w}, v \in Q_h \times U_h \qquad (4.71)$$

where Q_h and U_h are two finite element spaces,

$$Q_h = \{v : v \in Q(\Omega_j) \forall \Omega_j \in \Omega\} \; ; \; U_h = \{w : w \in U(\Omega_j) \forall \Omega_j \in \Omega\} \qquad (4.72)$$

and the bilinear forms are defined as follows:

$$a(\mathbf{q},\mathbf{r}) = \sum_j \int_{\Omega_j} \mathbf{q} \cdot \mathbf{w} dV + \int_{\partial \Omega_j} C_{22} \underline{[\mathbf{q}][\mathbf{w}]} dS \qquad (4.73a)$$

$$b(T,\mathbf{r}) = \sum_j \int_{\Omega_j} T \nabla \cdot \mathbf{w} dV - \int_{\partial \Omega_j} (\{T\} + \mathbf{C}_{12} \cdot \underline{[T][\mathbf{w}]} dS \qquad (4.73b)$$

$$c(T,v) = \sum_j \int_{\partial \Omega_j} C_{11} \underline{[T]} \cdot \underline{[v]} dS + \int_{\partial \Omega} C_{11} T v dS \qquad (4.73c)$$

$$F(\mathbf{r}) = \int_{\Omega} f v dV \qquad (4.73d)$$

where $\partial \Omega$ means integrating once along the boundary interface. As a consequence, the corresponding matrix equation may be arranged to take the following form, which we have seen in Section 4.2:

$$\begin{pmatrix} \mathbf{A} & \mathbf{B} \\ \mathbf{B} & \mathbf{C} \end{pmatrix} \begin{pmatrix} \underline{\mathbf{q}} \\ \underline{\mathbf{T}} \end{pmatrix} = \begin{pmatrix} \mathbf{0} \\ \mathbf{F} \end{pmatrix} \qquad (4.74)$$

This matrix form is typical of the stabilized mixed finite element methods [1, 21]; and the "stabilizing" form $c(\cdot,\cdot)$, usually associated with residuals, is introduced to meet the stability condition. Thus the relation between the discontinuous and the mixed finite element formulations is immediately clear. For discontinuous methods, the "stabilizing" form $c(\cdot,\cdot)$ solely depends on the parameter C_{11}, and the jumps across elements of the function in U_h. Thus, the discontinuous finite element formulation stabilizes the numerical scheme by *penalizing the jumps*, with C_{11} being the penalization parameter [5, 6, 21]. A detailed derivation of discontinuous formulation from the standpoint of the mixed finite element formulation is given in [21].

We note that the matrix form similar to Equation 4.72 also results from a mixed finite element formulation and the inf-sup (or *LBB*) condition needs to be satisifed to ensure stability, which in turn requires the use of non-equal order interpolation functions for the temperature and its fluxes [19, 22]. Different types of stabilization have been applied to circumvent the inf-sup condition so that equal order interpolation polynomials may be applied to approximate both the temperature and its gradient components [23, 24].

For discontinuous finite formulations, penalizing the jumps is equivalent to introducing stabilization by using residuals. To see that the residuals are related to the jumps, we set $\mathbf{R}_1 = \mathbf{q}_h - \nabla u_h$ and $\mathbf{R}_2 = \nabla \cdot \mathbf{q}_h - Q$ and use the weak formulation of the discontinuous method, and the numerical fluxes to get

$$\int_{\Omega_j} \mathbf{R}_1 \cdot \mathbf{w} dx = \int_{\partial\Omega_j} \left((-\tfrac{1}{2}\mathbf{n} - \mathbf{C}_{12}) \cdot [T_h] + C_{22}[\mathbf{q}_h] \right) \mathbf{w} \cdot \mathbf{n} \, dS \qquad (4.75a)$$

$$\int_{\Omega_j} R_2 v dx = \int_{\partial\Omega_j} \left((-\tfrac{1}{2}\mathbf{n} + \mathbf{C}_{12})[\mathbf{q}_h] + C_{11}[T_h] \right) \cdot \mathbf{n} v dS \qquad (4.75b)$$

for all $\mathbf{w}, v \in Q(\Omega_j) \times U(\Omega_j)$.

The above discussion indicates that the relaxation on the inter-element continuity allows a variety of different stabilization schemes to be developed. The available schemes published in the literature are all developed based on the stability requirements illustrated above. In Castillo *et al.* [2], a comprehensive study of these schemes is carried out in a unified approach. In Table 4.1, the function $a'([T_h])$ is a special stabilization term introduced by Bassi and Rebay [9] and later studied by Brezzi *et al.* [7].

As mentioned above, the steady state heat conduction problems fall into the category of elliptic problems, whose governing equations do not have real characteristics; thus there are no characteristic curves as there are for hyperbolic equations (or wave equations) to carry the data into a region from the boundary. Therefore boundary conditions must be imposed on an elliptic equation. In general, elliptic problems are more difficult than their hyperbolic and parabolic counterparts because a solution must exist over an entire domain, whereas for the latter, it still may be of interest to obtain a local solution in some small interval of time. These characteristic differences are also reflected in their stability results.

4.6.2 Stability for Time Dependent Problems

A numerical scheme for time dependent problems needs to satisfy the consistency and stability requirements in the same way as described for steady state problems. We present L^2-integral analysis, matrix analysis and Fourier analysis for time dependent problems.

4.6.2.1 Numerical Fluxes for Transient Problems
Here, we discuss the selection of numerical fluxes based on the L^2-integral analysis. For this, the procedure discussed in Section 4.6.1.1 is used to obtain the L^2-stability for transient heat conduction problems. That is, we first derive the L^2-stability results for the continuous problem, and then enforce them on the discretized equations. To do that, we consider a somewhat simpler form of the transient heat conduction equation with properties set to unity,

$$\frac{\partial T}{\partial t} = \nabla^2 T \qquad \in \Omega \qquad\qquad (4.76a)$$

$$T = 0 \qquad \in \partial\Omega \qquad\qquad (4.76b)$$

$$T(\mathbf{r},0) = T_0(\mathbf{r}) \qquad \in \Omega \tag{4.76c}$$

The above equation is multiplied by $T(\mathbf{r},t)$ and integrated in time and in space to give the following expression:

$$\int_\Omega \int_0^\tau T \frac{\partial T}{\partial t} dt\, dV = \int_0^\tau \int_\Omega T \nabla^2 T\, dV dt \tag{4.77}$$

Carrying out the integration, and substituting the definition $\mathbf{q} = \nabla T$, we have the final result for the L^2-stability,

$$\frac{1}{2}\int_\Omega T^2(\mathbf{r},\tau)\, dV + \int_0^\tau \int_\Omega |\,\mathbf{q}(\mathbf{r},t)\,|^2\, dV dt = \frac{1}{2}\int_\Omega T^2(\mathbf{r},0)\, dV \tag{4.78}$$

A procedure similar to that used for the stability analysis of steady state problems can then be applied. The resultant discretized solutions satisfy the following inequality:

$$\frac{1}{2}\int_\Omega T_h^2(\mathbf{r},\tau)\, dV + \int_0^\tau \int_\Omega \mathbf{q}_h^2(\mathbf{r},t) dV dt + \int_0^\tau \Theta_h dt$$

$$= \frac{1}{2}\int_\Omega T_h^2(r,0)\, dV \le \frac{1}{2}\int_\Omega T^2(r,0)\, dV \tag{4.79}$$

By forcing the third term on the right hand side to be non-negative, one can show that the stability of the scheme requires C_{11} and $\mathbf{C}_{12}$ to be non-negative. This procedure is similar to, but more involved than, that for the steady state case. This is detailed by Cockburn and Shu [6], who further proved that for transient heat conduction, $C_{11} = 0$ and $\mathbf{C}_{12} = \mathbf{0}$ can also be applied. This is different from the steady state case. Their analyses and numerical experiments indicate, however, that the order of the accuracy is sub-optimal for this choice.

Numerical fluxes that pass the stability requirement posed by the above equation are given in Table 4.2.

4.6.2.2 Stability Analysis Using Matrix

The L^2- integral analysis given above is closely related to Lyapunov's theory on the differential equations [18]. As a result, the matrix equations, resulting from the discontinuous finite element discretization, can also be used to carry out the stability analysis. To see that, we rewrite Equation 4.77 without time integration. The stability requirement is such that the solution of the resulting differential equation needs to meet the following condition if the solution is stable and unique [20]:

$$\frac{1}{2}\frac{d}{dt}\int_\Omega T^2 dV = \int_\Omega T \nabla^2 T\, dV \le 0 \tag{4.80}$$

The discretized form of the equation takes the following form:

$$\frac{d\mathbf{T}}{dt} = \mathbf{AT} \tag{4.81}$$

where $\mathbf{A}$ is an $N \times N$ matrix independent of t and $\mathbf{T}$ is the vector of the N-dimension. The solution is subject to the initial condition $\mathbf{T}_0$. Assuming that $\mathbf{A}$ has eigenvalues $\lambda_1, \lambda_2, \cdots, \lambda_N$, and C is a positive constant, i.e., $C > 0$, then we have the following stability theorem [25]:

a. If $Re(\lambda_k) < 0$, $k = 1, \cdots, N$, then for each $\mathbf{T}_0 \in R^N$ and suitably chosen positive constant μ, we have

$$\|\mathbf{T}(t)\| \leq C\|\mathbf{T}_0\| e^{-\mu t} \quad \text{and} \quad \lim_{t \to \infty} \mathbf{T}(t) = 0 \tag{4.82}$$

b. If $Re(\lambda_k) \leq 0$, $k = 1, \ldots, N$, where the eigenvalues with $Re(\lambda_k) = 0$ are distinct, then $\mathbf{T}(t)$ is bounded for $t \geq 0$,

$$\|\mathbf{T}(t)\| \leq C\|\mathbf{T}_0\| \tag{4.83}$$

c. If there exists an eigenvalue λ_k with $Re(\lambda_k) > 0$, then in each neighborhood of $\mathbf{T}_0 = 0$ there are initial values such that the corresponding solutions behave as follows:

$$\lim_{t \to \infty} \|\mathbf{T}(t)\| = +\infty \tag{4.84}$$

In case a, the solution $\mathbf{T}_0 = \mathbf{0}$ is exponentially stable, in case b $\mathbf{T}_0 = \mathbf{0}$ is Lyapunov-stable, and in case c, it is unstable. In all the above, $\|\cdot\|$ denotes the L^2-norm.

In addition to the above, we also have the following stability theorem that can be very useful in our analysis. For a matrix equation given below,

$$\frac{d\mathbf{T}}{dt} = \mathbf{AT} + \mathbf{B}(t)\mathbf{T} \tag{4.85}$$

the solution is considered Lyapunov stable (or L^2-stable), that is, for a constant C, it satisfies the following inequality,

$$\|\mathbf{T}(t)\| \leq C\|\mathbf{T}_0\| \tag{4.86}$$

if all the eigenvalues of $\mathbf{A}$ are such that $Re(\lambda_k) < 0$, $k = 1, \cdots, N$; and if $\int_0^\infty \|\mathbf{B}(t)\| \, dt \leq \infty$, that is, $\mathbf{B}(t)$ is bounded.

In particular, for an equation of the form,

$$\frac{d\mathbf{T}}{dt} = \mathbf{AT} + \mathbf{f}(t, \mathbf{T}) \tag{4.87}$$

the solution is L^2-stable, that is,

$$\|\mathbf{T}(t)\| \leq C\|\mathbf{T}_0\| \tag{4.88}$$

if all the eigenvalues of $\mathbf{A}$ are such that $Re(\lambda_k) < 0$, $k = 1, \cdots, N$, and if the following condition is met on $\mathbf{f}(t, \mathbf{T})$:

$$\lim_{\|T\| \to 0} \frac{\|\mathbf{f}(t, \mathbf{T})\|}{\|\mathbf{T}\|} = 0 \text{ uniformly in } t \tag{4.89}$$

which means that $\|\mathbf{f}(t, \mathbf{T})\| \leq c\|\mathbf{T}\|$, with c being a positive constant [26].

The above theorem provides the basis for the Nuemann stability analysis, which is often employed to determine the time steps for transient calculations. The theorem also will be used in later chapters for stability analysis.

4.6.3 Fourier Analysis

Let us consider the Fourier analysis of the error and stability of the numerical schemes for the solution of transient problems. This type of analysis is based on Neumann's spectrum analysis and gives the criterion for critical time steps required for the solution. This can be particularly useful for the discontinuous finite element methods in that the discontinuous schemes often use explicit methods for time marching. As usual, the analysis is applied to a 1-D problem,

$$\frac{\partial u}{\partial t} = D\frac{\partial^2 u}{\partial x^2}, \qquad\qquad x \in [a, b] \tag{4.90}$$

with periodic conditions and the initial condition $u(x,0) = \sin(x)$ and D denotes the thermal diffusivity.

The above diffusion equation can be formulated using the discontinuous schemes presented in Section 4.3.1. The stability analysis of these schemes has been made recently by Zhang and Shu using the Fourier expansion technique [4].

Here we perform the stability analysis of different, but more general, forms of the discontinuous finite element formulation using the Neumann stability analysis method. As discussed in Chapter 2, discontinuous formulations can come in various forms. The various forms presented so far have been based on the unified numerical flux approach. Here, as a variation, we consider the formulation based on the double integration approach or through weakly imposing the cross-element continuity (see Section 2.1.1).

Towards this end, we split the governing equation above into two first order differential equations and integrate them over element $j \in [x_j, x_{j+1}]$ with respect to weighting functions,

$$\frac{\partial T}{\partial t} + D\frac{\partial q}{\partial x} = 0 \; ; \qquad\qquad q + \frac{\partial T}{\partial x} = 0 \qquad (4.91)$$

$$\int_{x_j}^{x_{j+1}} \left[\frac{\partial T}{\partial t} + D\frac{\partial q}{\partial x}\right] v(x)\,dx = 0 \; ; \quad \int_{x_j}^{x_{j+1}} \left[q + \frac{\partial T}{\partial x}\right] w(x)\,dx = 0 \qquad (4.92)$$

We integrate by parts twice and apply the appropriate upwinding scheme at the element boundaries to obtain the following expressions:

$$\int_{x_j}^{x_{j+1}} v_h \left(\frac{\partial T_h}{\partial x} + q_h\right) dx + \alpha_1 [T]_j\, v_h(x_j) + (1-\alpha_1)[T]_{j+1}\, v_h(x_{j+1}) = 0 \quad (4.93a)$$

$$\int_{x_j}^{x_{j+1}} v_h \left(\frac{\partial T_h}{\partial t} + D\frac{\partial q_h}{\partial x}\right) dx$$

$$+\alpha_2 D[q]_j\, v_h(x_j) + (1-\alpha_2)D[q]_{j+1}\, v_h(x_{j+1}) = 0 \qquad (4.93b)$$

Here, values of the constant α_1 and α_2 are chosen to be 1 or ½, which correspond to different schemes of flux calculations. That is, it is the full upwind scheme if $\alpha_1 = \alpha_2 = 1$, the central scheme if $\alpha_1 = \alpha_2 = $ ½, and the alternating upwinding if $\alpha_1 = 0$ and $\alpha_2 = 1$ (or $\alpha_1 = 1$ and $\alpha_2 = 0$). Also, $[\bullet]_j$ denotes the jump condition at the element boundaries,

$$[q]_j = q_j^+ - q_j^- \; ; \quad [T]_j = T_j^+ - T_j^- \qquad (4.94)$$

To demonstrate that the above formulation is consistent with the alternating upwinding, we note that in the double integration approach, the values of q across one boundary of the element are the same, and reflects a jump condition on $[T]$ at the same boundary. We do just the opposite to the other boundary of the element.

The discontinuous Galerkin solution procedure for Equation 4.88 yields the matrix equations,

$$\sum_{m=1}^{N_e} S_{(j)}^{(km)} q_{(j)}^{(m)} + \sum_{m=1}^{l} L_{(j)}^{(km)} T_{(j)}^{(m)}$$

$$+\alpha_1 [T]_j\, \phi_k(x_j) + (1-\alpha_1)[T]_{j+1}\, \phi_k(x_{j+1}) = 0 \qquad (4.95a)$$

$$\sum_{m=1}^{N_e} S_{(j)}^{(km)} \dot{T}_{(j)}^{(m)} + D\sum_{m=1}^{Ne} L_{(j)}^{(km)} q_{(j)}^{(m)}$$

$$+\alpha_2 D[q]_j \phi_k(x_j) + (1-\alpha_2) D[q]_{j+1} \phi_k(x_{j+1}) = 0 \qquad (4.95b)$$

for $j = 1, 2, \cdots, N$, N being the number of elements, and $k = 1, 2, \cdots, N_e$, N_e being the number of nodes of element j. Also, subscript (m) on q and T with $m = 1, 2, \ldots,$ Ne refers to the node number local to the element and $S_{(j)}^{(km)}$ and $L_{(j)}^{(km)}$ are matrices calculated by

$$S_{(j)}^{(km)} = \int_{x_j}^{x_{j+1}} \phi_k \phi_m dx \quad \text{and} \quad L_{(j)}^{(km)} = \int_{x_j}^{x_{j+1}} \phi_k \frac{d\phi_m}{dx} dx \qquad (4.96)$$

where $\phi_k(x)$ is the shape function. Since $\mathbf{q} = (q_1, q_2, \ldots, q_{Ne})^T$ is an auxiliary variable, it is eliminated and the actual scheme for $\{T\}$ takes a form similar to a continuous finite element scheme,

$$\dot{\mathbf{T}}_{(j)} = \mathbf{A}\underline{T}_{(j-2)} + \mathbf{B}\underline{T}_{(j-1)} + \mathbf{C}\underline{T}_{(j)} + \mathbf{D}\underline{T}_{(j+1)} + \mathbf{E}\underline{T}_{(j+2)} \qquad (4.97)$$

where $\mathbf{A}, \mathbf{B}, \mathbf{C}, \mathbf{D},$ and $\mathbf{E}$ denote the coefficient matrices and again the subscript $()$ refers to the element. The time step is selected based on the criterion for the numerical scheme to avoid numerical instability. We now perfom the linear stability analysis for the discontinuous finite element formulations with the Runge–Kutta time integration schemes for time marching.

By Neumann's theory, the round-off error may be expanded in a Fourier series,

$$\varepsilon(x,t) = T(x,t) - T_{exact}(x,t) = \sum_{m=-\infty}^{\infty} \varepsilon_m(t) e^{ik_m x} = \sum_{m=-\infty}^{\infty} e^{at} e^{ik_m x} \qquad (4.98)$$

where $T(x,t)$ is the numerical solution to Equation 4.90.

It is often sufficient to consider a typical component of the series at a nodal point j,

$$\varepsilon_j^n = e^{at} e^{ik_m x_j} = e^{at_n} e^{ik_m jh} \qquad (4.99)$$

where $h = \Delta x$, $x_j = jh$, and n is the nth time step. To simplify the notation, we will set $k = k_m$. The amplification factor is defined by

$$G = \frac{\varepsilon_j^{n+1}}{\varepsilon_j^n} = \frac{e^{a(t_n + \Delta t)} e^{ikjh}}{e^{at_n} e^{ikjh}} = e^{a\Delta t} \qquad (4.100)$$

To prevent the error from growing in time, it is required that

$$|G| \le 1 \quad \text{or} \quad Re(a) \le 0 \qquad (4.101)$$

This concept applies to any component of a polynomial of order l defined on an element,

$$\boldsymbol{\varepsilon}_j(t) = \begin{pmatrix} \varepsilon_{(j)}^{(0)}(t) \\ \vdots \\ \varepsilon_{(j)}^{(l)}(t) \end{pmatrix} e^{ikjh} \qquad (4.102)$$

where as usual the subscript () refers to the element.

Setting $\mathbf{T} = \boldsymbol{\varepsilon}$ in Equation 4.97, one has the matrix equation for errors,

$$\dot{\boldsymbol{\varepsilon}}_{(j)} = \mathbf{A}\boldsymbol{\varepsilon}_{(j-2)} + \mathbf{B}\boldsymbol{\varepsilon}_{(j-1)} + \mathbf{C}\boldsymbol{\varepsilon}_{(j)} + \mathbf{D}\boldsymbol{\varepsilon}_{(j+1)} + \mathbf{E}\boldsymbol{\varepsilon}_{(j+2)}$$

$$= (\mathbf{A}e^{-i2kh} + \mathbf{B}e^{-ikh} + \mathbf{C} + \mathbf{D}e^{ikh} + \mathbf{E}e^{i2kh})\boldsymbol{\varepsilon}_{(j)}$$

$$= \underline{\mathbf{C}}(k,h)\,\boldsymbol{\varepsilon}_{(j)} \qquad (4.103)$$

Clearly, the resultant matrix $\underline{\mathbf{C}}(k,h)$ plays the role of amplification factor. One is then able to use the theory of matrix stability to perform the L^2-stability analysis of the numerical scheme ([4], also Section 4.6.2.2). Using Equation 4.103, we are also able to determine the critical time step for a particular time integration scheme.

We illustrate the use of Equations 4.97 and 4.103 to study the numerical stability of the scheme, and to determine the time step for the explicit Runge–Kutta integration scheme of up to the third order for a piecewise constant approximation. Setting $l = 0$, Equation 4.97 takes the following form:

$$\dot{T}_j = \frac{D}{h^2}(\alpha_1\alpha_2 T_{j-2} + (\alpha_1 + \alpha_2 - 4\alpha_1\alpha_2)T_{j-1} + (1 - 3\alpha_1 - 3\alpha_2 + 6\alpha_1\alpha_2)T_j$$

$$+ (-2 + 3\alpha_1 + 3\alpha_2 - 4\alpha_1\alpha_2)T_{j+1} + (1 - \alpha_1)(1 - \alpha_2) \qquad (4.104)$$

Thus, we have the differential equation for the error,

$$\dot{\varepsilon}_j = \underline{C}(k,h)\varepsilon_j \qquad (4.105)$$

where the amplification matrix has only one term, $\underline{C}(k,h) = (D/h^2)C$, and C is given by

$$C = \alpha_1\alpha_2 e^{-2ikh} + (\alpha_1 + \alpha_2 - 4\alpha_1\alpha_2)e^{-ikh} + (1 - 3\alpha_1 - 3\alpha_2 + 6\alpha_1\alpha_2)$$

$$+ (-2 + 3\alpha_1 + 3\alpha_2 - 4\alpha_1\alpha_2)e^{ikh} + (1 - \alpha_1)(1 - \alpha_2)e^{2ikh} \qquad (4.106)$$

From the theory of stability presented in the last section, the coefficient $Re(C)$ needs to be negative to ensure that Equation 4.104 has a stable solution. For example, if $\alpha_1 = \alpha_2 = 0$ is taken, $C = (1 - e^{ikh})^2 \geq 0$; and thus the solution will be unstable. This is confirmed by numerical experiments, as shown in Table 4.3a. Similarly, if $\alpha_1 = \alpha_2 = 1$ is taken, $C = (1 - e^{-ikh})^2 \geq 0$.

We now obtain information on the time step required for a stable solution. Let us begin by considering the first order Runge–Kutta scheme, which is the Euler forward method, for time integration. By this scheme, we have from Equation 4.99

$$\frac{\varepsilon_j^{n+1} - \varepsilon_j^n}{\Delta t} = \frac{D}{h^2} C \varepsilon_j^n \tag{4.107}$$

Other discretization using the Runge–Kutta schemes can be obtained similarly. With the definition of the amplification factor, $G = \varepsilon_j^{n+1} / \varepsilon_j^n = e^{a\Delta t}$, these discretizations yield the following relations:

$$G = 1 + z \quad \text{for first order (Euler forward)} \tag{4.108a}$$

$$G = 1 + z + \tfrac{1}{2} z^2 \quad \text{for second order} \tag{4.108b}$$

$$G = 1 + z + \tfrac{1}{2} z^2 + \tfrac{1}{6} z^3 \quad \text{for third order} \tag{4.108c}$$

where $z = \lambda D \Delta t / h^2$. Here, λ is the eigenvalue of the matrix C.

To avoid numerical instability, that is, to make the error smaller as time marching continues, it is required that $|G| \leq 1$. Therefore, the stability condition for the discontinuous finite element scheme, with the Runge–Kutta time integration schemes of difference order, is such that the time step Δt is chosen to satisfy

$$r = \frac{D \Delta t}{h^2} \leq f \tag{4.109}$$

where f is determined by the most critical eigenvalue of C, which satisfies $|G| \leq 1$. The results using Equations 4.108 and 4.109 are given in Table 4.3.

Table 4.3a. Stability criterion f for the first order Runge–Kutta scheme

α_2 \ α_1	0	0.5	1.0
0	0.0	2.0	0.5
0.5	2.0	2.0	2.0
1.0	0.5	2.0	0.0

Note: $f = 0$ means that the scheme is unconditionally unstable.

Table 4.3b. Stability criterion f for the second order Runge–Kutta scheme

α_2 \ α_1	0	0.5	1.0
0	0.0	1.3333	0.5
0.5	1.3333	2.0	1.3333
1.0	0.5	1.3333	0.0

Note: $f = 0$ means that the scheme is unconditionally unstable.

Table 4.3c. Stability criterion f for the third order Runge–Kutta scheme

α_2 \ α_1	0	0.5	1.0
0	0.0	1.6	0.6282
0.5	1.6	2.5128	1.6
1.0	0.6282	1.6	0.0

Note: $f = 0$ means that the scheme is unconditionally unstable.

Exercises

1. Consider the problem defined below,

$$\frac{\partial T}{\partial t} = D\frac{\partial^2 T}{\partial x^2} + g(T) \qquad \in [0,1], \; t > 0$$

$$T(x,0) = T_0(x) \qquad \in [0,1]$$

$$\frac{\partial T(0,t)}{\partial x} = \frac{\partial T(1,t)}{\partial x} = 0 \qquad t > 0$$

with $D > 0$ and $\sup|dg(T)/dT| = M < \infty$. Defining the energy and its gradient at time t as

$$E(t) = \int_0^1 T^2(x,t)dx \; ; \qquad F(t) = \int_0^1 \left(\frac{\partial T(x,t)}{\partial x}\right)^2 dx$$

Show that

$$\frac{dF(t)}{dt} \leq -2(M - \pi^2)\int_0^1 \left(\frac{dT}{dt}\right)^2 dx$$

and further show that if $M < \pi^2$,

$$\lim_{t \to \infty} F(t) = \lim_{t \to \infty} \int_0^1 \left(\frac{\partial T(x,t)}{\partial x}\right)^2 dx = 0$$

Hint: to prove the above relations, the Poincaré inequality may prove convenient. The Poincaré inequality is stated as follows:

$$\int_0^1 \left(\frac{du(x)}{dx}\right)^2 dx \geq \pi^2 \int_0^1 u(x)^2 dx, \; \text{if} \; u(0) = u(1) = 0$$

$$\int_0^1 \left(\frac{d^2 u(x)}{dx^2}\right)^2 dx \geq \pi^2 \int_0^1 \left(\frac{du(x)}{dx}\right)^2 dx, \quad \text{if} \quad \frac{du(0)}{dx} = \frac{du(1)}{dx} = 0$$

where u is a function that is continuously difrferentiable.

2. Consider a problem given by the following set of equations,

$$\frac{\partial T}{\partial t} = D\frac{\partial^2 T}{\partial x^2} + g(T), \quad \in [0,1],\ t > 0$$

$$T(x,0) = T_0(x), \quad \in [0,1]$$

$$T(0,t) = T(0,t) = 0, \quad t > 0$$

with $D > 0$ and $\sup|dg(T)/dT| = \pi^2 < \infty$. Show that if the solution to the above problem if unique, then

$$\lim_{t\to\infty} E(t) = \lim_{t\to\infty} \int_0^1 T^2(x,t)dx = 0$$

3. A wall 0.12 m thick having a thermal diffusivity of 1.5×10^{-6} m²/s is initially at a uniform temperature of 85°C. Suddenly one face is lowered to a temperature of 20°C, while the other face is perfectly insulated. (a) Using the explicit discontinuous finite element technique with space and time increments of 30 mm and 300 s, respectively, determine the temperature distribution at $t = 45$ min. (b) With $\Delta x = 30$ mm and $\Delta t = 300$ s, compute $T(x,t)$ for $0\leq t\leq t_{ss}$, where t_{ss} is the time required for the temperature at each nodal point to reach a value that is within 1°C of the steady state temperature. Repeat the foregoing calculation for $\Delta t = 75$ s. For each value of Δt, plot temperature histories for each face and the mid-plane.

4. Consider steady state heat conduction in a square region of side a (Figure 4.1p). Assume that the medium has a thermal conductivity of $k = 30$ W/(m K) and a uniform heat generation of $Q_0 = 10^7$ W/m³. For the boundary conditions shown in Figure 4.1p, form the discontinuous finite element

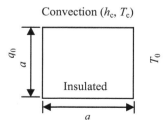

Figure 4.1p.

matrix equations for two different discretizations: (a) the domain discretized by one quadrilateral element and (b) the domain is discretized into two triangles. Compare the results obtained by these two different discretizations. Take h_c = 60 W/(m^2 °C), T_c = 0.0°C, T_0 = 100°C, q_0 = 2×10^5 W/m^2, and a = 1 cm.

5. Consider a steady state heat transfer in a 2-D fin. The fin shown in Figure 4.2p has its base maintained at 300°C and is exposed to convection on its remaining boundary. Write a discontinuous finite element formulation and develop a computer code to solve the problem. Use quadrilateral elements to calculate the temperature distribution. Take h_c = 40 W/(m^2 K), T_c =20°C, and k = 5 W/(m K). Compare the results using constant, linear and quadratic elements. Compare the results obtained using different numerical fluxes.

6. For Problem 5 above, develop a transient discontinuous finite element program and solve the temperature distribution history, assuming $T(x,t=0)$ = 300°C.

7. Consider the square channel shown in the sketch (Figure 4.3p) operating under steady state conditions. The inner surface of the channel is at a uniform temperature of 600 K, while the outer surface is exposed to convection with a fluid at 300 K and a convection coefficient of 50 W/m^2 K. From a symmetrical element of the channel, a two-dimensional grid has been constructed and the nodes labeled. The temperatures for nodes 1, 3, 6, 8, and 9 are identified.

Convection (h_c, T_c)

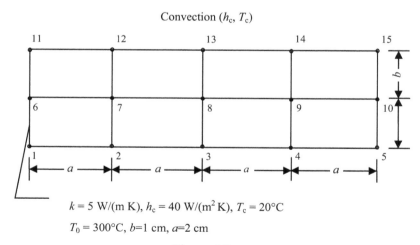

k = 5 W/(m K), h_c = 40 W/(m^2 K), T_c = 20°C

T_0 = 300°C, b=1 cm, a=2 cm

Figure 4.2p.

(a) Beginning with properly defined control volumes, and following Example 4.3, derive a discontinuous finite element code for the solution of the problem using a triangular mesh with nodes shown in Figure 4.3p and determine the temperature T_2, T_4 and T_7 (K).

(b) Calculate the heat loss per unit length from the channel.

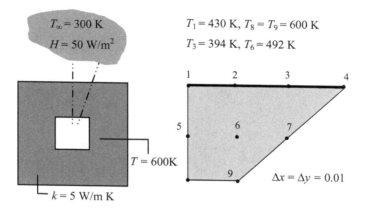

Figure 4.3p.

8. Apply the Fourier analysis to determine the critical time step for a 1-D transient heat conduction with the Runge–Kutta scheme of second order for time integration and with linear elements for spatial discretization.

References

[1]   Zienkiewicz OC, Taylor RL. The Finite Element Method, 4th Ed, Volume 1. New York: McGraw-Hill, 1989.

[2]   Castillo P, Cockburn B, Perugia I, Schotzau D. An a priori Error Analysis of the Local Discontinuous Galerkin Method for Elliptic Problems. SIAM J. Numer. Anal. 2000; 38: 1676–1706.

[3]   Arnold D, Brezzi F, Cockburn B, Marini D. Unified Analysis of Discontinuous Galerkin Methods for Elliptic Problems. SIAM J. Numer. Anal. 2001; 39: 1749–1779.

[4]   Zhang M, Shu CW. An Analysis of Three Different Formulations of the Discontinuous Galerkin Method for Diffusion Equations. Math. Models Meth. Appl. Sci. 2003; 13: 395–413.

[5]   Cockburn B. Discontinuous Galerkin Methods. School of Mathematics, University of Minnesota, 2003; 1–25.

[6]   Cockburn B, Shu CW. The Local Discontinuous Galerkin Method for Time-Dependent Convection-Diffusion Systems. SIAM J. Numer. Anal. 1998; 35: 2440–2463.

[7]   Brezzi F, Manzini G, Marini D, Pietra P, Russo A. Discontinuous Galerkin Approximations for Elliptic Problems. Numer. Methods for Partial Differ. Equ. 2000; 16: 365–378.

[8]   Douglas J, DuPont T. Interior Penalty Procedures for Elliptic and Parabolic Galerkin Methods. Lecture Notes in Physics, 58. Berlin: Springer-Verlag, 1976.

[9]   Bassi F, Rebay S, Mariotti G, Pedinotti S, Savini M. A High-Order Accurate Discontinuous Finite Element Method for Inviscid and Viscous Turbomachinery Flows. In: Decuypere R, Dibelius G, editors. 2nd European Conference on Turbomachinery Fluid Dynamics and Thermodynamics. Belgium: Antwerpen Technologisch Institut, 1997; 5–7: 99–108.

[10] Riviere B, Wheeler M, Girault V. Improved Energy Estimates for Interior Penalty, Constrained and Discontinuous Galerkin Methods for Elliptic Problems. Part I. Comp. Geo. 1999; 3: 337–360.

[11] Shu CW. Different Formulations of the Discontinuous Galerkin Method for the Viscous Terms. In: Shi ZC, Mu M, Xue W, Zou J, editors. Advances in Scientific Computing. Beijing: Science Press, 2001: 144–155.

[12] Bassi F, Rebay S. A High-Order Accurate Discontinuous Finite Element Method for the Numerical Solution of the Compressible Navier–Stokes Equations. J. Comput. Phys. 1997; 131: 267–279.

[13] Kirby RM, Karniadakis GE. Selecting the Numerical Flux in Discontinuous Galerkin Methods for Diffusion Problems. J. Sci. Comp. 2005; 22(1): 385–411.

[14] Babuska I, Zl´amal M. Nonconforming Elements in the Finite Element Method with Penalty. SIAM J. Numer. Anal. 1973; 10: 863–875.

[15] Brezzi F, Fortin M. Mixed and Hybrid Finite Element Methods. New York: Springer, 1991.

[16] Baumann C, Oden J. A Discontinuous hp Finite Element Method for Convection-Diffusion Problems. Comput. Meth. Appl. Mech. Eng. 1999; 175: 311–341.

[17] Riviére, B. Analysis of a Discontinuous Finite Element Method for the Coupled Stokes and Darcy Problems. J. Sci. Comp. 2005; 22(1): 479–500.

[18] Brezzi F, Hughes TJR, Marini LD. Mixed Discontinuous Galerkin Methods for Darcy Flow. J. Sci. Comp. 2005; 22(1): 119–145.

[19] Bathe KJ. Finite Element Procedures. Englewood Cliffs: Prentice Hall, 1996.

[20] Evans LC. Partial Differential Equations. Philadelphia: American Society of Mathematics, 2000.

[21] Zienkiewicz OC, Taylor RL, Sherwin SJ, Peiro J. On Discontinuous Galerkin Methods. Int. J. Numer. Meth. Eng. 2003; 58: 1119–1148.

[22] Chapelle D, Bathe KJ. The Finite Element Analysis of Shells – Fundamentals. New York: Springer, 2003.

[23] Franca LP, Hughes TJR, Loula AFD, Miranda I. A New Family of Stable Elements for Nearly Incompressible Elasticity Based on a Mixed Petrov–Galerkin Finite Element Method. Numer. Math. 1988; 53: 123–141.

[24] Hughes TJR. Multiscale Phenomena: Green's Functions, the Dirichlet-to-Neumann Formulation, Subgrid Scale Models, Bubbles and the Origins of Stabilized Methods. Comp. Meth. Appl. Mech. Engrg. 1995; 127: 387–401.

[25] Verhust F. Nonlinear Differential Equations and Dynamical Systems. New York: Springer, 1991.

[26] Brauer R, Nohel JA. The Qualitative Theory of Ordinary Differential Equations. New York: Dover Publications, 1989.

5

Convection-dominated Problems

Convection is the second mode for heat transfer. It plays a dominant role in determining the overall behavior of the fluid flow field and the redistribution of thermal energy in a wide range of thermal and fluids systems. The mathematical description of convection problems involves first order derivatives in the spatial coordinates, which differs from pure conduction, where the first order derivatives do not exist. These first order derivative terms are often the origin of spurious oscillations that pollute the numerical solution, regardless of the types of numerical techniques used. This is particularly true whenever there is not enough diffusion in the system. Many techniques have been designed to suppress these oscillations, and the discontinuous finite elements provide a natural formulation to minimize the oscillations while maintaining a high accuracy. The purpose of this chapter is to discuss the application of the discontinuous finite element method to the solution of the convection and convection-diffusion problems.

The chapter starts with pure convection and then moves on to study the convection-diffusion problems. Various discontinuous formulations and their numerical implementation for these problems are presented. An important issue concerning the effective use of the discontinuous finite element methods is the choice of effective numerical fluxes. The selection of these fluxes is discussed in detail and the L^2-stability analyses used for selecting these numerical fluxes are also presented for these problems. For transient problems, the von Neumann analysis is presented, which is a powerful tool for determining the critical time steps, or the so-called *CFL* conditions, for explicit time marching solutions using the discontinuous formulations. The subject of non-physical oscillation is discussed in detail for 1-D steady state convection-diffusion problems, and its origin is investigated in terms of the eigenvalues of the resultant matrix. Often in the study of convection problems of practical importance, the convective terms are nonlinear in nature. The subject of nonlinear convection is discussed in the context of Burgers' problems. Various numerical algorithms, designed to minimize the oscillations associated with the discontinuous finite element solution of the Burgers' equations, are presented. These algorithms include the *TVD* scheme and its variants for higher order approximations, along with the appropriate use of slope limiters or flux limiters.

5.1 Pure Convection Problems

Pure convection problems are idealized systems where viscous effects are neglected. Without viscosity, the system develops sharp fronts and discontinuities in the mathematical solution. In the real world, thermal fluids systems in general are dissipative, meaning that the viscosity, however small it is, is present and plays an important role in smoothing out the sharp fronts. Nonetheless, a pure convection problem offers a system for analysis, and for the understanding of the nature of convection problems. In this section, we first develop a discontinuous formulation for a 1-D pure convection equation, and then generalize the formulation for multidimensional problems.

5.1.1 1-D Pure Convection

For the purpose of understanding the nature of this type of problem, the method of characteristics is discussed before the discontinuous finite element formulation is presented.

5.1.1.1 Method of Characteristics
The following linear partial differential equation describes the pure convection effect on the temperature field $T(x,t)$:

$$\frac{\partial T}{\partial t} + u \frac{\partial T}{\partial x} = 0 \tag{5.1}$$

where u is the convection velocity and is taken to be a positive constant for the sake of simplicity. We consider a particular curve $x = x(t)$ in the $x-t$ plane. Then the total derivative of $T(x,t)$ along the curve is governed by the chain rule,

$$\frac{dT(x(t),t)}{dt} = \frac{\partial T(x(t),t)}{\partial t} + \frac{\partial T(x(t),t)}{\partial x} \frac{dx(t)}{dt} = 0 \tag{5.2}$$

Comparison with Equation 5.1 indicates that the curve has a characteristic of $dx(t)/dt = u$. Thus we have the solution for the temperature $T(x, t)$,

$$\frac{dT}{dt} = 0 \text{ along the characteristic curve } \frac{dx(t)}{dt} = u \tag{5.3}$$

Integrating the above equation yields the following results:

$$T(x,t) = \text{const} ; \quad x(t) = ut + x_0 \tag{5.4}$$

where $x_0 = x(0)$. Since T is constant along the characteristic curve, the solution may be written as

$$T(x,t) = T(x_0,0) = T(x - ut) \tag{5.5}$$

We see that $T(x, t)$ remains a constant along the curves $x(t) - ut = x_0$, which are called characteristic curves. The constant x_0 is a parameter. The solution $T(x,t)$ carries the initial data $T(x_0, 0)$ at the boundary into the x–t domain. A set of characteristic curves in the x–t plane is called the characteristic diagram. Figures 5.1(a) and (b) show the characteristic curves in the x–t plane.

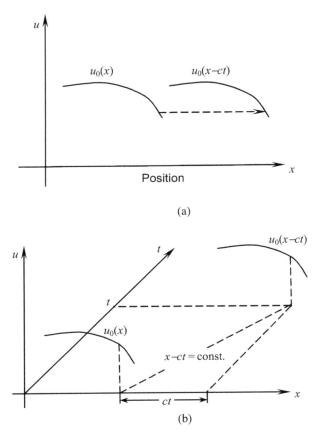

Figure 5.1. Characteristic diagrams for pure convection problems: (a) 2-D view and (b) 3-D view

In the fluids literature, the above solution method is called the method of characteristics [1, 2]. The method can also be employed to obtain a solution to a system of equations. Let us consider such a system,

$$\frac{\partial \mathbf{T}}{\partial t} + \mathbf{A}\frac{\partial \mathbf{T}}{\partial x} = \mathbf{B} \tag{5.6}$$

where $\mathbf{A}$ is an $N \times N$ matrix, $\mathbf{T} = [T_1, T_2, ..., T_N]^T$ is a vector of dimension N and $\mathbf{B}$ $= [B_1, B_2, ..., B_N]^T$ is the source vector of dimension N. The superscript T denotes the transpose. If the above equation is to be written in the form of Equation 5.1, then we need to first find the eigenvalues λ of matrix $\mathbf{A}$,

$$\mathbf{Aa} = \lambda\mathbf{Ia} \tag{5.7}$$

where $\mathbf{a}$ is an arbitrary vector of dimension N and $\mathbf{I}$ is the identity matrix. For a pure convection (or hyperbolic) problem, the eigenvalues are distinct and the following eigenvalue–eigenvector relation exists:

$$\mathbf{AP} = \mathbf{P\Lambda} \tag{5.8}$$

where Λ is a diagonal matrix with terms on the diagonal being the eigenvalues, Λ_{ij} $= \lambda_j \delta_{ij}$. $\mathbf{P}$ is the eigenvector matrix, $\mathbf{P} = [\boldsymbol{p}_1, \boldsymbol{p}_2, ..., \boldsymbol{p}_N]$, where $\boldsymbol{p}_i$ of dimension N being the eigenvector corresponding to the eigenvalue λ_i.

Premultiplying Equation 5.6 by $\mathbf{P}^{-1}$ gives

$$\mathbf{P}^{-1}\frac{\partial \mathbf{T}}{\partial t} + \mathbf{P}^{-1}\mathbf{A}\frac{\partial \mathbf{T}}{\partial x} = \mathbf{P}^{-1}\frac{\partial \mathbf{T}}{\partial t} + \mathbf{\Lambda P}^{-1}\frac{\partial \mathbf{T}}{\partial x} = \mathbf{P}^{-1}\mathbf{B} \tag{5.9}$$

and defining two new vectors $\mathbf{w}$ and $\mathbf{g}$, both of dimension N,

$$\mathbf{w} = \mathbf{P}^{-1}\mathbf{T}\,;\ \mathbf{g} = \mathbf{P}^{-1}\mathbf{B} \tag{5.10}$$

we can write Equation 5.9 in the following form:

$$\frac{\partial \mathbf{w}}{\partial t} + \mathbf{\Lambda}\frac{\partial \mathbf{w}}{\partial x} = \mathbf{g} \tag{5.11}$$

In component form, Equation 5.11 becomes

$$\frac{\partial w_i}{\partial t} + \lambda_i \frac{\partial w_i}{\partial x} = g_i\,, \qquad i = 1, 2, ..., N \tag{5.12}$$

Following the steps leading from Equation 5.1 to Equation 5.3, we identify the characteristic curves for the system of equations as

$$\frac{dx}{\partial t} = \lambda_i\,, \quad i = 1, 2, ..., N \tag{5.13a}$$

along which w_i is determined by the equation,

$$\frac{dw_i}{dt} = g_i, \quad i = 1, 2, ..., N \tag{5.13b}$$

With **w** so obtained, **T** is calculated by

$$\mathbf{T} = \mathbf{Pw} \tag{5.14}$$

It is noted that the above equation is derived based on the assumption that **A** is a constant. The same procedure also applies when $\mathbf{A} = \mathbf{A}(x, t, T_1, T_2, ..., T_N)$. In this case, one can show that the above equations remain the same, except that **g** will be replaced by **g'**

$$\mathbf{g}' = \mathbf{g} + [\partial(\mathbf{P}^{-1})/\partial t + [\Lambda]\partial(\mathbf{P}^{-1})/\partial x]\mathbf{Pw} \tag{5.15}$$

The solution procedure remains the same as above and thus needs no elaboration. Let us now discuss an example of using the method of characteristics for the solution of a pure convection problem.

Example 5.1. Obtain the solution of the temperature distribution governed by the pure convection equation,

$$\frac{\partial T}{\partial t} + u\frac{\partial T}{\partial x} = 0 \quad \in [-\infty, \infty] \times [0, \infty] \tag{5.1e}$$

$$T(x, t = 0) = T_0(x) \quad x \in [-\infty, \infty] \tag{5.2e}$$

where u is the known velocity, which is taken to be a positive constant here, and T is the temperature.

Solution. The general solution of the above equation can be obtained using the method of characteristics,

$$T(x - ut) = T(x, ut) = T_0(x - ut) \tag{5.3e}$$

The solution represents a wave pack propagating with speed u. The solution (x_1, t_1) is then related to $t = 0$ in the following way:

$$T(x_1, ut_1) = T_0(x_1 - ut_1) = T(x_1 - ut_1, 0) \tag{5.4e}$$

For this type of problem, the solution is a set of straight lines in the x–t plane originating from the boundary of $t=0$, and therefore depends strongly on the initial data. Two of these straight lines are plotted in Figure 5.1e. Both of the lines have the same slope, $1/u$. Also, the solution carries the data from the boundary directly to the x–t domain. Moreover, along the characteristics, $T(x,t)$ remains constant, and is determined by its value at the boundary.

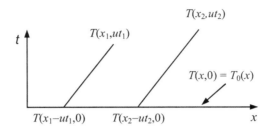

Figure 5.1e. Illustration of the analytic solution of the temperature field

5.1.1.2 Discontinuous Finite Element Formulation
Having understood the basic nature of pure convection problems, let us now turn our attention to the discontinuous finite element solution of the 1-D problem given by Equation 5.1. As usual, we start with discretizing the domain into a collection of N elements, and then integrating the above equation over element $j \in [x_j, x_{j+1}]$ with respect to a weighting function $v(x)$,

$$\int_{x_j}^{x_{j+1}} \left(v\frac{\partial T_h}{\partial t} - uT_h \frac{\partial v}{\partial x} \right) dx + v(x_{j+1})\overleftrightarrow{uT}(T_{j+1}^-, T_{j+1}^+)$$

$$- v(x_j)\overleftrightarrow{uT}(T_j^-, T_j^+) = 0 \qquad (5.16)$$

where we have replaced the convective temperature values at the element boundaries by the numerical flux expressions (see Figure 5.2 for geometric definitions). For a constant u or a linear problem, an effective numerical flux is given by the Lax–Friedrichs flux:

$$\overleftrightarrow{uT}(a,b) = u\frac{a+b}{2} - |u|\frac{b-a}{2} \qquad (5.17)$$

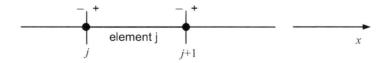

Figure 5.2. Illustration of boundary interfacial quantities for element j

which in essence is an upwinding scheme. The unknowns T_h may be approximated with a polynomial of order $k = N_e - 1$ as local basis functions,

$$T_h(x,t) = \sum_{i=1}^{N_e} T_i(t)\phi_i(x) \tag{5.18}$$

with N_e being the number of nodes associated with an element.

Substituting the above equations into Equation 5.16 and taking the Galerkin approximation, followed by numerical integration, we have the following matrix equation:

$$\mathbf{M}\frac{d\mathbf{U}_{(j)}}{dt} + \mathbf{K}\mathbf{U}_{(j)} + \mathbf{K}_B\mathbf{U}_{(j)} + \mathbf{N}_{B,1}\mathbf{U}_{(j-1)} + \mathbf{N}_{B,2}\mathbf{U}_{(j+1)} = \mathbf{0} \tag{5.19}$$

where subscript () refers to the element. For example, $\mathbf{U}_{(j)}$ denotes the unknowns belonging to element j, subscript B refers to the boundary, and $\mathbf{U}$ is the vector containing unknowns at the nodal points of the elements,

$$\mathbf{U}_{(j)} = [T_1, T_2, \cdots, T_{N_e}]^T_{(j)}; \qquad \mathbf{U}_{(j-1)} = [T_{N_e}, 0, \cdots, 0]^T_{(j-1)};$$

$$\mathbf{U}_{(j+1)} = [0, 0, \cdots, T_1]^T_{(j+1)}$$

if the Lagrangian interpolation functions are used. The matrices, all of dimensions $N_e \times N_e$, are calculated by

$$M_{mn} = \int_{x_j}^{x_{j+1}} \phi_m\phi_n dx, \quad K_{mn} = -u\int_{x_j}^{x_{j+1}} \phi_n\frac{\partial\phi_m}{\partial x}dx, \quad m,n = 1,2,\cdots,N_e$$

$$K_{B,11} = -\frac{u-|u|}{2}, \quad K_{B,N_eN_e} = \frac{u+|u|}{2}$$

$$N_{B,1,11} = -\frac{u+|u|}{2}, \quad N_{B,2,N_eN_e} = \frac{u-|u|}{2}$$

with all other elements in $\mathbf{K}_B$, $\mathbf{N}_{B,1}$, and $\mathbf{N}_{B,2}$ set to zero.

In particular, if $k = 0$, that is, if the constant element approximation is used, we recover the finite volume formulation. In addition, if $u > 0$, then we have the following matrices ($m=n=1$),

$$M_{mn} = \Delta x = x_{j+1} - x_j; \quad K_{mn} = 0; \quad K_{B,11} = N_{B,2,N_eN_e} = 0;$$

$$K_{B,N_eN_e} = -N_{B,1,11} = u$$

The matrix equation is then simplified as

$$\Delta x \frac{dT_{(j)}}{dt} + uT_{(j)} - uT_{(j-1)} = 0 \qquad (5.20)$$

This is the simple upwinding finite difference (or finite volume) scheme.

The above equations are ordinary differential equations, which can be integrated in time using various time integrators. For convenience, the matrix equation may be written in a generic form,

$$\frac{d\mathbf{U}_{(j)}}{dt} = L(\mathbf{U}) \qquad (5.21a)$$

with the operator L defined by

$$L(\mathbf{U}) = -\mathbf{M}^{-1}(\mathbf{K}\mathbf{U}_{(j)} + \mathbf{K}_B\mathbf{U}_{(j)} + \mathbf{N}_{B,1}\mathbf{U}_{(j-1)} + \mathbf{N}_{B,2}\mathbf{U}_{(j+1)}) \qquad (5.21b)$$

If a simple Euler forward scheme is used, then we have the following time-discretized form for Equation 5.21a:

$$\mathbf{U}_{(j)}^{n+1} = \mathbf{U}_{(j)}^n + \Delta t^n L(\mathbf{U}^n) \qquad (5.22)$$

or explicitly,

$$\mathbf{U}_{(j)}^{n+1} = \mathbf{U}_{(j)}^n - \Delta t^n \mathbf{M}^{-1}(\mathbf{K}\mathbf{U}_{(j)}^n + \mathbf{K}_B\mathbf{U}_{(j)}^n + \mathbf{N}_{B,1}\mathbf{U}_{(j-1)}^n + \mathbf{N}_{B,2}\mathbf{U}_{(j+1)}^n) \quad (5.23)$$

where superscript n denotes the nth time step and Δt^n represents the time step, where the critical time step must be chosen to enforce the *CFL* condition. It is seen that the step involves the inversion of the mass matrix. For practical applications, the mass matrix $\mathbf{M}$ may be diagonalized, either through numerical quadrature or by constructing orthogonal basis functions, as discussed in Chapter 3, to speed up the calculations.

Other high order explicit schemes, such as the Runge–Kutta integrator, may also be used. An implicit time scheme is also possible. It is noted that when an implicit time scheme is used, a global matrix needs to be assembled, which for this case would be larger than the global matrix generated by the continuous finite element formulation. In fact, for a 1-D problem, the matrix would be almost twice as big if linear elements are used, because every node is shared by two elements.

The complete computational procedure for the solution of Equation 5.21 may now be described. To start, all the values are set to the initial data and a time step is selected. Then the calculation starts with a boundary where the boundary condition is prescribed and sweeps from this element forward until the entire domain is covered. After a sweep, the convergence is checked. If convergence is achieved, then the same procedure is applied for the next time step. This calculation is repeated for every time step until the time elapsed reaches the total time for simulation. It is noted that in this element-by-element approach, the terms associated with the adjacent elements are moved to the right of the equation and treated as source terms. These terms are updated with the U values as soon as they become available during the course of solution.

It is obvious that the above procedure can be directly applied to the system of equations (i.e., Equation 5.6), when the system is appropriately decomposed, and the details are thus not elaborated.

Example 5.2. Consider the following convection problem:

$$\frac{\partial T}{\partial t} + u \frac{\partial T}{\partial x} = 0 \qquad x \in [0,1] \tag{5.5e}$$

with $u = 0.1$ and the initial conditions,

$$T(x,0) = \begin{cases} \sin(10\pi x) & \text{for } 0 \le x \le 0.1 \\ 0 & \text{for } 0.1 \le x \le 1.0 \end{cases} \tag{5.6e}$$

Solution. For illustrative purposes, only the linear elements are used. Following the procedure discussed above, we take the linear elements to discretize the domain equally, and to obtain the following matrix for a typical element:

$$\mathbf{M} = \frac{h}{3} \begin{bmatrix} 2 & 1 \\ 1 & 2 \end{bmatrix}; \ \mathbf{K} = -\frac{uh}{2} \begin{bmatrix} -1 & -1 \\ 1 & 1 \end{bmatrix}; \ \mathbf{K}_B = \begin{bmatrix} 0 & 0 \\ 0 & u \end{bmatrix};$$

$$\mathbf{N}_{B,1} = \begin{bmatrix} -u & 0 \\ 0 & 0 \end{bmatrix}; \ \mathbf{N}_{B,2} = \begin{bmatrix} 0 & 0 \\ 0 & 0 \end{bmatrix}$$

where $h = \Delta x$. The explicit forward time integration is employed to march the solution in time.

The calculations used 200 elements and a time step of 1.0×10^{-4}. The computed results are plotted in Figure 5.2e, along with the analytic solution. For comparison, the finite difference solution with upwinding is also given and plotted in the same graph. We can see that there is considerable numerical dissipation associated with the finite volume scheme with upwinding, and the peak is subdued substantially. The discontinuous finite element solution, even with linear element approximation, almost reproduces the analytic solution. A little overshoot near the two tails, however, is noticed.

Comments. The finite difference (or finite volume) formulation based on the traditional control volume approach is discussed by Fletcher [3]. For this problem, the finite volume formulation was obtained from the above discontinuous formulation by using the piecewise constant approximation over an element. Here, the discontinuous finite element method is just an extension of the finite volume method, which is limited to the step function approximation. The discontinuous finite element method, on the other hand, easily employs the higher order approximation.

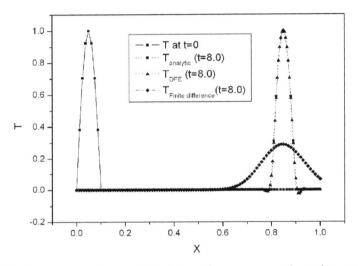

Figure 5.2e. Comparison of numerical solutions for pure convection using constant and linear elements. In both cases, the upwinding scheme is used

5.1.2 Pure Convection in Multidimensions

As with the 1-D pure convection problems, a multidimensional pure convection problem takes a simple form,

$$\frac{\partial T}{\partial t} + \mathbf{u} \cdot \nabla T = 0 \quad \in \Omega \times (0, \tau_T) \tag{5.24}$$

where, for simplicity, $\mathbf{u}$ is taken as a constant vector. To develop a discontinuous formulation, the domain is first discretized into a tessellation of finite elements, say, triangular elements for a 2-D geometry, and then the equation is integrated over an element j with the result,

$$\int_{\Omega_j} \left(v \frac{\partial T}{\partial t} - \mathbf{u} T \cdot \nabla v \right) dV + \int_{\partial \Omega_j} v \overrightarrow{\mathbf{u} T} \cdot \mathbf{n}_j dS = 0 \tag{5.25}$$

where $\mathbf{n}_j$ is the outward normal of element j, and Ω_j and $\partial\Omega_j$ are its domain and boundary, as shown in Figure 5.3. The numerical fluxes now are needed to complete the formulation. For this purpose, the following definitions are used (see also Figure 5.4):

$$\{T\} = 0.5(T^+ + T^-) \; ; \qquad [T] = T^+\mathbf{n}^+ + T^-\mathbf{n}^- \tag{5.26}$$

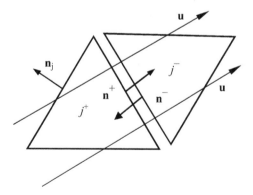

Figure 5.3. Schematic of the triangularization of computational domain and element arrangements.

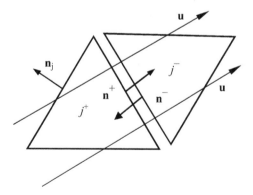

Figure 5.4. Element j (j^+), its neighbor j^- and other quantities used to define numerical fluxes

Again, the underscored brackets are associated with these definitions only. With the above equation, the consistent numerical fluxes are given by

$$\overrightarrow{\mathbf{u}T} = \mathbf{u}\{T\} + \mathbf{C}_u \cdot [T] \tag{5.27}$$

where $\mathbf{C}_u$ is a non-negative definite matrix depending on the value of $\mathbf{u \cdot n}$. The relative relation of the quantities is schematically shown in Figure 5.4.

With a choice of $\mathbf{C}_u = 0.5|\mathbf{u \cdot n}|\mathbf{I}$, the numerical flux is calculated by the classical upwinding scheme,

$$\overrightarrow{\mathbf{u}T_h} = \mathbf{u}\{T\} + \tfrac{1}{2}|\mathbf{u \cdot n}|[T] \qquad (5.28)$$

where $\mathbf{I}$ is the unit matrix. If, on the other hand, $\mathbf{C}_u = 0.5|\mathbf{u}|\mathbf{I}$, then we have the local Lax–Friedrichs numerical flux,

$$\overrightarrow{\mathbf{u}T_h} = \mathbf{u}\{T\} + \tfrac{1}{2}|\mathbf{u}|[T] \qquad (5.29)$$

Once the numerical fluxes are selected, the unknowns can be interpolated using a polynomial basis function. With the Galerkin procedure, the element matrix can be calculated and the resultant matrix equation has the following form:

$$\mathbf{M}\frac{d\mathbf{U}}{dt} + \left(\mathbf{K} + \sum_{i=1}^{NS}\mathbf{K}_{B,i}\right)\mathbf{U} + \sum_{i=1}^{NS}\mathbf{N}_{B,i}\mathbf{U}_{(NB,i)} = \mathbf{0} \qquad (5.30)$$

where $\mathbf{M}$ is the mass matrix, $\mathbf{K}$ is the volume integral, $\mathbf{K}_{B,i}$ and $\mathbf{N}_{B,i}$ represent the boundary integral contribution associated with element j, $\mathbf{U}$ is the unknown vector for element j, NS is the number of sides of the element, and $\mathbf{U}_{(NB,i)}$ is the unknown vectors associated with the neighboring elements. The above equation may be further written in the same generic form as Equation 5.21.

As discussed for the 1-D case, time integration can now be applied to obtain a solution from Equation 5.27 in time. Restriction on the time step applies when explicit numerical time integration schemes are used.

Aside from the direct solution given above, another approach may also be applied to solve this problem. Since $\mathbf{u}$ is constant, the equation can be re-written as follows:

$$\frac{\partial T}{\partial t} + |u|\frac{\partial T}{\partial s_u} = 0 \qquad (5.31)$$

where s_u indicates that the directional derivative is along the $\mathbf{u}$ direction. This system is practically 1-D, and the above solution procedure for the 1-D problem may be applied directly.

We will see some of these applications in the radiative transfer processes (Chpater 9) and in the discontinuous finite element solution of the lattice Boltzmann equations (Chapter 11).

In passing, we note that selection of appropriate time steps and numerical fluxes for a meaningful numerical solution of pure convection equations requires requires stability analysis, which is to be discussed below.

5.1.3 Stability Analysis

As for the solution of heat conduction problems, stability is a critical issue for the discontinuous finite element method for the solution of pure convection equations. Numerical fluxes must be selected to satisfy the stability condition. There are different approaches to stability analysis. In this section, we present the integral analysis, the discretized analysis and the Fourier analysis; the last also being used for the determination of the critical time step for explicit time integration, and for the study of dissipation and dispersion behavior of the numerical schemes.

5.1.3.1 L^2-Stability – Integral Analysis
Integral analysis is based on the existence and uniqueness theory of partial differential equations and is a powerful tool for stability analysis. We have seen its use in considering the heat conduction problems in Chapter 4. Here again, we follow the approach given by Cockburn [4,5] and Evans [2]. Let us consider the L^2-stability for a pure convection problem,

$$\frac{\partial T}{\partial t} + \nabla \cdot (\mathbf{u}T) = 0 \qquad \in \Omega \times (0, \tau_T] \tag{5.32}$$

with periodic boundary conditions. The stability result is first obtained for the problem of a continuous case. To do that, we multiply the above equation by T and integrate over space and time to get

$$\frac{1}{2}\int_\Omega T^2(\mathbf{r},t)\,dV + \frac{1}{2}\int_0^{\tau_T}\int_\Omega (\nabla \cdot \mathbf{u})T^2(\mathbf{r},t)\,dVdt = \frac{1}{2}\int_\Omega T^2(\mathbf{r},0)\,dV \tag{5.33}$$

where use has been made of the following vector identity:

$$\nabla \cdot (\mathbf{u}T^2(\mathbf{r},t)) + (\nabla \cdot \mathbf{u})T^2(\mathbf{r},t) = 2T(\mathbf{r},t)\nabla \cdot (\mathbf{u}T(\mathbf{r},t)) \tag{5.34}$$

Also the periodic condition causes the following integral on the boundary to vanish:

$$\int_\Omega \nabla \cdot (\mathbf{u}T^2(\mathbf{r},t))\,dV = \int_{\partial\Omega} \mathbf{n} \cdot (\mathbf{u}T^2(\mathbf{r},t))\,dS = 0 \tag{5.35}$$

In particular, if $\nabla \cdot \mathbf{u} = 0$ or $\mathbf{u}$ is a constant vector, a stability result is obtained as follows:

$$\frac{1}{2}\int_\Omega T^2(\mathbf{r},t)\,dV = \frac{1}{2}\int_\Omega T^2(\mathbf{r},0)\,dV \tag{5.36}$$

The same procedure used for stability studies on the discontinuous finite element solution of heat conduction equations may be employed here to obtain the same estimate. In essence, we compare the discrete system with the above analytic weak form solution, and enforce stability upon the discrete system. Towards this end, we set $v = T_h$ in the weak formulation (i.e., Equation 5.25) and then sum over all the elements with the result,

$$\frac{1}{2}\int_\Omega T^2(\mathbf{r},t)\,dV + \int_0^{\tau_T} \Theta_h(t)\,dt = \frac{1}{2}\int_\Omega T^2(\mathbf{r},0)\,dV \tag{5.37}$$

Equation 5.37 is required to converge to Equation 5.33 when the limit of the element size approaches zero. This requires that $\Theta_h(t) \geq 0$, and goes to zero as the limit is taken to ensure compatibility. The term $\Theta_h(t)$ collects all the boundary contributions:

$$\begin{aligned}
\Theta_h(t) &= \sum_j \left(-\frac{1}{2}\int_{\Omega_j} \nabla\cdot(\mathbf{u}T_h^2)\,dV + \int_{\partial\Omega_j} \mathbf{u}\hat{T}_h \cdot \mathbf{n}T_h\,dS \right) \\
&= \sum_j \int_{\partial\Omega_j} \left(\mathbf{u}\hat{T}_h \cdot \mathbf{n}T_h - \tfrac{1}{2}\mathbf{u}T_h^2 \cdot \mathbf{n} \right)dS \\
&= \sum_{j_e} \int_{\partial\Omega_{j_e}} [\mathbf{u}\hat{T}_h T_h - \tfrac{1}{2}\mathbf{u}T_h^2]\,dS \\
&= \sum_{j_e} \int_{\partial\Omega_{j_e}} \mathbf{u}\hat{T}_h[T_h] - \tfrac{1}{2}[\mathbf{u}T_h^2]\,dS \\
&= \sum_{j_e} \int_{\Omega_{j_e}} (\mathbf{u}\hat{T}_h - \mathbf{u}\{T_h\})\cdot[T_h]\,dS \tag{5.38}
\end{aligned}$$

where the second equation in Equation 5.26 has been used in the third step, and subscript j_e means integrating only once along the element boundary shared by two elements. It is important to stress that the subscript j means integration along the element boundary for every element and thus integration is carried out twice along the bondary of the shared elements. Consequently, if a numerical flux is defined as follows:

$$\overrightarrow{\mathbf{u}T} = \mathbf{u}\{T\} + \mathbf{C}\cdot[T] \tag{5.39}$$

which is Equation 5.27, then $\Theta_h(t)$ satisfies the stability condition,

$$\Theta_h(t) = \sum_j \int_{\partial\Omega_j} \mathbf{C}\cdot[T_h]\cdot[T_h]\,dS \geq 0 \tag{5.40}$$

provided that $\mathbf{C}$ is a non-negative matrix.

At this point, we note that for a piecewise constant approximation, Equation 5.25 reduces to the familiar finite volume formulation,

$$\int_{\Omega_j} \frac{\partial T_h}{\partial t} dV + \int_{\partial \Omega_j} \overleftrightarrow{\mathbf{u} T_h} \cdot \mathbf{n} \, dS = 0 \tag{5.41}$$

In this regard, the finite volume method is just a subclass of the discontinuous finite element method, which is consistent with the previous discussions for 1-D problems.

The discontinuous finite element method is a higher order method. Cockburn [4] shows that an order of convergence of $k + \frac{1}{2}$ can be obtained with polynomials of degree k at most. Also Equation 5.37 provides a dissipative effect (equivalently the artificial viscosity) for the numerical scheme, where the dissipation is related to the across-element jumps.

As we have discussed for the heat conduction problems, the residuals are also related to the stabilization. To see that, Equation 5.25 is integrated once again to produce

$$\int_{\Omega_j} v \left(\frac{\partial T}{\partial t} + \nabla \cdot (\mathbf{u}T) \right) dV = \int_{\Omega_j} R v \, dV = \int_{\partial \Omega_j} (\mathbf{u} T_h \cdot \mathbf{n} v - \overleftrightarrow{\mathbf{u} T_h} \cdot \mathbf{n} v) \, dS \tag{5.42}$$

In the case of upwinding, the above equation becomes

$$\int_{\Omega_j} R v \, dV = \int_{\partial \Omega_j} \mathbf{u} \cdot [\underline{T_h^2}] \, dS \tag{5.43}$$

which shows that the residuals are directly related to the inflow jump at the element boundary [4]. We have seen a similar role of residuals in the case of heat conduction problems (see Section 4.7).

5.1.3.2 L^2-Stability – Discretized Analysis
The stability of a discretized system can also be carried out either by the Fourier series method, as seen for the transient heat conduction problems in Chapter 4 (see Section 4.7), or by the local method, which is discussed in this section. Let us illustrate this point by a 1-D example,

$$\frac{\partial T}{\partial t} + u \frac{\partial T}{\partial x} = 0 \qquad \in [0,1] \tag{5.44}$$

For simplicity, we choose $u > 0$. We discretize the system into N equations, with nodes denoted by $x_1, x_2, \cdots, x_{N+1}$, as shown in Figure 5.5. Over each element, the discontinuous procedure yields the following weak form solution:

$$\int_{x_j}^{x_{j+1}} v \left(\frac{\partial T_h}{\partial t} + u \frac{\partial T_h}{\partial x} \right) dx = v(x_j) u \left(T_h^-(x_j) - T_h^+(x_j) \right),$$

$$\forall T_h, v \in P^k (x_j, x_{j+1}) \tag{5.45}$$

where $P^k(x_j, x_{j+1})$ is a polynomial of order k defined over element j, $x \in [x_j, x_{j+1}]$. Note that if $u < 0$, the upwinding point would be at $j + 1$. In the jth element, an approximation of T is denoted by T_h.

Figure 5.5. Discretization of the 1-D domain for a pure convection problem

By choosing $k = 0$ (i.e., the space of piecewise constant functions) the discontinuous formulation reduces to the finite volume formulation,

$$\int_{x_j}^{x_{j+1}} v \frac{\partial T_h}{\partial t} dx = vu\left(T_{h,(j-1)} - T_{h,(j)}\right) \tag{5.46}$$

that is,

$$h v \frac{\partial T_h}{\partial t} = vu\left(T_{h,j-1} - T_{h,j}\right) \tag{5.47a}$$

$$\frac{\partial T_h}{\partial t} = \frac{u}{h}\left(T_{h,j-1} - T_{h,j}\right) \tag{5.47b}$$

Let us now examine the stability condition of the above discrete formulation. To do that, the Galerkin method is applied, that is, $v(x) = T_h(x)$, whence Equation 5.45 becomes

$$\int_{x_j}^{x_{j+1}} T_h\left(\frac{\partial T_h}{\partial t} + u \frac{\partial T_h}{\partial x}\right) dx = T_h(x_j)u\left(T_h^-(x_j) - T_h^+(x_j)\right) \tag{5.48}$$

This equation can be further written as

$$\int_{x_j}^{x_{j+1}} T_h\left(\frac{\partial T_h}{\partial t} + u \frac{\partial T_h}{\partial x}\right) dx = T_h^+(x_j)u\left(T_h^-(x_j) - T_h^+(x_j)\right) \tag{5.49}$$

which is integrated over the element to give

$$\frac{d}{dt}\int_{x_j}^{x_{j+1}} \frac{T_h^2}{2} dx = -\left(u \frac{T_h^2}{2}\right)_{x_j}^{x_{j+1}} + T_h^+(x_j)u\left(T_h^-(x_j) - T_h^+(x_j)\right)$$

$$= u \frac{-\left(T_h^+(x_j)\right)^2 - \left(T_h^-(x_{j+1})\right)^2}{2} + u T_h^+(x_j) T_h^-(x_j) \tag{5.50}$$

Summing over all the elements, one has the following result:

$$\frac{d}{dt} \sum_{j=1}^N \int_{x_j}^{x_{j+1}} \frac{T_h^2}{2} dx$$

$$= \sum_{j=1}^N \left\{ u \frac{-T_h^{+2}(x_j) - T_h^{-2}(x_{j+1})}{2} + u T_h^+(x_j) T_h^-(x_j) \right\}$$

$$\leq \sum_{j=1}^N \left\{ u \frac{-T_h^{+2}(x_j) - T_h^{-2}(x_{j+1})}{2} + u \frac{T_h^{+2}(x_j) + T_h^{-2}(x_{j-1})}{2} \right\}$$

$$\leq \frac{u}{2} \sum_{j=1}^N \left(T_h^{-2}(x_{j-1}) - T_h^{-2}(x_{j+1}) \right)$$

$$\leq \frac{u}{2} \left(T_h^{-2}(x_1) - T_h^{-2}(x_{N+1}) \right) \tag{5.51}$$

where use has been made of the inequality $2ab \leq a^2 + b^2$ to arrive at the first inequality.

As we have seen before, the analysis of the discontinuous method uses the L^2-norm stability theory. Defining the L^2-norm,

$$\| T_h \|_0 = \left(\sum_{j=1}^N \int_{x_j}^{x_{j+1}} T_h^2 \, dx \right)^{1/2} \tag{5.52}$$

and using Equation 5.51, we have the following stability result:

$$\frac{d}{dt} \frac{\| T_h \|_0^2}{2} = \frac{d}{dt} \sum_{j=1}^N \int_{x_j}^{x_{j+1}} \frac{T_h^2}{2} dx \leq \frac{u}{2} \left(T_h^2(x_1) - T_h^2(x_{N+1}) \right) \tag{5.53}$$

where $\|\bullet\|_0 = \|\bullet\|_{L^2}$ denotes the L^2-norm. The above formula is useful to check for consistency. If two solutions are initially close with the same boundary condition imposed weakly at x_0, then these two solutions are bounded by the following condition:

$$\frac{d \| T_{h,1} - T_{h,2} \|_0^2}{dt} \leq -u \left(T_{h,1}(x_{N+1}) - T_{h,2}(x_{N+1}) \right)^2 \tag{5.54}$$

which means that if the time derivative is computed exactly, then the distance between the two solutions (measured in the L^2-norm) will decrease in time.

The above stability condition serves also as a basis for error analysis. For this purpose, it is often useful to project the solution into a polynomial space. Let us further assume that such a projection function exists. Then we have the following three equations:

$$\int_{x_j}^{x_{j+1}} vQ^k \left(\frac{\partial T}{\partial t} + u \frac{\partial T}{\partial x} \right) dx = 0 \qquad (5.55a)$$

$$\int_{x_j}^{x_{j+1}} vQ^k \left(\frac{\partial T}{\partial t} + u \frac{\partial T}{\partial x} \right) dx = \int_{x_j}^{x_{j+1}} vR \, dx$$

$$+ v(x_j)u \left(Q^k T^-(x_j) - Q^k T^+(x_j) \right) \qquad (5.55b)$$

$$\int_{x_j}^{x_{j+1}} v \left(\frac{\partial T_h}{\partial t} + u \frac{\partial T_h}{\partial x} \right) dx = v(x_j)u \left(T_h^-(x_j) - Q^k T_h^+(x_j) \right) \qquad (5.55c)$$

where Q^k is the projection function,

$$Q^k : H^1(x_0, x_N) \rightarrow \bigcup_{j=1}^{N} P^k(x_j, x_{j+1})$$

Note that Equation 5.55a projects the exact equation, Equation 5.55b means that the projected exact solution is substituted into the weak formulation, and Equation 5.55c is the basic numerical scheme used for numerical solution. We note that Equation 5.55b has the error term associated with the residual R.

Subtracting Equation 5.55c from Equation 5.55b and making use of the relation, $T^-(x_j) = T^+(x_j)$, that is, the exact solution is continuous for a smooth solution, we obtain the following equation:

$$\int_{x_j}^{x_{j+1}} v \left(\frac{\partial (Q^k T - T_h)}{\partial t} + u \frac{\partial (Q^k T - T_h)}{\partial x} \right) dx = \int_{x_j}^{x_{j+1}} vR(x,t)dx$$

$$+ v(x_j)u \left(Q^k T^-(x_j) - T_h^-(x_j) + T_h^+(x_j) - Q^k T^+(x_j) \right) \qquad (5.56)$$

By setting $v(x) = Q^k T(x) - T_h(x)$, the preceding equation becomes

$$\int_{x_j}^{x_{j+1}} (Q^k T - T_h) \left(\frac{\partial (Q^k T - T_h)}{\partial t} + u \frac{\partial (Q^k T - T_h)}{\partial x} \right) dx$$

$$= \left(Q^k T^+(x_j) - T_h^+(x_j)\right) u \left(Q^k T^-(x_j) - T_h^-(x_j) + T_h^+(x_j) - Q^k T^+(x_j)\right)$$

$$+ \int_{x_j}^{x_{j+1}} R(x,t)\left(Q^k T - T_h\right) dx \tag{5.57}$$

Use of the relation $2ab \le a^2 + b^2$ allows the first term on the right and the second on the left to be combined,

$$-\frac{1}{2}\left(Q^k T - T_h\right)^2 \Big|_{x_j}^{x_{j+1}}$$

$$+ \left(Q^k T^+(x_j) - T_h^+(x_j)\right)\left(Q^k T^-(x_j) - T_h^-(x_j) + T_h^+(x_j) - Q^k T^+(x_j)\right)$$

$$= -\frac{1}{2}\left(Q^k T^-(x_{j+1}) - T_h^-(x_{j+1})\right)^2 - \frac{1}{2}\left(Q^k T^+(x_j) - T_h^+(x_j)\right)^2$$

$$+ \left(Q^k T^+(x_j) - T_h^+(x_j)\right)\left(Q^k T^-(x_j) - T_h^-(x_j)\right)$$

$$\le \frac{1}{2}\left(Q^k T^-(x_j) - T_h^-(x_j)\right)^2 - \frac{1}{2}\left(Q^k T^-(x_{j+1}) - T_h^-(x_{j+1})\right)^2 \tag{5.58}$$

Combining the above two equations yields the following estimate:

$$\frac{d}{dt}\left\|Q^k T - T_h\right\|_0^2$$

$$\le u \sum_{j=1}^{N} \left\{ \frac{1}{2}\left(Q^k T^-(x_j) - T_h^-(x_j)\right)^2 - \frac{1}{2}\left(Q^k T^-(x_{j+1}) - T_h^-(x_{j+1})\right)^2 \right\}$$

$$+ \sum_{j=1}^{N} \int_{x_j}^{x_{j+1}} R(x,t)\left(Q^k T - T_h\right) dx$$

$$\le \sum_{j=1}^{N} \int_{x_j}^{x_{j+1}} R(x,t)\left(Q^k T - T_h\right) dx$$

$$\le \left\|Q^k T - T_h\right\|_0 \left\|R(x,t)\right\|_0 \tag{5.59}$$

where use has been made of the Schwarz inequality,

$$\int_{x_j}^{x_{j+1}} ab\, dx \le \int_{x_j}^{x_{j+1}} a^2 dx \int_{x_j}^{x_{j+1}} b^2 dx \tag{5.60}$$

The consistency analysis indicates that the numerical solution T_h, and some projection of the exact solution T to the same space of polynomials that we use to represent T_h, grow apart slowly in time, i.e.,

$$\frac{d}{dt}\left\|Q^k T - T_h\right\|_0 \leq \left\|R\right\|_0 \leq Ch^{k+1}\,|T(0)|_{H^{k+2}(0,1)} \tag{5.61}$$

where we have used the basic estimate given by Cockburn [5], the subscript H^{k+2} (0,1) denotes a Hilbert space and $T(0) = T(t = 0) \in H^{k+2}$ (0,1). Also, C is a constant depending solely on k, $|u|$ and T. Integrating over time, we have the error $T - T_h$ bounded by the following expression:

$$
\begin{aligned}
\left\|T - T_h\right\|_0 &= \left\|T - Q^k T + Q^k T - T_h\right\|_0 \\
&\leq \left\|T - Q^k T\right\|_0 + \left\|Q^k T - T_h\right\|_0 \\
&\leq \left\|T - Q^k T\right\|_0 + \left\|Q^k T(t = 0) - T_h(t = 0)\right\|_0 + \tau_T Ch^{k+1}\,|T(0)|_{H^{k+2}(0,1)}
\end{aligned}
\tag{5.62}
$$

Here the first term represents the error between the exact solution and the projected solution in the finite polynomial space. The second term represents the error in the approximation of the initial data. The third term is the accumulation of truncation errors in time, which depend on the discretization and the total time τ_T [4].

5.1.3.3 Fourier Analysis

Fourier analysis is useful for determining critical time steps for transient calculations; and for the pure convection problems, it is also a valuable tool to analyze the numerical dispersion and dissipation phenomena associated with wave propagation. In pure convection problems, the system of equations is hyperbolic. The solution is characterized by a train of waves propagating with little or no loss of amplitude. It is important that the numerical solutions do not introduce non-physical dissipation, which shows up as a broadening of the wave pack and reduced amplitude, that is, artificial diffusion. It is equally important that the numerical schemes do not introduce artificial dispersion. Dispersion refers to the change of the speed at which waves propagate, and it often shows up in a numerical solution as numerical oscillation [3].

Dissipation and Dispersion. Information on the numerical dissipation and dispersion introduced by a computational scheme can be obtained by comparing the Fourier representations of the exact and numerical solutions. For the problem given by Equation 5.44, the initial condition may be expanded in a Fourier series,

$$T(x,0) = \sum_{m=-\infty}^{\infty} T_m e^{imx} \tag{5.63}$$

and the exact solution to the problem at any instant in time is also represented by a Fourier series,

$$T_{ex}(x,t) = \sum_{m=-\infty}^{\infty} T_m e^{im(x-ut)} \tag{5.64}$$

Clearly, all the Fourier components in the above equation convect with the same velocity u, and are not subject to any reduction in amplitude. This is simply a statement that there is no diffusion effect (or even order derivative) in the equation. Similarly, a numerical algorithm can be represented by a Fourier series. Taking into consideration the errors involved in numerical schemes, a numerical solution may have the following form of the Fourier expansion:

$$T_h(x,t) = \sum_{m=-\infty}^{\infty} T_m e^{-i\omega t} e^{imx} = \sum_{m=-\infty}^{\infty} T_m e^{-p(m)t} e^{-imq(m)t} e^{imx} \tag{5.65}$$

where we have taken $\omega(m) = -ip(m) + mq(m)$. For an exact solution, $p(m) = 0$ and $q(m) = u$; thus there is no attenuation of the amplitude of T_m as the wave propagates with a speed u. In general, $p(m \neq 0$ and $q(m) \neq u$, that is, the amplitude is attenuated and the propagation speed is altered. For the latter case, if waves of more than one wavelength are present, they propagate at different speeds, i.e., they disperse. The change in wave propagation is often more pronounced for short waves (large m).

In the computational literature, dissipation is defined as the attenuation of the amplitude of the waves, and dispersion is defined as the propagation of waves of different wave numbers m at different speeds $q(m)$. Dissipation is often associated with even order derivatives, and dispersion with odd order derivatives.

In numerical analysis, the dissipation and dispersion behaviors of the system are studied using the amplification factor and the phase angle for a typical Fourier component. The amplification factor is defined by the ratio of the Fourier component at two consecutive time steps [6],

$$G = \frac{T_m \exp(-p(m)(t+\Delta t)) \exp(-imq(m)(t+\Delta t)) \exp(imx)}{T_m \exp(-p(m)t) \exp(-imq(m)t) \exp(imx)}$$

$$= \exp(-p(m)\Delta t) \exp(-imq(m)\Delta t) \tag{5.66}$$

By this definition, the amplitude of G, $|G|$, is related to the dissipation at x and the phase angle φ is associated with dispersion,

$$|G| = \exp(-p(m)\Delta t) = \sqrt{(\mathrm{Re}(G))^2 + (\mathrm{Im}(G))^2} \tag{5.67a}$$

$$\varphi = -mq(m)\Delta t = \tan^{-1}(\mathrm{Im}(G)/\mathrm{Re}(G)) \tag{5.67b}$$

To use the Fourier analysis to understand the basics of numerical dissipation and dispersion, we consider the 1-D pure convection problem,

$$\frac{\partial T}{\partial t} + u\frac{\partial T}{\partial x} = 0 \tag{5.68}$$

with $u > 0$ and periodic boundary conditions. For the Galerkin approximation, we have the following discretized matrix equations:

$$\mathbf{M}\frac{d\mathbf{U}_{(j)}}{dt} + \mathbf{K}\mathbf{U}_{(j)} + \mathbf{K}_B\mathbf{U}_{(j)} + \mathbf{N}_{B,1}\mathbf{U}_{(j-1)} = \mathbf{0} \tag{5.69}$$

where we have incorporated the upwinding scheme. For a three-element discretization with periodicity, the matrix system has the form of

$$\begin{bmatrix} \mathbf{M} & \mathbf{0} & \mathbf{0} \\ \mathbf{0} & \mathbf{M} & \mathbf{0} \\ \mathbf{0} & \mathbf{0} & \mathbf{M} \end{bmatrix}\frac{d\mathbf{U}}{dt} + \begin{bmatrix} \mathbf{K}+\mathbf{K}_B & \mathbf{0} & \mathbf{N}_{B,1} \\ \mathbf{N}_{B,1} & \mathbf{K}+\mathbf{K}_B & \mathbf{0} \\ \mathbf{0} & \mathbf{N}_{B,1} & \mathbf{K}+\mathbf{K}_B \end{bmatrix}\mathbf{U} = \mathbf{0} \tag{5.70}$$

This is a generalized eigenvalue problem for $\mathbf{U}$, since we know $\mathbf{K}$, $\mathbf{M}$, $\mathbf{K_B}$, and $\mathbf{N}_{B,1}$. The dimension of the matrix is $(k+1)\times(k+1)$ for a polynomial of order k.

We consider the dissipation and dispersion analyses below. Assuming that a Fourier component takes the following form:

$$T_{j,m} = T_m e^{-i\omega(m)t}e^{-imx} \tag{5.71}$$

and substituting the above equation into the matrix equations, one has an eigenvalue problem for ω,

$$-\left(i\omega(m)\mathbf{M} + u\mathbf{D} + u\mathbf{F} + u\mathbf{G}e^{-im\Delta x}\right)\mathbf{T}_m = 0 \tag{5.72}$$

If a piecewise constant approximation is taken to evaluate the matrices, then the following result is obtained:

$$-\left(i\omega(m)h + u(1 - e^{-imh})\right)\mathbf{T}_m = 0 \tag{5.73}$$

with $\Delta x = h$. The solution to the eigenvalue problem for a non-trivial $\mathbf{T}_m$ is given by the following expression:

$$i\omega(m) = \frac{u(1 - e^{-imh})}{h} = \frac{u}{h}\left(1 - \cos(mh) + i\sin(mh)\right) \tag{5.74}$$

which yields the values for $p(m)$ and $q(m)$ as follows:

$$p(m) = \frac{u}{h}(1 - \cos(mh)), \quad mq(m) = \frac{u}{h}\sin(mh) \tag{5.75}$$

The dissipation is measured by the amplitude of the amplification factor,

$$|G| = \exp(-p(m)\Delta t) = \exp(-c(1 - \cos(mh))) \tag{5.76}$$

where $c = u\Delta t/h$ is the Courant number, and the dispersion behavior is given by the phase φ of the amplification factor,

$$\varphi = -mq(m)\Delta t = -c\sin(mh) \tag{5.77}$$

For this problem, the exact solution can be easily obtained and its mth Fourier component has the form of

$$T_{ex,m}(x,t) = T_m e^{im(x-ut)} \tag{5.78}$$

which has a phase over a consecutive time step $t \to t+\Delta t$,

$$\varphi_{ex} = -um\Delta t \tag{5.79}$$

The change of the phase for the mth wave component is given by

$$\frac{\varphi}{\varphi_{ex}} = \frac{-mq(m)\Delta t}{-mu\Delta t} = \frac{1}{mh}\sin(mh) \tag{5.80}$$

From the above analysis, it is clear that if the time integration is exact, then the spatial discretization would have attenuation and dispersion given by Equations 5.76 and 5.80. In particular, we have from these two equations,

$$|G| \to 1, \quad \varphi/\varphi_{ex} \to 1, \text{ as } m \to 0 \tag{5.81}$$

Critical Time Step for Time Integration. As for the heat conduction problems, the Fourier analysis is also used to determine the critical time step for a time marching scheme for pure convection problems. In this use, the analysis is intended to prevent the round-off error from growing during time marching, which is considered as an important stability issue for time integration schemes. The round-off error $\varepsilon(x,t)$ is defined as the difference between the exact solution, computed with infinite accuracy, and the numerical solution with the actual machine. For illustrative purposes, we analyze an explicit time integration scheme applied to Equation 5.70,

$$
\begin{bmatrix} \mathbf{M} & 0 & 0 \\ 0 & \mathbf{M} & 0 \\ 0 & 0 & \mathbf{M} \end{bmatrix} \frac{\mathbf{U}^{n+1} - \mathbf{U}^{n}}{\Delta t} + \begin{bmatrix} \mathbf{K} + \mathbf{K}_{B} & 0 & \mathbf{N}_{B,1} \\ \mathbf{N}_{B,1} & \mathbf{K} + \mathbf{K}_{B} & 0 \\ 0 & \mathbf{N}_{B,1} & \mathbf{K} + \mathbf{K}_{B} \end{bmatrix} \mathbf{U}^{n} = 0 \quad (5.82)
$$

where n denotes the nth time step. The round-off error $\varepsilon(x,t)$ is expanded in terms of a Fourier series,

$$
\varepsilon(x,t) = \sum_{m=-\infty}^{\infty} \varepsilon_m(t)e^{ik_m x} \quad (5.83)
$$

Again, if the piecewise constant is used and the mth Fourier component is substituted for $\mathbf{U}$ in Equation 5.82, then the equation for error becomes

$$
\varepsilon_m(t_n + \Delta t) = \left(h - u\Delta t(1 - e^{-imh}) \right) \varepsilon_m(t_n) \quad (5.84)
$$

where subscript n denotes the nth time step. Following Neumann's analysis, we assume $\varepsilon_m(t) = e^{at}$, and we have the following relation upon substitution:

$$
G = \frac{\varepsilon_m(t_n + \Delta t)}{\varepsilon_m(t_n)} = e^{a\Delta t} = \left(1 - c(1 - e^{-imh})\right)h \quad (5.85)
$$

where G is the amplification factor and $c = u\Delta t/h$. To prevent the error from growing in time, it is necessary that

$$
|G| = |e^{a\Delta t}| = |h - u\Delta t(1 - e^{-imh})| = 1 - 2c(1 - c)(1 - \cos mh) \leq 1 \quad (5.86)
$$

This gives the well known Courant–Friedrichs–Lewy (*CFL*) criterion for stability,

$$
c = \frac{u\Delta t}{h} \leq 1 \quad (5.87)
$$

which is a restriction to be respected. It it noted that while the above restriction is obtained from the stability analysis, its physical meaning is such that the distance a particle travels over a time step can not be greater than the mesh size; otherwise the information about the traveling particle may be lost.

While the Fourier analyses have been used to study different behaviors associated with a transient numerical scheme, the equation for the amplification factor (i.e., Equation 5.85) is basically the same as that obtained from Equation 5.76, the two differing only in higher order terms [3, 6]. In fact, in numerical analysis literature, Equation 5.85 is often used as the amplification factor for dissipation and dispersion analysis [3].

5.2 Steady State Convection-diffusion

The above discussion has been concerned with pure convection problems where the viscosity is neglected, and their numerical solutions found using the discontinuous finite element approach. Pure convection is an idealization, since in the real world thermal fluids systems are dissipative. In this section, we focus on the numerical aspects of the discontinuous solution of convection-diffusion problems. Convection-diffusion problems are known to exhibit oscillations in their numerical solutions and suppression of these types of oscillations requires consideration of various factors. The origin of numerical oscillation associated with a convection term, and its analysis, is also discussed.

5.2.1 1-D Problem

The steady state convection diffusion equation is a useful system to illustrate the oscillatory behavior of the numerical solution, when the exact solution changes rapidly across a thin boundary layer, over which the dissipative mechanism is significant. For a 1-D steady state convection-diffusion problem with temperatures fixed at the two boundaries, the mathematical statement is given below,

$$u\frac{dT}{dx} - D\frac{d^2T}{dx^2} = 0 \qquad \in [0,1] \tag{5.88a}$$

$$T(0) = 0; \quad T(1) = 1.0 \tag{5.88b}$$

From the physical point of view, the above equation represents a balance between the convection and diffusion mechanisms in the system. The exact solution for the system is simple and has the following form:

$$T(x) = \frac{1 - e^{ux/D}}{1 - e^{u/D}} \tag{5.89}$$

To develop a discontinuous finite element formulation, the equation is split into two first order equations,

$$q - \frac{dT}{dx} = 0 \; ; \; u\frac{dT}{dx} - D\frac{dq}{dx} = 0 \tag{5.90}$$

The domain is then discretized into N elements as shown in Figure 5.5, and the above equations are integrated over element j. After integration by parts, one has the following results:

$$\int_{x_j}^{x_{j+1}} q_h w \, dx + \int_{x_j}^{x_{j+1}} T_h w_x \, dx - \hat{T}_{j+1}^- w_{j+1}^- + \hat{T}_j^+ w_j^+ = 0 \tag{5.91a}$$

$$\int_{x_j}^{x_{j+1}} (Dq_h - uT_h)v_x dx - (D\hat{q}_{j+1}^- - (\overrightarrow{uT})_{j+1}^-)v_{j+1}^- + (D\hat{q}_j^+ - (\overrightarrow{uT})_j^+)v_j^+ = 0$$

(5.91b)

where two types of numerical fluxes are used, and subscript x denotes the derivative, e.g., $w_x = dw/dx$. The numerical fluxes associated with diffusion are given by Equation 4.20, which, for 1-D problems, takes the following form:

$$\hat{q} = 0.5(q^+ + q^-) - C_{11}(T^- - T^+) + C_{12}(q^- - q^+)$$

(5.92)

$$\hat{T} = 0.5(T^+ + T^-) - C_{12}(T^- - T^+)$$

(5.93)

For convection flux, a simple upwinding scheme is used, and one has the following convective fluxes:

$$(\overrightarrow{uT})_j^+ = \begin{cases} uT_j^- & u > 0 \\ uT_j^+ & u \le 0 \end{cases}$$

(5.94)

$$(\overrightarrow{uT})_{j+1}^- = \begin{cases} uT_{j+1}^- & u > 0 \\ uT_{j+1}^+ & u \le 0 \end{cases}$$

(5.95)

The unknowns (q_h, T_h) may be approximated with a polynomial as local basis functions,

$$T_h(x,t) = \sum_{i=1}^{k+1} T_i(t)\phi_i(x), \quad q_h(x,t) = \sum_{i=1}^{k+1} q_i(t)\phi_i(x)$$

(5.96)

where k is the order of the polynomial and for 1-D problems $k+1 = N_e$, with N_e being the number of nodes per element. With the numerical fluxes for diffusion and convection defined above, and making use of the Galerkin approximation procedures, we obtain the following matrix equation:

$$\begin{bmatrix} \mathbf{E} & \mathbf{H} \\ \mathbf{J} & \mathbf{G} \end{bmatrix} \begin{pmatrix} \mathbf{q} \\ \mathbf{T} \end{pmatrix}_{(j)} + \begin{bmatrix} \mathbf{E}_B & \mathbf{H}_B \\ \mathbf{J}_B & \mathbf{G}_B \end{bmatrix} \begin{pmatrix} \mathbf{q} \\ \mathbf{T} \end{pmatrix}_{(j)} + \begin{bmatrix} \mathbf{E}_{B,1} & \mathbf{N}_{B,1} \\ \mathbf{J}_{B,1} & \mathbf{G}_{B,1} \end{bmatrix} \begin{pmatrix} \mathbf{q} \\ \mathbf{T} \end{pmatrix}_{(j-1)}$$

$$+ \begin{bmatrix} \mathbf{E}_{B,2} & \mathbf{N}_{B,2} \\ \mathbf{J}_{B,2} & \mathbf{G}_{B,2} \end{bmatrix} \begin{pmatrix} \mathbf{q} \\ \mathbf{T} \end{pmatrix}_{(j+1)} = \begin{pmatrix} \mathbf{0} \\ \mathbf{0} \end{pmatrix}$$

(5.97)

where, as usual, subscript () refers to the element; subscript B denotes the matrices associated with element boundary integrals; and $\mathbf{E}$, $\mathbf{G}$, $\mathbf{J}$, and $\mathbf{N}$ are matrices associated with element calculations. Note also that for this problem, $\mathbf{J}$ is related to $\mathbf{H}$ such that $\mathbf{J} = D\mathbf{H}^{\mathsf{T}}$. This suggests that appropriate scaling would render $\mathbf{J}$ to be $\mathbf{H}$

transposed [9]. The above matrix equation can be further written in a simplified form:

$$\mathbf{KU} = \mathbf{F} \tag{5.98}$$

where $\mathbf{K}$ is the final matrix including the contributions from both the domain and boundary integrals, $\mathbf{U} = (\mathbf{q}^T, \mathbf{T}^T)$, and $\mathbf{F}$ is the source term. The unknowns $\mathbf{U}$ are obtained by inversion of the matrix equations above.

The numerical procedure for the computations is essentially the same as for steady state heat conduction calculations discussed in Chapter 4. Again, it is worth emphasizing that the element-wise sweep, coupled with successive substitution, provides perhaps the most convenient, though arguably the most efficient, algorithm for a numerical solution. By this approach, the field is initialized as zero to start, and an element is selected at the boundary. The calculations are performed element by element to sweep through the entire domain. The unknowns obtained for an element are immediately available to the neighboring elements, and are applied in boundary source terms. The procedure is iterative and iteration is considered *converged* if the successive solutions are within a preset tolerance measured in a norm.

Example 5.3. Consider the 1-D convection diffusion problem defined below,

$$u\frac{dT}{dx} - D\frac{d^2T}{dx^2} = 0 \quad \in [0,1] \tag{5.7e}$$

$$T(0) = 0, \quad T(1) = 1.0 \tag{5.8e}$$

with u being the convection velocity. Calculate the temperature distribution using the discontinuous finite element method.

Solution. In the following equations, $C_{12}^{(1)}$ and $C_{12}^{(2)}$ are the values of C_{12} at the start and end points of each element. The following values were used for calculations: $C_{12}^{(1)} = C_{12}^{(2)} = 0$ at inner nodes (central scheme); at the boundary $j = 1$, we set $C_{12}^{(1)} = 1/2$ and at $j = N+1$, $C_{12}^{(2)} = -1/2$. Also, we set $C_{11} \sim O(1/h)$. The matrix equations for $(k = 0)$ and $(k = 1)$ are given below.

For a constant element approximation $(k = 0)$, the matrix equation takes the following form:

for $u > 0$,

$$\begin{bmatrix} h_j & -(C_{12}^{(1)} + C_{12}^{(2)}) \\ D(C_{12}^{(1)} + C_{12}^{(2)}) & 2DC_{11} + u \end{bmatrix} \begin{bmatrix} q_j \\ T_j \end{bmatrix}$$

$$= \begin{pmatrix} (1/2 - C_{12}^{(2)})T_{j+1} - (1/2 + C_{12}^{(1)})T_{j-1} \\ D(1/2 + C_{12}^{(2)})q_{j+1} + D(-1/2 + C_{12}^{(1)})q_{j-1} + DC_{11}(T_{j-1} + T_{j+1}) + uT_{j-1} \end{pmatrix}$$

$$(5.9e)$$

and for $u \leq 0$,

$$\begin{bmatrix} h_j & -(C_{12}^{(1)} + C_{12}^{(2)}) \\ D(C_{12}^{(1)} + C_{12}^{(2)}) & 2DC_{11} - u \end{bmatrix} \begin{bmatrix} q_j \\ T_j \end{bmatrix}$$

$$= \begin{pmatrix} (1/2 - C_{12}^{(2)})T_{j+1} - (1/2 + C_{12}^{(1)})T_{j-1} \\ D(1/2 + C_{12}^{(2)})q_{j+1} + D(-1/2 + C_{12}^{(1)})q_{j-1} + DC_{11}(T_{j-1} + T_{j+1}) - uT_{j+1} \end{pmatrix}$$

$$(5.10e)$$

For the linear elements ($k=1$), the matrix equations have the forms below:

for $u > 0$,

$$\begin{bmatrix} h_j/3 & h_j/6 & -C_{12}^{(1)} & -1/2 \\ h_j/6 & h_j/3 & 1/2 & -C_{12}^{(2)} \\ DC_{12}^{(1)} & -D/2 & u/2 + DC_{11} & u/2 \\ D/2 & DC_{12}^{(2)} & -u/2 & u/2 + DC_{11} \end{bmatrix} \begin{pmatrix} q_j^+ \\ q_{j+1}^- \\ T_j^+ \\ T_{j+1}^- \end{pmatrix}$$

$$= \begin{pmatrix} -(1/2 + C_{12}^{(1)})T_j^- \\ (1/2 - C_{12}^{(2)})T_{j+1}^+ \\ -D(1/2 - C_{12}^{(1)})q_j^- + (DC_{11} + u)T_j^- \\ D(1/2 + C_{12}^{(2)})q_{j+1}^+ + DC_{11}T_{j+1}^+ \end{pmatrix}$$

$$(5.11e)$$

and for $u \leq 0$,

$$\begin{bmatrix} h_j/3 & h_j/6 & -C_{12}^{(1)} & -1/2 \\ h_j/6 & h_j/3 & 1/2 & -C_{12}^{(2)} \\ DC_{12}^{(1)} & -D/2 & -u/2 + DC_{11} & u/2 \\ D/2 & DC_{12}^{(2)} & -u/2 & -u/2 + DC_{11} \end{bmatrix} \begin{pmatrix} q_j^+ \\ q_{j+1}^- \\ T_j^+ \\ T_{j+1}^- \end{pmatrix}$$

$$= \begin{pmatrix} -(1/2 + C_{12}^{(1)})T_j^- \\ (1/2 - C_{12}^{(2)})T_{j+1}^+ \\ -D(1/2 - C_{12}^{(1)})q_j^- + DC_{11}T_j^- \\ D(1/2 + C_{12}^{(2)})q_{j+1}^+ + (DC_{11} - u)T_{j+1}^+ \end{pmatrix}$$

$$(5.12e)$$

Diffusion flux: $\hat{\mathbf{q}} = \{\mathbf{q}\} - C_{11}[T] - \mathbf{C}_{12}[\mathbf{q}]$ (5.111a)

Diffusion flux: $\hat{T} = \{T\} - C_{22}[\mathbf{q}] + \mathbf{C}_{12} \cdot [T]$ (5.111b)

Convection flux: $\overrightarrow{\mathbf{u}T} = \mathbf{u}\{T\} + \mathbf{C}_u \cdot [T]$ (5.111c)

Note that $\mathbf{C}_u$ is a matrix for convection flux.

With the approximation of unknowns using the polynomial basis functions and the Galerkin approach taken, the following matrix equations are obtained:

$$
\begin{bmatrix}
\mathbf{E} & \mathbf{0} & \mathbf{0} & \mathbf{H}_x \\
\mathbf{0} & \mathbf{E} & \mathbf{0} & \mathbf{H}_y \\
\mathbf{0} & \mathbf{0} & \mathbf{E} & \mathbf{H}_z \\
\mathbf{J}_x & \mathbf{J}_y & \mathbf{J}_z & \mathbf{G}_{T,u}
\end{bmatrix}
\begin{Bmatrix}
\underline{\mathbf{q}}_x \\
\underline{\mathbf{q}}_y \\
\underline{\mathbf{q}}_z \\
\underline{\mathbf{T}}
\end{Bmatrix}
$$

$$
+ \sum_{i=1}^{NS}
\begin{bmatrix}
\mathbf{E}_{xx,i} & \mathbf{E}_{xy,i} & \mathbf{E}_{xz,i} & \mathbf{H}_{x,i} \\
\mathbf{E}_{yx,i} & \mathbf{E}_{yy,i} & \mathbf{E}_{yz,i} & \mathbf{H}_{y,i} \\
\mathbf{E}_{zx,i} & \mathbf{E}_{zy,i} & \mathbf{E}_{zx,i} & \mathbf{H}_{z,i} \\
\mathbf{J}_{x,i} & \mathbf{J}_{y,i} & \mathbf{J}_{z,i} & \mathbf{G}_{T,i} + \mathbf{G}_{T,u,i}
\end{bmatrix}
\begin{Bmatrix}
\underline{\mathbf{q}}_x \\
\underline{\mathbf{q}}_y \\
\underline{\mathbf{q}}_z \\
\underline{\mathbf{T}}
\end{Bmatrix}
$$

$$
+ \sum_{i=1}^{NS}
\begin{bmatrix}
\mathbf{E}_{xx,B,i} & \mathbf{E}_{xy,B,i} & \mathbf{E}_{xz,B,i} & \mathbf{H}_{x,B,i} \\
\mathbf{E}_{yx,B,i} & \mathbf{E}_{yy,B,i} & \mathbf{E}_{yz,B,i} & \mathbf{H}_{y,B,i} \\
\mathbf{E}_{zx,B,i} & \mathbf{E}_{zy,B,i} & \mathbf{E}_{zz,B,i} & \mathbf{H}_{z,B,i} \\
\mathbf{J}_{x,B,i} & \mathbf{J}_{y,B,i} & \mathbf{J}_{z,B,i} & \mathbf{G}_{T,B,i} + \mathbf{G}_{T,u,B,i}
\end{bmatrix}
\begin{Bmatrix}
\underline{\mathbf{q}}_x \\
\underline{\mathbf{q}}_y \\
\underline{\mathbf{q}}_z \\
\underline{\mathbf{T}}
\end{Bmatrix}_{(NB,i)}
=
\begin{Bmatrix}
0 \\
0 \\
0 \\
\mathbf{S}_T
\end{Bmatrix}
$$

$$
(5.112)
$$

where NS is the number of sides of the element, the matrices with subscript i under summation are those from the boundaries of element j, and those marked with subscript B represent the contributions from the neighboring elements. The subscript (NB, i) represent the ith neighbor element that shares the ith side of element j. The vector $\mathbf{S}$ on the right includes the contribution from the source and boundary conditions, if the element shares its element boundary with the domain boundary. See Figure 5.3 for element j and its geometric relation to its surrounding elements.

The above equation is very similar to its counterpart of the pure conduction matrix equation in Equation 4.24 except for the convection term, which comes from the convection effect on temperature only. This portion of the contribution to the matrix is calculated as

$$
G_{T,u,km} = - \int_{\partial\Omega_{j,i}} (\mathbf{u} \cdot \nabla \phi_k) \phi_m \, dS
$$

$$
(5.113a)
$$

$$G_{T,u,i,km} = -G_{T,u,B,i,km} = \int_{\partial\Omega_{j,i}} (\mathbf{u}\cdot\mathbf{n}+\mathbf{n}\cdot\mathbf{C}_u\cdot\mathbf{n})\phi_k\phi_m dS \qquad (5.113b)$$

where the definition of the shape functions ϕ_k, and other matrix quantities in Equation 5.122, were given in Section 4.3.

For actual computations, the matrices may be assembled such that all the unknowns of element j are stored in the column vector of $\mathbf{U} = [\mathbf{q}_x^T, \mathbf{q}_y^T, \mathbf{q}_z^T, \mathbf{T}^T]^T = [q_{x,1}, q_{x,2}, \ldots, q_{x,N_e}, q_{y,1}, q_{y,2}, \ldots, q_{y,N_e}, q_{z,1}, q_{z,2}, \ldots, q_{z,N_e}, T_1, T_2, \ldots, T_{N_e}]^T$ with N_e being the number of the nodes of element j, and other available information from the neighbor elements is included in the force term $\mathbf{F}$, which also includes the contribution from $\mathbf{S}$. The final matrix equation takes the following form:

$$\mathbf{KU} = \mathbf{F} \qquad (5.113c)$$

where $\mathbf{K}$ is the resultant matrix having contributions from both the domain and the boundary. The computational procedure is exactly the same as used for the discontinuous finite element solution of the 1-D steady convection-diffusion problems discussed above, and thus is not elaborated upon here.

We emphasize that the oscillatory behavior associated with the 1-D convection-diffusion problems occurs also in multidimensional convection-diffusion problems. The analysis of this issue in multidimensional geometry, however, is much more difficult, because the matrix in general becomes much larger and more complex. The basic theorem governing the behavior of the matrix still holds, that is, the eigenvalues of the matrix must remain non-negative and real in order to avoid spurious oscillation in the numerical solutions.

The above algorithm has been applied to obtain the solution for 2-D steady state heat convection-conduction problems. One of these results is plotted in Figure 5.7. The solution shown is for the 2-D convection-diffusion problem defined by

$$u\frac{\partial T}{\partial x} = D\left(\frac{\partial^2 T}{\partial x^2} + \frac{\partial^2 T}{\partial y^2}\right) \qquad (5.114a)$$

with the boundary conditions

$$T(x,0) = 0 \; ; \; \frac{\partial T}{\partial y}(x,1) = 0 \; ; \; T(0,y) = \begin{cases} 1 & y \le 0.5 \\ 0 & y > 0.5 \end{cases} \text{ and } T(1,y) = 0 \quad (5.114b)$$

The discontinuous finite element discretization used an unstructured triangular mesh. The convection flux (Equation 5.111c) is used to model the convective effect. From Figure 5.7, it is apparent that as the ratio of convection over diffusion progressively increases, convection plays a more important role and the temperature contour becomes more distorted, which is consistent with the theory of transport processes associated with this problem.

5.3 Transient Convection-diffusion

We now study the time dependent problems of convection and diffusion. Since much of the framework is the same as for the steady state convection-diffusion problems, only the general discontinuous finite element formulation for multidimensional problems is presented here. This is then followed by stability analyses.

5.3.1 Multidimensional Problem

A transient convection-diffusion problem defined in a multidimensional domain takes the following form:

$$\frac{\partial T}{\partial t} + \mathbf{u} \cdot \nabla T = \nabla \cdot k\nabla T + Q \quad \in \Omega \times (0, \tau_T] \tag{5.115a}$$

$$T = T_D \quad \in \partial\Omega_D \times (0, \tau_T] \tag{5.115b}$$

$$-\mathbf{n} \cdot k\nabla T = h(T - T_\infty) \quad \in \partial\Omega_N \times (0, \tau_T] \tag{5.115c}$$

$$T(\mathbf{r}, t = 0) = T_0(\mathbf{r}) \quad \in \Omega \tag{5.115d}$$

To develop a discontinuous finite element formulation for the problem, the governing differential equation is split into a system of first order differential equations,

$$\mathbf{q} = k\nabla T \ ; \ \frac{\partial T}{\partial t} + \mathbf{u} \cdot \nabla T - \nabla \cdot \mathbf{q} - Q = 0 \tag{5.116}$$

The domain is now divided into a set of elements, and we consider a typical element, say, the jth element as shown in Figure 5.3. Multiplying the first and second equations in Equation 5.116 by test functions $\mathbf{w}$ and v and integrating over the element, one has the following form of solution:

$$\int_{\Omega_j} \mathbf{q} \cdot \mathbf{w} \, dV = \int_{\Omega_j} k\nabla T \cdot \mathbf{w} \, dV \tag{5.117a}$$

$$\int_{\Omega_j} v\frac{\partial T}{\partial t} \, dV + \int_{\Omega_j} v\mathbf{u} \cdot \nabla T \, dV - \int_{\Omega_j} v\nabla \cdot \mathbf{q} \, dV = \int_{\Omega_j} vQ \, dV \tag{5.117b}$$

Integration-by-parts and approximating the exact solution $(\mathbf{q}, T)$ with functions $(\mathbf{q}_h, T_h)$ in the finite element broken space, one obtains the desired discontinuous finite element solution,

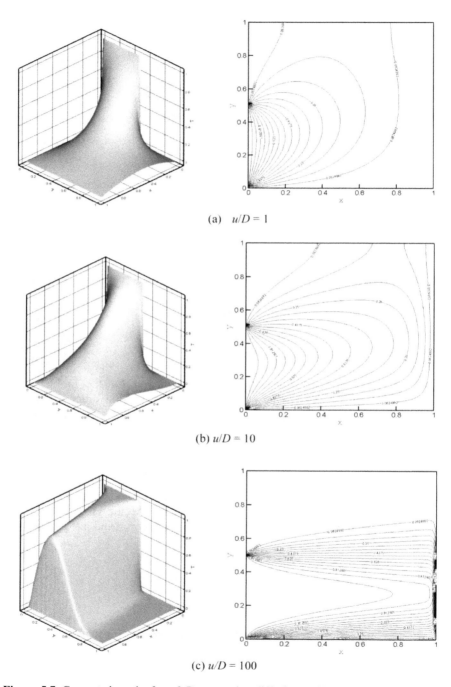

(a) $u/D = 1$

(b) $u/D = 10$

(c) $u/D = 100$

Figure 5.7. Computed results for a 2-D convection-diffusion problem showing the effect of convection on the temperature distribution in the system

$$\int_{\Omega_j} \mathbf{q}_h \cdot \mathbf{w}\, dV = -\int_{\Omega_j} T_h \nabla \cdot (k\mathbf{w})dV + \int_{\partial\Omega_j} k\hat{T}_h \mathbf{n}_j \cdot \mathbf{w}dS \qquad (5.118a)$$

$$\int_{\Omega_j} v\frac{\partial T}{\partial t}dV + \int_{\Omega_j} \mathbf{q}_h \cdot \nabla v\, dV - \int_{\Omega_j} T_h \mathbf{u} \cdot \nabla v\, dV$$

$$= \int_{\Omega_j} vQ dV + \int_{\partial\Omega_j} \hat{\mathbf{q}}_h \cdot \mathbf{n}_j v dS - \int_{\partial\Omega_j} \overrightarrow{\mathbf{u}T_h} \cdot \mathbf{n}_j v dS \qquad (5.118b)$$

where two types of numerical fluxes are used. One type is the diffusion numerical fluxes, ($\hat{\mathbf{q}}_h, \hat{T}_h$), which can be taken from Table 4.2. The convection numerical fluxes are of the form given by Equation 5.27. With the approximation of unknowns using the polynomial basis functions and taking the Galerkin approach, the following matrix equations are obtained:

$$\begin{bmatrix} 0 & 0 & 0 & 0 \\ 0 & 0 & 0 & 0 \\ 0 & 0 & 0 & 0 \\ 0 & 0 & 0 & \mathbf{M}_T \end{bmatrix} \begin{pmatrix} \underline{\dot{q}}_x \\ \underline{\dot{q}}_y \\ \underline{\dot{q}}_z \\ \underline{\dot{T}} \end{pmatrix} + \begin{bmatrix} \mathbf{E} & 0 & 0 & \mathbf{H}_x \\ 0 & \mathbf{E} & 0 & \mathbf{H}_y \\ 0 & 0 & \mathbf{E} & \mathbf{H}_z \\ \mathbf{J}_x & \mathbf{J}_y & \mathbf{J}_z & \mathbf{G}_{T,u} \end{bmatrix} \begin{pmatrix} \underline{q}_x \\ \underline{q}_y \\ \underline{q}_z \\ \underline{T} \end{pmatrix}$$

$$+ \sum_{i=1}^{NS} \begin{bmatrix} \mathbf{E}_{xx,B,i} & \mathbf{E}_{xy,B,i} & \mathbf{E}_{xz,B,i} & \mathbf{H}_{x,B,i} \\ \mathbf{E}_{yx,B,i} & \mathbf{E}_{yy,B,i} & \mathbf{E}_{yz,B,i} & \mathbf{H}_{y,B,i} \\ \mathbf{E}_{zx,B,i} & \mathbf{E}_{zy,B,i} & \mathbf{E}_{zz,B,i} & \mathbf{H}_{z,B,i} \\ \mathbf{J}_{x,B,i} & \mathbf{J}_{y,B,i} & \mathbf{J}_{z,B,i} & \mathbf{G}_{T,B,i}+\mathbf{G}_{T,u,B,i} \end{bmatrix} \begin{pmatrix} \underline{q}_x \\ \underline{q}_y \\ \underline{q}_z \\ \underline{T} \end{pmatrix}_{(NB,i)}$$

$$+ \sum_{i=1}^{NS} \begin{bmatrix} \mathbf{E}_{xx,i} & \mathbf{E}_{xz,i} & \mathbf{E}_{xy,i} & \mathbf{H}_{x,i} \\ \mathbf{E}_{yx,i} & \mathbf{E}_{yy,i} & \mathbf{E}_{yz,i} & \mathbf{H}_{y,i} \\ \mathbf{E}_{zx,i} & \mathbf{E}_{zy,i} & \mathbf{E}_{zz,i} & \mathbf{H}_{z,i} \\ \mathbf{J}_{x,i} & \mathbf{J}_{y,i} & \mathbf{J}_{z,i} & \mathbf{G}_{T,i}+\mathbf{G}_{T,u,i} \end{bmatrix} \begin{pmatrix} \underline{q}_x \\ \underline{q}_y \\ \underline{q}_z \\ \underline{T} \end{pmatrix} = \begin{pmatrix} 0 \\ 0 \\ 0 \\ \mathbf{S}_T \end{pmatrix} \qquad (5.119)$$

where $\mathbf{M}_T$ is the mass matrix associated with the temperature field, and the definitions of the other terms are the same as for the steady state parts (see Section 5.2.3 and Equation 5.112). In principle, one could write the above equation in terms of matrices and vectors,

$$\mathbf{M}\frac{d\mathbf{U}}{dt} + \mathbf{K}\mathbf{U} = \mathbf{F} \qquad (5.120)$$

where $\mathbf{M}$ is the resultant mass matrix, which, in this case, has a contribution from the mass matrix associated with temperature.

From Equations 5.119–5.120, the matrix $\mathbf{M}$ is a singular matrix. Thus in applications, it is often useful to separate the matrix equations into two subsystems with one for fluxes and the other for temperature:

$$
\begin{bmatrix} \mathbf{E} & \mathbf{0} & \mathbf{0} \\ \mathbf{0} & \mathbf{E} & \mathbf{0} \\ \mathbf{0} & \mathbf{0} & \mathbf{E} \end{bmatrix} \begin{pmatrix} \underline{\mathbf{q}}_x \\ \underline{\mathbf{q}}_y \\ \underline{\mathbf{q}}_z \end{pmatrix} + \begin{pmatrix} \mathbf{H}_x \\ \mathbf{H}_y \\ \mathbf{H}_z \end{pmatrix} \underline{\mathbf{T}} + \sum_{i=1}^{NS} \begin{bmatrix} \mathbf{E}_{xx,i} & \mathbf{E}_{xy,i} & \mathbf{E}_{xz,i} \\ \mathbf{E}_{yx,i} & \mathbf{E}_{yy,i} & \mathbf{E}_{yz,i} \\ \mathbf{E}_{zx,i} & \mathbf{E}_{zy,i} & \mathbf{E}_{zz,i} \end{bmatrix} \begin{pmatrix} \underline{\mathbf{q}}_x \\ \underline{\mathbf{q}}_y \\ \underline{\mathbf{q}}_z \end{pmatrix} + \begin{pmatrix} \mathbf{H}_{x,i} \\ \mathbf{H}_{y,i} \\ \mathbf{H}_{z,i} \end{pmatrix} \underline{\mathbf{T}}
$$

$$
+ \sum_{i=1}^{NS} \begin{bmatrix} \mathbf{E}_{xx,B,i} & \mathbf{E}_{xy,B,i} & \mathbf{E}_{xz,B,i} \\ \mathbf{E}_{yx,B,i} & \mathbf{E}_{yy,B,i} & \mathbf{E}_{yz,B,i} \\ \mathbf{E}_{zx,B,i} & \mathbf{E}_{zy,B,i} & \mathbf{E}_{zz,B,i} \end{bmatrix} \begin{pmatrix} \underline{\mathbf{q}}_x \\ \underline{\mathbf{q}}_y \\ \underline{\mathbf{q}}_z \end{pmatrix}_{(NB,i)} + \begin{pmatrix} \mathbf{H}_{x,B,i} \\ \mathbf{H}_{y,B,i} \\ \mathbf{H}_{z,B,i} \end{pmatrix} \underline{\mathbf{T}}_{(NB,i)} = \begin{pmatrix} \mathbf{S}_x \\ \mathbf{S}_y \\ \mathbf{S}_z \end{pmatrix}
$$

$$(5.121a)$$

$$
\mathbf{M}_T \frac{d\underline{\mathbf{T}}}{dt} + \mathbf{G}_{T,u}\underline{\mathbf{T}} + \sum_{i=1}^{NS}(\mathbf{G}_{T,i} + \mathbf{G}_{T,u,i})\underline{\mathbf{T}} + \sum_{i=1}^{NS}(\mathbf{G}_{T,B,i} + \mathbf{G}_{T,u,B,i})\underline{\mathbf{T}}_{(NB,i)}
$$

$$
+ \sum_{l=x,y,z} \mathbf{J}_l \underline{\mathbf{q}}_l + \sum_{i=1}^{NS}\left(\sum_{l=x,y,z} \mathbf{J}_{l,i}\underline{\mathbf{q}}_l + \sum_{l=x,y,z} \mathbf{J}_{l,B,i}\underline{\mathbf{q}}_{l(NB,i)} \right) = \mathbf{S}_T \quad (5.121b)
$$

The computational procedures may now be described as follows. For a given time step, the calculations start from an element located at the domain boundary and sweep element by element through the entire domain. The unknowns obtained for the element are immediately available to the neighboring elements, and are used in boundary source terms. This way the terms with subscript (NB,i) can be moved to the right-hand side of the equation and added to the source. At every time step, the two equations above need to be solved iteratively. A typical procedure is as follows. The temperature is first calculated using Equation 5.121b with the $\underline{\mathbf{q}}$ values available, and then $\underline{\mathbf{q}}$ is updated using Equation 5.121a. The procedure is iterative and iteration is considered converged if the successive solutions are within a preset tolerance (see Equation 4.14). Then the next time step is selected, and the above procedure is repeated until the total time is equal to the preset value.

As an alternative treatment, the vector $\underline{\mathbf{q}}$ may be eliminated by combining the two equations above. If this is done, then we have the matrix equation in terms of $\underline{\mathbf{T}}$ only,

$$
\mathbf{M}_T \frac{d\underline{\mathbf{T}}}{dt} + \mathbf{K}\underline{\mathbf{T}} = \mathbf{F} \tag{5.122}
$$

where $\mathbf{K}$ is the combined element stiffness matrix and $\mathbf{F}$ is the combined source vector. Equation 5.122 can then be integrated using a time integrator.

Whichever procedure one chooses, the oscillatory behavior associated with steady convection-diffusion problems may also appear in transient solutions. In particular, they may occur near the sharp fronts if appropriate conditions are not satisfied. To perform the analysis of spurious oscillations in the solution, it is often more convenient to assemble the element equations shown in Equation 5.122 into a global matrix equation,

$$\frac{d\mathbf{T}_g}{dt} = -\mathbf{M}_g^{-1}\mathbf{K}_g\mathbf{T}_g + \mathbf{M}_g^{-1}\mathbf{F}_g = -\mathbf{A}\mathbf{U} + \mathbf{B} \tag{5.123}$$

where subscript g denotes global quantities, $\mathbf{A} = \mathbf{M}_g^{-1}\mathbf{K}_g$ and $\mathbf{B} = \mathbf{M}_g^{-1}\mathbf{F}_g$. Carey and Oden [10] analyzed the eigenvalues for the above equation system and show that, as a rule of thumb, the eigenvalues of the matrix $-\mathbf{A}$ must be negative and real in order to eliminate unphysical oscillations in the numerical solution.

The additional integral and Fourier analysis of the transient convection-diffusion problems will be discussed in the next section. An important point to remember is that if an explicit time step is used, the critical time step needs to be selected. Here, let us consider some calculated results.

The above algorithm, based on Equation 5.121, has been applied to solve a 2-D convection-diffusion problem. The governing equation for the problem has the following statement:

$$\frac{\partial T}{\partial t} + u\frac{\partial T}{\partial x} = D\left(\frac{\partial^2 T}{\partial x^2} + \frac{\partial^2 T}{\partial y^2}\right) \qquad \in [0,1]\times[0,1]\times(0,\infty] \tag{5.124a}$$

with the boundary conditions

$$T(x,0,t) = 0 \;;\; \frac{\partial T}{\partial y}(x,1,t) = 0 \;;\; T(0,y,t) = \begin{cases} 1 & y \le 0.5 \\ 0 & y > 0.5 \end{cases} \text{ and } T(1,y,t) = 0$$

$$\tag{5.124b}$$

and the initial data

$$T(x,y,t) = \begin{cases} 1 - x^2 & y \le 0.5 \\ 0 & y > 0.5 \end{cases} \tag{5.124c}$$

An unstructured mesh and linear triangular elements are used to carry out the calculations. An explicit time integration is used with a time step selected to satisfy the restricted *CLF* condition. The computed results are given in Figure 5.8.

5.3.2 Stability Analysis

5.3.2.1 L^2-Stability – Integral Analysis
A complete L^2-stability analysis, including error bounds, has been given by Cockburn and Shu [11] for general nonlinear convection-diffusion problems. For a

basic understanding, a linear problem is analyzed here to show how the stability criterion from the theory of partial differential equations [2] can be used to enforce stability on a numerical scheme, and thus to provide guidance on the choice of numerical fluxes.

The linear convection-diffusion problem to be analyzed is mathematically stated as

$$\frac{\partial T}{\partial t} + \nabla \cdot (\mathbf{u}T) = D\nabla^2 T \tag{5.125}$$

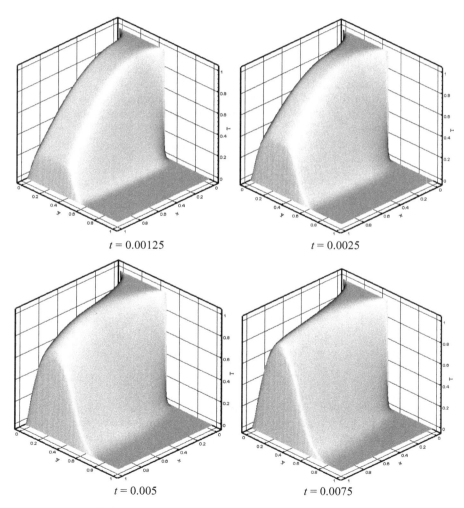

$t = 0.00125$ $t = 0.0025$

$t = 0.005$ $t = 0.0075$

Figure 5.8. Evolution of temperature distribution in a convection-diffusion problem calculated using the discontinuous finite element method ($D/u = 0.01$)

with a periodic boundary condition. Here **u** is a constant vector. As usual, we first work out the L^2-stability result for the continuous case, and enforce the condition on the discontinuous formulation. Thus, the above equation is integrated with respect to T over the spatial domain,

$$\int_\Omega T\frac{\partial T}{\partial t}dV + \int_\Omega T\nabla\cdot(\mathbf{u}T)dV = \int_\Omega T\nabla\cdot(\nabla T)dV \qquad (5.126)$$

Integrating by parts, one has

$$\frac{d}{dt}\int_\Omega \frac{T^2}{2}dV + \frac{1}{2}\int_{\partial\Omega} T^2\mathbf{u}\cdot\mathbf{n}dS = \int_{\partial\Omega} TD\nabla T\cdot\mathbf{n}dS - \int_\Omega D\nabla T\cdot\nabla TdV \qquad (5.127)$$

Applying the periodic boundary conditions in Equation 5.127 yields the following equation,

$$\frac{d}{dt}\int_\Omega \frac{T^2}{2}dV + \int_\Omega \mathbf{q}\cdot\mathbf{q}dV = 0 \quad\text{with}\quad \mathbf{q}=\sqrt{D}\nabla T \qquad (5.128)$$

Here, q is defined following Cockburn and Shu [11]. Integrating over $t\in[0,\tau_T]$, one has the stability result for the continuous case,

$$\frac{1}{2}\int_\Omega T^2(\mathbf{r},\tau_T)dV + \int_0^{\tau_T}\int_\Omega \mathbf{q}\cdot\mathbf{q}dV = \frac{1}{2}\int_\Omega T^2(\mathbf{r},0)dV \qquad (5.129)$$

For the discontinuous finite element formulation, we obtain the following equation by summing up the contributions from all elements:

$$\frac{1}{2}\int_\Omega T_h^2(\mathbf{r},\tau_T)dV + \int_0^{\tau_T}\int_\Omega q_h^2 dV + \Theta_{T,C}([\mathbf{w}]) = \frac{1}{2}\int_\Omega T_h^2(\mathbf{r},0)dV \qquad (5.130)$$

where $\Theta_{T,C}([\mathbf{w}])$ represents the errors due to the jumps across the inter-element boundaries. It is calculated by

$$\Theta_{T,C}([\mathbf{w}]) = \sum_{j=1}^N \int_0^{\tau_T}\int_{\partial\Omega_j} [\mathbf{w}(t)]^T\mathbf{C}_E[\mathbf{w}(t)]dS\,dt \qquad (5.131a)$$

where $[\mathbf{w}(t)]$ is defined as

$$\mathbf{w}(t) = \left(T_h, q_{x,h}, q_{y,h}, q_{z,h}\right)^T \qquad (5.131b)$$

and $\mathbf{C}_E$ is a matrix:

$$\mathbf{C}_E = \begin{pmatrix} |u|/2 & -\sqrt{D}/2 & -\sqrt{D}/2 & -\sqrt{D}/2 \\ \sqrt{D}/2 & 0 & 0 & 0 \\ \sqrt{D}/2 & 0 & 0 & 0 \\ \sqrt{D}/2 & 0 & 0 & 0 \end{pmatrix} \qquad (5.131c)$$

From Equation 5.130, it transpires that one needs to have $\Theta_{T,C}([\mathbf{w}]) \geq 0$ to ensure the stability.

5.3.2.2 L^2-Stability – Discretized Analysis

Here we consider the stability analysis of discretized equations. Toward this end, we consider the 1-D convection-diffusion problem as follows:

$$\frac{\partial T}{\partial t} + u\frac{\partial T}{\partial x} = D\frac{\partial^2 T}{\partial x^2} \qquad (5.132)$$

To simplify the notation below, an inner product operator is introduced,

$$(a,b) = \int a\,b\,dV \qquad (5.133)$$

For an element, e.g., element j, we have the following discretized discontinuous finite element formulation:

$$\left(v,\frac{\partial T}{\partial t}\right)_{\Omega_j} + \left(v,a(\tfrac{u+|u|}{2},\tfrac{u-|u|}{2})T\right)_{\Omega_j} = \left(v,a(1,1)q\right)_{\Omega_j} \qquad (5.134a)$$

$$\left(v,q\right)_{\Omega_j} = \left(v,Da(1,1)C\right)_{\Omega_j} \qquad (5.134b)$$

where v is the weighting function, and the operator $a(s_1, s_2)$ is defined by the following expression:

$$(v,a(s_1,s_2)w)_{\Omega_j} = (v,w)_{\Omega_j} + s_1 v(x_{j+1})\left(\frac{w_{j+1}(x_{j+1}) - w_j(x_{j+1})}{2}\right)$$

$$+ s_2 v(x_j)\left(\frac{w_j(x_j) - w_{j-1}(x_j)}{2}\right) \qquad (5.135)$$

For a 1-D case, Warburton [12] shows that the above operator may be written in a general form,

$$(v,a(s_1,s_2)w)_{\Omega_j} = s_1\left(v_j(x_{j+1})+v_{j+1}(x_{j+1})\right)\left(\frac{w_{j+1}(x_{j+1})-w_j(x_{j+1})}{2}\right)$$

$$+(v,w)_{\Omega_j}+s_2\left(v_j(x_j)+v_{j-1}(x_j)\right)\left(\frac{w_j(x_j)-w_{j-1}(x_j)}{2}\right) \qquad (5.136)$$

This will equip the operator with the following property:

$$\left(v,a(1,1)T\right)_{\Omega_j} = -\left(a(1,1)v,T\right)_{\Omega_j} \qquad (5.137)$$

Consequently, we have the following relation:

$$\left(v,\frac{\partial T}{\partial t}\right)_{\Omega_j}+\left(v,ua(\tfrac{u+|u|}{2},\tfrac{u-|u|}{2})T\right)_{\Omega_j} = (v,Da(1,1)q)_{\Omega_j} \qquad (5.138)$$

$$(v,q)_{\Omega_j} = (v,Da(1,1)T)_{\Omega_j} = -\left(Da(1,1)v,T\right)_{\Omega_j} \qquad (5.139)$$

A combination of the two equations above yields

$$\left(v,\frac{\partial T}{\partial t}\right)_{\Omega_j} = -\left(v,ua(\tfrac{u+|u|}{2},\tfrac{u-|u|}{2})T\right)_{\Omega_j}-\left(a(1,1)v,Da(1,1)T\right)_{\Omega_j} \qquad (5.140)$$

Setting $v = T$, and summing up over all elements, one has the following stability result:

$$\frac{1}{2}\frac{d}{dt}\|T\|_2^2 = -\sum_{j=1}^{N}\left(T,ua(\tfrac{u+|u|}{2},\tfrac{u-|u|}{2})T\right)_{\Omega_j}-\sum_{j=1}^{N}\left(a(1,1)T,Da(1,1)T\right)_{\Omega_j}$$

$$= -\frac{|u|}{2}\sum_{j=1}^{N}(T_j(x_j)-T_{j-1}(x_j))^2-D\sum_{j=1}^{N}\left(a(1,1)T,a(1,1)T\right)_{\Omega_j} \leq 0 \quad (5.141)$$

5.3.2.3 Fourier Analysis
We present a generalized Fourier analysis for the transient convection-diffusion problems in this section, and will use the analysis to determine the critical time steps for explicit time integration for the solution of the problem. Toward this end, we consider the discretized ordinary different equation system of the form,

$$\frac{d\mathbf{U}}{dt} = -\mathbf{A}\mathbf{U}+\mathbf{B} \qquad (5.142)$$

If the simple Euler forward scheme is used for time integration, then the following numerical implementation is obtained:

$$\mathbf{U}(t + \Delta t) = (\mathbf{I} - \mathbf{A}\Delta t)\mathbf{U}(t) + \mathbf{B}\Delta t \tag{5.143}$$

For the purpose of determining the critical time step, $\mathbf{B} = \mathbf{0}$ may be set. As a result, the above equation can be written for errors,

$$\varepsilon(t + \Delta t) = (\mathbf{I} - \mathbf{A}\Delta t)\varepsilon(t) \tag{5.144}$$

where $\varepsilon(t) = \mathbf{U}(t) - \mathbf{U}_{exact}(t)$. Repeated application of the above recursive relation gives the following equation:

$$\varepsilon((m + 1)\Delta t) = (\mathbf{I} - \mathbf{A}\Delta t)^{m+1}\varepsilon(0) \tag{5.145}$$

For a stable time marching scheme, the amplification factor should decrease,

$$\frac{|\varepsilon((m + 1)\Delta t)|}{|\varepsilon(0)|} < 1 \tag{5.146}$$

This condition requires that $|\mathbf{I} - \mathbf{A}\Delta t|^{m+1} < 1$, which means

$$|1 - \lambda_i \Delta t| \le 1 \quad \text{or} \quad \Delta t \le 2/\max \lambda_i \tag{5.147}$$

where λ_i is an eigenvalue of the matrix $\mathbf{A}$. The convection often causes matrix $\mathbf{A}$ to be asymmetric and the eigenvalues may be complex. Assuming $\lambda_i = \upsilon + j\mu$, we then have the stability constraint that defines the critical time step used for the explicit time marching,

$$\Delta t \le \frac{2\upsilon}{\mu^2 + \upsilon^2} \tag{5.148}$$

Here the imaginary part μ provides the oscillation, but not growth. The above critical time step is obtained from the time marching stability consideration, which requires that the errors at every ensuing step are not larger than the previous error. For convection-diffusion problems, the convection term causes numerical oscillation, as we have seen in the steady state case. For the transient case, the oscillation levies another time step constraint on the time marching scheme.

Let us determine the time step from the requirement that unphysical oscillation be suppressed. Using the exponential matrix $\exp(\mathbf{A}t)$, the exact solution to the differential equation (Equation 5.142) may be written as [13]

$$\frac{d(\exp(\mathbf{A}t)\mathbf{U})}{dt} = \exp(\mathbf{A}t)\mathbf{B} \tag{5.149}$$

On integration over $[t_0, t]$, one has the following result:

$$\mathbf{U}(t) = \exp(-\mathbf{A}(t-t_0))\mathbf{U}(t_0) + (1 - \exp(-\mathbf{A}(t-t_0)))\mathbf{A}^{-1}\mathbf{B} \qquad (5.150)$$

Given the initial data $\mathbf{U}(0) = \mathbf{0}$, the above recursive relation will become, after m time steps,

$$\mathbf{U}(t) = \{1 - \exp(-m\mathbf{A}\Delta t)\}\mathbf{A}^{-1}\mathbf{B} \qquad (5.151)$$

where we have used $t = t_0 + \Delta t$. To avoid numerical oscillation, we need to have the condition,

$$\exp(-\mathbf{A}\Delta t) > 0 \qquad (5.152)$$

Otherwise, $\exp(-m\mathbf{A}\Delta t)$ will change sign from one time step to the next, which causes oscillation. Carey and Oden [10] showed that the exponential matrix can be calculated using the Pade approximations,

$$\exp(-\mathbf{A}\Delta t) \approx \frac{P(\mathbf{A}\Delta t)}{Q(\mathbf{A}\Delta t)} \qquad (5.153)$$

where $P(\mathbf{A}\Delta t)$ and $Q(\mathbf{A}\Delta t)$ are polynomials of $\mathbf{A}\Delta t$. In a simple Euler forward scheme, $P(\mathbf{A}\Delta t) = \mathbf{I} - \mathbf{A}\Delta t$ and $Q(\mathbf{A}\Delta t) = \mathbf{I}$. Thus the solution will oscillate if

$$\frac{P(\lambda_i \Delta t)}{Q(\lambda_i \Delta t)} < 0 \quad \text{or} \quad 1 - \lambda_i \Delta t < 0 \qquad (5.154)$$

Therefore, if no numerical oscillation occurs for this time scheme, then the time step must be such that

$$\Delta t \leq \frac{1}{\max \lambda_i} \qquad (5.155)$$

In comparison with Equation 5.147, we see that the time step limit to suppress oscillations due to integration is half that required for stability of the forward time integration scheme.

Additional analysis of the problem is also given in the context of finite volume methods, where the time step required for explicit time integration is discussed for various schemes [3]. Warburton [12] argues that the critical time step should take into account both the convection and diffusion effects, and he showed that $\Delta t = \min(c_1 u h / k^2, c_2 D h^2 / k^4)$ to ensure stability, where h is the mesh size, k is the order of polynomial, and c_1 and c_2 are two Courant constants. The first term accounts for convection, while the second accounts for diffusion. This argument is consistent with the analysis above.

5.4 Nonlinear Problems

Thus far, our attention has been on the linear convection and diffusion problems. Another major class is the nonlinear convection-diffusion problems, where nonlinear convection plays a crucial role in generating and propagating sharp fronts and discontinuities such as shock waves in compressible flows. Benchmark problems for nonlinear convection and diffusion are Burgers' equations. We consider both inviscid and viscous forms of Burgers' equations.

5.4.1 1-D Inviscid Burgers' Equation

5.4.1.1 Basic Considerations
Let us start with a 1-D inviscid Burgers' equation, which is an idealized case for shock wave and rarefaction wave phenomena in compressible fluid flow and heat transfer systems. The 1-D inviscid Burgers' equation has the following statement:

$$\frac{\partial u}{\partial t} + u\frac{\partial u}{\partial x} = 0 \qquad (5.156a)$$

with the initial data given by

$$u(x,0) = \begin{cases} u_L & x < -\delta \\ \frac{u_L}{2\delta}(\delta - x) + \frac{u_R}{2\delta}(\delta + x) & -\delta < x < \delta \\ u_R & \delta < x \end{cases} \qquad (5.156b)$$

For this problem, the characteristic curve along which u is constant is given by

$$\frac{dx}{dt} = u = C; \quad \frac{du}{dt} = 0 \qquad (5.157)$$

where C is a constant. By integrating the above equation, the following expressions for the characteristic curves are obtained:

$$x = Ct + x_0; \quad u = C \qquad (5.158)$$

The two constant regions in the initial data are carried into the domain from the boundary, and remain constant along the characteristic curves, which become, on applying the initial data,

$$x_0 = x - Ct < -\delta; \quad x - Ct > \delta \qquad (5.159)$$

Over the range $-\delta < x_0 \le \delta$, the characteristic curves are dependent upon the relative values of u_L and u_R. In the case of $u_L < u_R$,

$$C - \frac{\delta}{t} < \frac{x}{t} < C + \frac{\delta}{t}; \quad u = C = \frac{x - x_0}{t} \tag{5.160}$$

If the limiting case is taken such that $x_0 \to \delta \to 0$, then we have the following limits for u:

$$C(x_0^-) = \lim_{\substack{\delta \to 0 \\ x_0 \to \delta-}} \frac{u_L}{2\delta}(\delta - x_0) + \frac{u_R}{2\delta}(\delta + x_0) = u_L \tag{5.161a}$$

$$C(x_0^+) = \lim_{\substack{\delta \to 0 \\ x_0 \to \delta+}} \frac{u_L}{2\delta}(\delta - x_0) + \frac{u_R}{2\delta}(\delta + x_0) = u_R \tag{5.161b}$$

$$u = \lim_{x_0 \to \delta} \frac{x - x_0}{t} = \frac{x}{t} \tag{5.161c}$$

where the first equation is in the region $-\delta < x_0 < 0$ and the second equation is taken in the region $0 < x_0 < -\delta$. Thus, the solution becomes

$$u_L < \frac{x}{t} < u_R; \quad u = \frac{x}{t} \tag{5.162}$$

If $u_L > u_R$, however, the above limiting process would imply that $u_L < u_R$, which would violate the given condition. Let us then consider the limiting process: $\delta \to x_0 \to 0$,

$$C(x_0^-) = \lim_{\substack{\delta \to 0 \\ x_0 \to \delta-}} \frac{u_L}{2\delta}(\delta - x_0) + \frac{u_R}{2\delta}(\delta + x_0) = \frac{u_R + u_L}{2} \tag{5.163a}$$

$$C(x_0^+) = \lim_{\substack{\delta \to 0 \\ x_0 \to \delta+}} \frac{u_L}{2\delta}(\delta - x_0) + \frac{u_R}{2\delta}(\delta + x_0) = \frac{u_R + u_L}{2} \tag{5.163b}$$

We thus have the following consistent expressions:

$$\frac{x}{t} = \frac{u_R + u_L}{2}; \quad u = C = \frac{u_R + u_L}{2} \tag{5.164}$$

To summarize the above results, for initial data,

$$u(x,0) = \begin{cases} u_L & x < 0 \\ \frac{u_L + u_R}{2} & x = 0, \\ u_R & 0 < x \end{cases} \qquad u_L > u_R \tag{5.165}$$

we have the following solution for $u(x,t)$:

$$u(x,t) = \begin{cases} u_L & \text{if } x/t < (u_L + u_R)/2 \\ (u_L + u_R)/2 & \text{if } x/t = (u_L + u_R)/2 \\ u_R & \text{if } x/t > (u_L + u_R)/2 \end{cases} \qquad (5.166)$$

and for initial data,

$$u(x,0) = \begin{cases} u_L & x < 0 \\ \frac{u_L + u_R}{2} & x = 0, \\ u_R & 0 < x \end{cases} \qquad u_L < u_R \qquad (5.167)$$

the solution for $u(x,t)$ is given by

$$u(x,t) = \begin{cases} u_L & \text{if } x/t \le u_L \\ x/t & \text{if } u_L < x/t < u_R \\ u_R & \text{if } x/t \ge u_R \end{cases} \qquad (5.168)$$

These results are plotted in Figures 5.9 and 5.10, where rarefaction waves and shock waves are generated depending on the initial data. For the case of initial data $u_L < u_R$, rarefaction waves are generated and the fan of characteristics populate the area that emanates from the origin, as shown in Figure 5.9. For the case of initial data $u_L > u_R$, a shock wave is degenerated and the discontinuity present in the initial data propagate from the left to the right, as sketched in Figure 5.10. This basic understanding will guide us to develop various discontinuous finite element algorithms, which allow appropriate handling of these discontinuities.

5.4.1.2 Discontinuous Finite Element Formulation
We now consider the discontinuous finite element formulation for the inviscid Burgers' equation,

$$\frac{\partial u}{\partial t} + u \frac{\partial u}{\partial x} = 0 \qquad u(x,0) = u_0(x) \qquad (5.169)$$

with the boundary conditions $u(0) = 1$ and $u(1) = 0$. The above equation may be written in a generic form,

$$\frac{\partial u}{\partial t} + \frac{\partial f(u)}{\partial x} = 0 \qquad (5.170)$$

where $f(u) = 0.5u^2$.

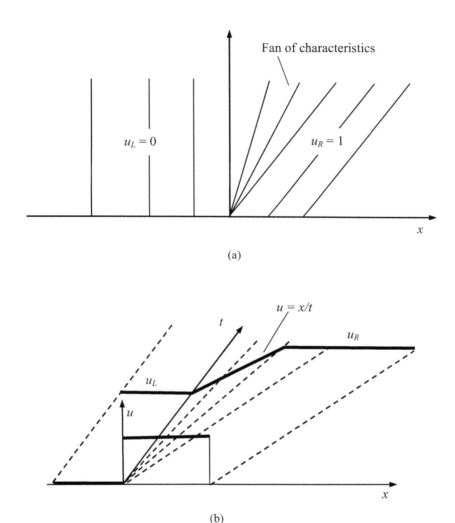

Figure 5.9. Solution of an inviscid Burgers' equation with initial data: $u_L < u_R$, which gives rise to the rarefaction waves: (a) characteristic curves – a fan of characteristics emanating from the boundary singularity as if a fluid flows out of a source, and (b) a 3-D view of a wave profile at time t evolving from that at $t = 0$: u varies from 0 to 1 along the characteristic fan indicating expansion (or rarefaction) waves spreading out as time increases

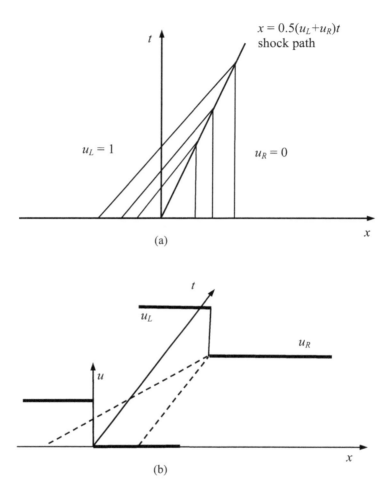

Figure 5.10. Solution of an inviscid Burgers' equation with the initial data: $u_L > u_R$. Shock waves are generated and propagate from left to right: (*a*) characteristics of shock waves and (*b*) 3-D view of shock wave in motion

To apply the discontinuous finite element method, the domain is first discretized into N elements. The above equation is then integrated with respect to the weighting function v over element j,

$$\int_{x_j}^{x_{j+1}} \left(v \frac{\partial u_h}{\partial t} - f(u_h) \frac{\partial v}{\partial x} \right) dx + v(x_{j+1}) \hat{h}(u_{j+1}^-, u_{j+1}^+)$$

$$-v(x_j)\hat{h}(u_j^-, u_j^+) = 0 \qquad (5.171)$$

where the numerical fluxes $\hat{h}(a, b)$ have been used at the element boundaries.

Cockburn [4, 5] showed that a well behaved numerical flux needs to satisfy the following conditions: (*i*) locally Lipschitz and consistent with the flux $f(u)$, i.e., $\hat{h}(u, u) = f(u)$, (*ii*) a non-decreasing function of its first argument, and (*iii*) a non-increasing function of its second argument. Some of the numerical fluxes have been widely used in the definite difference approximations, and are found also to satisfy these consistency conditions. These fluxes are listed below for convenience:

(i) the Godunov flux:

$$\hat{h}^G(a, b) = \begin{cases} \min_{a \le T \le b} f(u), & \text{if } a \le b \\ \max_{b \le T \le a} f(u), & \text{otherwise.} \end{cases} \qquad (5.172)$$

(ii) the Engquist–Osher flux:

$$\hat{h}^{EO}(a, b) = \int_0^b \min(f'(s), 0)ds + \int_0^b \max(f'(s), 0)ds + f(0) \qquad (5.173)$$

(iii) the Lax–Friedrichs flux:

$$\hat{h}^{LF}(a, b) = \frac{1}{2}[f(a) + f(b) - C(b - a)] \qquad (5.174a)$$

$$C = \max_{\inf u^0(x) \le s \le \sup u^0(x)} |f'(s)| \qquad (5.174b)$$

(iv) the local Lax–Friedrichs flux:

$$\hat{h}^{LLF}(a, b) = \frac{1}{2}[f(a) + f(b) - C(b - a)] \qquad (5.175a)$$

$$C = \max_{\min(a,b)(x) \le s \le \max(a,b)} |f'(s)| \qquad (5.175b)$$

(v) the Roe flux with "entropy fix":

$$\hat{h}^{Reo}(a, b) = \begin{cases} f(a), & \text{if } f'(u) \ge 0 \text{ for } u \in [\min(a, b), \max(a, b)] \\ f(b), & \text{if } f'(u) \le 0 \text{ for } u \in [\min(a, b), \max(a, b)] \\ \hat{h}^{LLF}(a, b), & \text{otherwise} \end{cases} \qquad (5.176)$$

For an inviscid Burgers' problem, the Godunov flux $\hat{h}^G$ may be used, which is known to produce the smallest amount of artificial viscosity. The local Lax–Friedrichs flux produces more artificial viscosity than the Godunov flux, with similar numerical performance. Cockburn [4] shows that as the degree k of the approximate solution increases, the choice of the numerical flux does not have a significant impact on the quality of the approximations.

The calculations can begin once the numerical flux is selected. The unknowns may be approximated with a polynomial as local basis functions,

$$u_h(x,t) = \sum_{i=1}^{k+1} u_i(t)\phi_i(x) \tag{5.177}$$

where $k + 1 = Ne$, with Ne being the number of nodes per element. Substituting the above equations into Equation 5.171 and taking the Galerkin approximation by setting $v = u_h$, followed by the numerical manipulations, we have the following matrix equation:

$$\mathbf{M}\frac{d\mathbf{U}_{(j)}}{dt} = L(\mathbf{U}) \tag{5.178}$$

where the subscript $(\cdot)$ refers to the element. For example, $\mathbf{U}_{(j)}$ denotes the unknowns belonging to element j, and $\mathbf{U}$ is the vector containing unknowns at the nodal points of the elements,

$$\mathbf{U}_{(j)} = [u_1, u_2, \cdots, u_{k+1}]_{(j)}^T \tag{5.179}$$

$$\mathbf{U}_{(j-1)} = [u_{k+1}, 0, \cdots, 0]_{(j-1)}^T , \; \mathbf{U}_{(j+1)} = [0, 0, \cdots, u_1]_{(j+1)}^T \tag{5.180}$$

The operator L depends on the choice of the numerical functional flux and it can be written in general as

$$L(\mathbf{U}) = \int_{x_j}^{x_{j+1}} \frac{\partial \mathbf{\Phi}}{\partial x} f(\mathbf{\Phi}^T(x)\mathbf{U}) \, dx + \begin{pmatrix} \hat{h}(u_j^-, u_j^+) \\ -\hat{h}(u_{j+1}^-, u_{j+1}^+) \end{pmatrix} \tag{5.181}$$

Time integration is now applied to solve the ordinary differential equations resulting from the spatial discretization. If a simple Euler forward scheme is used, then we have the following time-discretized form for Equation 5.178:

$$\mathbf{U}_{(j)}^{n+1} = \mathbf{U}_{(j)}^n + \Delta t^n \mathbf{M}^{-1} L(\mathbf{U}^n) \tag{5.182}$$

where the superscript n denotes the nth time stepping and Δt^n is the time step, for which the critical time step must be chosen to enforce the *CFL* condition. Various integration schemes, discussed for linear convection-diffusion equations, can also be employed to integrate the above system of equations in time. It is noted that the above approach is linear in accuracy, because of the use of the numerical fluxes presented above. Higher order schemes require additional considerations, which will be discussed in Section 5.4.3.

5.4.2 Multidimensional Inviscid Burgers' Equation

The discontinuous finite element method for 1-D pure convection problems discussed above is extended to solve multidimensional inviscid Burgers' problems in this section. We discuss below the formulation and the use of the chracteristics decomposition method for the computation of numerical fluxes required to complete a discontinuous finite element formulation.

5.4.2.1 Discontinuous Finite Element Formulation
Let us consider the generalized Burgers' equation in a multidimensional domain,

$$\mathbf{u}_t + \nabla \cdot \mathbf{F}(\mathbf{u}) = \mathbf{G}(\mathbf{r},t,\mathbf{u}) \quad \in \Omega \quad t > 0 \tag{5.183a}$$

$$\mathbf{u}(\mathbf{r},0) = \mathbf{0} \quad \in \Omega \cup \partial\Omega \tag{5.183b}$$

where $\mathbf{u}_t = \partial\mathbf{u}/\partial t$, the nonlinear function $\mathbf{F}(\mathbf{u})$ and its derivative are defined as

$$\mathbf{F}(\mathbf{u}) = [\mathbf{f}(\mathbf{u}),\mathbf{g}(\mathbf{u}),\mathbf{h}(\mathbf{u})] ; \quad \nabla \cdot \mathbf{F}(\mathbf{u}) = \frac{\partial \mathbf{f}(\mathbf{u})}{\partial x} + \frac{\partial \mathbf{g}(\mathbf{u})}{\partial y} + \frac{\partial \mathbf{h}(\mathbf{u})}{\partial z} \tag{5.184}$$

As usual, to develop a discontinuous formulation for the problem, the domain is first partitioned into a set of finite elements and a weak form solution is constructed over an element, say, element j. The procedure involves integrating the differential equation with respect to a weighting function $\mathbf{v}$, and integration by parts, to generate a flux term at the boundary,

$$\int_{\Omega_j} \mathbf{v} \cdot \mathbf{u}_t dV + \int_{\partial\Omega_j} \mathbf{v} \cdot \mathbf{F} \cdot \mathbf{n}\, dS - \int_{\Omega_j} \mathbf{F} : \nabla\mathbf{v} dV = \int_{\Omega_j} \mathbf{v} \cdot \mathbf{G} dV \tag{5.185a}$$

where $\mathbf{n}$ is the outward normal of the element and $\mathbf{F}:\nabla\mathbf{v}$ is defined by

$$\mathbf{F} : \nabla\mathbf{v} = \mathbf{f}(\mathbf{u}) \cdot \partial_x\mathbf{v} + \mathbf{g}(\mathbf{u}) \cdot \partial_y\mathbf{v} + \mathbf{h}(\mathbf{u}) \cdot \partial_z\mathbf{v} \tag{5.185b}$$

and $\mathbf{v} \cdot \mathbf{F} \cdot \mathbf{n}$ is calculated by

$$\mathbf{v} \cdot \mathbf{F} \cdot \mathbf{n} = \mathbf{v} \cdot \mathbf{f}(\mathbf{u})n_x + \mathbf{v} \cdot \mathbf{g}(\mathbf{u})n_y + \mathbf{v} \cdot \mathbf{h}(\mathbf{u})n_z \tag{5.185c}$$

Since only the normal fluxes are involved in surface integration, the 1-D forms of numerical fluxes discussed in the previous section may be used in the normal direction for this problem. These normal numerical fluxes $\mathbf{H}_n$ can be constructed using the values available from adjacent elements during computation and, as in other cases considered so far, replace the term $\mathbf{F} \cdot \mathbf{n}$ in the above formulation. This is very similar to the procedure discussed above for 1-D inviscid Burgers' equations, and thus all the flux expressions (such as the Godunov flux, etc.) presented in the previous section are candidates for multidimensional numerical fluxes, $\mathbf{H}_n$.

Let (NB,k) $(k = 1, 2, 3, \ldots, NS)$ denote the indices of the NS elements that are neighboring elements of element j and let $\partial\Omega_{j,k}$, $k = 1, 2, 3, \ldots, NS$, be the faces of Ω_j, which are shared with the neighboring elements. See Figure 5.3 for the geometric arrangement of element j and its neighbors. Then the weak form integral is written as

$$\int_{\Omega_j} \mathbf{v} \cdot \mathbf{u}_t \, dV + \sum_{k=1}^{NS} \int_{\partial\Omega_{j,k}} \mathbf{v} \cdot \mathbf{H}_n(\mathbf{u}_{h,j}, \mathbf{u}_{h,(NB,k)}) \, dS$$

$$-\int_{\Omega_j} \mathbf{F} : \nabla\mathbf{v} \, dV = \int_{\Omega_j} \mathbf{v} \cdot \mathbf{G} \, dV \qquad (5.186)$$

where $\mathbf{u}_{h,j}$ is the nodal values of element j.

For discontinuous formulations, inter-element continuity is not required. Thus, virtually any polynomial basis functions can be used to construct the approximate solutions to $\mathbf{u}_h$ on Ω_j. For multidimensional problems, the use of tensor products is a common approach to construct local interpolation functions. Various forms of interpolation functions were discussed in Chapter 3. The discontinuous finite element solution has the usual form over a canonical 3-D element,

$$\mathbf{u}_h(x,y,t) = \sum_{m=1}^{n_k} u_m \phi_m(\xi,\eta,\zeta) \qquad (5.187)$$

where n_k is the number of terms in a complete polynomial of degree k. With these substituted into the formulation, and applying the Galerkin approximation, followed by extensive but routine algebra, we obtain the final matrix form for Equation 5.187,

$$\mathbf{M} \frac{d\mathbf{U}_{(j)}}{dt} + \mathbf{K}\mathbf{U} = \mathbf{S} \qquad (5.188)$$

where $\mathbf{U}$ is the unknown vector, $\mathbf{M}$ is the mass matrix, $\mathbf{K}$ is the stiffness matrix, including the portion of normal numerical flux, and $\mathbf{S}$ is the source vector, including contributions from the neighboring elements. The matrix equation, and hence the unknowns, can be solved using the same procedure as described in the previous sections.

5.4.2.2 Characteristic Decomposition

For a multidimensional problem, the characteristic decomposition method, also called the flux-vector splitting method, may be used to assist in constructing numerical fluxes needed for discontinuous finite element formulations. In general, a flux vector may be expressed as

$$\mathbf{F(u)} = \mathbf{Au} = \mathbf{F_u(u)u} \tag{5.189}$$

For a hyperbolic system, the Jacobian $\mathbf{A}$ may be diagonalized as shown in Section 5.1 to yield

$$\mathbf{F(u)} = \mathbf{P}^{-1}\mathbf{\Lambda}\,\mathbf{Pu} \tag{5.190}$$

where the diagonal matrix $\mathbf{\Lambda}$ contains the eigenvalues of $\mathbf{A}$,

$$\mathbf{\Lambda} = \begin{bmatrix} \lambda_1 & & & \\ & \lambda_2 & & \\ & & \cdot & \\ & & & \cdot & \\ & & & & \lambda_N \end{bmatrix} \tag{5.191}$$

The matrix may be further decomposed into two components,

$$\mathbf{\Lambda} = \mathbf{\Lambda}^+ + \mathbf{\Lambda}^- \tag{5.192}$$

where $\mathbf{\Lambda}^+$ and $\mathbf{\Lambda}^-$ are composed of non-negative and non-positive parts of $\mathbf{\Lambda}$,

$$\lambda_i^{\pm} = \frac{\lambda_i \pm |\lambda_i|}{2} \qquad i = 1, 2, \dots, N \tag{5.193}$$

The flux vectors are now written as two components,

$$\mathbf{F(u)} = \mathbf{P}^{-1}(\mathbf{\Lambda}^+ + \mathbf{\Lambda}^-)\mathbf{Pu} = \mathbf{F(u)}^+ + \mathbf{F(u)}^- \tag{5.194}$$

with the positive and negative fluxes calculated by

$$\mathbf{F(u)}^+ = \mathbf{P}^{-1}\mathbf{\Lambda}^+\mathbf{Pu} = \mathbf{A}^+\mathbf{u} \ ; \quad \mathbf{F(u)}^- = \mathbf{P}^{-1}\mathbf{\Lambda}^-\mathbf{Pu} = \mathbf{A}^-\mathbf{u} \tag{5.195}$$

where $\mathbf{A}^+$ contains only rightward-moving characteristic information, and $\mathbf{A}^-$ carries only leftward-moving characteristic information: $\mathbf{A} = \mathbf{A}^+ + \mathbf{A}^-$. For scalar linear advection this is just the upwinding flux.

The characteristic decomposition of the flux vector allows us to construct the numerical fluxes at the boundary more effectively. Consider a constant element approximation. Then we can write

$$\mathbf{F}_{j+1/2} = \mathbf{A}^+\mathbf{u}_j + \mathbf{A}^-\mathbf{u}_{j+1} \qquad (5.196)$$

Thus the flux at the interface can be evaluated using the upwinding methods, and at the interface $\mathbf{x}_j$, $\mathbf{F}(\mathbf{u})^+$ is calculated by $\mathbf{u}(\mathbf{x}_j, t)$ and $\mathbf{F}(\mathbf{u})^-$ by $\mathbf{u}(\mathbf{x}_{j+1}, t)$.

In passing, we note that the above formulation and decomposition method can also be used to solve a system of hyperbolic problems, in which case there is a system of variables defined in either one spatial dimension or multiple dimensions.

5.4.3 Higher Order Approximations and TVD Formulations

In many applications, the linear accuracy approximations presented above are not adequate, and higher order interpolations are needed. For nonlinear inviscid convection problems, a higher order discontinuous method is particularly attractive. It not only provides higher accuracy, but also is more efficient in suppressing the spurious oscillations appearing in the numerical results, especially around discontinuities, provided that the higher order scheme is constructed correctly. For the use of higher order spatial approximations, the order of the time integration scheme has to be compatible to maintain accuracy. Merely increasing the spatial resolution may not eliminate these oscillations, as the numerical schemes may not satisfy the Total Variation Diminishing (*TVD*) property. To avoid these oscillatory problems, a numerical scheme needs to be constructed that satisfies the *TVD* property, by using a second order accurate numerical scheme on smooth solutions and adding diffusion to the numerical scheme near discontinuities. Such numerical schemes, which are often referred to as high resolution schemes in the literature, are at least second order accurate on smooth solutions, and minimize the spurious oscillations present near discontinuities. Cockburn and Shu [14–17] show that for a polynomial of order k, the order of an explicit temporal integration needs to be $k+1$ to achieve the desired effects.

To be fully consistent with the literature on the subject, we use $\Omega_j = [x_{j-1/2}, x_{j+1/2}]$ to define the domain of element j for 1-D problems for both constant and higher order polynomial approximations. This use will be exclusively for this section (Section 5.4.1). As such, u_j refers to the value at the center of the element j and $\bar{u}_j$ is the averaged value over element j. Note that u_j is the same as $\bar{u}_j$ for a constant element approximation formulation. The difference between u_j and $\bar{u}_j$ is important for constructing slope limiters for higher order approximations.

5.4.3.1 Concept of Total Variation Diminishing
Much of the work on higher order approximations has been discussed along with the *TVD* scheme. Cockburn and Shu [14] recommend a total variation diminishing Runge–Kutta scheme; however, Biswar *et al.* [18] point out that the classical Runge–Kutta was equally satisfactory. For most applications, an explicit time integration is used. Since *TVD* schemes are only first order accurate at the local

extrema, alternative reconstruction procedures, for which some growth of the total variation is allowed, have also been developed. Among those, we mention the total variation bounded (*TVB*) schemes [14–17], the essentially non-oscillatory (*ENO*) schemes [20], and the least extremum diminishing (*LED*) schemes [19].

To illustrate the concept of a total variation diminishing (*TVD*) scheme, we consider a 1-D scalar conservation law, or inviscid Burgers' equation,

$$\frac{\partial u}{\partial t} + \frac{\partial f(u)}{\partial x} = 0 \tag{5.197}$$

where $f(u) = 0.5u^2$ is the flux function. The total variation (*TV*) of the solution to the above problem is defined as

$$TV(u) = \int_{-\infty}^{+\infty} \left| \frac{\partial u}{\partial x} \right| dx \tag{5.198}$$

and the total variation for the discrete case is given by

$$TV(u) = \sum_j \left| u_{j+1} - u_j \right| \tag{5.199}$$

A numerical method is considered total variation diminishing, or *TVD*, if the following condition is satisfied:

$$TV(u^{n+1}) \le TV(u^n) ; \ TV(u^n) = \sum_j \left| u_{j+1}^n - u_j^n \right| \tag{5.200}$$

where n is the nth time step and subscript j refers to nodal point j. Harten [21–24] proved that a *TVD* scheme is monoticity preserving and a montonic scheme is *TVD*.

Harten [21–24] studied the central finite difference (or constant element) scheme for the above equation,

$$\frac{du_j}{dt} + \frac{f_{j+1/2} - f_{j-1/2}}{\Delta x} = 0 \tag{5.201}$$

Here subscript j denotes nodal point j, which, for a piecewise constant element approximation, is the same as the element j. Also, subscripts $j+1/2$ and $j-1/2$ represent the values at the two boundaries of element j. These geometric relations are shown in Figure 5.11. Harten proved that the scheme is total variation diminishing, provided that it can be written in the form of

$$\frac{du_j}{dt} = c_{j-1/2}(u_{j-1} - u_j) + c_{j+1/2}(u_{j+1} - u_j) \tag{5.202}$$

with non-negative coefficients $c_{j-1/2} \geq 0$, $c_{j+1/2} \geq 0$, and for an explicit scheme $\Delta t(c_{j-1/2} + c_{j+1/2}) \leq 1$. The coefficiens $c_{j-1/2}$ and $c_{j+1/2}$ may be nonlinear. Furthermore, an explicit time integration scheme is required to satisfy the *CLF* condition.

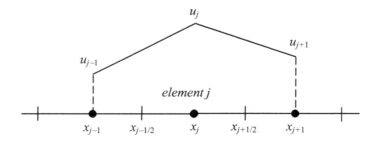

Figure 5.11. Construction of a flux limiter, using constant element approximation. The nodal value is at the center of the element, and the boundaries of element j are marked by $x_{j-1/2}$ and $x_{j+1/2}$

The spatial and explicit time discreitization may also be written in the following form:

$$u_j^{n+1} = u_j^n - C_{j-1/2}^n (u_j^n - u_{j-1}^n) + C_{j+1/2}^n (u_{j+1}^n - u_j^n) \tag{5.203}$$

where C is an arbitrary constant. Harten's theorem states that the algorithm given in a general form of the above equation is *TVD*, i.e., the criterion expressed by Equation 5.200 is satisifed if the constants are such that

$$C_{j-1/2}^n \geq 0; \qquad C_{j+1/2}^n \geq 0; \qquad C_{j-1/2}^n + C_{j+1/2}^n \leq 1 \quad \forall j \tag{5.204}$$

One thus needs to construct a numerical flux that satisfies the above *TVD* criterion. It is important to note that an explicit time scheme may not be *TVD* even if it satisfies the *CFL* condition. For example, one can show that an explicit Lax–Wendroff scheme does not satisfy the *TVD* conditions in the range of time steps that satisfy the *CFL* condition [6]. It is worth noting that the corrections making a method *TVD* are always associated with nonlinear limiting even for linear convection problems. We consider these limiting corrections below.

5.4.3.2 Flux Limiters

The flux limiter approach is based on the idea of approximating flux expressions to obtain higher accuracy, while maintaining the *TVD* property. We limit the flux of u between elements and subsequently limit spurious growth in u near discontinuities. Godunov [6] further proved that a linear *TVD* scheme is doomed to be first order accurate. To overcome this difficulty, numerical fluxes are constructed by combining the high and low order approximations,

$$f_{j\pm1/2} = f^{L}_{j\pm1/2} + \Phi_{j\pm1/2}[f^{H}_{j\pm1/2} - f^{L}_{j\pm1/2}]$$ (5.205)

where superscripts L and H are the low and high order approximations of f, and $\Phi_{j\pm1/2}$ is a correction factor referred to as the flux limiter, and is designed in a numerical algorithm to satisfy Harten's *TVD* condition. Here, to make the presentation clear, we consider the constant element approximation, which is equivalent to the finite volume scheme. The purpose of the limiter is to regulate the diffusion to the solution. It acts as a nonlinear anti-diffusion factor to the lower order flux approximation, in order to improve its accuracy without generating spurious oscillations and violating the *TVD* property. All the numerical fluxes using flux limiters are constructed similarly and consist of two pieces: a high order flux (e.g., the Lax–Wendroff flux) for smooth regions of the flow, and a low order flux (e.g., the flux from some monotone method) near discontinuities.

We illustrate this point through a simple case study. Consider a pure convection problem with a constant velocity $c>0$,

$$\frac{\partial u}{\partial t} + c\frac{\partial u}{\partial x} = 0$$ (5.206)

The flux functions for a constant element scheme may be approximated by the following flux functions:

$$f^{L}_{j+1/2} = cu_j \qquad\qquad \text{for upwinding} \qquad\qquad (5.207a)$$

$$f^{H}_{j+1/2} = c0.5(u_j + u_{j+1}) \qquad\qquad \text{for central scheme} \qquad\qquad (5.207b)$$

If a combined approximation with a flux limiter is used, the numerical flux takes the form of

$$f_{j+1/2} = cu_j + 0.5c\Phi_{j+1/2}[u_{j+1} - u_j]$$ (5.208)

and the other term $f_{j-1/2}$ is constructed by substituting $j = j-1$ into the above equation. Equation 5.208 shows that the flux at the element boundary is constructed by the jump in u there by a limiting function. To control the anti-diffusion correction, the flux limiter should vary depending upon the local condition of the solution. A suitable choice is to examine the solutions at the adjacent elements (see Figure 5.11). It is suggested that the limiter be designed to be a function of the slope ratio of the velocities near the element boundary $x_{j+1/2}$, $r_{j+1/2}$,

$$r_{j+1/2} = \frac{u_j - u_{j-1}}{u_{j+1} - u_j}$$ (5.209)

With this definition, $\Phi(r) = 0$ if $r \leq 0$ and $\Phi_{j+1/2} = \Phi(r_{j+1/2})$. Jameson [25] suggests that the flux limiter be expressed in terms of a special operator Ξ, which is a function of two variables and defines $\Xi(1, r) = \Phi(r)$. The operator needs to have the following four properties:

$$1. \; \Xi(a,b) = \Xi(b,a) \tag{2.110}$$

$$2. \; \Xi(ca,cb) = c\Xi(b,a) \tag{2.211}$$

$$3. \; \Xi(a,a) = a \tag{2.212}$$

$$4. \; \Xi(a,b) = 0 \; \text{ if } ab \leq 0 \tag{2.213}$$

As shown from Example 5.4 below, the flux limiter is related to the operator,

$$\Phi(r) = \Xi(1,r) = r\Xi(1,1/r) = r\Phi(1/r) \tag{5.214}$$

$$\Phi(r_{j+1/2})(u_{i+1} - u_j) = \Xi(u_{j+1} - u_j, u_j - u_{j-1}) = \Phi(1/r_{j+1/2})(u_j - u_{j-1}) \tag{5.215}$$

Using the operator, the problem of dividing by zero can be eliminated. With these relations, one can show that the constant element scheme satisfies the *TVD* condition if the coefficients are calculated using the following expressions:

$$c_{j-1/2} = \frac{c}{2\Delta x}\left[2 - \Phi(r_{j-1/2}) + \frac{\Phi(r_{j+1/2})}{r_{j+1/2}}\right]; \; c_{j+1/2} = 0 \tag{5.216}$$

with $c_{j-1/2} \geq 0$ and $\Delta t c_{j-1/2} \leq 1$.

Example 5.4. Prove the relations given in Equations 5.214, 5.215 and 5.216.

Solution. The proofs can be made using the definitions of the operator Ξ and its properties:

$$\Phi(r) = \Xi(1,r) = \Xi(r,1) = r\Xi(1/r,1) = r\Xi(1,1/r) = r\Phi(1/r) \tag{5.13e}$$

$$\Phi(r_{j+1/2})(u_{j+1} - u_j) = (u_{j+1} - u_j)\Xi(1, r_{j+1/2}) = \Xi(u_{j+1} - u_j, u_j - u_{j-1})$$

$$= (u_j - u_{j-1})\Xi(1/r_{+1/2},1) = (u_j - u_{j-1})\Phi(1/r_{j+1/2}) \tag{5.14e}$$

To prove Equation 5.216, we start with the constant element (or finite difference) equation,

$$\frac{du_j}{dt} = -\frac{f_{j+1/2} - f_{j-1/2}}{\Delta x} \tag{5.15e}$$

Substituting the numerical flux, combined from the upwinding and central differencing, one has

$$\frac{du_j}{dt} = -\frac{c}{\Delta x}\left(u_j + 0.5\Phi_{j+1/2}(u_{j+1} - u_j) - u_{j-1} - 0.5\Phi_{j-1/2}(u_j - u_{j-1})\right)$$

$$= -\frac{c}{2\Delta x}\{2(u_j - u_{i-1}) + \Phi_{j+1/2}(1/r_{j+1/2})(u_j - u_{j-1})$$

$$- \Phi_{j-1/2}(r_{j-1/2})(u_j - u_{j-1})\}$$

$$= \frac{c}{2\Delta x}\left(2 + \Phi_{j+1/2}(r_{j+1/2})/r_j - \Phi_{j-1/2}(r_{j-1/2})\right)(u_{j-1} - u_j) \tag{5.16e}$$

Comparison with Equation 5.216 gives the needed result immediately.

Further analysis shows that the discretized form using a flux limiter can also be written in the general form given in Equation 5.203 (see Example 5.5 below) with the following coefficients:

$$C^n_{j-1/2} = \delta + \frac{1}{2}\delta(1-\delta)\left(\frac{\Phi(r^n_{j+1/2})}{r^n_{j+1/2}} - \Phi(r^n_{j-1/2})\right); \quad C^n_{j+1/2} = 0 \tag{5.217}$$

where $\delta = c\Delta t/\Delta x$. In this case, the Harten *TVD* conditions reduce to $0 \leq C^n_{j-1} \leq 1$,

$$0 \leq C^n_{j-1} = \delta + \frac{1}{2}\delta(1-\delta)\left(\frac{\Phi(r^n_{j+1/2})}{r^n_{j+1/2}} - \Phi(r^n_{j-1/2})\right) \leq 1 \quad \forall j \tag{5.218}$$

This will hold true if

$$0 \leq \frac{\Phi(r)}{r} \leq 2; \qquad 0 \leq \Phi(r) \leq 2 \quad \forall r \geq 0 \tag{5.219}$$

In addition,we require that $\Phi(r) = 0$ if $r < 0$, as stated above.

A variety of flux limiters is devised in the literature and some of the popular ones are plotted in Figures 5.12 and 5.13. They are given below also for convenience:

For linear methods ($\Xi(1, r) = \Phi(r)$):

Godunov slope: $\Phi(r) = 0$ (5.220)

Centered slope (Fromm): $\Phi(r) = 0.5(1 + r)$ (5.221)

Upwinding slope (Beam–Warming): $\Phi(r) = r$ (5.222)

Downwinding slope (Lax–Wendroff): $\Phi(r) = 1$ (5.223)

For higher order methods:

Minmod: $\Xi(a, b) = S(a, b) \min\{|a|, |b|\}$ (5.224)

Van Leer: $\Xi(a, b) = S(a, b) \dfrac{2|a||b|}{|a| + |b|}$ (5.225)

Monotonized central: $\Xi(a, b) = S(a, b) \min\left\{\dfrac{|a| + |b|}{2}, 2|a|, 2|b|\right\}$ (5.226)

Superbee: $\Xi(a, b) = S(a, b) \max\{\min\{2|a|, |b|\}, \min\{|a|, 2|b|\}$ (5.227)

where $S(a, b) = 0.5(\text{sgn}(a) + \text{sgn}(b))$.

As can be seen above, the linear methods do not give rise to a *TVD* scheme. To guarantee second order accuracy and avoid excessive compression of solutions, Sweby suggested a reduced portion of the *TVD* region as a suitable range for the flux limiting function. The Sweby *TVD* region and the nonlinear flux limiters are illustrated in Figure 5.13.

5.4.3.3 Slope Limiters

As shown above, the way the element boundary fluxes are computed determines the spatial order of accuracy of the numerical algorithm and controls the amplitude of the local jumps at an element interface. If these jumps are monotonically reduced, the scheme provides more accurate initial guesses for the solution of the local Riemann problems (the average values give only first order accuracy). Besides the flux limiters for reconstructing numerical fluxes, the slope limiters are also commonly, maybe more commonly, used for the *TVD* schemes. Van Leer [26] discussed the second order, piecewise-linear reconstruction in the design of the *MUSCL* (Monotonic Upstream Scheme for Conservation Laws) scheme. The third order, piecewise parabolic reconstruction scheme was developed by Colella and Woodward [27] in their Piecewise Parabolic Method (*PPM*). We consider below slope limiters for discontinuous finite element applications.

A slope limiter method is based on a geometric approach to construct a higher order approximation for the fluxes in Equation 5.201. The idea is to reconstruct the variable u at the element boundaries from the values in the element and its neighbors and then use the variable in the flux expressions. We illustrate the concept using the piecewise constant approximation. Knowing the numerical data $u(x_i, t^n)$, we can construct a piecewise approximation to u,

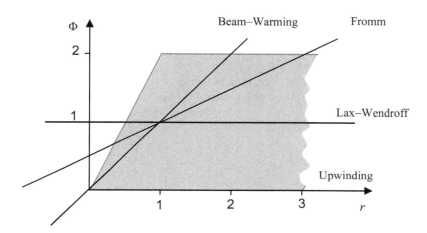

Figure 5.12. The shaded region is the *TVD* region and four slope limiters for linear methods

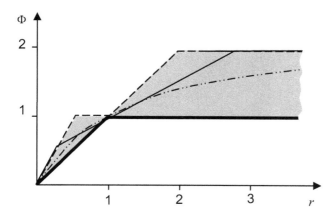

Figure 5.13. The Sweby *TVD* region (marked by shading) and nonlinear slope limiters: thicker solid line – minmod, double-dot dash curve – the monotonized central limiter, dash line – superbee, and lighter solid line – van Lear

$$\tilde{u}(x,t^n) = \overline{u}(x_j,t^n) + \sigma_j^n(x - x_j), \quad x_{j-1/2} < x < x_{j+1/2} \tag{5.228}$$

and the approximation $\tilde{u}(x,t^n)$ is then used to calculate the numerical fluxes at the element boundaries at $x = x_{j-1/2}$ and $x = x_{j+1/2}$. In Equation 5.228, $\overline{u}(x_j,t^n)$ is the averaged value over element j (see Figure 5.11 and also Equation 5.230 below) and the slope is interpolated using the averages at the element boundary from adjacent elements,

$$\sigma_j^n = \Delta_j u / \Delta_j x \tag{5.229}$$

One could choose the central flux or upwinding flux for σ_j^n. Whatever the choice may be, it needs to produce a polynomial for $u(x,t^n)$ such that the average over the element is the same,

$$\overline{u}(x_j,t^n) = \frac{1}{\Delta x}\int_{x_{j-1.2}}^{x_{j+1.2}} u(x,t^n)dx \tag{5.230}$$

Thus, one has the freedom of constructing any slope as needed, so long as the above condition is satisfied. Figure 5.14 shows the procedure by which a piecewise approximation is made from the element-averaged values.

Some of the standard formulae for slopes (assuming same-size elements) are given as follows:

Godunov slope: $\sigma_j^n = 0$ $\qquad\qquad\qquad\qquad\qquad$ (5.231)

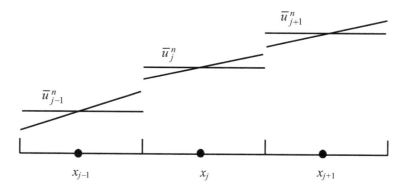

Figure 5.14. Reconstruction of $u(x)$ in an element from element-averaged values (horizontal lines) and extratpolate u at the element boundaries. Over each element an approximation to the slope is made. The reconstructed solution passes through the element average at the element center and has the computed slope (slant lines)

Centered slope (Fromm): $\sigma_j^n = \dfrac{\overline{u}_{j+1}^n - \overline{u}_{j-1}^n}{2\Delta x_j}$ (5.232)

Upwinding slope (Beam–Warming): $\sigma_j^n = \dfrac{\overline{u}_j^n - \overline{u}_{j-1}^n}{\Delta x_j}$ (5.233)

Downwinding slope (Lax–Wendroff): $\sigma_j^n = \dfrac{\overline{u}_{j+1}^n - \overline{u}_j^n}{\Delta x_j}$ (5.234)

By the Taylor expansion, one can show that the latter three methods (Fromm, Beam–Warning, Lax–Wendroff) are second order accurate for sufficiently smooth solutions. This is one order better that the straight upwind version (assuming zero slope). Numerical experience indicates that the use of the above standard numerical fluxes sometimes would actually introduce oscillation near the discontinuity as shown in Figure 5.15. This is because the values used for numerical flux estimation are extrapolated to the element boundaries without error control. Thus special care has to be taken to construct a slope limiter that is both higher order accurate and oscillation free.

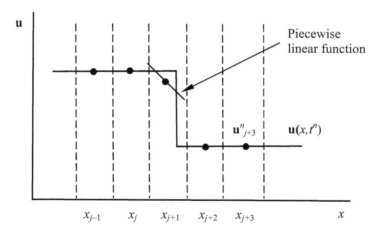

Figure 5.15. A linear function causes oscillation at the shock front. The solid line is the exact solution and the dots are numerical solutions

Several choices of slopes that satisfy the *TVD* condition are reported in the literature, which include:

Setting slope to zero: Godunov method: $\sigma_j^n = 0$ (5.235)

Minmod slope limiter: $\sigma_j^n = \min \mathrm{mod}(a,b) = S(a,b)\min\{|a|,|b|\}$ (5.236)

Monotonized central: $\sigma_j^n = S(a,b)\min\left\{\dfrac{|a|+|b|}{2},\, 2|a|,\, 2|b|\right\}$ (5.237)

where a and b are the upwinding and downwinding fluxes, respectively, which are calculated by

$$a = \frac{\overline{u}_j^n - \overline{u}_{j-1}^n}{\Delta x};\quad b = \frac{\overline{u}_{j+1}^n - \overline{u}_j^n}{\Delta x}$$ (5.238)

The concepts for both slope limiters and flux limiters can be extended to multidimensional calculations in regular grids with ease. For unstructured meshes, however, a procedure can be tedious [19]. In general, it involves: (1) reconstruction of a local 1-D stencil, by inserting equidistant dummy nodes on the continuation of each mesh edge, (2) interpolation/extrapolation using the adjacent elements containing node j, (3) interpolation using the actual triangle T containing the dummy node k, and (4) extrapolation, using a least square reconstruction (L^2-projection) for the gradient at node j. The detailed procedures for each of these steps are discussed in Kuzmin and Turek [19].

Example 5.5. Consider a pure convection problem with a constant convective velocity c,

$$\frac{\partial u}{\partial t} + c\frac{\partial u}{\partial x} = 0 \quad \in \Omega,\, t > 0$$ (5.17e)

$$u(x,0) = u_0(x) \quad \in \Omega$$ (5.18e)

Develop discontinuous finite element formulations using the constant element approximation incorporating (1) a slope limiter and (2) a flux limiter.

Solution. The spatial discretization using a constant element gives the following conservation expression:

$$\frac{d\overline{u}_j}{dt} + \frac{f(u(x_{j+1/2},t)) - f(u(x_{j-1/2},t))}{\Delta x} = 0$$ (5.19e)

where the overbar denotes the average over an element and $f = cu$. We construct the slope limiter from the calculated values in the elements,

$$\tilde{u}(x,t^n) = \overline{u}(x_j,t^n) + (x - x_j)\sigma_j^n$$ (5.20e)

Since the propagation speed c is a constant, we can use a method of characteristics to evaluate the flux integral (for the reconstructed solution as it propagates through the end points of each element). A fairly standard approach is to linearize the flux function about some appropriate mean state, decompose the solution into characteristic variables of the linearized system and perform the same kind of linear wave propagation analysis.

In the space-time plane we can plot the characteristics for the advection equation for the outflow condition (see Figure 5.4e). If the discretization for the slopes only includes the jth and $(j+1)$th elements, then the time step must satisfy $c\Delta t \leq \Delta x$. Otherwise we would have to use the reconstruction from the $(j-1)$th element. Similarly, we can plot the characteristics for the advection equation for the inflow condition (see Figure 5.5e) and we back-track into the $(j-1)$th element to evaluate the inflow flux for element j.

We now use values of the polynomial reconstructed u at the nth time level to evaluate the flux integral at the inflow and outflow boundaries. Given a piecewise linear reconstructed solution each of the flux integrals can be evaluated. Thus, we have the flux expression at $x_{j+1/2}$,

$$\int_{t^n}^{t^{n+1}} f(u(x_{j+1/2},t))dt = \int_{t^n}^{t^{n+1}} f(\tilde{u}(x_{j+1/2},t))dt$$

$$= \int_{t^n}^{t^{n+1}} f(\tilde{u}(x_{j+1/2} - c(t-t^n)))dt$$

$$= \int_{t^n}^{t^{n+1}} c(\overline{u}_j^n + (x_{j+1/2} - c(t-t^n) - x_j)\sigma_j^n)dt$$

$$= c\Delta t\overline{u}_j^n + c\frac{\Delta x\Delta t}{2}\sigma_j^n - \frac{c^2\Delta t^2}{2}\sigma_j^n \qquad (5.21e)$$

In deriving the above, we back-track along the characteristics to find the value $\tilde{u}$ at $(x_{j+1/2}, t)$ for (t^n, t^{n+1}). A similar expression can be obtained for the inflow integral using the information from the neighboring elements.

Thus, one can express approximations for the right hand integrals in terms of element averages:

$$\overline{u}_j^{n+1} = \overline{u}_j^n + \frac{1}{\Delta x}\left\{\int_{t^n}^{t^{n+1}} f(u(x_{j-1/2},t))dt - \int_{t^n}^{t^{n+1}} f(u(x_{j+1/2},t))dt\right\}$$

$$= \overline{u}_j^n + \frac{1}{\Delta x}\left\{c\Delta t\overline{u}_{j-1}^n + c\frac{\Delta x\Delta t}{2}\sigma_{j-1}^n - \frac{c^2\Delta t^2}{2}\sigma_{j-1}^n\right.$$

$$\left. - c\Delta t\overline{u}_j^n - c\frac{\Delta x\Delta t}{2}\sigma_j^n + \frac{c^2\Delta t^2}{2}\sigma_j^n\right\}$$

$$= \overline{u}_j^n - \frac{1}{\Delta x}\left\{c\Delta t(\overline{u}_j^n - \overline{u}_{j-1}^n) + \left(c\frac{\Delta x\Delta t}{2} - \frac{c^2\Delta t^2}{2}\right)(\sigma_j^n - \sigma_{j-1}^n)\right\} \quad (5.22e)$$

Re-arranging, one has the final expression,

$$\overline{u}_j^{n+1} = \overline{u}_j^n - \frac{c\Delta t}{\Delta x}\left\{\overline{u}_j^n - \overline{u}_{j-1}^n + \frac{\Delta x}{2}\left(1 - \frac{c\Delta t}{\Delta x}\right)\left(\sigma_j^n - \sigma_{j-1}^n\right)\right\} \qquad (5.23e)$$

where σ is the slope limiter, which one can choose from the list of the slope limiters discussed in Section 5.4.3.2.

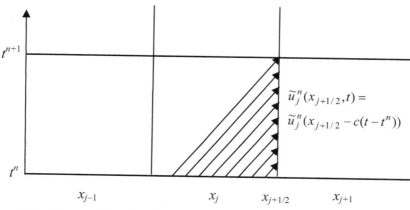

Figure 5.4e. Characteristics plots in the x–t plane. Initial data at t at element j is carried over into the domain and leave out the out-flow boundary of element j.

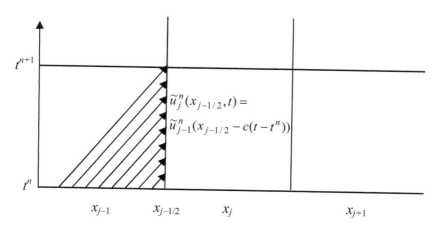

Figure 5.5e. Characteristics plots in the x–t plane. Initial data at t at element j–1 is carried over into the domain and enter through the left-hand boundary of element j

We now turn to the derivation of a flux limiter expression. We start with Equation (5.19e) and integrate it using a Euler forward scheme to obtain the following result:

$$\bar{u}_j^{n+1} = \bar{u}_j^n + \frac{\Delta t}{\Delta x}\left(F_{j-1/2}^n - F_{j+1/2}^n\right) \tag{5.24e}$$

where F is the averaged boundary flux and is calculated by (see also Equation 5.21e)

$$F_{j-1/2}^n = \frac{1}{\Delta t}\int_{t^n}^{t^{n+1}} f(\tilde{u}(x_{j-1/2},t))dt = c\bar{u}_{j-1}^n + \frac{c}{2}(\Delta x - c\Delta t)\sigma_{j-1}^n \tag{5.25e}$$

$$F_{j+1/2}^n = \frac{1}{\Delta t}\int_{t^n}^{t^{n+1}} f(\tilde{u}(x_{j+1/2},t))dt = c\bar{u}_j^n + \frac{c}{2}(\Delta x - c\Delta t)\sigma_j^n \tag{5.26e}$$

The two equations above are nothing but the flux formulations with piecewise reconstruction.

So far we have assumed $c > 0$ but we can also obtain the flux expression for the condition of $c < 0$ using the same approach. The results for both cases may be summarized as follows:

$$F_{j-1/2}^n = \begin{cases} c\bar{u}_{j-1}^n + 0.5c(\Delta x - c\Delta t)\sigma_{j-1}^n & \text{if } c \geq 0 \\ c\bar{u}_j^n - 0.5c(\Delta x + c\Delta t)\sigma_j^n & \text{if } c \leq 0 \end{cases} \tag{5.27e}$$

The flux expression may be further written in a simplified version,

$$F_{j-1/2}^n = c^-\bar{u}_j^n + c^+\bar{u}_{j-1}^n + \frac{|c|}{2}\left(1 - \left|\frac{c\Delta t}{\Delta x}\right|\right)\Lambda_{j-1/2}^n \tag{5.28e}$$

where $\Lambda = \Delta x g(\sigma_{j-1}^n, \sigma_j^n)$ with g being a function of $\sigma_{j-1}^n, \sigma_j^n$ is the flux limiter, and c^- and c^+ are calculated by the following expressions:

$$c^- = \frac{c - |c|}{2}; \qquad c^+ = \frac{c + |c|}{2} \tag{5.29e}$$

By writing the time interval averaged flux function in this way, we have changed the numerical strategy from a local element reconstruction approach towards controlling the flux contribution from jumps in the averages between elements. In particular, we use the flux limiter, intead of the slope limiter, to approximate the numerical fluxes at the element boundaries. Equations 5.27e and 5.28e also reveal the relation between a slope limiter and a flux limiter.

From the analysis of the units of Λ, we devise the flux limiting function, which allows us to control the boundary flux by multiplying the jump in element averages by a flux limiting function,

$$\Lambda_{j-1/2}^n = \Phi(r_{j-1/2}^n)(\overline{u}_j^n - \overline{u}_{j-1}^n) \tag{5.30e}$$

where the flux limiter is calculated by

$$\Phi(r_{j-1/2}^n) = \frac{\overline{u}_j^n - \overline{u}_{J-1}^n}{\overline{u}_j^n - \overline{u}_{j-1}^n} \tag{5.31e}$$

with

$$J = \begin{cases} j-1 & \text{if } c > 0 \\ j+1 & \text{if } c < 0 \end{cases} \tag{5.32e}$$

The ratio $r_{j-1/2}$ can be thought of as a smoothness indicator near the element interface at $x_{j-1/2}$. If the data is smooth we expect the ratio to be approximately 1 (except at extrema). Near a discontinuity we expect the ratio to be far away from 1. The flux limiter, Φ, will range between 0 and 2. The smaller it is, the more limiting is applied to a jump in element averages. Above 1 it is being used to steepen the effective reconstruction.

Using this notation one can write down the scheme in terms of the flux limiter $(\delta = c\Delta t/\Delta x)$,

$$\overline{u}_j^{n+1} = \overline{u}_j^n - \delta(\overline{u}_j^n - \overline{u}_{j-1}^n) - \frac{\delta(1-\delta)}{2}\left\{ \Phi(r_{j+1/2}^n)(\overline{u}_{j+1}^n - \overline{u}_j^n) \right.$$

$$\left. - \Phi(r_{j-1/2}^n)(\overline{u}_j^n - \overline{u}_{j-1}^n) \right\} \qquad \text{for } c > 0 \tag{5.33e}$$

$$\overline{u}_j^{n+1} = \overline{u}_j^n - \delta(\overline{u}_{j+1}^n - \overline{u}_j^n) - \frac{\delta(1-\delta)}{2}\left\{ \Phi(r_{j-1/2}^n)(\overline{u}_j^n - \overline{u}_{j-1}^n) \right.$$

$$\left. - \Phi(r_{j+1/2}^n)(\overline{u}_{j+1}^n - \overline{u}_j^n) \right\} \qquad \text{for } c < 0 \tag{5.34e}$$

In the above equations, the second term represents the upwinding scheme flux contributions, the third term the limited downwinding element interface flux contribution, and the fourth term the upwinding element interface flux contribution. The above equation can be written in a general form,

$$\overline{u}_j^{n+1} = \overline{u}_j^n - C_{j-1/2}^n(\overline{u}_j^n - \overline{u}_{j-1}^n) + C_{j+1/2}^n(\overline{u}_{j+1}^n - \overline{u}_j^n) \tag{5.35e}$$

which has been used in the discussion of *TVD* with flux limiters in Section 5.4.3.2.

5.4.3.4 TVD-Runge–Kutta Schemes
In a series of recent papers [14–17], Cockburn and Shu presented a *TVD*-Runge–Kutta scheme for the discontinuous finite element solution of nonlinear hyperbolic conservation equations. Their procedure essentially involves three steps: (1) discontinuous finite element space discretization, (2) explicit Runge–Kutta time integration and (3) generalized slope limiters to ensure the *TVD* properties. Their schemes are, strictly speaking, *TVB* (total variation bounded), which is a modification of *TVD*.

These steps may be better illustrated through an example, for which the Burgers' equation should serve the purpose,

$$\frac{\partial u}{\partial t} + \frac{\partial f(u)}{\partial x} = 0 \tag{5.239}$$

where $f(u) = 0.5u^2$ is the flux function.

The first step involves the space discretization of the domain into a collection of elements and integrating locally over an element (say, element j) to obtain the weak form solution,

$$\int_{\Omega_j} v \cdot u_t dV + \int_{\partial\Omega_j} vh(u)\hat{i} \cdot \mathbf{n}_j \, dS - \int_{\Omega_j} f(u)\frac{\partial v}{\partial x} dV = 0 \tag{5.240}$$

where integration by parts has been taken, and the boundary fluxes have been replaced by the numerical fluxes $h(u)$.

For this type of problem, the numerical fluxes can be taken as the approximate Riemann solver. Some examples of the numerical fluxes that satisfy the stability criteria are the following:

(1) the Godunov flux:

$$h(a,b) = \begin{cases} \min_{a \le u \le b} f(u) & \text{if } a \le b \\ \max_{b \le u \le a} f(u) & \text{otherwise} \end{cases} \tag{5.241}$$

(2) the Engquist–Osher flux:

$$h(a,b) = \int_0^b \min(f'(s),0)ds + \int_0^b \max(f'(s),0)ds + f(0) \tag{5.242}$$

(3) the Lax–Friedrichs flux:

$$h(a,b) = \tfrac{1}{2}[f(a) + f(b) - C(b-a)],$$

$$C = \max_{\inf u^0(x) \le s \le \sup u^0(x)} |f'(s)| \tag{5.243}$$

These fluxes are part of the list discussed in Section 5.4.2.1, but given here for convenience. The above fluxes are useful for piecewise constant approximations and, when used with an explicit Euler forward time integrator, give linear order accuracy. These fluxes are not suitable for high order *TVD* schemes without slope or flux limiters. By interpolating unknowns using polynomial basis functions, and carrying out the necessary element calculations, one generally arrives at a system of ordinary differential equations,

$$\frac{d\mathbf{U}_{(j)}}{dt} = L(\mathbf{U}) \tag{5.244}$$

where L is the operator and $\mathbf{U}_{(j)} = [u(x_{j-1/2},t), \ldots, u(x_j,t), \ldots, u(x_{j+1/2},t)]^T$ are the unknowns, or coefficients of the polynomial basis functions, associated with element j. Also, $\mathbf{U}$ includes unknowns from neighboring elements. Note that for hyperbolic problems, the discontinuous finite element method has order $k + \frac{1}{2}$ accuracy when polynomials of degree k are used.

The second step for the calculations involves the Runge–Kutta time integration. This is an explicit Runge–Kutta method specially modified for the discontinuous finite element solutions [14]. If $[0, T]$ is partitioned into N time steps with $\Delta t^n = t^{n+1} - t^n$, $n = 0, 1, \ldots, N-1$, then the time-marching algorithm is numerically implemented as follows:

Set $\mathbf{U}^0 = \mathbf{U}_0$

For $n = 0, 1, \ldots, N-1$, calculate $\mathbf{U}^{n+1}$ from $\mathbf{U}^n$ by the following procedures:

(1) Set $\mathbf{U}^{[0]} = \mathbf{U}^n$

(2) For $i = 1, 2, \ldots, K (= k+1)$ compute the intermediate functions,

$$\mathbf{U}^{[i]} = \sum_{l=0}^{i-1} \alpha_{il} \mathbf{W}^{il}; \quad \mathbf{W}^{il} = \mathbf{U}^{[l]} + \frac{\beta_{il}}{\alpha_{il}} \Delta t L\left(\mathbf{U}^{[l]}\right)$$

(3) Set $\mathbf{U}^{n+1} = \mathbf{U}^{[K]}$

where k is the order of the spatial polynomial.

Gottlieb and Shu [28] show that the following properties are required of the coefficients α_{il} and β_{il} in order for the scheme to be *TVD*:

if $\beta_{il} \neq 0$ then $\alpha_{il} \neq 0$,

$\alpha_{il} \geq 0$

$\sum_{i=0}^{i-1} \alpha_{ij} = 1$

and if the single Euler forward time step (that is, $\mathbf{W}^{il} = \mathbf{U}^{[l]} + (\beta_{il} / \alpha_{il})\Delta t L(\mathbf{U}^{[l]})$) satifies the *CFL* condition [4, 5],

$$\delta_0 = \max_{0 \le n \le N} |\Delta t^n \max\{\beta_{il} / \alpha_{il}\}|$$

Thus, the stability of the Runge–Kutta schemes follows from the stability of the intermediate steps for the mapping $\mathbf{U}^{[l]} \mapsto \mathbf{W}^{il} : \mathbf{W}^{il} = \mathbf{U}^{[l]} + (\beta_{il}/\alpha_{il})\Delta t(\mathbf{U}^{[l]})$. In other words, to render a Runge–Kutta scheme stable, every single Euler forward intermdiate step needs to be stable. This is a crucial aspect of the TVD-Runge–Kutta scheme.

Since a TVD-Runge–Kutta scheme is explicit, the round-off errors must be fenced in. For a discontinuous formulation using polynomials of order k, a $(k+1)$-stage Runge–Kutta scheme (i.e., order $k+1$) must be used. A von Neumann analysis shows that for a 1-D problem with $f(u) = cu$ or ($\partial u / \partial t + c\partial u / \partial x = 0$), c being a constant, the following CLF condition is required to ensure the stability:

$$|c|\frac{\Delta t}{\Delta x} \le \frac{1}{2k+1} \tag{5.245}$$

where k is the order of polynomials for space discretization [5]. This is much smaller than a Euler forward scheme alone, which is easily recovered from the above equation by setting $k = 0$.

For discontinuous finite element space discretizations, Cockburn [4, 5] shows that the Euler forward mapping $\mathbf{U}^{[l]} \mapsto \mathbf{W}^{il}$ is not stable in the L^2-norm, except in the case where polynomials of degree 0 are used. If the polynomials of degree k are used, the intermediate single Euler forward step may not be stable. For example, for $k = 1$, that is, a piecewise linear approximate solution, the single Euler forward step is unconditionally unstable for any fixed ratio of $\Delta t/\Delta x$. On the other hand, if a Runge–Kutta method of second order ($k + 1 = 2$) is used, it is conditionally stable for $|c|\Delta t/\Delta x \le 1/3$. This means that even though a single Euler forward step is not stable, the TVD-Runge–Kutta scheme is stable according to Equation 5.221. Thus, the stability of the complete Runge–Kutta scheme cannot be deduced from the stability of the single Euler forward step.

It is shown that with the use of a piecewise constant approximation the TVD property means

$$TV(u^n) = \sum_{1 \le j \le N} |\bar{u}_{j+1} - \bar{u}_j| \tag{5.246}$$

where the overbar represents the element-averaged value. Note that for a constant element approximation, $\mathbf{U}_{(j)} = u_j$. For discontinuous solutions that are not piecewise constant, the above result still holds if the following sign conditions are met for a single Euler forward step mapping $u \mapsto w$ [4, 5]:

$$\operatorname{sgn}(u_{j+1/2}^+ - u_{j-1/2}^+) = \operatorname{sgn}(\bar{u}_{j+1} - \bar{u}_j);$$

$$\text{sgn}(u_{j+1/2}^{-} - u_{j-1/2}^{-}) = \text{sgn}(\overline{u}_j - \overline{u}_{j-1})$$ (5.247)

For higher-order approximations, these properties need to be ensured by the generalized slope limiter. Note that the result expressed by Equation 5.246 is, strictly speaking, the total variation diminishing in means (*TVDM*) for a higher order approximation [5].

The generalized slope limiter is constructed in a procedure similar to that discussed above for slope limiters for constant elements. For a piecewise linear approximation of u over the element for the purpose of evaluating fluxes,

$$\tilde{u}(x,t^n) = \overline{u}_j^n + (x - x_j)\frac{\partial u(x,t^n)}{\partial x}, \qquad x \in \Omega_j = [x_{j-1/2}, x_{j+1/2}]$$ (5.248)

The generalized slope limiter $SP(\cdot)$ is then defined as

$$u(x,t^n) = SP(u) = \overline{u}_j^n$$

$$+ (x - x_j)\min \text{mod}\left\{\frac{\partial u^n}{\partial x}, 2\frac{\overline{u}_{j+1}^n - \overline{u}_j^n}{\Delta_j}, 2\frac{\overline{u}_j^n - \overline{u}_{j-1}^n}{\Delta_j}\right\}$$ (5.249)

where $\Delta_j = x_{j+1/2} - x_{j+1/2}$ and also

$$\min \text{mod}(a,b,c) = \begin{cases} \text{sgn}(a)\min(|a|,|b|,|c|), & \text{if } \text{sgn}(a) = \text{sgn(b)} = \text{sgn(c)} \\ 0 & \text{otherwise} \end{cases}$$ (5.250)

At the boundary of element j, the projection has the following form:

$$u_{j+1/2}^{-,n} = \overline{u}_j^n + \min \text{mod}\left(v_{j+1/2}^{-,n} - \overline{u}_j^n, \overline{u}_{j+1}^n - \overline{u}_j^n, \overline{u}_j^n - \overline{u}_{j-1}^n\right)$$ (5.251a)

$$u_{j-1/2}^{+,n} = \overline{u}_j^n - \min \text{mod}\left(\overline{u}_j^n - v_{j-1/2}^{+,n}, \overline{u}_{j+1}^n - \overline{u}_j^n, \overline{u}_j^n - \overline{u}_{j-1}^n\right)$$ (5.251b)

where $\overline{u}_j^n = \overline{u}(x_j,t^n)$ and the values at the boundary are calculated by the following expression:

$$v^n = v(x,t^n) = \overline{u}(x_j,t^n) + (x - x_j)\frac{\partial u^n}{\partial x}\bigg|_{x_j}, \qquad x \in [x_{j-1/2}, x_{j+1/2}]$$ (5.252)

Two of the slope limiters used for discontinuous formulations are given in Figure 5.16. The limiting procedure prevents the overshooting at the boundaries: the overshooting causes the oscillation in solution near the discontinuities.

Thus, we have a complete *TVD*-Runge–Kutta scheme for the numerical solution of inviscid Burgers' equation using a discontinuous Galerkin formulation. Let $[0, T]$ be partitioned into N time steps with $\Delta t^n = t^{n+1} - t^n$ ($n = 0, 1, ..., N - 1$) and let $SP(\cdot)$ be the generalized slope limiter. Then the time-marching algorithm is numerically implemented as follows:

Set $\mathbf{U}^0 = SP(\mathbf{U}_0)$

For $n = 0, 1, ..., N - 1$, calculate $\mathbf{U}^{n+1}$ from $\mathbf{U}^n$ by the following procedures:

(1) Set $\mathbf{U}^{[0]} = \mathbf{U}^n$

(2) For $i = 1, 2, ..., K (= k + 1)$ compute the intermediate functions,

$$\mathbf{U}^{[i]} = SP\left(\sum_{l=0}^{i-1} \alpha_{il} \mathbf{W}^{il}\right); \ \mathbf{W}^{il} = \mathbf{U}^{[l]} + \frac{\beta_{il}}{\alpha_{il}} \Delta t L\left(\mathbf{U}^{[l]}\right)$$

(3) Set $\mathbf{U}^{n+1} = \mathbf{U}^{[K]}$

where k is the order of the spatial polynomial.

We consider below an example illustrating the difference between a *TVD*-Runge–Kutta scheme and a non-*TVD*-Runge–Kutta scheme.

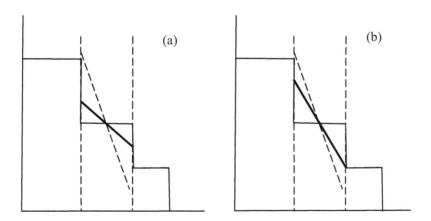

Figure 5.16. Slope limters: the MUSCL limiter (a) and the less restrictive SP limiter (b). Local means are denoted by the horizontal solid line. Also shown are the linear function u in the element of the middle before limiting (dashed diagonal line) and the resulting function after limiting (diagonal solid line)

Example 5.6. Consider the 1-D inviscid Burgers' equation,

$$\frac{\partial u}{\partial t} + \frac{\partial}{\partial x}\left(\frac{u^2}{2}\right) = 0 \tag{5.36e}$$

with discontinuous initial data

$$u(x,0) = \begin{cases} 1.0, & \text{if } x \le 0 \\ -0.5, & \text{if } x > 0 \end{cases} \tag{5.37e}$$

Develop a discontinuous finite element formulation with the *TVD*-Runge–Kutta time integration, with the MUSCL slope limiter, and compare the results obtained without the use of a *TVD* scheme.

Solution. Let $f(u) = 0.5u^2$. Then the discontinuous formulation has the form,

$$\int_{\Omega_j} v \cdot \frac{\partial u}{\partial t} dV + \int_{\partial\Omega_j} v h(u)\hat{i} \cdot \mathbf{n}_j \, dS - \int_{\Omega_j} f(u)\frac{\partial v}{\partial x} dV = 0 \,,$$

$$\Omega_j = [x_{j-1/2}, x_{j+1/2}] \tag{5.38e}$$

For a piecewise constant approximation, the above can be integrated to yield the result,

$$\frac{du}{dt} = \frac{1}{\Delta x}\left(h_{j+1/2} - h_{j-1/2}\right) = L(u) \tag{5.39e}$$

where to simplify the notation, the subscript on u has been dropped. We take the slope limiter as follows:

$$u^-_{j+1/2} = \bar{u}_j + \frac{1}{2}\min \text{mod}\left(\bar{u}_{j+1} - \bar{u}_j, \bar{u}_j - \bar{u}_{j-1}\right) \tag{5.40e}$$

$$u^+_{j+1/2} = \bar{u}_{j+1} - \frac{1}{2}\min \text{mod}\left(\bar{u}_{j+2} - \bar{u}_{j+1}, \bar{u}_{j+1} - \bar{u}_j\right) \tag{5.41e}$$

and use the Godunov flux as the monotone numerical flux,

$$h(u^-_{j+1/2}, u^+_{j+1/2}) = \begin{cases} \min_{u^-_{j+1/2} \le u \le u^+_{j+1/2}} 0.5u^2, & \text{if } u^-_{j+1/2} \le u^+_{j+1/2} \\ \max_{u^+_{j+1/2} \le u \le u^-_{j+1/2}} 0.5u^2, & \text{otherwise} \end{cases} \tag{5.42e}$$

For this choice, the time step is restricted to

$$\Delta t \le \frac{\Delta x}{a \max |u_j^n|} \tag{5.43e}$$

The *TVD* second order accurate Runge–Kutta (RK) scheme is given by

$$u^{[1]} = u^n + \Delta t L(u^n) \tag{5.44e}$$

$$u^{n+1} = 0.5u^n + 0.5u^{[1]} + 0.5\Delta t L(u^{[1]}) \tag{5.45e}$$

The non-*TVD* method is

$$u^{[1]} = u^n - 20\Delta t L(u^n) \tag{5.46e}$$

$$u^{n+1} = u^n + \tfrac{41}{40}\Delta t L(u^n) - \tfrac{1}{40}\Delta t L(u^{[1]}) \tag{5.47e}$$

Gottlieb and Shu [28] solved the program using the above algorithm and the results are given in Figure 5.6e. Clearly, the *TVD*-RK scheme is superior, and shows no oscillation near the discontinuity.

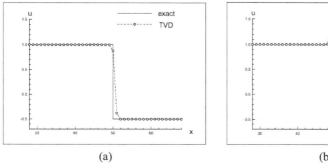

(a)   (b)

Figure 5.6e. Comparison of numerical solutions to the inviscid Burgers' equation obtained from the *TVD* and non-*TVD*-Runge–Kutta discontinuous methods: (a) with *TVD* and (b) without *TVD* [28]

For multidimensional problems, the construction of a slope limiter can be complex and difficult. Cockburn and Shu [17] give the general procedure for constructing such a limiter for various elements. Here we consider the triangular elements only, as they are the most frequently used for discontinuous finite element analyses. As for the 1-D case, the linear slope needs to satisfy the following conditions:

1. If u_h is linear, then $SP(u_h)$ is linear,

2. For every element of the triangularization, the mass is conserved,

$$\int_{\Omega_j} SP(u_h)\,dV = \int_{\Omega_j} u_h\,dV$$

3. On each element, the gradient of the $SP(u_h)$ is not bigger than that of u_h.

To construct the slope limiter for a triangular element that satisfies the above properties, the following procedure is taken. First, the element and its neighbors are defined as shown in Figure 5.17, when m is the middle point of the edge on the boundary of element j, and b_i denotes the center of the triangle Ω_i, $i = 0, 1, 2, 3$. Then, the averaged value of u is calculated by

$$\bar{u}_{\Omega_i} = \frac{1}{\Omega_i}\int_{\Omega_j} u_h\,dV = u_h(b_i),\ i = 0, 1, 2, 3. \tag{5.253}$$

and

$$\tilde{u}(m_1,\Omega_0) = u_h(m_1) - \bar{u}_{\Omega_0} = \alpha_1(\bar{u}_{\Omega_1} - \bar{u}_{\Omega_0}) + \alpha_2(\bar{u}_{\Omega_2} - \bar{u}_{\Omega_0})$$

$$= \Delta\bar{u}(m_1,\Omega_0) \tag{5.254}$$

For a piecewise linear function,

$$u_h(x, y) = \sum_{i=1}^{3} u_h(m_i)\phi_i(x, y) = \bar{u}_{\Omega_0} + \sum_{i=1}^{3} \tilde{u}_h(m_i,\Omega_0)\phi_i(x, y) \tag{5.255}$$

where $m_i(i = 1, 2, 3)$ is the middle point of the ith edge of the triangle (see Figure 5.17).

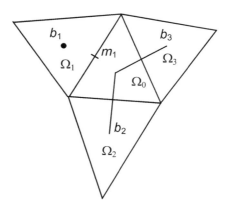

Figure 5.17. Geometric relations used to construct general slope limiters in a triangular mesh

The slope limiter can be constructed for the triangle as follows:

$$SP(u_h) = \bar{u}_{\Omega_0} + \sum_{i=1}^{3} (\theta^+ \max(0, \Delta_i) - \theta^- \max(0, -\Delta_i)) \phi_i(x, y) \qquad (5.256)$$

where the following definitions are used:

$$\theta^+ = \min\left(1, \frac{\sum_{i=1}^{i=3} \max(0, -\Delta_i)}{\sum_{i=1}^{i=3} \max(0, \Delta_i)}\right); \quad \theta^- = \min\left(1, \frac{\sum_{i=1}^{i=3} \max(0, \Delta_i)}{\sum_{i=1}^{i=3} \max(0, -\Delta_i)}\right) \qquad (5.257)$$

$$\Delta_i = \bar{m}(\tilde{u}_h(m_i, \Omega_0), \nu \Delta \bar{u}(m_i, \Omega_0)) \qquad (5.258)$$

In the above a modified minmod function is used,

$$\bar{m}(a, b) = \begin{cases} a, \text{ if } |a| \le Mh^2 \\ \min \operatorname{mod}(a, b) \end{cases} \qquad (5.259)$$

where h is the size of the element, and M is a given constant.

Cockburn and Shu [17] developed a discontinuous finite element method for the solution of a double Mach problem, a benchmark problem used for fluid dynamics calculations. The results are presented in Figure 5.18.

5.5 Viscous Burgers' Equations

The preceding section has been focused on inviscid Burgers' equations when the system exhibits no dissipation, because the viscosity is neglected. In the real world, viscosity is always present, and plays an important role in both providing stabilizing effects for the Galerkin based numerical algorithms, and smearing the sharp discontinuity in the solution.

5.5.1 1-D Burgers' Equation

Let us consider the 1-D Burgers' equation,

$$\frac{\partial u}{\partial t} + u \frac{\partial u}{\partial x} = D \frac{\partial^2 u}{\partial x^2}, \quad x \in [0,1] \qquad (5.260)$$

with the boundary conditions $u(0, t) = 1$ and $u(1, t) = 0$ and the initial condition $u(x, 0) = 0$.

To develop a discontinuous formulation, the above equations are first split into two first order differential equations,

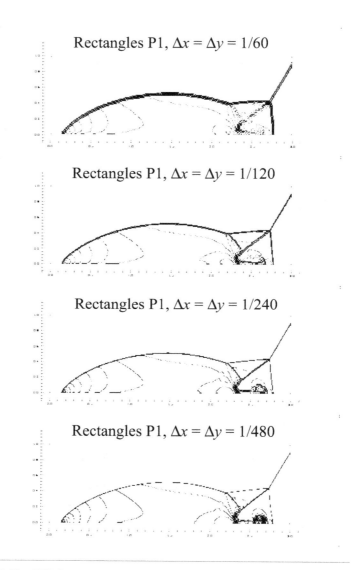

Figure 5.18. The *TVD*-Runge–Kutta discontinuous finite element solution of a double Mach benchmark problem [17]

$$q = \frac{\partial u}{\partial x} \; ; \quad \frac{\partial u}{\partial t} + u \frac{\partial u}{\partial x} = D \frac{\partial q}{\partial x} \tag{5.261}$$

The domain is now discretized and integration is carried out over element $j \in [x_j, x_{j+1}]$. After integration by parts, we arrive at the weak form integrals (see Figure 5.2 for geometric relations),

$$\int_{x_j}^{x_{j+1}} q_h w dx + \int_{x_j}^{x_{j+1}} u_h w_x dx - \hat{u}_{j+1}^- w_{j+1}^- + \hat{u}_j^+ w_j^+ = 0 \qquad (5.262a)$$

$$\int_{x_j}^{x_{j+1}} \frac{\partial u_h}{\partial t} v dx + \int_{x_j}^{x_{j+1}} (Dq_h - u_h^2) v_x dx - (D\hat{q}_{j+1}^- - \hat{h}_{j+1}^-) v_{j+1}^-$$
$$+ (D\hat{q}_j^+ - \hat{h}_j^+) v_j^+ = 0 \qquad (5.262b)$$

where two types of numerical fluxes are used. The numerical fluxes associated with diffusion are given by Equation 4.20,

$$\hat{q} = 0.5(q^+ + q^-) - C_{11}(u^- - u^+) - C_{12}(q^- - q^+) \qquad (5.263a)$$

$$\hat{u} = 0.5(u^+ + u^-) + C_{12}(u^- - u^+) \qquad (5.263b)$$

The convective numerical flux $\hat{h}$ may be approximated in two ways: one is to use the convective numerical flux given by Equation 5.27, and the other is to choose one from the function fluxes defined by Equations 5.172–5.175 [19]. If Equation 5.27 is used, $u \partial u / \partial x$ is treated approximately as $u_c \partial u / \partial x$, where u_c is assumed a constant, but changes its value during each iteration. This should give the first approximation as shown in the example below.

The unknowns (q_h, u_h) may be approximated with a polynomial as local basis functions,

$$u_h(x,t) = \sum_{i=1}^{k+1} u_i(t)\phi_i(x) = \Phi^T \underline{\mathbf{u}} \; ; \; q_h(x,t) = \sum_{i=1}^{k+1} q_i(t)\phi_i(x) = \Phi^T \underline{\mathbf{q}} \qquad (5.264)$$

where k is the order of the polynomial and $k+1 = N_e$, N_e being the number of nodes per element. With the numerical fluxes for diffusion and convection defined above, we may start to carry out calculations. If the convective numerical flux is used and the Galerkin approximation is invoked, then we have the following matrix equation:

$$\begin{bmatrix} \mathbf{0} & \mathbf{0} \\ \mathbf{0} & \mathbf{M} \end{bmatrix} \begin{pmatrix} \dot{\underline{\mathbf{q}}} \\ \dot{\underline{\mathbf{u}}} \end{pmatrix}_{(j)} + \begin{bmatrix} \mathbf{E} & \mathbf{H} \\ \mathbf{J} & \mathbf{G} \end{bmatrix} \begin{pmatrix} \underline{\mathbf{q}} \\ \underline{\mathbf{u}} \end{pmatrix}_{(j)} + \begin{bmatrix} \mathbf{E}_B & \mathbf{H}_B \\ \mathbf{J}_B & \mathbf{G}_B \end{bmatrix} \begin{pmatrix} \underline{\mathbf{q}} \\ \underline{\mathbf{u}} \end{pmatrix}_{(j)}$$
$$+ \begin{bmatrix} \mathbf{E}_{B,1} & \mathbf{N}_{B,1} \\ \mathbf{J}_{B,1} & \mathbf{G}_{B,1} \end{bmatrix} \begin{pmatrix} \underline{\mathbf{q}} \\ \underline{\mathbf{u}} \end{pmatrix}_{(j-1)} + \begin{bmatrix} \mathbf{E}_{B,2} & \mathbf{N}_{B,2} \\ \mathbf{J}_{B,2} & \mathbf{G}_{B,2} \end{bmatrix} \begin{pmatrix} \underline{\mathbf{q}} \\ \underline{\mathbf{u}} \end{pmatrix}_{(j+1)} = \begin{pmatrix} \mathbf{0} \\ \mathbf{0} \end{pmatrix} \qquad (5.265)$$

where the overdot denotes the time derivative, subscript () refers to the element, subscript B denotes the matrices associated with element boundary integrals, and $\mathbf{E}$, $\mathbf{G}$, $\mathbf{J}$, and $\mathbf{N}$ are matrices associated with element calculations, but they can be a function of u because of the need to update the variable during each iteration.

The above equation is very similar to that given in Section 5.4.2 and thus the same method can be applied to solve it, the difference being that, in this case, the variable u (hence u_c) needs to be updated during each iteration, since the problem is nonlinear.

Example 5.7. Develop a discontinuous formulation for the 1-D viscous Burgers' equation defined by Equation 5.260 and discuss the results. Use linear elements and the convective numerical flux approximation.

Solution. The domain is discretized to N elements. For linear element approximations, we have

$$u_h(x,t) = u_1(t)\phi_1(x) + u_2(t)\phi_2(x) ; \quad q_h(x,t) = q_1(t)\phi_1(x) + q_2(t)\phi_2(x) \tag{5.29e}$$

The diffusion numerical fluxes then become

$$\hat{u}_j^+ = \begin{cases} u_0 & j=1 \\ (1/2 - C_{12}^{(1)})u_j^+ + (1/2 + C_{12}^{(1)})u_j^- & j \neq 1 \end{cases} \tag{5.48e}$$

$$\hat{u}_{j+1}^- = \begin{cases} u_{N+1} & j = N \\ (1/2 - C_{12}^{(2)})u_{j+1}^+ + (1/2 + C_{12}^{(2)})u_{j+1}^- & j \neq N \end{cases} \tag{5.49e}$$

$$\hat{q}_j^+ = \begin{cases} q_j^+ - C_{11}(u_0 - u_j^+) & j = 1 \\ (1/2 + C_{12}^{(1)})q_j^+ + (1/2 - C_{12}^{(1)})q_j^- - C_{11}(u_j^- - u_j^+) & j \neq 1 \end{cases} \tag{5.50e}$$

$$\hat{q}_{j+1}^- = \begin{cases} q_{j+1}^- - C_{11}(u_{j+1}^- - u_{N+1}) & j = N \\ (1/2 + C_{12}^{(2)})q_{j+1}^+ + (1/2 - C_{12}^{(2)})q_{j+1}^- - C_{11}(u_{j+1}^- - u_{j+1}^+) & j \neq N \end{cases}$$
$$\tag{5.51e}$$

If the convective numerical fluxes are used, then we have

$$\hat{h}_j^+ = \begin{cases} u_j^+ u_j^- & u_j^+ > 0 \\ u_j^+ u_j^+ & u_j^+ \leq 0 \end{cases} ; \quad \hat{h}_{j+1}^- = \begin{cases} u_{j+1}^+ u_{j+1}^- & u_{j+1}^- > 0 \\ u_{j+1}^- u_{j+1}^+ & u_{j+1}^- \leq 0 \end{cases} \tag{5.52e}$$

The final matrix equations, after some algebra, take the following forms:

$$\frac{h_{(j)}}{6} \begin{bmatrix} 2 & 1 \\ 1 & 2 \end{bmatrix} \begin{pmatrix} q_j^+ \\ q_{j+1}^- \end{pmatrix} + \begin{bmatrix} -C_{12}^{(1)} & -1/2 \\ 1/2 & -C_{12}^{(2)} \end{bmatrix} \begin{pmatrix} u_j^+ \\ u_{j+1}^- \end{pmatrix} = \begin{pmatrix} -(1/2 + C_{12}^{(1)})u_j^- \\ (1/2 - C_{12}^{(2)})u_{j+1}^+ \end{pmatrix}$$
$$\tag{5.53e}$$

$$\frac{h_{(j)}}{6}\begin{bmatrix}2 & 1\\ 1 & 2\end{bmatrix}\begin{pmatrix}\dot{u}_j^+\\ \dot{u}_{j+1}^-\end{pmatrix}+\begin{bmatrix}DC_{12}^{(1)} & -D/2\\ D/2 & DC_{12}^{(2)}\end{bmatrix}\begin{pmatrix}q_j^+\\ q_{j+1}^-\end{pmatrix}$$

$$+\begin{bmatrix}1/2 & 1/2\\ -1/2 & -1/2\end{bmatrix}\begin{bmatrix}u_j^+ & 0\\ 0 & u_{j+1}^-\end{bmatrix}\begin{pmatrix}u_j^+\\ u_{j+1}^-\end{pmatrix}+\begin{bmatrix}DC_{11} & 0\\ 0 & DC_{11}\end{bmatrix}\begin{pmatrix}u_j^+\\ u_{j+1}^-\end{pmatrix}$$

$$=\begin{pmatrix}-D(1/2-C_{12}^{(1)})q_j^- +DC_{11}u_j^- +\hat{h}_j^+\\ D(1/2+C_{12}^{(2)})q_{j+1}^+ +DC_{11}u_{j+1}^+ -\hat{h}_{j+1}^-\end{pmatrix} \qquad (5.54e)$$

For the calculations, $C_{12} = 0$ in the interior. To incorporate the boundary conditions, we set $C_{12}^{(1)} = 1/2$ when $j = 1$ and $C_{12}^{(1)} = -1/2$ when $j = N+1$. For the results shown below in Figure 5.7e, the Euler forward time scheme is used and the time step was $\Delta t = 10^{-4}$. From these results, it is clear that upwinding seems to give the best results and downwinding is simply a disaster. The mesh size is important, as a coarse mesh generates considerable oscillation, particularly near the discontinuity front. As the mesh size progressively reduces, the resolution becomes better. For this problem, a mesh of 100 linear elements provides the best results.

5.5.2 2-D Viscous Burgers' Equation

Just as the one-dimensional heat conduction equation has a multidimensional counterpart, so can the one-dimensional Burgers' equation be extended to multidimensions. The two-dimensional Burgers' equation is written as

$$\frac{\partial u}{\partial t}+u\frac{\partial u}{\partial x}+v\frac{\partial u}{\partial y}-D\left(\frac{\partial^2 u}{\partial x^2}+\frac{\partial^2 u}{\partial y^2}\right)=0 \qquad (5.266a)$$

$$\frac{\partial v}{\partial t}+u\frac{\partial v}{\partial x}+v\frac{\partial v}{\partial y}-D\left(\frac{\partial^2 v}{\partial x^2}+\frac{\partial^2 v}{\partial y^2}\right)=0 \qquad (5.266b)$$

The two-dimensional Burgers' equations are basically two-dimensional momentum equations for incompressible laminar flow, with the pressure terms dropped.

Like the one-dimensional Burgers' equation, exact solutions can be constructed [3] using the Cole–Hopf transformation. In two dimensions, the Cole–Hopf transformation introduces a single function Φ, to which u and v are related as

$$u=\frac{-2D\dfrac{\partial\Phi}{\partial x}}{\Phi}\ ;\ \overline{v}=\frac{-2D\dfrac{\partial\Phi}{\partial x}}{\Phi} \qquad (5.267)$$

As a result, Equations 5.266a–b are transformed to a single equation,

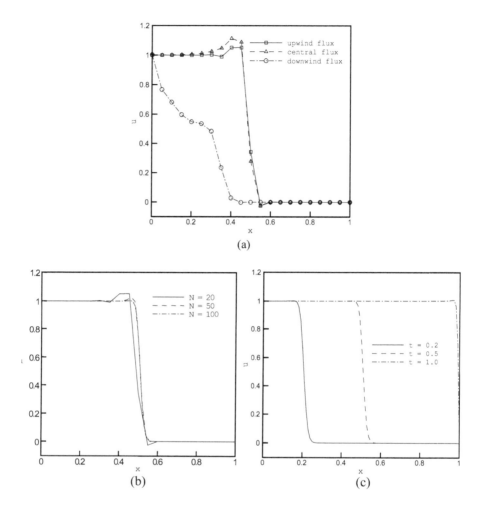

Figure 5.7e. Calculations for the viscous Burgers' equation ($D = 0.001$): (a) effect of upwinding schemes ($N = 20$, $t = 0.5$), (b) effect of mesh size (upwinding, $t = 0.5$) and (c) evolution of variable u at different times ($N = 100$, upwinding)

$$\frac{\partial \Phi}{\partial t} - \left(\frac{\partial^2 \Phi}{\partial x^2} + \frac{\partial^2 \Phi}{\partial y^2} \right) = 0 \qquad (5.268)$$

which is the two-dimensional diffusion equation. With appropriate boundary conditions, the above equation can be solved and the velocities can be derived using Equation 5.267. Of course, Equation 5.266 can be solved directly using a discontinuous fintie element scheme presented in this chapter.

If convective numerical fluxes are used to approximate the convection term, then the final matrix equation would be very similar to Equation 5.119 and the computational procedure then follows from the discussion ensuing there. However, better approximations are expected if the viscous effects are very small and if the *TVD*-based approach is used. In deriving the discontinuous finite element algorithm for the solution of the multidimensional Burgers' equation, the *TVD*-Runge–Kutta discontinuous scheme should be an attractive approach, especially when the viscous term is small in comparison with the convection. Kuzmin and Turek [19] show that for a convection-dominant Burgers' equation, the slope limiter is required to obtain a higher resolution scheme for a numerical solution.

Exercises

1. Starting from Equation 5.6 and assuming **A** is a function of **T**, derive Equation 5.14.
2. Consider the first order hyperbolic equation,

$$\frac{\partial u}{\partial t} + u\frac{\partial u}{\partial x} = 0$$

$$u(x,0) = \begin{cases} 1 & x < 0 \\ 1-x & 0 \le x < 1 \\ 0 & 1 \le x \end{cases}$$

Calculate the characteristic curves and plot the characteristic curves emanating from $x = 0.5$ and $x = 1.5$. Calculate the time at which the two curves meet.

3. Develop a discontinuous finite element formulation for the two-dimensional wave problem,

$$\frac{\partial u}{\partial t} + c\frac{\partial u}{\partial x} + c\frac{\partial u}{\partial y} = 0$$

and compare the results with those reported in Biswas *et al.* [18].

4. Consider a convection-diffusion equation with the initial and boundary data given below:

$$\frac{\partial T}{\partial t} + c\frac{\partial T}{\partial x} = D\frac{\partial^2 T}{\partial x^2} + g(T), \quad \in [0,1], \, t > 0$$

$$T(x,0) = T_0(x), \quad \in [0,1]$$

$$T(0,t) = T(1,t) = 0, \quad t > 0$$

with $D > 0$ and $\sup|g(T)| = M < \infty$. Defining the energy at time t as

$$E(t) = \int_0^1 T^2(x,t)dx ,$$

we then have the derivative of the above term,

$$\frac{dE(t)}{dt} = 2\int_0^1 T\frac{dT}{dt}dx = 2\int_0^1 T\left(D\frac{\partial^2 T}{\partial x^2} + g(T) - c\frac{\partial T}{\partial x}\right)dx$$

Show that the energy for this system is bounded by the following expression:

$$\frac{dE(t)}{dt} + D\pi^2 E(t) \le \frac{M^2}{D\pi^2}$$

or

$$E(t) \le E(0)\exp\left(-\pi^2 Dt\right) + \frac{M^2}{D^2\pi^4}\left(1 - \exp\left(-\pi^2 Dt\right)\right)$$

Hint: to prove the above relation, the inequalities of Hölder, Poincaré and Young may be needed.
   a) Holder's inequality: if $p,q \in R$, $p,q > 0$, $1/p + 1/q = 1$, then

$$\int_a^b |uv|dx \le \left|\int_a^b |u|^p dx\right|^{1/p}\left|\int_a^b |v|^q dx\right|^{1/q}$$

   b) Young's inequality: if $p,q \in R$, $p,q > 0$, $1/p + 1/q = 1$, f and g are real positive quantities, then

$$fg \le \frac{f^p}{p} + \frac{g^q}{q}$$

The Poincaré inequality is given in Problem 1 in Chapter 4.
5.   Consider the pure convection system,

$$\frac{\partial \mathbf{u}}{\partial t} = A\frac{\partial \mathbf{u}}{\partial x}, \quad \in [0,1], \ t > 0$$

$$\mathbf{u}(x,0) = \mathbf{u}_0(x), \quad \in [0,1]$$

$$\mathbf{u}(0,t) = \mathbf{u}(1,t) = 0, \quad t > 0$$

where A is a constant matrix with $A = A^T$. Show that

$$\cdot \quad \frac{d}{dt}\int_0^1 \mathbf{u}^T\mathbf{u}\,dx = 0$$

where $\mathbf{u}$ is a vector of n dimensions.

6. Consider Burgers' equation with the initial and boundary data:

$$\frac{\partial u}{\partial t} + u\frac{\partial u}{\partial x} = D\frac{\partial^2 u}{\partial x^2},\ \in[0,1],\ t>0$$

$$u(x,0) = u_0(x),\qquad \in[0,1]$$

$$u(0,t) = u(1,t) = 0,\quad t>0$$

with $D>0$. Let $w = u_1 - u_2$, where u_1 and u_2 are two solutions to the above problem. Define the energy term,

$$E(t) = \int_0^1 w^2(x,t)dx,$$

and show that

$$\frac{dE(t)}{dt} \le -\frac{1}{4}\int_0^1 \frac{\partial(u_1+u_2)}{\partial x}w^2(x,t)dx \le CE(t)$$

Further show that since $E(t)=0$, then $E(t)=0$, so long as $|C|<\infty$.

7. Develop a discontinuous finite element code to solve a 1-D pure convection problem. Compare the results obtained with linear and higher order elements. Use different time marching schemes.

8. Develop a discontinuous finite element code to solve a 1-D convection-duffision problem. Compare the results obtained with linear and higher order elements. Perform the stability analysis. Compare the results obtained using different numerical fluxes.

9. Develop a discontinous finite element code to solve a 1-D viscous Burgers' equation and study the solution as a function of diffusion coefficient. Compare the results obtained using constant, linear and quadratic elements and using linear and higher order Runge–Kutta time integrators.

References

[1]   Lagon JD. An Introduction to Nonlinear Partial Differential Equations. New York: John Wiley & Sons, 1994.

[2]   Evans LC. Partial Differential Equations. Philadelphia: American Society of Mathematics, 2000.

[3]   Fletcher CAJ. Computational Techniques for Fluid Dynamics. Berlin: Springer-Verlag, 1991.

[4]   Cockburn B. Discontinuous Galerkin Methods. School of Mathematics, University of Minnesota, 2003: 1–25.

[5]   Cockburn B. Discontinuous Galerkin Methods for Convection Dominated Problems. NATO Lecture Notes. School of Mathematics, University of Minnesota, 2001.

[6]   Tannehill JC, Anderson DA, Pletcher RH. Computational Fluid Mechanics and Heat Transfer. New York: Taylor & Francis, 1997.

[7]   Bender CM, Orszag SA. Advanced Mathematical Methods for Scientists and Engineers. Berlin: Springer-Verlag, 1999.

[8]   Moon TK, Stirling WC. Mathematical Methods and Algorithms for Signal Processing. Englewood Cliffs: Prentice Hall, 2000.

[9]   Zienkiewicz OC, Taylor RL, Sherwin SJ, Peiro I. On Discontinuous Galerkin Methods. Int. J. Numer. Meth. Eng. 2003; 58: 1119–1148.

[10]   Carey, GF, Oden JT. Finite Elements, Volume VI: Fluid Mechanics. Englewood Cliffs: Prentice Hall, 1986

[11]   Cockburn B, Shu CW. The Local Discontinuous Galerkin Method for Time-Dependent Convection-Diffusion Systems. SIAM J. Numer. Anal. 1998; 35: 2440–2463.

[12]   Warburton T. Lecture Notes on Numerical Solution of Partial Differential Equations. University of New Mexico, 2003.

[13]   Boye WE, DiPrima RC. Elementary Differential Equations and Boundary Value Problems, 7th Ed. New York: John Wiley, 2004.

[14]   Cockburn B, Shu CW. TVB Runge–Kutta Local Projection Discontinuous Finite Element Method for Conservation Laws II: General Framework. Math. Comput. 1989; 52: 411–435.

[15]   Cockburn B, Lin SY, Shu CW. TVB Runge–Kutta Local Projection Discontinuous Finite Element Method for Conservation Laws III: One-Dimensional Systems. J. Comput. Phys. 1989; 84: 90–113.

[16]   Cockburn B, Hou S, and Shu CW. TVB Runge–Kutta Local Projection Discontinuous Galerkin Finite Element Method for Conservation Laws IV: The Multidimensional Case. Math. Comput. 1990; 54: 545–581.

[17]   Cockburn B, Shu CW. The Runge–Kutta Discontinuous Galerkin Method for Conservation Laws V: Multidimensional Systems. J. Comput. Phys. 1998; 141: 199–224.

[18]   Biswas R, Devine K, Flaherty J. Parallel Adaptive Finite Element Methods for Conservation Laws. Appl. Numer. Math. 1994; 14: 255–283.

[19]   Kuzmin D, Turek S. High-Resolution FEM-TVD Schemes Based on a Fully Multidimensional Flux Limiter. J. Comput. Phys. 2004; 198: 131–158.

[20]   Shu CW, Osher S. Efficient Implementation of Essentially Non-oscillatory Shock Capture Schemes. J. Comp. Phys. 1988; 77: 439–471.

[21]   Harten A. On the Symmetric Form of Systems of Conservation Laws with Entropy. J. Comp. Phys. 1983; 49: 151–164.

[22]   Harten A, Lax PD, Van Leer B. On Upstream Differencing and Godunov-type Schemes for Hyperbolic Conservation Laws. SIAM Review. 1983; 25: 35–61.

[23]   Harten A. On a Class of High Resolution Total-Variation-Stable Finite-Difference Schemes. SIAM J. Numer. Anal. 1984; 21: 1–23.

[24]   Harten A, Engquist B, Osher S, Chakravarthy S. Uniformly High Order Accurate Essentially Non-oscillatory Schemes III. J. Comput. Phys. 1987; 71: 231–303.

[25]   Jameson A. Essential Elements of Computational Algorithms for Aerodynamic Analysis and Design. Tech. Rep. 97–68, Inst. Comput. Appl. in Sci. and Eng., NASA Langley Research Center, Hampton, Virginia, December 1997.

[26] Leer BV. Towards the Ultimate Conservative Difference Scheme, V. A Second-Order Sequel to Godunov's Method. J. Comput. Phys. 1979; 32: 101–136.
[27] Colella P, Woodward PR. The Piecewise Parabolic Method (ppm) for Gas-Dynamics Simulations. J. Comput. Phys. 1984; 54: 174–201.
[28] Gottlieb, Shu CW. Total Variation Diminishing Runge–Kutta Schemes. Math. Comput. 1998; 67: 73–85.

6

Incompressible Flows

Incompressible flows occur in a wide range of thermal and fluids systems, where the transport of energy and species is dominated by the convection mechanism. Understanding of the flow field distribution and its effect on thermal and species transport is thus of critical importance in these systems. The general mathematical description of fluid motion is given in Chapter 1, which consists of balance equations for momentum, mass and energy, the latter being required for a non-isothermal fluid flow system. These equations consist of a set of coupled, nonlinear, partial differential equations in terms of the velocity components, pressure and temperature. The momentum balance equations are also referred to as the Navier–Stokes equations. For an isothermal flow system, or a system in which the thermal field has a negligible effect on flow, the energy equation may be decoupled from the Navier–Stokes equations and the continuity equation.

For the purpose of numerical computations, the Navier–Stokes equations are often solved using two approaches: the primitive variable (e.g., velocity–pressure) approach and the vorticity–stream function (or derived variable) approach. While the derived variable approach is popular for 2-D problems, its extension to a general 3-D description can be rather complex and thus there have been limited applications of a 3-D derived variable approach. Probably the majority of numerical simulations have employed the primitive variable approach. Most of the computational fluid dynamics codes commercially available today are also developed on the primitive variable formulation. An important advantage of the primitive variable formulation is that the extension of a 2-D formulation to its 3-D counterpart is straightforward.

In practical applications, flows often are in the turbulence regime. The effects of turbulence on incompressible viscous flows are usually modeled using the concept of eddy viscosity, which is computed from a k–ε engineering turbulence model or its variations. The differential equations for k and ε are structurally similar to the energy transport equation and are usually discretized and calculated in the same way, except in the near wall region, where a special treatment is required for the standard k–ε model. Of course, for stratified flows, additional terms need to be added due to buoyancy effects. The actual implementation for flow calculations involves the use of an effective viscosity in place of the laminar viscosity. Consequently, the computational algorithms used for laminar flow simulations can be modified relatively easily for incompressible turbulent flow calculations.

This chapter starts with a discussion of the discontinuous finite element solution of the Navier–Stokes equations for isothermal incompressible flows using the primitive variable approach. Other variations of the formulation are then discussed. This is followed by the discontinuous formulation, using the derived variable approach. The chapter ends with a discussion on the discontinuous formulation for non-isothermal fluid flow problems.

6.1 Primitive Variable Approach

The primitive variable approach refers to the direct solution of the Navier–Stokes equations in terms of velocity and pressure. This is in contrast with the derived variable approach by which the velocity field is derived from other field variables. The primitive variable approach is perhaps the most widely used method for the numerical solution of the Navier–Stokes equations. A major advantage of this approach is that the formulations and numerical algorithms developed and tested for 2-D calculations can be directly extended to 3-D calculations. The incompressibility constraint, however, poses a considerable difficulty for numerical algorithm development.

In the finite element solution of incompressible flow problems, the study of the divergence-free constraint leads to the well known *LBB* (or inf-sup) condition, by which the interpolation order for the pressure field is required to be one order lower than that for the velocity field, in order to have a stable numerical solution [1, 2]. The *LBB* condition is, in essence, a modification of the coersiveness condition of the Lax–Milgram theory [3] for a mixed finite element setting, and is a key condition required for a mixed finite element procedure to be well-posed and optimal [2, 4, 5]. The inf-sup condition may also be derived from convergence considerations or the finite element matrix equations [6, 7]. The use of the *LBB* condition for the stability analysis of various mixed finite element formulations can be found in [6–10]. From the constrained optimization point of view, the pressure field is a Lagrangian multiplier and the continuity is the constraint. The saddle point of the augmented dual optimization, or variational, functional gives the solution to the Navier–Stokes equations. Strictly speaking, this saddle point is derived for the Stokes flow problems [1]. Since the discontinuous finite element method is in some sense a stabilized mixed finite element method, it inherits certain features of the continuous finite element method, for the incompressible fluid flow calculations in particular.

The development of a discontinuous finite element formulation for incompressible fluid problems is based on the ideas and methodologies discussed in Chapters 4 and 5. We consider the continuity and the Navier–Stokes equations, written in terms of velocity and pressure [11],

$$\nabla \cdot \mathbf{u} = 0 \quad \in \Omega \qquad\qquad (6.1)$$

$$\frac{\partial \mathbf{u}}{\partial t} + (\mathbf{u} \cdot \nabla)\mathbf{u} = -\nabla p + \frac{1}{\mathrm{Re}}\nabla^2 \mathbf{u} + \mathbf{f} \quad \in \Omega \tag{6.2}$$

$$\mathbf{u} = \mathbf{u}_a \quad \in \partial\Omega \tag{6.3}$$

where $\mathbf{u} = (u_x, u_y, u_z)$ is the velocity field, p is the pressure, $\mathbf{f}$ is the external body force, and Re is the Reynolds number, defined by $\mathrm{Re} = U\rho L/\mu$ where ρ is the density of fluids, μ is the viscosity, and U and L are the characteristic scales of velocity and length associated with the flow fields. Here we take Ω to be the bounded domain with its boundary denoted by $\partial\Omega$, where for illustrative purposes a simple Dirichlet boundary condition is imposed. It is noted that other types of boundary conditions can be readily incorporated.

To develop an integral formulation suitable for a discontinuous finite element solution, the governing equations (Equations 6.1–6.2) are split into a set of first order partial differential equations,

$$\nabla \cdot \mathbf{u} = 0 \tag{6.4}$$

$$\underline{\tau} = \frac{1}{\mathrm{Re}}\nabla \mathbf{u} \tag{6.5}$$

$$\mathbf{u}_t + (\mathbf{u} \cdot \nabla)\mathbf{u} = -\nabla p + \nabla \cdot \underline{\tau} + \mathbf{f} \tag{6.6}$$

where $\mathbf{u}_t = \partial \mathbf{u}/\partial t$.

The Dirichlet boundary condition remains unchanged. The computational domain is now discretized into a finite number of elements. Multiplying the above equations by smooth test functions v, $\underline{\sigma}$ and $\mathbf{v}$ respectively, and integrating by parts over an arbitrary element Ω_j (see Figure 4.3), we have the following integral representation:

$$-\int_{\Omega_j} \mathbf{u} \cdot \nabla v\, dV + \int_{\partial\Omega_j} \mathbf{u} \cdot \mathbf{n}v\, dS = 0 \tag{6.7}$$

$$\int_{\Omega_j} \underline{\tau} : \underline{\sigma}\, dV = -\frac{1}{\mathrm{Re}}\int_{\Omega_j} \mathbf{u} \cdot \nabla \cdot \underline{\sigma}\, dV + \frac{1}{\mathrm{Re}}\int_{\partial\Omega_j} \mathbf{u} \cdot \underline{\sigma} \cdot \mathbf{n}\, dS \tag{6.8}$$

$$\int_{\Omega_j} \mathbf{u}_t \cdot \mathbf{v}\, dV - \int_{\Omega_j} \mathbf{u} \cdot \nabla \cdot (\mathbf{v} \otimes \mathbf{u})\, dV + \int_{\partial\Omega_j} \mathbf{u} \cdot \mathbf{n}\mathbf{u} \cdot \mathbf{v}\, ds - \int_{\Omega_j} p\nabla \cdot \mathbf{v}\, dV$$

$$+ \int_{\partial\Omega_j} p\mathbf{v} \cdot \mathbf{n}\, ds + \int_{\Omega_j} \underline{\tau} : \nabla \mathbf{v}\, dV - \int_{\partial\Omega_j} \underline{\tau} : (\mathbf{v} \otimes \mathbf{n})\, ds = \int_{\Omega_j} \mathbf{f} \cdot \mathbf{v}\, dV \tag{6.9}$$

where $\otimes$ is the dyadic operator, $(\mathbf{u}\otimes\mathbf{v})\cdot\mathbf{w} = (\mathbf{w}\cdot\mathbf{v})\mathbf{u}$. In some literature, $\mathbf{u}\otimes\mathbf{v}$ is also written as $\mathbf{uv}$, i.e., $\mathbf{u}\otimes\mathbf{v} = \mathbf{uv}$. Also, use has been made of the following relation: $\mathbf{v}\cdot(\mathbf{u}\cdot\nabla\mathbf{u}) = \nabla\cdot(\mathbf{u}\cdot(\mathbf{v}\otimes\mathbf{u})) - \mathbf{u}\cdot\nabla\cdot(\mathbf{v}\otimes\mathbf{u})$ and of the boundary condition. The above equations define the weak form of the incompressible Navier–Stokes equations. To develop numerical solutions, the exact solution $(p\times\underline{\tau}\times\mathbf{u})$ is approximated with functions $(p_h\times\underline{\tau}_h\times\mathbf{u}_h)$ in the finite element broken space and the function fluxes at the interface of the elements are replaced by the numerical fluxes. We then have the discontinuous finite element formulation of the incompressible fluid flow problems,

$$-\int_{\Omega_j} \mathbf{u}_h\cdot\nabla v\, dV + \int_{\partial\Omega_j} \hat{\mathbf{u}}_h^p\cdot\mathbf{n}v\, dS = 0 \tag{6.10}$$

$$\int_{\Omega_j} \underline{\tau}_h : \underline{\sigma}\, dV = -\frac{1}{\mathrm{Re}}\int_{\Omega_j} \mathbf{u}_h\cdot\nabla\cdot\underline{\sigma}\, dV + \frac{1}{\mathrm{Re}}\int_{\partial\Omega_j} \hat{\mathbf{u}}_h^{\sigma}\cdot\underline{\sigma}\cdot\mathbf{n}dS \tag{6.11}$$

$$\int_{\Omega_j} \mathbf{u}_t\cdot\mathbf{v}\,dV - \int_{\Omega_j} \mathbf{u}_h\cdot\nabla\cdot(\mathbf{v}\otimes\mathbf{u}_h)dV + \int_{\partial\Omega_j} \mathbf{u}_h\cdot\mathbf{n}\hat{\mathbf{u}}_h^c\cdot\mathbf{v}dS$$

$$-\int_{\Omega_j} p_h\nabla\cdot\mathbf{v}dV + \int_{\partial\Omega_j} \hat{p}_h\mathbf{v}\cdot\mathbf{n}dS + \int_{\Omega_j} \underline{\tau}_h : \nabla\mathbf{v}dV$$

$$-\int_{\partial\Omega_j} \hat{\underline{\tau}}_h : (\mathbf{v}\otimes\mathbf{n})ds = \int_{\Omega_j} \mathbf{f}\cdot\mathbf{v}dV \tag{6.12}$$

Here, $\hat{\mathbf{u}}_h^c$, $\hat{\mathbf{u}}_h^p$, $\hat{\mathbf{u}}_h^{\sigma}$, $\hat{\underline{\tau}}_h$ and $\hat{p}_h$ are the numerical fluxes, which are discrete approximations to traces on the boundary of elements, where $\hat{\mathbf{u}}_h^{\sigma}$ and $\hat{\underline{\tau}}_h$ are referred to as the diffusive numerical fluxes, $\hat{\mathbf{u}}_h^c$ is referred to as the convective numerical flux, and $\hat{\mathbf{u}}_h^p$ and $\hat{p}_h$ are referred to as the incompressible numerical fluxes which are related to the incompressibility condition on the velocity. Also, $\mathbf{n}$ is the outward normal of element j.

With this discontinuous finite element formulation, the same order approximation can be used for the pressure and the velocity fields [12]. To ensure the numerical stability of the discontinuous finite element method, the numerical fluxes must be chosen carefully. These numerical fluxes are defined in terms of the jump operators introduced in Chapters 4 and 5. Thus, on the interface of the interior elements $\partial\Omega_{j,i}\in\partial\Omega_j^+\cap\partial\Omega_j^-$, the mean values $\{\!\{\}\!\}$ and jumps $[\![\,]\!]$ for p, $\underline{\sigma}$ and $\underline{\mathbf{u}}$ are defined as follows:

$$\{\!\{p\}\!\} = \tfrac{1}{2}(p^+ + p^-)\,;\ [\![p]\!] = p^+\mathbf{n}^+ + p^-\mathbf{n}^- \tag{6.13a}$$

$$\{\!\{\underline{\tau}\}\!\} = \tfrac{1}{2}(\underline{\tau}^+ + \underline{\tau}^-)\,;\ [\![\underline{\tau}]\!] = \underline{\tau}^+\cdot\mathbf{n}^+ + \underline{\tau}^-\cdot\mathbf{n}^- \tag{6.13b}$$

$$\{\mathbf{u}\} = \tfrac{1}{2}(\mathbf{u}^+ + \mathbf{u}^-) \; ; \; [\![\mathbf{u}]\!] = \mathbf{u}^+ \otimes \mathbf{n}^+ + \mathbf{u}^- \otimes \mathbf{n}^- \qquad (6.13c)$$

Note that the jumps $\{p\}$ and $[\![\tau]\!]$ are both vectors, and jump $[\![\mathbf{u}]\!]$ is a tensor of rank two. Also, $+$ and $-$ refer to the inside and outside element j (see Figure 4.4).

For incompressible fluid flow calculations, Cockburn et al. [12 – 16] suggest the use of three different numerical fluxes. The first one accounts for viscous diffusion, which is based on the study of heat conduction problems in Chapter 4; the second for convection, which was discussed in Chapter 5 for convection-diffusion problems; and the third for the incompressibility condition.

The diffusive numerical fluxes are constructed as follows. If $\partial\Omega_j$ lies inside the domain Ω, the diffusive fluxes $\hat{\mathbf{u}}_h^\sigma$ and $\underline{\tau}_h$ are taken in the form of

$$\hat{\mathbf{u}}_h^\sigma = \{\mathbf{u}\} + [\![\mathbf{u}]\!] \cdot \mathbf{C}_{12} \; ; \; \hat{\underline{\tau}}_h = \{\underline{\tau}\} - C_{11}[\![\mathbf{u}]\!] - [\![\underline{\tau}]\!] \otimes \mathbf{C}_{12} \qquad (6.14a)$$

If $\partial\Omega_j$ lies on the boundary, the diffusive fluxes are taken in the form of

$$\hat{\mathbf{u}}_h^\sigma = \mathbf{u}_a \; ; \; \hat{\underline{\tau}}_h = \{\underline{\tau}\} - C_{11}(\mathbf{u} - \mathbf{u}_a) \otimes \mathbf{n} \qquad (6.14b)$$

The purpose of parameter C_{11} and $\mathbf{C}_{12}$ is to enhance the stability and accuracy of the discontinuous finite element method. It is worth noting that since the numerical flux $\hat{\mathbf{u}}_h^\sigma$ is independent of the variable $\underline{\tau}$, it is possible to eliminate it from the equations by using Equation 6.11 to solve $\underline{\sigma}$ in terms of $\mathbf{u}$ in an element-by-element manner.

The convective numerical fluxes are constructed based on the local flow conditions. For the convective flux $\hat{\mathbf{u}}_h^c$ in Equation 6.12, the standard upwinding flux scheme is used, namely,

$$\hat{\mathbf{u}}_h^c = \lim_{\varepsilon \to 0} \mathbf{u}(\mathbf{x} - \varepsilon\mathbf{u}(\mathbf{x})) \qquad (6.15)$$

The numerical fluxes $\hat{\mathbf{u}}_h^p$ and $\hat{p}_h$ are related to the incompressibility condition on the velocity. They are constructed as follows. If $\partial\Omega_j$ lies inside the domain Ω, $\hat{\mathbf{u}}_h^p$ and $\hat{p}_h$ are taken in the form of

$$\hat{\mathbf{u}}_h^p = \{\mathbf{u}\} + D_{11}[\![p]\!] + \mathbf{D}_{12} tr([\![\mathbf{u}]\!]) \; ; \; \hat{p}_h = \{p\} - \mathbf{D}_{12} \cdot [\![p]\!] \qquad (6.16a)$$

where D_{11} and $\mathbf{D}_{12}$ are constant and constant matrix respectively. On the boundary, they can be defined by

$$\hat{\mathbf{u}}_h^p = \mathbf{u}_a \; ; \; \hat{p}_h = p^+ \qquad (6.16b)$$

As usual, the unknowns can be approximated using the local polynomial basis functions,

$$\mathbf{u}_h = \sum_{i=1}^{N_e} \phi_i \mathbf{u}_i = \Phi^T \underline{\mathbf{u}}; \quad p_h = \sum_{i=1}^{N_e} \phi_i p_i = \Phi^T \underline{p}; \quad \underline{\tau}_h = \sum_{i=1}^{N_e} \phi_i \underline{\tau}_i = \Phi^T \underline{\Sigma}$$

where N_e is the number of nodes associated with the element, and the unknown functions are found as follows:

$$\underline{\mathbf{u}} = (\underline{\mathbf{u}}_x, \underline{\mathbf{u}}_y, \underline{\mathbf{u}}_z) = \underline{\mathbf{u}}_x \hat{i} + \underline{\mathbf{u}}_y \hat{j} + \underline{\mathbf{u}}_z \hat{k}$$

$$\underline{\mathbf{u}}_k = (u_{k,1}, u_{k,2}, \cdots, u_{k,N_e})^T \text{ with } k = x, y, z$$

$$\underline{p} = (p_1, p_2, \cdots, p_{N_e})^T$$

$$\underline{\Sigma} = \begin{vmatrix} \underline{\Sigma}_{xx} & \underline{\Sigma}_{xy} & \underline{\Sigma}_{xz} \\ \underline{\Sigma}_{yx} & \underline{\Sigma}_{yy} & \underline{\Sigma}_{yz} \\ \underline{\Sigma}_{zx} & \underline{\Sigma}_{zy} & \underline{\Sigma}_{zz} \end{vmatrix}$$

$$\underline{\Sigma}_{km} = (\sigma_{km,1}, \sigma_{km,2}, \cdots, \sigma_{km,N_e})^T \text{ with } k, m = x, y, z$$

$$\Phi = (\phi_1, \phi_2, \cdots, \phi_{N_e})^T$$

With these numerical fluxes substituted, and making use of the Galerkin approximation, Equations 6.10–6.12 can be calculated. The discretized form of each of these equations is presented below.

To obtain the discretized form for Equation 6.10, that is, the continuity equation, we set $\mathbf{D}_{12} = \mathbf{0}$ and obtain from Equation 6.16

$$\hat{\mathbf{u}}_h^p = \{\underline{\mathbf{u}}\} + D_{11}[\underline{p}] = 0.5(\mathbf{u}^+ + \mathbf{u}^-) + D_{11}(p^+\mathbf{n}^+ + p^-\mathbf{n}^-) \tag{6.17}$$

With this choice substituted into Equation 6.10, one has the following integrals:

$$-\left(\int_{\Omega_j} (\nabla\Phi)\Phi^T dV \right) \cdot \underline{\mathbf{u}} + \left(\int_{\partial\Omega_j} 0.5\Phi\Phi^T \mathbf{n} dS \right) \cdot \underline{\mathbf{u}} + \left(\int_{\partial\Omega_j} D_{11}\Phi\Phi^T dS \right) \underline{p}$$

$$+ \left(\int_{\partial\Omega_j} 0.5\Phi\Phi^T \mathbf{n} dS \right) \cdot \underline{\mathbf{u}}_{(NB)} + \left(-\int_{\partial\Omega_j} D_{11}\Phi\Phi^T dS \right) \underline{p}_{(NB)} = 0 \tag{6.18}$$

The above can be further written in matrix form,

$$\left(\mathbf{K}_x \quad \mathbf{K}_y \quad \mathbf{K}_z\right)\begin{pmatrix}\underline{\mathbf{u}}_x\\\underline{\mathbf{u}}_y\\\underline{\mathbf{u}}_z\end{pmatrix} - \sum_{i=1}^{NS}\left(\mathbf{E}_{x,i} \quad \mathbf{E}_{y,i} \quad \mathbf{E}_{z,i}\right)\begin{pmatrix}\underline{\mathbf{u}}_x\\\underline{\mathbf{u}}_y\\\underline{\mathbf{u}}_z\end{pmatrix} - \sum_{i=1}^{NS}\mathbf{H}_{1,i}\underline{\mathbf{p}}$$

$$+ \sum_{i=1}^{NS}\left(\mathbf{E}_{x,i} \quad \mathbf{E}_{y,i} \quad \mathbf{E}_{z,i}\right)\begin{pmatrix}\underline{\mathbf{u}}_x\\\underline{\mathbf{u}}_y\\\underline{\mathbf{u}}_z\end{pmatrix}_{(NB,i)} + \sum_{i=1}^{NS}\mathbf{H}_{1,i}\underline{\mathbf{p}}_{(NB,i)} = 0 \qquad (6.19)$$

where NS is the number of boundary nodes per element, (NB, i) represents the neighboring element sharing the ith side of element j, and other quantities are defined by the following expressions:

$$\mathbf{K}_x = \int_{\Omega_j}\frac{\partial\Phi}{\partial x}\Phi^T dV; \quad \mathbf{K}_y = \int_{\Omega_j}\frac{\partial\Phi}{\partial y}\Phi^T dV; \quad \mathbf{K}_z = \int_{\Omega_j}\frac{\partial\Phi}{\partial z}\Phi^T dV$$

$$\mathbf{E}_{x,i} = \int_{\partial\Omega_{j,i}}0.5\Phi\Phi^T(\mathbf{n}_{j,i}\cdot\hat{i})dS; \quad \mathbf{E}_{y,i} = \int_{\partial\Omega_{j,i}}0.5\Phi\Phi^T(\mathbf{n}_{j,i}\cdot\hat{j})dS;$$

$$\mathbf{E}_{z,i} = \int_{\partial\Omega_{j,i}}0.5\Phi\Phi^T(\mathbf{n}_{j,i}\cdot\hat{k})dS; \quad \mathbf{H}_{1,i} = \int_{\partial\Omega_{j,i}}D_{11}\Phi\Phi^T dS$$

Here $\mathbf{n}_{j,i}$ denotes the normal pointing outward from the ith side of element j. Also, $\partial\Omega_{j,i}$ represents the interface between element j and its neighbor element i: $\partial\Omega_j = \cup_{i=1}^{NS}\partial\Omega_{j,i}$. The above matrices can also be written in terms of the component form. For example, an element in matrix $\mathbf{E}_{z,i}$ is calculated by

$$E_{z,i,qr} = \int_{\partial\Omega_{j,i}}0.5\phi_q\phi_r(\mathbf{n}_{j,i}\cdot\hat{k})dS, \qquad q, r = 1, 2, \dots, N_e$$

We now turn to the calculation of Equation 6.11. By selecting the following numerical flux,

$$\hat{\mathbf{u}}_h^\sigma = \{\underline{\mathbf{u}}\} + [\underline{\mathbf{u}}]\cdot\mathbf{C}_{12} = 0.5(\mathbf{u}^+ + \mathbf{u}^-) + C_{12}\mathbf{n}^+\cdot(\mathbf{u}^+\otimes\mathbf{n}^+ + \mathbf{u}^-\otimes\mathbf{n}^-)$$

$$= (0.5 + C_{12})\mathbf{u}^+ + (0.5 - C_{12})\mathbf{u}^- \qquad (6.20)$$

and with $C_{12} = \mathbf{C}_{12}\cdot\mathbf{n}^+$, Equation 6.11 becomes

$$\left(\int_{\Omega_j}\Phi\Phi^T dV\right)\Sigma_{km} = \left(-\frac{1}{Re}\int_{\Omega_j}(\hat{e}_k\cdot\nabla\Phi)\Phi^T dV\right)\underline{\mathbf{u}}_m$$

$$+ \left(\frac{1}{\mathrm{Re}} \int_{\partial \Omega_j} (0.5 + C_{12}) \Phi \Phi^T (\mathbf{n} \cdot \hat{e}_k) dS \right) \underline{\mathbf{u}}_m$$

$$+ \left(\frac{1}{\mathrm{Re}} \int_{\partial \Omega_j} (0.5 - C_{12}) \Phi \Phi^T (\mathbf{n} \cdot \hat{e}_k) dS \right) \underline{\mathbf{u}}_{m(NB)}, \qquad (k, m = x, y, z) \quad (6.21)$$

where $\hat{e}_x = \hat{i}$, $\hat{e}_y = \hat{j}$, and $\hat{e}_z = \hat{k}$. The result can also be written in the following matrix form:

$$\mathbf{K}\underline{\Sigma}_{km} + \frac{1}{\mathrm{Re}} \mathbf{K}_k \underline{\mathbf{u}}_m - \frac{1}{\mathrm{Re}} \sum_{i=1}^{NS} \mathbf{E}_{21,k,i} \underline{\mathbf{u}}_m - \frac{1}{\mathrm{Re}} \sum_{i=1}^{NS} \mathbf{E}_{22,k,i} \underline{\mathbf{u}}_{m,(NB,i)} = 0$$

$$(k, m = x, y, z) \qquad (6.22)$$

where the matrices are calculated by

$$\mathbf{K} = \int_{\Omega_j} \Phi \Phi^T dV$$

$$\mathbf{E}_{21,k,i} = \int_{\partial \Omega_{j,i}} (0.5 + C_{12}) \Phi \Phi^T (\mathbf{n}_{j,i} \cdot \hat{e}_k) dS$$

$$\mathbf{E}_{22,k,i} = \int_{\partial \Omega_{j,i}} (0.5 - C_{12}) \Phi \Phi^T (\mathbf{n}_{j,i} \cdot \hat{e}_k) dS$$

At last, we consider Equation 6.12. To obtain a discretized form, the appropriate numerical fluxes are selected,

$$\hat{p}_h = \{p_h\} = 0.5(p_h^+ + p_h^-); \qquad \{\underline{\tau}_h\} = \frac{1}{2}(\underline{\tau}_h^+ + \underline{\tau}_h^-)$$

$$[\underline{\tau}_h] = \underline{\tau}_h^+ \cdot \mathbf{n}^+ + \underline{\tau}_h^- \cdot \mathbf{n}^-; \qquad [\underline{\mathbf{u}}_h] = \mathbf{u}_h^+ \otimes \mathbf{n}^+ + \mathbf{u}_h^- \otimes \mathbf{n}^-$$

$$\hat{\mathbf{u}}_h^c = \alpha \mathbf{u}_h^+ + (1 - \alpha) \mathbf{u}_h^- \quad \text{with} \quad \alpha = \mathrm{sgn}(1, \mathbf{u}_h^+ \cdot \mathbf{n}^+) = \begin{cases} 1 & \text{if } \mathbf{u}_h^+ \cdot \mathbf{n}^+ \geq 0 \\ -1 & \text{if } \mathbf{u}_h^+ \cdot \mathbf{n}^+ < 0 \end{cases};$$

$$\hat{\underline{\tau}}_h = \{\underline{\tau}_h\} - C_{11}[\underline{\mathbf{u}}_h] - [\underline{\tau}_h] \otimes \mathbf{C}_{12} = (0.5 - C_{12})\underline{\tau}_h^+ + (0.5 + C_{12})\underline{\tau}_h^- \quad (6.23)$$

Substituting the above numerical fluxes into Equation 6.12 yields the following integrals:

$$\left(\int_{\Omega_j} \Phi\Phi^T dV\right)\underline{\dot{\mathbf{u}}}_k - \left(\int_{\Omega_j} (\nabla \cdot (\hat{e}_k\mathbf{u}_h\Phi))\Phi^T dV\right)\cdot\underline{\mathbf{u}}$$

$$+\left(\int_{\partial\Omega_j} \alpha(\mathbf{u}_h \cdot \mathbf{n}_j)\Phi\Phi^T dS\right)\underline{\mathbf{u}}_k + \left(\int_{\partial\Omega_j} (1-\alpha)(\mathbf{u}_h \cdot \mathbf{n})\Phi\Phi^T dS\right)\underline{\mathbf{u}}_{k(NB)}$$

$$-\left(\int_{\Omega_j} (\hat{e}_k \cdot \nabla\Phi)\Phi^T dV\right)\underline{\mathbf{p}} + \left(\int_{\partial\Omega_j} 0.5\Phi\Phi^T\mathbf{n}\cdot\hat{e}_k dS\right)\underline{\mathbf{p}}$$

$$+\left(\int_{\partial\Omega_j} 0.5\Phi\Phi^T\mathbf{n}\cdot\hat{e}_k dS\right)\underline{\mathbf{p}}_{(NB)} + \sum_{m=1}^{3}\left(\int_{\Omega_j} (\hat{e}_m \cdot \nabla\Phi)\Phi^T dV\right)\underline{\Sigma}_{km}$$

$$-\sum_{m=1}^{3}\left(\int_{\partial\Omega_j} (0.5 - C_{12})\Phi\Phi^T\mathbf{n}\cdot\hat{e}_m dS\right)\underline{\Sigma}_{km}$$

$$-\sum_{m=1}^{3}\left(\int_{\partial\Omega_j} (0.5 + C_{12})\Phi\Phi^T\mathbf{n}\cdot\hat{e}_m dS\right)\underline{\Sigma}_{km(NB)} = \int_{\Omega_j} \Phi f_k dV ,$$

$$m, k = x, y, z; \quad \hat{e}_1 = \hat{e}_x = \hat{i}, \quad \hat{e}_2 = \hat{e}_y = \hat{j}, \quad \hat{e}_3 = \hat{e}_z = \hat{k} \qquad (6.24)$$

with $\dot{u} = \partial u/\partial t$. The corresponding matrix equation is given by

$$\begin{bmatrix} \mathbf{M} & & \\ & \mathbf{M} & \\ & & \mathbf{M} \end{bmatrix}\begin{pmatrix} \underline{\dot{\mathbf{u}}}_x \\ \underline{\dot{\mathbf{u}}}_y \\ \underline{\dot{\mathbf{u}}}_z \end{pmatrix} - \begin{bmatrix} \mathbf{K}_{xx} & \mathbf{K}_{xy} & \mathbf{K}_{xz} \\ \mathbf{K}_{xy} & \mathbf{K}_{yy} & \mathbf{K}_{yz} \\ \mathbf{K}_{zx} & \mathbf{K}_{zy} & \mathbf{K}_{zz} \end{bmatrix}\begin{pmatrix} \underline{\mathbf{u}}_x \\ \underline{\mathbf{u}}_y \\ \underline{\mathbf{u}}_z \end{pmatrix} + \sum_{i=1}^{NS}\begin{pmatrix} \mathbf{E}_{31,i}\underline{\mathbf{u}}_x \\ \mathbf{E}_{31,i}\underline{\mathbf{u}}_y \\ \mathbf{E}_{31,i}\underline{\mathbf{u}}_z \end{pmatrix}$$

$$+\sum_{i=1}^{NS}\begin{pmatrix} \mathbf{E}_{32,i}\underline{\mathbf{u}}_x \\ \mathbf{E}_{32,i}\underline{\mathbf{u}}_y \\ \mathbf{E}_{32,i}\underline{\mathbf{u}}_z \end{pmatrix}_{(NB,i)} - \begin{pmatrix} \mathbf{L}_{3,x} \\ \mathbf{L}_{3,y} \\ \mathbf{L}_{3,z} \end{pmatrix}\underline{\mathbf{p}} + \sum_{i=1}^{NS}\begin{pmatrix} \mathbf{H}_{31,x,i} \\ \mathbf{H}_{31,y,i} \\ \mathbf{H}_{31,z,i} \end{pmatrix}\underline{\mathbf{p}} + \sum_{i=1}^{NS}\begin{pmatrix} \mathbf{H}_{32,x,i} \\ \mathbf{H}_{32,y,i} \\ \mathbf{H}_{32,z,i} \end{pmatrix}\underline{\mathbf{p}}_{(NB,i)}$$

$$-\sum_{i=1}^{NS}\left\{\mathbf{J}_{31,x,i}\begin{pmatrix} \underline{\Sigma}_{xx} \\ \underline{\Sigma}_{yx} \\ \underline{\Sigma}_{zx} \end{pmatrix} + \mathbf{J}_{31,y,i}\begin{pmatrix} \underline{\Sigma}_{xx} \\ \underline{\Sigma}_{yy} \\ \underline{\Sigma}_{zy} \end{pmatrix} + \mathbf{J}_{31,z,i}\begin{pmatrix} \underline{\Sigma}_{xz} \\ \underline{\Sigma}_{yz} \\ \underline{\Sigma}_{zz} \end{pmatrix}\right\}$$

$$-\sum_{i=1}^{NS}\left\{\mathbf{J}_{32,x,i}\begin{pmatrix} \underline{\Sigma}_{xx} \\ \underline{\Sigma}_{yx} \\ \underline{\Sigma}_{zx} \end{pmatrix}_{(NB,i)} + \mathbf{J}_{32,y,i}\begin{pmatrix} \underline{\Sigma}_{xx} \\ \underline{\Sigma}_{yy} \\ \underline{\Sigma}_{zy} \end{pmatrix}_{(NB,i)} + \mathbf{J}_{32,z,i}\begin{pmatrix} \underline{\Sigma}_{xz} \\ \underline{\Sigma}_{yz} \\ \underline{\Sigma}_{zz} \end{pmatrix}_{(NB,i)}\right\}$$

$$+ \mathbf{G}_x \begin{pmatrix} \Sigma_{xx} \\ \Sigma_{yx} \\ \Sigma_{zx} \end{pmatrix} + \mathbf{G}_y \begin{pmatrix} \Sigma_{xx} \\ \Sigma_{yy} \\ \Sigma_{zy} \end{pmatrix} + \mathbf{G}_z \begin{pmatrix} \Sigma_{xz} \\ \Sigma_{yz} \\ \Sigma_{zz} \end{pmatrix} = \begin{pmatrix} \underline{\mathbf{s}}_x \\ \underline{\mathbf{s}}_y \\ \underline{\mathbf{s}}_z \end{pmatrix} \tag{6.25}$$

where the source term and the matrices are calculated by

$$\mathbf{M} = \int_{\Omega_j} \Phi \Phi^T dV$$

$$\mathbf{K}_{qr} = \int_{\Omega_j} \frac{\partial (\Phi u_r)}{\partial q} \Phi^T dV , \quad q, r = x, y, z$$

$$\mathbf{E}_{31,i} = \int_{\partial \Omega_{j,i}} \alpha \mathbf{u}_h \cdot \mathbf{n}_{j,i} \Phi \Phi^T dS$$

$$\mathbf{E}_{32,i} = \int_{\partial \Omega_{j,i}} (1 - \alpha) \mathbf{u}_h \cdot \mathbf{n}_{j,i} \Phi \Phi^T dS$$

$$\mathbf{L}_{3,q} = \int_{\Omega_j} \frac{\partial \Phi}{\partial q} \Phi^T dV , \quad q = x, y, z;$$

$$\mathbf{H}_{31,q,i} = \mathbf{H}_{32,q,i} = \int_{\partial \Omega_{j,i}} 0.5 \Phi \Phi^T \mathbf{n}_{j,i} \cdot \hat{e}_q dS , \quad q = x, y, z$$

$$\mathbf{G}_q = \int_{\Omega_j} \frac{\partial \Phi}{\partial q} \Phi^T dV , \quad q = x, y, z$$

$$\mathbf{J}_{31,q,i} = \int_{\partial \Omega_{j,i}} (0.5 - C_{12}) \Phi \Phi^T \mathbf{n}_{j,i} \cdot \hat{e}_q dS , \quad q = x, y, z$$

$$\mathbf{J}_{32,q,i} = \int_{\partial \Omega_{j,i}} (0.5 + C_{12}) \Phi \Phi^T \mathbf{n}_{j,i} \cdot \hat{e}_q dS , \quad q = x, y, z$$

$$\underline{\mathbf{s}}_q = \int_{\Omega_j} \Phi f_q dV , \quad q = x, y, z$$

Equations 6.19, 6.22 and 6.25 represent the discretized discontinuous finite element formulation for the Navier–Stokes equations with the continuity constraint. The unknown variables for this system include three components of velocity field, pressure and nine stress components. An iterative computational procedure is required to solve these equations. In practice, the successive substitution method works well for this type of problem. A typical computational procedure starts with the initial condition for velocity and then the pressure and stresses are calculated using the element-by-element sweep. To carry out a time marching, Equation 6.25 is solved using an explicit time scheme, with a time step selected to satisfy the *CFL* condition, to advance the velocity field ($\underline{u}^{n+1}$) with the known pressure ($\underline{p}^{n}$) and the stress components ($\underline{\Sigma}^{n}$) and the velocity field ($\underline{u}^{n}$) at the previous time steps. The pressure ($\underline{p}^{n+1}$) and the stresses ($\underline{\Sigma}^{n+1}$) are then calculated from this velocity field $\underline{u}^{n+1}$ using Equations 6.19 and 6.22. Because of the nonlinearity, ($\underline{p}^{n+1}$), ($\underline{\Sigma}^{n+1}$) and $\underline{u}^{n+1}$ at the time step ($n+1$) need to be determined iteratively using, say, the successive substitution method. The iteration continues until convergence is achieved at the time step ($n+1$). This procedure repeats for every element and every time step.

The above discontinuous finite element formulation has been applied to simulate incompressible fluid flows driven by a moving lid in a 2-D cavity. The dimensionless governing equations for this problem take the following component form:

$$\frac{\partial u}{\partial x} + \frac{\partial v}{\partial y} = 0 \tag{6.26a}$$

$$u\frac{\partial u}{\partial x} + v\frac{\partial u}{\partial y} = -\frac{\partial p}{\partial x} + \frac{1}{Re}\left(\frac{\partial^2 u}{\partial x^2} + \frac{\partial^2 u}{\partial y^2}\right) \tag{6.26b}$$

$$u\frac{\partial v}{\partial x} + v\frac{\partial v}{\partial y} = -\frac{\partial p}{\partial y} + \frac{1}{Re}\left(\frac{\partial^2 v}{\partial x^2} + \frac{\partial^2 v}{\partial y^2}\right) \tag{6.26c}$$

with the following boundary conditions:

$$u(x=0,y) = v(x=0,y) = 0 , \ u(x=1,y) = v(x=1,y) = 0 \tag{6.26d}$$

$$u(x,y=0) = v(x,y=0) = 0 , \ u(x,y=1) = 1 , \ v(x,y=1) = 0 \tag{6.26e}$$

where the lid velocity is normalized to one and the motion is from the left to right or in the positive x direction.

The computations were made using an unstructured mesh with linear triangular element approximation. Since this is a steady state case, the transient terms are set to zero. The iterative procedure starts with an initial guess of the field variables, followed by the element-by-element computation of the velocity, the stress and the pressure, with imposed boundary conditions. The successive substitution method is applied to perform the iteration until the variables converge within a preset tolerance. The finite element mesh and computed results are given in Figure 6.1 for different Re numbers.

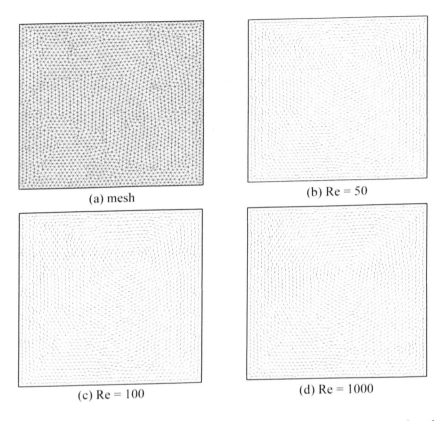

Figure 6.1 Computed results of a velocity field in a lid-driven cavity: (a) unstructured mesh, (b–d) velocity at (b) Re = 50, (c) Re = 100 and (d) Re = 1000

6.2 Fractional Step (Projection) Approach

The fractional step method is based on the operator splitting concept and has been widely used in the framework of finite difference approximations. The method was first proposed by Chorin [17] and is also known as the projection method. This method is considered useful for the solution of transient Navier–Stokes equations. There are many variations of the projection method. However, it essentially involves a two-step approximation for any time step. Two time steps are used in the formulation. In the first time step, a provisional velocity $\mathbf{u}^*$ is obtained. In the second step, the velocity then is corrected by accounting for a pressure gradient and an equation of continuity.

We consider splitting the velocity into two parts,

$$\mathbf{u} = \mathbf{u}^* + \mathbf{u}''$$

(6.27)

such that $\mathbf{u}^*$ is made to satisfy the momentum equation without the pressure,

$$\frac{\partial \mathbf{u}^*}{\partial t} = -\nabla \cdot (\mathbf{u}\mathbf{u}) + \left(\frac{1}{Re}\right)\nabla^2 \mathbf{u} \qquad (6.28)$$

$\mathbf{u}''$ is made to satisfy a portion of the pressure p,

$$\mathbf{.}'' = -\nabla p \qquad (6.29)$$

time marching scheme is used, then the provisional velocity field
ith information available at the nth time step by integrating

$$\left(1 - \left(\frac{\Delta t}{Re}\right)\nabla^2 \mathbf{u}^* = \mathbf{u}^n \qquad (6.30)\right.$$

sion and reaction equation. The provisional velocity
t for the pressure using Equation 6.29,

$$(6.31)$$

ulated using the Poisson equation:

$$(6.32)$$

$\mathbf{u}^{n+1}$ satisfy the continuity condition,

$$(6.33)$$

method, the pressure field is
oming from the divergence
n the momentum involves
tep. Consequently, this
and an equal order
sure fields.
venient to define a

$$(6.34)$$

$$(6.35)$$

With this definition, we can show that

$$\int_{\partial\Omega} \frac{\partial\phi}{\partial n} dS = \int_\Omega \nabla^2\phi\, dV = -\int_\Omega \nabla\cdot \mathbf{u}^* dV \qquad (6.36)$$

In practice, $\partial\phi/\partial n = 0$ is imposed at a solid boundary but the values on an open boundary are adjusted so that the above relation is satisfied. To ensure the no-slip condition, it is often necessary to impose the following relation:

$$\mathbf{u}^* = -\left[\frac{\partial\phi}{\partial s}\right]^n \qquad (6.$$

where s is the tangential direction.

The governing equations developed above using the fractional step a? can be solved using the discontinuous finite element. In fact, we treat equation (Equation 6.30) in the sequence as a convection-diffusion and equation, for which the discontinuous Galerkin formulation has been (see Chapter 5); and the Poisson equation (Equation 6.34) can be solved discontinuous schemes that have been discussed in Chapter 4. For a m? problem, the continuous finite element method may be used to s? Poisson equation, whereas the discontinuous finite element method is ? the solution of the convection-diffusion and reaction equation in vect combined approach has been presented in the context of the s incompressible flow problem over a 2-D domain using the de approach.

6.3 Vorticity and Stream Function Approach

The Navier–Stokes equations can also be written in terms of v functions. This approach is one of the most popular method? incompressible fluid flow problems in 2-D geometries. This a? the derived variable method, because the velocity field is r from field equations, but derived from the stream fun advantage of this approach is that the staggered-grid arran when the finite difference method is used. In the context unequal order interpolation becomes obviated. The stag order interpolations, which sometimes complicate progra to be used when the primitive variable approaches are en the Navier–Stokes equations. Another major advant approach is that the pressure field, which can be numerical scheme, is eliminated from the formatio? solved. This can be particularly useful for therm? information on a pressure distribution is not need method include the convective heat and mass tra? systems.

In general 3-D formulations, the stream function is generalized as a vector potential, because the concept of a stream function does not apply. First, we consider the general 3-D vorticity–vector potential formulation and its variation, and then deduce from them, the 2-D dimensional formulation.

In the fluid dynamics literature, the vorticity $\boldsymbol{\omega}$ and a vector potential $\boldsymbol{\psi}$ are related to the velocity field $\mathbf{u}$ through the following two relations:

$$\boldsymbol{\omega} = \nabla \times \mathbf{u} \tag{6.38}$$

$$\mathbf{u} = \nabla \times \boldsymbol{\psi} \tag{6.39}$$

Note that the second equation comes directly from the incompressibility (or divergence free) condition,

$$\nabla \cdot \mathbf{u} = \nabla \cdot (\nabla \times \boldsymbol{\psi}) = 0 \tag{6.40}$$

With the above definitions, the Navier–Stokes equations can now be written in terms of the vorticity and the potential vector,

$$\nabla^2 \boldsymbol{\psi} = -\boldsymbol{\omega} \tag{6.41}$$

$$\frac{\partial \boldsymbol{\omega}}{\partial t} + \nabla \cdot (\mathbf{u}\boldsymbol{\omega}) - (\boldsymbol{\omega} \cdot \nabla)\mathbf{u} - \frac{1}{Re} \nabla^2 \boldsymbol{\omega} = 0 \tag{6.42}$$

The first equation above is simply a vector Poisson equation, and is obtained by combining Equations 6.38 and 6.39, with $\nabla \cdot \boldsymbol{\psi} = 0$ imposed on the vector function. Apparently, this imposed divergence free condition is consistent with Equation 6.40 can be used freely for an arbitrary vector $\boldsymbol{\psi}$. To derive the second equation, which is the momentum balance equation, one needs to take the curl of the Navier–Stokes equations (Equation 6.2) and substitute into it the definition of vorticity (i.e., Equation 6.38). Note that the pressure has been eliminated from the equation, because of the vector identity, $\nabla \times \nabla p = 0$. For confined flows, boundary conditions for both vector potential and vorticity may also be derived by substituting vorticity and vector potential functions into the no-slip conditions along the walls [18 – 20]:

$$\frac{\partial \psi_x}{\partial x} = \psi_y = \psi_z = \omega_x = 0 \,; \; \omega_y = -\frac{\partial w}{\partial x} \,; \; \omega_z = \frac{\partial v}{\partial x} \quad \text{on surface } x = \text{const}$$

$$\frac{\partial \psi_y}{\partial y} = \psi_x = \psi_z = \omega_y = 0 \,; \; \omega_x = \frac{\partial w}{\partial y} \,; \; \omega_z = -\frac{\partial u}{\partial y} \quad \text{on surface } y = \text{const}$$

$$\frac{\partial \psi_z}{\partial z} = \psi_x = \psi_y = \omega_z = 0 \,; \; \omega_x = -\frac{\partial v}{\partial z} \,; \; \omega_y = \frac{\partial u}{\partial z} \quad \text{on surface } z = \text{const}$$

We note that Equation 6.42 has three components, but only two of them are independent. This approach requires at least two components of the vorticity equation and three components of the vector Poisson equation. However, Aziz and Hellums [20] reported that the vorticity–potential approach was faster and more accurate than a method based on the primitive velocity–pressure formulation.

For unconfined flows, such as flows involving inlets and outlets, Harasaki and Hellums [21,22] showed that the boundary conditions for the inflow and outflow can be simplified, if two potentials are used in lieu of ψ. Their formulation is based on the basic vector decomposition theorem that an arbitrary vector can be split into a curl free and a divergence-free (or solenoidal) part. Thus, with a dual potential approach, the velocity is defined by

$$\mathbf{u} = \nabla \times \boldsymbol{\psi} + \nabla \phi \tag{6.43}$$

Clearly, this definition of velocity satisfies Equation 6.36. The continuity condition on the velocity field immediately leads to the following Laplace equation for ϕ:

$$\nabla^2 \phi = 0 \tag{6.44}$$

Because $\nabla \times \nabla p = 0$ and $\nabla \cdot \boldsymbol{\psi} = 0$, other governing equations are the same as before. The boundary conditions for the potential are of the Neumann type,

$$\mathbf{n} \cdot \nabla \phi = -\mathbf{n} \cdot \mathbf{u} \tag{6.45}$$

On a solid boundary, $\mathbf{n} \cdot \nabla \phi = 0$. There were also other choices for the potential ϕ, which can also be applied to make it even easier to specify certain inflow and outflow boundary conditions [18 – 21]. It is noted that the potential ϕ satisfies the Laplace equation, which along with the preceding condition gives a solution $\phi = 0$, if there exists no flow throughout.

Other hybrid forms of the velocity and vorticity approach have been reported. One of the approaches uses the dependent variables as the vorticity and velocity components. The vorticity dynamics equations are the same as Equation 6.42 and the velocity components are solved from the following vector equation:

$$\nabla^2 \mathbf{u} = -\nabla \times \boldsymbol{\omega} \tag{6.46}$$

which is easily obtained by taking the curl of Equation 6.38 and using the condition $\nabla \cdot \mathbf{u} = 0$.

The use of the derived variable approaches described above, and the details of numerical implementation, including the boundary conditions, can be found in the works of Liu and Shu [19].

Compared with the limited use of the general 3-D vorticity–vector potential approach, the 2-D vorticity–stream function is far more popular among the computational fluid dynamics and computational heat transfer communities. The vorticity–stream function form of the Navier–Stokes equations can be rather easily deduced from the above vorticity–vector potential equations. For a 2-D geometry, only one component of the vorticity and stream function fields survives and other components vanish. Without loss of generality, we take the two velocity components in the x–y plane. Then we have $\omega = \omega_z$ and $\psi = \psi_z$,

$$\omega \hat{k} = \nabla \times \mathbf{u} \;; \qquad \mathbf{u} = -\frac{\partial \psi}{\partial y}\hat{i} + \frac{\partial \psi}{\partial x}\hat{j} \ \in \Omega \tag{6.47}$$

Equations 6.41 and 6.42 now reduce to two scalar equations for the variables ω and ψ,

$$\nabla^2 \psi = \omega \ \in \Omega \tag{6.48}$$

$$\frac{\partial \omega}{\partial t} + \nabla \cdot (\mathbf{u}\omega) = \frac{1}{Re}\nabla^2 \omega \ \in \Omega \tag{6.49}$$

$$\mathbf{u} = \mathbf{u}_0 \ \in \partial\Omega \tag{6.50}$$

Here, for an illustrative purpose, a simple velocity boundary condition is applied. Other boundary conditions can also be employed for computations. From the definition of stream functions, the no slip boundary condition puts the following constraint for ψ,

$$\mathbf{n} \cdot \nabla \psi = \mathbf{n} \cdot \mathbf{u}_0 \ \in \partial\Omega \tag{6.51}$$

The equations can be solved using the discontinuous finite element method. They can also be solved using the combined continuous and discontinuous finite element methods. The latter is used below, as this gives an opportunity to illustrate the combined approach to fluid flow problems. For this problem, the stream function field is solved using the continuous finite element method, whereas the vorticity field is solved using the discontinuous finite element method. To develop the required discontinuous finite element formulation, the higher order derivatives are split into first order derivatives,

$$\boldsymbol{\sigma} = \frac{1}{Re}\nabla \omega \tag{6.52}$$

$$\frac{\partial \omega}{\partial t} + \nabla \cdot (\mathbf{u}\omega) = \nabla \cdot \boldsymbol{\sigma} \tag{6.53}$$

We discretize the domain into a collection of elements of finite size, and integrate the above equations over a typical element with respect to weight functions v and s, followed by integration by parts to obtain the discontinuous finite element formulation,

$$\int_{\Omega_j} \mathrm{Re}\boldsymbol{\sigma}_h \cdot \mathbf{s}\, dV = -\int_{\Omega_j} \omega_h \nabla \cdot \mathbf{s}\, dV + \int_{\partial\Omega_j} \hat{\omega}_h \mathbf{s} \cdot \mathbf{n}\, dS \qquad (6.54)$$

$$\int_{\Omega_j} v\frac{\partial\omega_h}{\partial t}\, dV - \int_{\Omega_j} \omega_h \mathbf{u}_h \cdot \nabla v\, dV + \int_{\partial\Omega_j} \overline{\omega}_h \mathbf{u}_h \cdot \mathbf{n}v\, dS$$

$$= -\int_{\Omega_j} \boldsymbol{\sigma}_h \cdot \nabla v\, dV + \int_{\partial\Omega_j} \hat{\boldsymbol{\sigma}} \cdot \mathbf{n}v\, dS \qquad (6.55)$$

where n is the normal pointing outward from element j. In the above formulations, the numerical fluxes are used and they are defined as a central flux,

$$\hat{\omega}_h = 0.5(\omega^+ + \omega^-); \qquad \hat{\boldsymbol{\sigma}}_h = 0.5(\boldsymbol{\sigma}^+ + \boldsymbol{\sigma}^-) \qquad (6.56)$$

where $+$ and $-$ refer to element j and its neighbors (see Figure 4.4). The numerical fluxes for convection are modeled differently using the upwinding scheme,

$$\overline{\omega}_h = \begin{cases} \omega_h^+, & \text{for } \mathbf{u}_h \cdot \mathbf{n} \geq 0 \\ \omega_h^-, & \text{for } \mathbf{u}_h \cdot \mathbf{n} < 0 \end{cases} \qquad (6.57)$$

or the Lax–Friedrichs upwind biased flux,

$$\mathbf{u}_h \cdot \mathbf{n}\overline{\omega}_h = \frac{1}{2}\left(\mathbf{u}_h \cdot \mathbf{n}(\omega_h^- + \omega_h^+) - \alpha(\omega_h^- - \omega_h^+)\right) \qquad (6.58)$$

where α is the maximum of $|\mathbf{u}_h \cdot \mathbf{n}|$ either locally or globally.

For the stream function field, the continuous finite element formulation may be applied, which takes the following form:

$$-\int_{\Omega_j} \nabla\psi_h \cdot \nabla w\, dV = \int_{\Omega_j} \psi_h w\, dV \qquad (6.59)$$

Note that the inter-element jump terms vanish, since the continuity is enforced across element boundaries for the continuous finite element method. For this reason, C_0 elements are required for the finite element solution of the stream function ψ. It is also remarked that the equation for the stream function field may be solved using the discontinuous finite element method. In fact, the formulations presented in Chapter 4 for the steady state heat conduction equations with an applied source may be directly employed here. For a moderate sized 2-D pure diffusion problem, however, the continuous finite element method often performs better.

Once the stream function field is obtained, the local velocity is derived from the following relation:

$$\mathbf{u}_h = -\hat{i}\,\frac{\partial \psi}{\partial y} + \hat{j}\,\frac{\partial \psi}{\partial x} \tag{6.60}$$

Liu and Shu [13] showed that the above numerical scheme, combining the continuous and discontinuous formulations, has a stability property

$$\frac{d\,\|\,\omega_h\,\|^2}{dt} + 2\,\|\,\sigma_h\,\| \le 0 \tag{6.61}$$

The general solution procedure may be described as follows: (1) specify the values for ω and ψ at time $t = 0$, (2) solve the vorticity transport equation for ω at time $t + \Delta t$ using the discontinuous formulation and element-by-element sweep starting at the boundary, (3) iterate for new ψ values at all points by solving the Poisson equation, applying the new ω, using the continuous finite element method, (4) find the velocity components $u = \partial \psi / \partial y$ and $v = -\partial \psi / \partial x$, (5) determine the values of ω on the boundaries using ψ and ω values, and (6) return to Step 2 if convergence is not achieved.

Two of the calculated results derived from the approach above are given in Figure 6.2 for two different Re numbers [18]. In Figure 6.2a, a uniform rectangular mesh of 256×256 was used with the P^2/Q^2 method at $t = 8$ for Re $= 70000/2\pi$. In Figure 6.2b, a mesh doubling both the x and y directions was used with the P^1/Q^1 method at $t = 8$ for $20000/(2\pi)$. For both cases, the shear layers are dominated by thin layer structures and the flow develops roll-up structures. Further numerical simulations with various grids and conditions indicate that physical viscosity dominates the numerics at these high Reynolds numbers, suggesting that the built numerical dissipation of the discontinuous finite element method is rather small. In general, higher order methods have a better resolution and a minimized contribution from numerical viscosity.

6.4 Coupled Flow and Heat Transfer

In this section, a coupled fluid flow and heat transfer analysis is made of natural convection in a square cavity. The governing equations for this problem can be written in the dimensionless form as follows:

$$\frac{\partial u}{\partial x} + \frac{\partial v}{\partial y} = 0 \tag{6.62}$$

$$u\,\frac{\partial u}{\partial x} + v\,\frac{\partial u}{\partial y} = -\sqrt{\frac{Pr}{Ra}}\,\frac{\partial p}{\partial x} + \sqrt{\frac{Pr}{Ra}}\left(\frac{\partial^2 u}{\partial x^2} + \frac{\partial^2 u}{\partial y^2}\right) \tag{6.63}$$

$$u\,\frac{\partial v}{\partial x} + v\,\frac{\partial v}{\partial y} = -\sqrt{\frac{Pr}{Ra}}\,\frac{\partial p}{\partial y} + \sqrt{\frac{Pr}{Ra}}\left(\frac{\partial^2 v}{\partial x^2} + \frac{\partial^2 v}{\partial y^2}\right) - T \tag{6.64}$$

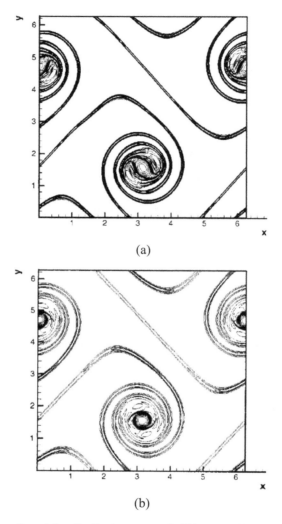

(a)

(b)

Figure 6.2 Contour of vorticity distribution at $t = 8$. Thirty equally spaced contour lines between $\omega = -15$ to $\omega = +15$. (a) Results are computed for Re $= 70000/2\pi$ using a 256 × 256 mesh with the P^2/Q^2 method. (b) Results are computed for Re $= 20000/2\pi$ using a 512 × 512 mesh with the P^1/Q^1 method [18]

$$u\frac{\partial T}{\partial x} + v\frac{\partial T}{\partial y} = \frac{1}{\sqrt{Ra\,Pr}}\left(\frac{\partial^2 T}{\partial x^2} + \frac{\partial^2 T}{\partial y^2}\right) \qquad (6.65)$$

where U $= \alpha\,(Ra\ Pr)^{1/2}/L$; L and L/U are used as the velocity, length and time scales. Also, Ra (Ra $= \rho\beta g(T_2^* - T_1^*)L^3/\mu\alpha$) is the Raleigh number, Pr (Pr $= v/\alpha$)

is the Prandtl number, α is the thermal diffusivity, μ is the molecular viscosity, ρ is the density, β is the thermal expansion coefficient, and g is the gravity acceleration. Moreover, the dimensionless temperature is written as $T = (T^* - T_1^*)/(T_2^* - T_1^*)$, where the superscript * denotes the dimensional temperature. The dimensionless boundary conditions on temperature are prescribed as follows:

$$T(x = 0, y) = 0 \; ; \; T(x = 1, y) = 1 \; ; \; \frac{\partial T}{\partial y}(x, y = 0) = \frac{\partial T}{\partial y}(x, y = 1) = 0 \qquad (6.66)$$

The velocity boundary conditions are such that all velocity components are set to zero on all walls.

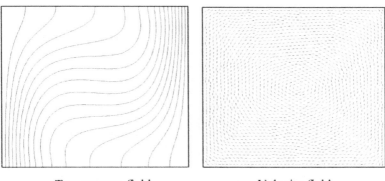

Temperature field   Velocity field

Pr = 1.0 and Ra = 1×10^4

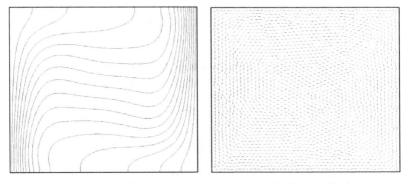

Temperature field   Velocity field

Pr = 1.0 and Ra = 5×10^4

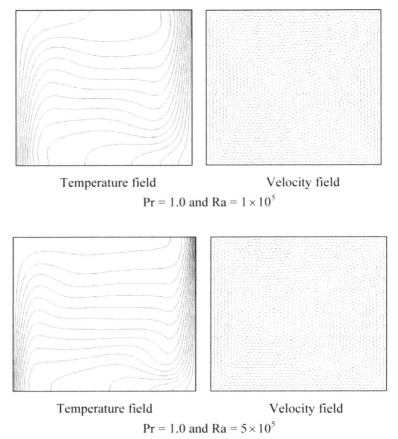

Temperature field   Velocity field
Pr = 1.0 and Ra = 1×10^5

Temperature field   Velocity field
Pr = 1.0 and Ra = 5×10^5

Figure 6.3. Numerical results obtained from the discontinuous finite element formulation for natural convection in a square cavity for different Ra numbers

For this coupled problem, the thermal and fluid flow fields need to be solved simultaneously. For the purpose of demonstrating the usefulness of the discontinuous algorithms, the discontinuous finite element formulation is used for the solution of both fluid flow and heat transfer equations. The flow is assumed incompressible and thermal effects are accounted for using the Boussinesq approximation. The discontinuous formulation for the fluid flow is given in Section 6.2, while that for convection is taken from Chapter 5. The solution procedure is iterative and requires updating of both fluid flow and thermal fields. The computations used unstructured mesh, with linear triangular elements used for fluid flow, pressure, and temperature fields. The computed results are shown in Figure 6.3. As can be seen, fluid flow has a very strong effect on temperature distribution. The calculations here compare very well with reported results for this problem.

Exercises

1. Consider an incompressible, inviscid fluid flow of constant density in an open domain Ω with smooth boundary. The flow is governed by the following continuity and momentum equations,

$$\nabla \cdot \mathbf{u} = 0 \quad \in \Omega, t > 0$$

$$\frac{\partial \mathbf{u}}{\partial t} + \mathbf{u} \cdot \nabla \mathbf{u} = -\nabla p \quad \in \Omega, t > 0$$

and is subject to the following boundary conditions:

$$\mathbf{n} \cdot \mathbf{u} = 0 \quad \in \partial\Omega, \quad t > 0$$

$$\mathbf{u}(\mathbf{r}, t) = 0 \quad \in \Omega, \quad t = 0$$

where $\mathbf{u}$ is the velocity and p is the pressure. Using the techniques discussed in Chapters 4 and 5, show that

$$E(t) = \int_\Omega \mathbf{u}(x,t) \cdot \mathbf{u}(x,t) dV = E(0), \qquad t > 0$$

2. Consider an incompressible, viscous fluid flow in an open, bounded region Ω. The flow is governed by the continuity and the Navier–Stokes equations,

$$\nabla \cdot \mathbf{u} = 0 \quad \in \Omega, t > 0$$

$$\frac{\partial \mathbf{u}}{\partial t} + \mathbf{u} \cdot \nabla \mathbf{u} = -\nabla p + \lambda \Delta \mathbf{u} + \mathbf{f} \quad \in \Omega, t > 0$$

and is subject to the following boundary conditions:

$$\mathbf{u} = 0 \quad \in \partial\Omega, t > 0$$

$$\mathbf{u}(\mathbf{r}, t) = 0 \quad \in \Omega, t = 0$$

where $\mathbf{u}$ is the velocity, $\mathbf{f}$ is the body force and p is the pressure. Using the techniques discussed in Chapters 4 and 5, show that

$$E(t) = \int_\Omega \mathbf{u}(x,t) \cdot \mathbf{u}(x,t) dV \le E(0)e^{-a\lambda t} + \frac{C}{\lambda}(1 - e^{-a\lambda t}), \quad t > 0$$

where C and a are two positive constants.

3. Develop a discontinuous finite element code for simulating incompressible flows in a channel. Perform the calculations using different orders of approximations and compare the results.
4. Develop a discontinuous finite element code for simulating natural convection and heat transfer in a 2-D cavity. Apply the code to study the fluid flow with various Rayleigh numbers.
5. Develop a discontinuous finite element code, using the projection method and the primitive variable formulation, for simulating transient flows in a 2-D cavity. Compare the results with those obtained using the derived variable approach presented in [13].

References

[1]   Carey GF, Oden JT. Finite Elements–Fluid Mechanics, Volume VI. Englewood Cliffs: Prentice Hall, 1986.
[2]   Babuska I. On the Inf-Sup (Babuska–Brezzi) Condition. Austin: TICAM Forum #5, October, 1996.
[3]   Evans LC. Partial Differential Equations. Philadelphia: American Society of Mathematics, 2000.
[4]   Babuska I, Aziz AK. Survery Lectures on the Mathematical Foundation of the Finite Element Method: In: Aziz, AK, editor. The Mathematical Foundation of the Finite Element Method with Applications to Partial Differential Equations. New York: Academic Press, 1972.
[5]   Chapelle D, Bathe KJ. The Finite Element Analysis of Shells – Fundamentals. New York: Springer, 2003.
[6]   Bathe KJ. Finite Element Procedures. Englewood Cliffs: Prentice Hall, 1996.
[7]   Brezzi F, Bathe KJ. A Discourse on the Stability Conditions for Mixed Finite Element Formulations. J. Comput. Meth. Appl. Mech. Eng. 1990; 82: 27–57.
[8]   Bathe KJ. The Inf-Sup Condition and its Evaluation for Mixed Finite Element Methods. Comput. & Struct. 2001; 79: 243–252.
[9]   Bathe KJ, Zhang H. A Flow-Condition-Based Interpolation Finite Element Procedure for Incompressible Fluid Flows. Comput. & Struct. 2002; 80: 1267–1277.
[10] Bathe KJ, Hendriana D, Brezzi F, Sangalli G. Inf-sup Testing of Upwind Methods. Int. J. Numer. Meth. Eng. 2000; 48: 745–760.
[11] Raris R. Vectors, Tensors and Basic Equations of Fluid Mechanics. New York: Dover, 1962.
[12] Cockburn B, Kanschat G, Schötzau D. A Locally Conservative LDG Method for the Incompressible Navier–Stokes Equations. Math. Comput. to appear.
[13] Cockburn B, Kanschat G, Schötzau D. The Local Discontinuous Galerkin Method for the Oseen Equations. Math. Comput. 2003: 73(2): 569–593.
[14] Cockburn B, Kanschat G, Schötzau D, Schwab C. Local Discontinuous Galerkin Methods for the Stokes System. SIAM J. Numer. Anal. 2002; 40(1): 319–343.
[15] Kanschat G, Rannacher R. Local Error Analysis of the Interior Penalty Discontinuous Galerkin Method for Second Order Elliptic Problems. J. Numer. Math. 2004; 10(4): 249–274.
[16] Cockburn B, Kanschat G, Schötzau D. The Local Discontinuous Galerkin Method for Linear Incompressible Fluid Flow: a Review. Comput. Fluids 2005; 34(4–5): 491–506.

[17] Chorin AJ. A Numerical Method for Solving Incompressible Viscous Flow Problems. J. Comput. Phys. 1967; 2: 12–26.

[18] Codina R. Pressure Stability in Fractional Step Finite Element Methods for Incompressible Flows. J. Comput. Phys. 2001; 29: 112–140.

[19] Liu JG, Shu CW. High-Order Discontinuous Galerkin Method for 2-D Incompressible Flows. J. Comput. Phys. 2000; 160: 577–596.

[20] Aziz AK, Hellums JD. Numerical Solution of the Three-Dimensional Equations of Motion for Laminar Natural Convection. Phys. Fluids 1967; 10: 314–324.

[21] Hirasaki GJ, Hellums JD. A General Formulation of the Boundary Conditions on the Vector Potential in Three Dimensional Hydrodynamics. Quart. App. Math. 1968; 26: 331–342.

[22] Hirasaki GJ, Hellums JD. Boundary Conditions on the Vector and Scalar Potentials in Viscous Three Dimensional Hydrodynamics. Quart. App. Math. 1970; 28: 293–296.

7

Compressible Fluid Flows

In this chapter, we discuss the application of the discontinuous finite element method for the solution of compressible fluid flow equations. One of the major challenges in compressible flow calculations is to devise a numerical scheme for an effective treatment of various discontinuities existing in the solution. In our study of inviscid Burgers' problems in Chapter 5, we showed through very simple examples how a discontinuity present in the initial data is being carried into the domain by the motion of a compressible fluid. This type of discontinuity occurs in various forms including shock waves, rarefaction waves and contact discontinuities. These discontinuities represent, mathematically, the singularities at which multiple solutions exist. Development of an effective algorithm in a general compressible flow setting requires, therefore, a careful consideration of the local behavior of the fluid flow field.

The discontinuous Galerkin finite element method has been developed to provide a higher order computational algorithm for the calculation of high speed compressible fluid flow problems with various discontinuities. The relaxation of the cross-element continuity requirement permits a variety of choices to incorporate different types of numerical fluxes to enhance the computational performance. Many of these numerical fluxes have their origin from finite difference approximations, which we know now are just the consequence of using lower order polynomials (i.e., the piecewise constant or zero order polynomial) in the discontinuous finite element formulation. These numerical fluxes have also been tested succesfully in the finite difference calculations for lower order approximations. An extension of the numerical fluxes to the higher order setting embedded in the discontinuous finite element formmulation has been a subject of recent research.

This chapter starts with the study of 1-D inviscid compressible fluid flow problems. The basic properties of the Euler equations, the exact and approximate Riemann solvers, and the low and high order discontinuous finite element formulations are discussed. We then extend the discontinuous formulation to the cases of the inviscid and viscous compressible flows in multidimensional geometries. The arbitrary Lagrangian–Eulerian (ALE) description of the compressible flow problems is then discussed. The discontinuous finite element formulations within the ALE framework and the computational procedures are presented.

7.1 1-D Compressible Flows

This section is concerned with some basic properties of the Euler equations, the Riemann solvers, and the discontinuous finite element solution of the equations in 1-D geometry. The shock tube problem is studied as a numerical example.

7.1.1 Governing Equations

General equations for compressible fluid flows were given in Section 1.8. In the absence of diffusive phenomena due to viscous stresses, a 1-D compressible flow problem is described by the following hyperbolic equation set, namely the Euler equations, representing the conservation of mass, momentum, and energy:

$$\frac{\partial U}{\partial t} + \frac{\partial F(U)}{\partial x} = 0 \tag{7.1}$$

where the U and F are vectors defined as

$$U = \begin{pmatrix} \rho \\ \rho u \\ \rho e \end{pmatrix}, \ F = \begin{pmatrix} \rho u \\ \rho u^2 + p \\ \rho u h \end{pmatrix} \tag{7.2}$$

In the above equation, ρ is the density, ρu is the momentum, e is the energy, p is the pressure, and h is the dynamic enthalpy. This last variable is related to the other quantities by the following relation:

$$h = e + \frac{p}{\rho}, \ e = E + \frac{1}{2} u^2 \tag{7.3}$$

with E being the internal energy. This system of three differential equations in four independent variables $(\rho, \rho u, \rho e \ \text{and} \ h)$ is closed by the equation of state, which is derived from thermodynamic principles. If the gas is calorically perfect and polytropic, then the pressure is related to the other variables by the relation,

$$e = \frac{p}{\rho(\gamma - 1)} + \frac{1}{2} u^2 \tag{7.4}$$

where γ is the ratio of specific heats and takes the value of 1.4 for an ideal gas–air is often approximated as an ideal gas in compressible flow studies [1, 2]. Unless otherwise indicated, the ideal gas law is assumed for the equation of state in this chapter.

7.1.2 Basic Properties of the Euler Equations

Understanding of the basic structure of the solution of the Euler equations is of great importance, both in developing an effective numerical scheme and in interpreting numerical solutions. To study the basic properties of the Euler equations, it is constructive to write Equation 7.1 in a quasi-linear form,

$$\frac{\partial U}{\partial t} + \mathbf{A}(U)\frac{\partial U}{\partial x} = 0 \tag{7.5a}$$

where $\mathbf{A}$ is the Jacobian matrix,

$$A_{ij} = \frac{\partial F_i}{\partial U_j} \tag{7.5b}$$

or explicitly,

$$\mathbf{A} = \frac{\partial F}{\partial U} = \begin{bmatrix} 0 & 1 & 0 \\ -\frac{1}{2}(\gamma-3)u & (3-\gamma)u & \gamma-1 \\ \frac{1}{2}(\gamma-2)u^3 - \frac{a^2}{\gamma-1} & \frac{3-2\gamma}{2}u^2 + \frac{a^2}{\gamma-1} & \gamma u \end{bmatrix} \tag{7.5c}$$

with $a = (\gamma p/\rho)^{1/2}$ being the speed of sound.

Further we assume that the initial data has a discontinuity at $x = 0$,

$$U(x,0) = \begin{cases} U_L & \text{if } x < 0 \\ U_R & \text{if } x > 0 \end{cases} \tag{7.5d}$$

This is the well known Riemann problem, which has played a fundamental role in compressible flow studies. Figure 7.1 schematically sketches the Riemann problem.

The Jacobian matrix $\mathbf{A}$ has three distinctive eigenvalues,

$$\lambda_1 = u - a \; ; \; \lambda_2 = u \; ; \; \lambda_3 = u + a \tag{7.6a}$$

which is sometimes written as an eigenvalue matrix $\mathbf{\Lambda}$,

$$\mathbf{\Lambda} = \begin{bmatrix} \lambda_1 & 0 & 0 \\ 0 & \lambda_2 & 0 \\ 0 & 0 & \lambda_3 \end{bmatrix} \tag{7.6b}$$

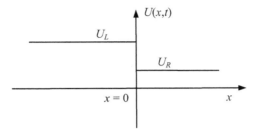

Figure 7.1. Schematic of the Riemann problem

Corresponding to these eigenvalues are the right eigenvector matrix **R** and its inverse $\mathbf{R}^{-1}$,

$$\mathbf{R} = (R^{(1)}, R^{(2)}, R^{(3)}) = \begin{bmatrix} 1 & 1 & 1 \\ u-a & u & u+a \\ h-ua & \frac{1}{2}u^2 & h+ua \end{bmatrix} \tag{7.7a}$$

$$\mathbf{R}^{-1} = \frac{\gamma-1}{2a^2} \begin{bmatrix} h+a(u-a)/\tilde{\gamma} & -u-(a/\tilde{\gamma}) & 1 \\ (4a^2/\tilde{\gamma})-2h & 2u & -2 \\ h-a(u+a)/\tilde{\gamma} & -u+(a/\tilde{\gamma}) & 1 \end{bmatrix} \tag{7.7b}$$

where $\tilde{\gamma} = \gamma - 1$ and each eigenvector $R^{(i)}$ is calculated by substituting each eigenvalue into the matrix equation,

$$\mathbf{A}R^{(i)} = \lambda_i R^{(i)} \tag{7.7c}$$

Similarly, one has the left eigenvector matrix **L**,

$$\mathbf{L} = \begin{pmatrix} L^{(i)} \\ L^{(i)} \\ L^{(i)} \end{pmatrix} \tag{7.8a}$$

where $L^{(i)}$ is the row vector and is obtained by the following relation:

$$L^{(i)}\mathbf{A} = \lambda_i L^{(i)} \tag{7.8b}$$

We further have the bi-orthonormal condition for $L^{(i)}$ and $R^{(j)}$,

$$L^{(i)}R^{(j)} = \begin{cases} 1 & \text{if } i = j \\ 0 & \text{if } i \neq j \end{cases} \tag{7.8c}$$

and the inverse relation,

$$\mathbf{L} = \mathbf{R}^{-1} \text{ and } \mathbf{L}^{-1} = \mathbf{R} \tag{7.8d}$$

From the study of the inviscid Burgers' equation, it is known that the three eigenvalues $(\lambda_1, \lambda_2, \lambda_3)$ correspond to three characteristics, which imply three different jump discontinuities in the solution of the Euler equations. The characteristics and the structure of the solution of the Euler equations are shown in Figure 7.2. These characteristics are useful in sampling the solution to the Riemann problem from data, and are discussed below

Associated with λ_2 is the contact discontinuity,

$$\lambda_2(U_{*L}) = \lambda_2(U_{*R}) = S_2 \tag{7.9a}$$

where S_2 is the speed of the movement of the contact discontinuity. This is a linear degenerate case. The rarefaction wave corresponds to the condition,

$$\lambda_1(U_L) \leq \lambda_1(U_{*L}) \tag{7.9b}$$

and the shock wave satisfies

$$\lambda_3(U_R) < S_3 < \lambda_3(U_{*R}) \tag{7.9c}$$

with S_3 being the shock wave speed. The cases of λ_1 and λ_2 are the nonlinear-genuine cases.

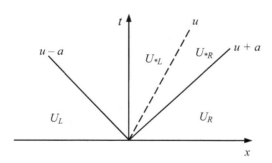

Figure 7.2. Structure of the solution of the Riemann problem in the x–t plane for the time dependent, 1-D Euler equations

7.1.3 The Rankine–Hugoniot Conditions

Conditions for the jump discontinuities may be obtained by a direct integration of the Euler equations. To do that, we define the variable and the operator,

$$\mathbf{G} = (F, U) \ ; \ \nabla = \hat{i}_x \frac{\partial}{\partial x} + \hat{i}_t \frac{\partial}{\partial t} \tag{7.10}$$

whence the 1-D Euler equations become

$$\nabla \cdot \mathbf{G} = 0 \tag{7.11}$$

Consider a control volume shown in Figure 7.3, which contains a jump discontinuity marked by S. Integration of Equation 7.11 over the control volume, followed by the use of the Gaussian theorem, yields

$$\int_{CV} \nabla \cdot \mathbf{G} dV = \int_{CS} \mathbf{G} \cdot \mathbf{n} dS = 0 \tag{7.12}$$

where CV and CS are the volume and surface of the control volume. As shown in Figure 7.3, the discontinuity S divides the control volume into two parts ($CS = CS_1 \cup CS_2$), which allows Equation 7.12 to be rewritten for each of the subcontrol volumes (i.e., CV_1 and CV_2),

$$\int_{CV_1} \nabla \cdot \mathbf{G} dV = \int_{(CS+S)_1} \mathbf{G} \cdot \mathbf{n} dS = \int_{CS_1} \mathbf{G} \cdot \mathbf{n} dS + \int_{S_1} \mathbf{G} \cdot \mathbf{n} dS = 0 \tag{7.13a}$$

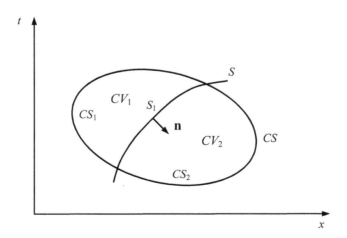

Figure 7.3. Control volume and control surfaces used to derive the jump conditions

$$\int_{CV_2} \nabla \cdot \mathbf{G} dV = \int_{(CS+S)_2} \mathbf{G} \cdot \mathbf{n} dS = \int_{CS_2} \mathbf{G} \cdot \mathbf{n} dS + \int_{S_2} \mathbf{G} \cdot \mathbf{n} dS = 0$$

(7.13b)

Adding the two equations together, noting that $S = CS_1 \cap CS_2$ and $\mathbf{n} = \mathbf{n}_{S_1} = -\mathbf{n}_{S_2}$, and then making use of Equation 7.12, one has the jump discondition across the discontinuity,

$$\int_S [\mathbf{G}] \cdot \mathbf{n} dS = 0$$

(7.14)

where [•] is used to denote the jump,

$$[\mathbf{G}] = \mathbf{G}_1 - \mathbf{G}_2$$

(7.15)

with subscripts 1 and 2 denoting regions 1 and 2, respectively.

Thus, for 1-D Euler equations, the jump conditions across a discontinuity are written as

$$\begin{bmatrix} \rho \\ \rho u \\ \rho e \end{bmatrix} n_t + \begin{bmatrix} \rho u \\ \rho u^2 + p \\ \rho u h \end{bmatrix} n_x = 0$$

(7.16)

If the discontinuity S is parameterized such that $S = x(t)$, $c = x(t)/dt$, and

$$\mathbf{n} = (n_x, n_t) = \frac{\hat{i}_x}{\sqrt{1+c^2}} - \frac{c\hat{i}_t}{\sqrt{1+c^2}}$$

(7.17)

then one has, upon substitution of the above equation into Equation 7.16,

$$-c[\rho] + [\rho u] = 0$$

(7.18a)

$$-c[\rho u] + [\rho u^2 + p] = 0$$

(7.18b)

$$-c[\rho e] + [\rho u h] = 0$$

(7.18c)

where $[g] = g_b - g_a$, with g_b being the data behind and g_a ahead of the discontinuity and c ($c = x(t)/dt$) is the velocity with which the discontinuity moves.

The above equations are written with the frame at rest. Consider that the discontinuity is a shock wave, which moves at a speed of S_3. Written in the frame that moves with the shock wave speed S_3, these jump conditions in the star region (see Figure 7.4: the star region is the region where quantities with subscript * reside) become

$$\rho_* \hat{u}_* = \rho_R \hat{u}_R \tag{7.19a}$$

$$\rho_* \hat{u}_*^2 + p_* = \rho_R \hat{u}_R^2 + p_R \tag{7.19b}$$

$$\hat{u}_*(\hat{E}_* + p_*) = \hat{u}_R(\hat{E}_R + p_R) \tag{7.19c}$$

with $\hat{u}_* = u_* - S_3$ and $\hat{u}_R = u_R - S_3$. From the above relations, it is straightforward to show that the shock wave speed S_3 is calculated by the following expression:

$$S_3 = u_R + a_R \sqrt{\left(\frac{\gamma+1}{2\gamma}\right)\left(\frac{p_*}{p_R}\right) + \left(\frac{\gamma-1}{2\gamma}\right)} \tag{7.20}$$

Then, the other variables of interest can also be determined [2].

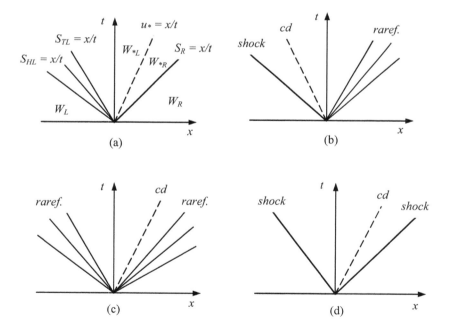

Figure 7.4. Four different solutions of a general Riemann problem (from left to right): (a) rarefaction, contact discontinuity and shock (RCS), (b) shock, contact discontinuity and rarefaction (SCR), (c) rarefaction, contact discontinuity and rarefaction (RCR), and (d) shock, contact discontinuity and shock (SCS). $W = (\rho, u, p)^T$

7.1.4 1-D Riemann Solver – Exact Solution

The above discussion can be extended to the general Riemann problem, for which possible wave patterns are shown in Figure 7.4. The start region, whose properties are marked with subscript *, contains a contact discontinuity and is the region between the left and right waves.

The solutions for the pressure and the particle velocity for the Riemann problem defined by Equations 7.8–7.9 are summarized in the equations below, assuming that the ideal gas law applies. The detailed derivation of these equations can be found in [3].

The solution of the pressure p_* is given by the root of the algebraic equation,

$$f_L(p,\rho_L,u_L,p_L)+f_R(p,\rho_R,u_R,p_R)+u_R-u_L=0 \qquad (7.21a)$$

with

$$f_L(p,\rho_L,u_L,p_L)=\begin{cases}(p-p_R)\left[\dfrac{A_L}{p+B_L}\right]^{1/2} & \text{if } p>p_L \ \text{(shock wave)}\\[2em] \dfrac{2a_L}{\gamma-1}\left[(p/p_L)^{\frac{\gamma-1}{2\gamma}}-1\right] & \text{if } p\le p_L \ \text{(Rarefaction wave)}\end{cases} \qquad (7.21b)$$

$$f_R(p,\rho_R,u_R,p_R)=\begin{cases}(p-p_R)\left(\dfrac{A_R}{p+B_R}\right)^{1/2} & \text{if } p>p_R \ \text{(shock wave)}\\[2em] \dfrac{2a_R}{\gamma-1}\left[(p/p_R)^{\frac{\gamma-1}{2\gamma}}-1\right] & \text{if } p\le p_R \ \text{(Rarefaction wave)}\end{cases} \qquad (7.21c)$$

$$A_L=\frac{2}{(\gamma+1)\rho_L}; \qquad B_L=\frac{(\gamma-1)}{(\gamma+1)}P_L \qquad (7.21d)$$

$$A_R=\frac{2}{(\gamma+1)\rho_R}; \qquad B_R=\frac{(\gamma-1)}{(\gamma+1)}P_R \qquad (7.21e)$$

and the solution for the particle velocity u_* is

$$u_*=\frac{1}{2}(u+u)+\frac{1}{2}(f_R(p_*)-f_L(p_*)) \qquad (7.22)$$

The solutions to other quantities are considered below, depending upon the specific conditions:

(i) Left shock wave (left region of Figure 7.4(b), $x < 0$),

$$\rho_{*L} = \rho_L \left(\frac{\frac{p_*}{p_L} + \frac{\gamma-1}{\gamma+1}}{\frac{\gamma-1}{\gamma+1}\frac{p_*}{p_L} + 1} \right); \quad S_L = u_L - a_L \left(\frac{\gamma+1}{2\gamma}\frac{p_*}{p_L} + \frac{\gamma-1}{2\gamma} \right)^{1/2} \tag{7.23a}$$

(ii) Right shock wave (right region of Figure 7.4(a), $x > 0$),

$$\rho_{*R} = \rho_R \left(\frac{\frac{p_*}{p_L} + \frac{\gamma-1}{\gamma+1}}{\frac{\gamma-1}{\gamma+1}\frac{p_*}{p_L} + 1} \right); \quad S_R = u_R + a_R \left(\frac{\gamma+1}{2\gamma}\frac{p_*}{p_R} + \frac{\gamma-1}{2\gamma} \right)^{1/2} \tag{7.23b}$$

(iii) Left rarefaction waves (left region of Figure 7.4(a), $x < 0$)

$$\rho_{*L} = \rho_L \left(\frac{p_*}{p_L} \right)^{1/\gamma}; \quad a_{*L} = a_L \left(\frac{p_*}{p_L} \right)^{\frac{\gamma-1}{2\gamma}} \tag{7.24a}$$

The rarefaction waves are enclosed by the head and tail, whose speeds are given by the following characteristics:

$$S_{HL} = \left. \frac{dx}{dt} \right|_{HL} = u_L - a_L; \quad S_{TL} = \left. \frac{dx}{dt} \right|_{TL} = u_{*L} - a_{*L} \tag{7.24b}$$

and the waves in the fan are given by

$$W_{Lfan} = \begin{cases} \rho = \rho_L \left[\frac{2}{\gamma+1} + \frac{\gamma-1}{(\gamma+1)a_L}\left(u_L - \frac{x}{t} \right) \right]^{\frac{2}{\gamma-1}} \\[2mm] u = \frac{2}{\gamma+1}\left[a_L + \frac{\gamma-1}{2}u_L + \frac{x}{t} \right] \\[2mm] p = p_L \left[\frac{2}{\gamma+1} + \frac{\gamma-1}{(\gamma+1)a_L}\left(u_L - \frac{x}{t} \right) \right]^{\frac{2\gamma}{\gamma-1}} \end{cases} \tag{7.24c}$$

(iv) Right rarefaction waves (right region of Figure 7.4(c), $x > 0$)

$$\rho_{*R} = \rho_R \left(\frac{p_*}{p_R} \right)^{1/\gamma}; \quad a_{*R} = a_R \left(\frac{p_*}{p_R} \right)^{\frac{\gamma-1}{2\gamma}} \tag{7.25a}$$

$$S_{HR} = \left. \frac{dx}{dt} \right|_{HR} = u_R + a_R; \quad S_{TR} = \left. \frac{dx}{dt} \right|_{TR} = u_{*R} + a_{*R} \tag{7.25b}$$

$$W_{Rfan} = \begin{cases} \rho = \rho_R \left[\frac{2}{\gamma+1} - \frac{\gamma-1}{(\gamma+1)a_R} \left(u_R - \frac{x}{t} \right) \right]^{\frac{2}{\gamma-1}} \\[2mm] u = \frac{2}{\gamma+1} \left[-a_R + \frac{\gamma-1}{2} u_R + \frac{x}{t} \right] \\[2mm] p = p_R \left[\frac{2}{\gamma+1} - \frac{\gamma-1}{(\gamma+1)a_R} \left(u_R - \frac{x}{t} \right) \right]^{\frac{2\gamma}{\gamma-1}} \end{cases} \tag{7.25c}$$

The above equations provide the exact solution of the complete wave structure of the Riemann problem at any point (x,t) in the relevant domain of interest $x_R < x < x_L$; $t > 0$, with $x_L < 0$ and $x_R > 0$. The numerical solution of the above equations is discussed in Toro and Reimann [3], and Gottlieb and Groth [4].

7.1.5 1-D Riemann Solver – Approximate Solution

Numerical schemes for compressible flow calculations require the solution of the Riemann problem, and for many schemes the solution is needed for every element boundary at each time step. The exact solution shown above requires a Newton iteration procedure and thus can be computationally expensive. This has motivated researchers to develop approximate Riemann solvers that can reduce computational time. One of the popular ideas in this category is the approximation proposed by Roe [5]. Roe's algorithm rests with the linearizing of the nonlinear term and thus the Roe solver is less expensive than the exact solver.

Roe's solver was derived based on the finite volume approach, which is in essence the constant element approximation used in discontinuous finite elements. According to Roe, the linearized Euler equation for the 1-D compressible flow problem is given by

$$\frac{\partial U}{\partial t} + \overline{A} \frac{\partial U}{\partial x} = 0 \tag{7.26}$$

with the matrix $\overline{A}$ evaluated at the interface $j+1/2$ (see Figure 7.5),

$$\overline{A} = \overline{A}_{j+1/2}(U_j, U_{j+1}) = \left. \frac{\partial F(U_j, U_{j+1})}{\partial U} \right|_{j+1/2} \tag{7.27}$$

where to be consistent with Roe's original derivation, the constant element approximation has been used. Note that for a piecewise-constant approximation, the node number coincides with the element number (see Figure 7.5). Three conditions are imposed on the matrix: (1) locally Lipshitz, $F_{j+1} - F_j = \overline{A}_{j+1/2}(U_{j+1} - U_j)$, (2) $\overline{A}_{j+1/2}$ is diagonalizable and all its eigenvalues are real, and (3)

$\overline{\mathbf{A}}_{j+1/2}(U,U) = \overline{\mathbf{A}}_j(U)$. These conditions will make the numerical scheme conservative and consistent with the original differential equation.

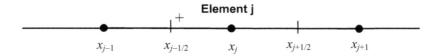

Figure 7.5. Constant element spatial discretization for the derivation of Roe's solver: (−) marks the element boundaries and (●) the center of the element

In the case of the Euler equations, the linearized Jacobian matrix $\overline{\mathbf{A}}_{j+1/2}$ may be calculated using an averaged state,

$$U^{Roe} = \overline{U}(U_L, U_R) \tag{7.28}$$

where subscripts L and R denote a left and right state, respectively. For a first order scheme, Roe's average is calculated by

$$\overline{U}_{j+1/2} = \left(\sqrt{\rho_L \rho_R}, \ \frac{(u\sqrt{\rho})_L + (u\sqrt{\rho})_R}{\sqrt{\rho_L} + \sqrt{\rho_R}}, \ \frac{(h\sqrt{\rho})_L + (h\sqrt{\rho})_R}{\sqrt{\rho_L} + \sqrt{\rho_R}} \right)^T \tag{7.29}$$

For the linearized Riemann problem, the flux can be calculated using the following expression:

$$F_{j+1/2} = 0.5[F(U_L) + F(U_R)] - 0.5\mathbf{R}|\mathbf{\Lambda}|\mathbf{R}^{-1}(U_L - U_R) \tag{7.30}$$

where subscripts L and R denote the values to the left and right of the interface $j+1/2$ and also we have the following decomposition relation for matrix $\mathbf{A} = \partial F / \partial U$,

$$\mathbf{\Lambda} = \mathbf{R}^{-1}\mathbf{A}\mathbf{R} \tag{7.31}$$

with $\mathbf{R}$ being the matrix of the right eigenvectors of $\mathbf{A}$.

7.1.6 Discontinuous Finite Element Formulation

Let us now consider the discontinuous finite element solution to the 1-D Euler equations. As usual, the domain is discretized into N elements and the governing equation is integrated with respect to a weighting function W over element $j \in [x_j, x_{j+1}]$,

$$\int_{x_j}^{x_{j+1}} \left(\frac{\partial U}{\partial t} + \frac{\partial F}{\partial x} \right) W \, dx = 0 \tag{7.32}$$

Integration by parts of the spatial derivative term yields the following weak form solution:

$$\int_{x_j}^{x_{j+1}} W \frac{\partial U}{\partial t} dx - \int_{x_j}^{x_{i+1}} F \frac{\partial W}{\partial x} dx$$

$$+ \hat{h}(U_{j+1}^-, U_{j+1}^+) W(x_{i+1}^-) - \hat{h}(U_j^-, U_j^+) W(x_j^-) = 0 \tag{7.33}$$

where, as usual, the element boundary function terms have been replaced by appropriate numerical fluxes,

$$F_{j+1}^- = \hat{h}(U_{j+1}^-, U_{j+1}^+) W(x_{j+1}^-) ; \qquad F_j^+ = \hat{h}(U_j^-, U_j^+) W(x_j^+) \tag{7.34}$$

By introducing the element basis functions $\phi^{(l)}(x)$ as follows:

$$U_h(x,t) = \sum_{l=1}^{N_e} U^{(l)}(t)\phi^{(l)}(x) ; \quad W_h(x,t) = \sum_{l=1}^{N_e} W^{(l)}(t)\phi^{(l)}(x) \tag{7.35}$$

with N_e being the number of nodes per element, Equation 7.33 can be written in a matrix form,

$$\frac{d\mathbf{U}_{(j)}}{dt} - \mathbf{K}\mathbf{U}_{(j)} + \hat{h}(U_{(j)}^{(N_e)}, U_{(j+1)}^{(1)})\mathbf{\theta}_{j+1} - \hat{h}(U_{(j-1)}^{(N_e)}, U_{(j)}^{(1)})\mathbf{\theta}_j = 0 \tag{7.36}$$

where $\mathbf{U} = (U^{(1)}, U^{(2)}, ..., U^{(N_e)})^T$ and the matrices are calculated by

$$\mathbf{K} = \mathbf{S}^{-1}\mathbf{L} \; ; \; \mathbf{\theta}_j = \mathbf{S}^{-1} \begin{pmatrix} \phi^{(1)}(x_j) \\ \phi^{(2)}(x_j) \\ \vdots \\ \phi^{(N_e)}(x_j) \end{pmatrix} \; ; \; \mathbf{\theta}_{j+1} = \mathbf{S}^{-1} \begin{pmatrix} \phi^{(1)}(x_{j+1}) \\ \phi^{(2)}(x_{j+1}) \\ \vdots \\ \phi^{(N_e)}(x_{j+1}) \end{pmatrix}$$

$$\mathbf{S} = \int_{x_j}^{x_{j+1}} W_h W_h^T \, dx \; ; \qquad \mathbf{L} = \int_{x_j}^{x_{j+1}} \left(\frac{\partial W_h}{\partial x} \right) W_h^T \, dx \tag{7.37}$$

Some terms of various orders are explicitly written below for the convenience of the subsequent discussions:

(1) $N_e = 1$:

$$\dot{U}_j + \frac{1}{h_j}[\hat{h}(U_j, U_{j+1}) - \hat{h}(U_{j-1}, U_j)] = 0 \tag{7.38a}$$

(2) $N_e = 2$:

$$\begin{pmatrix} \dot{U}_{(j)}^{(1)} \\ \dot{U}_{(j)}^{(2)} \end{pmatrix} - \frac{3}{h_j} \begin{pmatrix} -1 & -1 \\ 1 & 1 \end{pmatrix} \begin{pmatrix} U_{(j)}^{(1)} \\ U_{(j)}^{(2)} \end{pmatrix} + \frac{1}{h_{(j)}} \begin{pmatrix} -2 \\ 4 \end{pmatrix} \hat{h}(U_{(j)}^{(2)}, U_{(j+1)}^{(1)})$$

$$- \frac{1}{h_{(j)}} \begin{pmatrix} 4 \\ -2 \end{pmatrix} \hat{h}(U_{(j-1)}^{(2)}, U_{(j)}^{(1)}) = 0 \tag{7.38b}$$

(3) $N_e = 3$:

$$\begin{pmatrix} \dot{U}_{(j)}^{(1)} \\ \dot{U}_{(j)}^{(2)} \\ \dot{U}_{(j)}^{(3)} \end{pmatrix} - \frac{1}{h_{(j)}} \begin{pmatrix} -6 & -4 & 4 \\ \frac{5}{2} & 0 & -\frac{5}{2} \\ -4 & 4 & 6 \end{pmatrix} \begin{pmatrix} U_{(j)}^{(1)} \\ U_{(j)}^{(2)} \\ U_{(j)}^{(3)} \end{pmatrix} + \frac{1}{h_{(j)}} \begin{pmatrix} 3 \\ -\frac{3}{2} \\ 9 \end{pmatrix} \hat{h}(U_{(j)}^{(3)}, U_{(j+1)}^{(1)})$$

$$- \frac{1}{h_{(j)}} \begin{pmatrix} 9 \\ -\frac{3}{2} \\ 3 \end{pmatrix} \hat{h}(U_{(j-1)}^{(3)}, U_{(j)}^{(1)}) = 0 \tag{7.38c}$$

where the overdot denotes the time derivative, superscript () refers to node number local to the element and subscript (j) refers to the jth element. Note that in Equation 7.38a, subscript j refers to the jth node number, which is the same as the jth element for a constant polynomial approximation. Equation 7.36 can be expressed in the form of ordinary differential equations,

$$\frac{d\mathbf{U}_{(j)}}{dt} = L(\mathbf{U}) \tag{7.39}$$

where L is the operator and $\mathbf{U}$ without subscript includes the variables both in element j and its neighboring elements.

Time integration can be carried out numerically, for example, using the Runge–Kutta integration schemes. With the generalized slope limiter, the Runge–Kutta methods can be described as follows. Let $[0, T]$ be partitioned into N time steps with $\Delta t^n = t^{n+1} - t^n$ ($n = 0, 1, \ldots, N–1$) and $SP(\cdot)$ be the generalized slope limiter, then the time-marching algorithm is numerically implemented as follows:

Set $\mathbf{U}_{(j)}^0 = SP(\mathbf{U}_0)$

For $n = 0, 1, \ldots, N-1$, calculate $\mathbf{U}^{n+1}_{(j)}$ from $\mathbf{U}^{n}_{(j)}$ by the following procedures:

(1) Set $\mathbf{U}^{[0]}_{(j)} = \mathbf{U}^{n}_{(j)}$

(2) For $i = 1, 2, \ldots, K \,(= k+1)$ compute the intermediate functions,

$$\mathbf{U}^{[i]}_{(j)} = SP\left(\sum_{l=0}^{i-1} \alpha_{il} \mathbf{W}^{il}_{(j)} \right); \quad \mathbf{W}^{il}_{(j)} = \mathbf{U}^{[l]}_{(j)} + \frac{\beta_{il}}{\alpha_{il}} \Delta t L\left(\mathbf{U}^{[l]}\right)$$

(3) Set $\mathbf{U}^{n+1}_{(j)} = \mathbf{U}^{[K]}_{(j)}$

If $K = 1$, $\alpha_{10} = \beta_{10} = 1$

If $K = 2$, $\alpha_{10} = \beta_{10} = 1$, $\alpha_{20} = \alpha_{21} = 1/2$, $\beta_{20} = 0$ and $\beta_{21} = 1/2$

where k is the order of the spatial polynomial and superscript [•] denotes the intermediate time steps of the Runge–Kutta scheme. This process repeats for each element and marches to the next time step.

It is important to stress here that a Runge–Kutta scheme is explicit and thus the time step chosen for the simulations need to satisfy the *CFL* condition. For a spatial discretization using polynomials of degree k, a $(k+1)$-stage Runge–Kutta scheme of order $k+1$ needs to be used.

7.1.7 Low Order (Finite Volume) Approximations

It is easily shown that if the shape function is taken as a step function valid only within the element, we have from Equation 7.38a the classical finite volume formulation (or the piecewise constant discontinuous finite element formulation),

$$\frac{d\mathbf{U}_j}{dt} + \frac{1}{h_j}\hat{h}(\mathbf{U}^-_{j+1}, \mathbf{U}^+_{j+1}) - \hat{h}(\mathbf{U}^-_j, \mathbf{U}^+_j) = 0 \tag{7.40a}$$

$$\frac{d\mathbf{U}_j}{dt} = L(\mathbf{U}) = -\frac{1}{h_j}\hat{h}(\mathbf{U}^-_{j+1}, \mathbf{U}^+_{j+1}) - \hat{h}(\mathbf{U}^-_j, \mathbf{U}^+_j) \tag{7.40b}$$

where the node number and element number are the same, which is denoted by subscript j. For piecewise constant element approximations, the Euler forward scheme may be used. Thus, the above equation becomes

$$\mathbf{U}^{n+1}_j = \mathbf{U}^n_j - \frac{\Delta t}{h_j}\left(\hat{h}(\mathbf{U}^-_{j+1}(t^n), \mathbf{U}^+_{j+1}(t^n)) - \hat{h}(\mathbf{U}^-_j(t^n), \mathbf{U}^+_j(t^n)) \right) \tag{7.41}$$

where superscript n denotes the nth time step and $\Delta t = t^{n+1} - t^n$. We consider now how to incorporate the Godunov and Roe flux expressions into the above equation for a low order solution.

7.1.7.1 The Godunov Scheme

The variable update in Godunov-type schemes is done on the cell-averaged conservative variables [6–8]. The update requires an estimation of numerical fluxes at cell interfaces and a successive integration in time over a time step. Hence, the first step of a Godunov scheme approximates the point values of the solution at each interface by a piecewise-constant reconstruction. The values of the conservative variables at a grid point are considered to be a piecewise-constant approximation of the true solution over the cell, centered at that grid point, which is a cell average of the solution. So, the spatial error is of the same order as the cell size Δx, and the scheme is only first order accurate in space. High order accuracy generalizations of the scheme have been proposed based on high order polynomial reconstructions of pointwise values from cell-averaged values. Let us illustrate this step by applying the operator P to the ensemble of the cell-averaged state values, with $\bar{U}(t^n)$ representing the solution at time t^n. Hence, $P(\cdot, \bar{U}(t^n))$ will represent the high order accuracy polynomial approximation inside any cell of the computational domain at time t^n. This reconstruction produces a discontinuity in the state variables at each interface, which is taken as the initial condition for the local Riemann problem,

$$\frac{\partial U}{\partial t} + \frac{\partial F(U)}{\partial x} = 0, \qquad t \in [t^n, t^{n+1}] \tag{7.42}$$

$$U(x, t^n) = \begin{cases} U_j^n & x < x_{j+1/2} \\ U_{j+1}^n & x > x_{j+1/2} \end{cases} \tag{7.43}$$

This is shown in Figure 7.6.

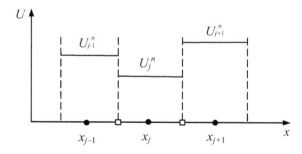

Figure 7.6. Schematic of the Godunov scheme for velocity update: open square (□)denotes the element interface or cell boundary and a dot denotes (•) the middle point of the element

Consider the point $x_{j+1/2}$ at time t^n, where U_j^n and U_{j+1}^n are the values to the left and right of the point. Within a neighborhood of $x_{j+1/2}$, the classical Riemann problem applies. This is schematically shown in Figure 7.6. For example, if $U_j^n >$

$U_{j+1}^n > 0$, we have a shock moving from the left to the right. Hence u_j^n is the solution at $(x_{j+1/2}, t^n)$. On the other hand, if $U_j^n < U_{j+1}^n < 0$, we have a rarefaction moving to the left and U_{j+1}^n is the solution at $(x_{j+1/2}, t^n)$. These quantities will be the solution until a shock/rarefaction moves in from adjacent cells. If $U_j^n < 0$ and $U_{j+1}^n > 0$, we have a rarefaction wave that will give a constant value at $x_{j+1/2}$. As long as the *CFL* condition for time stepping is satisfied, the value at $x_{j+1/2}$ is a function of U_j^n and U_{j+1}^n only. The solution to the Riemann problem is given by $\overline{U}(U_j^n, U_{j+1}^n)$. Thus the Godunov scheme is given by the following expression:

$$U_j^{n+1} = U_j^n - \frac{\Delta t}{h_j}[F(\overline{U}(U_j^n, U_{j+1}^n)) - F(\overline{U}(U_{j-1}^n, U_j^n))] \tag{7.44}$$

The requirement that the shocks or rarefaction waves from adjacent grid points do not touch leads to the *CFL* condition,

$$(\Delta t / h_j) \max_j | F(U_j^n) | \leq 1 \tag{7.45}$$

which needs to be respected during transient calculations.

7.1.7.2 The Roe Scheme
The Godunov scheme requires the solution of the Riemann problem locally at each grid point and is thus computationally intensive. In Roe's scheme, an approximate Riemann solver is applied. Applying Roe's approximate solver at $x_{j+1/2}$, we have the following Riemann problem:

$$\frac{\partial U}{\partial t} + \mathbf{A}U_x = 0, \qquad t \in [t^n, t^{n+1}] \tag{7.46a}$$

$$U(x, t^n) = \begin{cases} U_j^n & x < x_{j+1/2} \\ U_{j+1}^n & x > x_{j+1/2} \end{cases} \tag{7.46b}$$

The numerical fluxes are calculated using Equation 7.30 and thus the Roe scheme may be written as follows:

$$U_j^{n+1} = U_j^n - (\Delta t / h_j)[F_{i+1/2} - F_{i-1/2})] \tag{7.47}$$

Besides the two flux schemes described above, other flux schemes have been developed, and are summarized in Chapter 5. Some of these schemes, for example, the Lax–Friedrichs fluxes, do not require the solution of the local Riemann problem. These flux schemes can be extended to high order approximations.

7.1.8 High Order TVD Approximations

We take the quadratic approximation to illustrate how numerical fluxes and slope limiters can be applied in a *TVD* scheme for compressible flow calculations. As an alternative to other fluxes considered in Section 7.1.6 above, we employ the local Lax–Friedrichs flux [9–11],

$$\hat{h}^{LLF}(a,b) = \frac{1}{2}[F(a) + F(b) - \beta(b-a)] \tag{7.48a}$$

$$\beta = \max_{\min(a,b) \le u \le \max(a,b)} |F'(u)| \tag{7.48b}$$

$$\beta = \max(|F'(a)|, |F'(b)|) \text{ for convex } F(U) \tag{7.48c}$$

In the case of low order (FV) approximations, the Lax–Friedrichs fluxes take the following form:

$$\hat{h}^{LLF}(U_j, U_{j+1}) = \frac{1}{2}\left(F(U_j) + F(U_{j+1}) - \frac{\Delta t}{h_j}(U_{j+1} - U_j)\right) \tag{7.49}$$

Use of a *TVD*-Runge–Kutta scheme requires the construction of the general slope limiter, $U = SP(W)$. The limiter is generalized from the inviscid problems discussed in Chapter 5, and is calculated by the following procedures:

Compute an intermediate value W,

$$W(x,t) = \overline{U}_j + (\partial U_j/\partial x)(\overline{U}_j - \overline{U}_{j-1}), \quad x \in [x_{j-1/2}, x_{j+1/2}]$$

Compute $U^-_{j+1/2}$ and $U^+_{j-1/2}$ by

$$U^-_{j+1/2} = \overline{W}_j + m(W^-_{j+1/2} - \overline{W}_j, \overline{W}_j - \overline{W}_{j-1}, \overline{W}_{j+1} - \overline{W}_j)$$

$$U^+_{j-1/2} = \overline{W}_j - m(\overline{W}_j - W^+_{j-1/2}, \overline{W}_j - \overline{W}_{j-1}, \overline{W}_{j+1} - \overline{W}_j)$$

If $U^-_{j+1/2} = W^-_{j+1/2}$ and $U^+_{j-1/2} = W^+_{j-1/2}$, then $U = W$

If not, then $U = \overline{W}_j + m(W_j, \overline{W}_j - \overline{W}_{j-1}, \overline{W}_{j+1} - \overline{W}_j)(\partial U_j/\partial x)$

where the overbar on W and U denotes the averages over element j. The function $m()$ is the modified minmod function defined in Section 5.4.3.

With U evaluated at the boundaries of the element as described above, and substituted into Equation 7.48, the numerical fluxes are calculated. These numerical fluxes are then substituted into Equation 7.38 or 7.39, which is then integrated using the Runge–Kutta scheme as described in the paragraph below Equation 7.39.

Take as an example the second order accurate Runge–Kutta scheme for a linear spatial approximation ($k=1$), we then have the following four steps:

Step 1: $\mathbf{U}_{(j)}^0 = SP(\mathbf{U}_0)$

Step 2: For $n = 0, 1, ..., N-1$, calculate $\mathbf{U}_{(j)}^{n+1}$ from $\mathbf{U}_{(j)}^n$ as follows:

(1) Set $\mathbf{U}_{(j)}^{[0]} = \mathbf{U}_{(j)}^n$

(2) For $i = 1, 2 (= k+1)$ compute the intermediate functions,

$$\mathbf{W}_{(j)}^{10} = \mathbf{U}_{(j)}^{[0]} + \Delta t L\left(\mathbf{U}^{[0]}\right)$$

$$\mathbf{U}_{(j)}^{[1]} = SP(\mathbf{W}_{(j)}^{10})$$

$$\mathbf{W}_{(j)}^{20} = \mathbf{U}_{(j)}^{[0]}$$

$$\mathbf{W}_{(j)}^{21} = \mathbf{U}_{(j)}^{[1]} + \Delta t L\left(\mathbf{U}^{[1]}\right)$$

$$\mathbf{U}_{(j)}^{[2]} = SP(\alpha_{20}\mathbf{W}_{(j)}^{20} + \alpha_{21}\mathbf{W}_{(j)}^{21}) = \frac{1}{2}SP\left\{\mathbf{U}_{(j)}^{[0]} + \mathbf{U}_{(j)}^{[1]} + \Delta t L\left(\mathbf{U}^{[1]}\right)\right\}$$

(3) Set $\mathbf{U}_{(j)}^{n+1} = \mathbf{U}_{(j)}^{[2]}$

7.1.9 Numerical Examples

We consider three examples in this section. In the first example, the evolution of a scalar function is calculated. The problem is defined as follows:

$$\frac{\partial u}{\partial t} + \frac{\partial u}{\partial x} = 0, \; x \in [0,1] \tag{7.50a}$$

$$u(x, t = 0) = \begin{cases} 1, & x < 0.5 \\ 0, & \text{otherwise} \end{cases} \tag{7.50b}$$

with periodic boundary conditions at $x = 0$ and $x = 1$.

For this case, the local Lax–Friedrichs flux is used along with various orders of approximation. For constant elements, the first order Euler time integration method is used without the slope limiter. For linear elements, the second order Runge–Kutta method is used with the general slope limiter. For quadratic elements, the third order Runge–Kutta method is used with a linear slope limiter. The computed results are plotted in Figure 7.7, showing the effects of discretization and order of approximations.

Figure 7.7(a) compares the numerical results at $t = 0.5$ obtained using the different types of approximations but with the same number of elements ($N = 51$), N being the number of elements. Examination of these results indicates that considerable numerical dissipation occurs with the piecewise constant element (FV) approximation, as is evident by the strong flattening of the sharp edge of the square pulse. As discussed in Chapter 5, this dissipation comes from the numerical approximation and the behavior of the exact solution should be such that the initial rectangular-shaped wave remains its shape and the data is carried into the domain from the boundary. The use of linear elements improves the results substantially and shows considerably less dissipation.

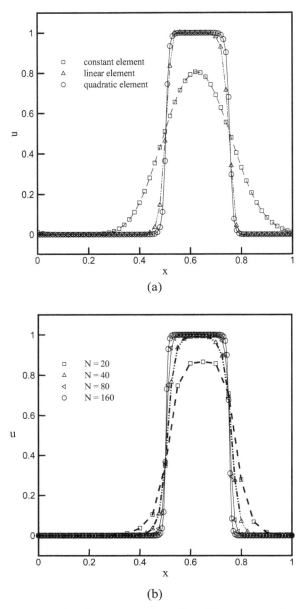

(a)

(b)

Figure 7.7. Computed results of pure convection using different types of approximations at t = 0.5: (a) different space discretizations (N=50) and (b) effects of the number of elements (t = 0.5 and with quadratic elements)

Figure 7.7(b) plots the results at $t = 0.5$ obtained using different numbers of linear elements. With $N = 21$, the results are already improvements over the constant element approximation as shown in Figure 7.7(a). Continued improvements are obtained as the discretization is refined. With $N = 160$, the results match almost exactly with the analytic solution.

As the second example, we consider the nonlinear transport of a scalar function, which is the typical inviscid Burgers' equation,

$$\frac{\partial u}{\partial t} + \frac{\partial}{\partial x}\left(\frac{u^2}{2}\right) = 0 \ , \ x \in [0,1] \tag{7.51a}$$

$$u(x, t = 0) = u_0(x) = 0.25 + 0.5\sin(\pi x) \tag{7.51b}$$

with periodic boundary conditions at $x = 0$ and $x = 1$.

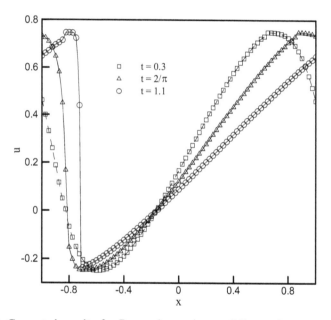

Figure 7.8. Computed results for Burgers' equation at different time steps using linear elements. Data used for calculations: $N = 80$, $t = 10^{-4}$, linear spatial approximation with a second order Runge–Kutta method

For this problem, the solution is smooth up to $t = 2/\pi$, and then it develops a moving shock wave, which interacts with a rarefaction wave. This results in a sonic wave. The calculations were made using a linear approximation with the second order Runge–Kutta time integration. The general slope limiter is used. Figure 7.8 shows the numerical results.

The third problem is the shock-tube problem. This is a very interesting test case because the exact time dependent solution is known and can be compared with the solution computed by applying numerical discretizations. The problem is schematically illustrated in Figure 7.9. The initial solution of the shock-tube problem is composed of two uniform states separated by a discontinuity that is usually located at the origin. This particular initial value problem is known as a Riemann problem. The initial left and right uniform states are usually introduced by giving the density, the pressure, and the velocity. This initial set represents a tube where the left and right regions are separated by a diaphragm, and filled by the same gas in two different physical states. If all the viscous effects are negligible along the tube walls and it is assumed that the tube is infinitely long in order to avoid reflections at the tube ends, then the exact solution of the full Euler equations can be obtained on the basis of a simple wave analysis. At the bursting of the diaphragm, the discontinuity between the two initial states breaks into leftward and rightward moving waves, which are separated by a contact surface. Each wave pattern is composed by a contact discontinuity (C) in the middle, and a shock (S) or a rarefaction wave (R) at the left and the right sides separating the uniform state solution. All the available combinations produce four wave patterns: RCR, RCS, SCR, and SCS, which are self-similar, that is they depend only on x/t. These four patterns are illustrated in Figure 7.4. A fifth pattern is possible in theory, and it contains a vacuum state between two central contact discontinuities, which occur between two rarefaction waves. This case is of theoretical interest only, because it expresses the limit of the perfect gas equations at zero pressure and temperature, but it can never occur in reality.

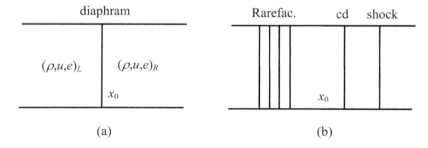

Figure 7.9. Schematic of a shock tube problem: (a) initial condition and (b) wave propagation

The shock tube test case corresponds to the Riemann problem. The governing equations for the above problem, with viscous effects neglected and the gas property of $\gamma = 1.4$, are the Euler equations,

$$\frac{\partial U}{\partial t} + \frac{\partial F(U)}{\partial x} = 0 \qquad (7.52a)$$

$$U(x,0) = \begin{cases} U_L, & x < x_0 \\ U_R, & x > x_0 \end{cases} \qquad (7.52b)$$

where initial conditions are given by

$$U_L = \begin{pmatrix} \rho \\ \rho u \\ \rho h \end{pmatrix} = \begin{pmatrix} 1 \\ 0 \\ 2.5 \end{pmatrix} ; \ U_R = \begin{pmatrix} \rho \\ \rho u \\ \rho h \end{pmatrix} = \begin{pmatrix} 0125 \\ 0 \\ 0.25 \end{pmatrix}$$ (7.52c)

The results are obtained using the TVD RK scheme with the slope limiter. The following data were used for calculations: $N=101$, $\Delta t = 10^{-4}$, t up to 0.5, and linear elements with a second order Runge–Kutta method. The results of the transient flow for density, velocity, energy and pressure at $t=0.5$ are shown in Figure 7.10, which compares well with the exact solution [2, 4].

The solution is composed of, from left to right, a constant undisturbed left state, then a continuous expansion wave moving to the left, followed by a constant state, a contact discontinuity moving to the right, followed by a constant state, and then a shock wave moving to the right in the undisturbed right state.

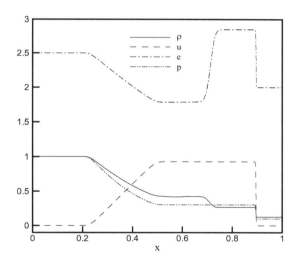

Figure 7.10. Computed results for the shock tube problem

7.2 Multidimensional Inviscid Compressible Flows

We now consider the discontinuous finite element formulation for computational compressible flows in multidimensions.

7.2.1 Governing Equations

The full three-dimensional Euler equations in a multidimensional domain are written in conservation form as follows:

$$\frac{\partial U}{\partial t} + \nabla \cdot \mathbf{F}(U) = 0 \tag{7.53}$$

where U is a vector of dimension 5 and $\mathbf{F}$ is a tensor,

$$U = \begin{Bmatrix} \rho \\ \rho u \\ \rho v \\ \rho w \\ \rho e \end{Bmatrix} ; \quad F_x(U) = \begin{Bmatrix} \rho u \\ \rho uu + p \\ \rho uv \\ \rho uw \\ \rho h u \end{Bmatrix} ;$$

$$F_y(U) = \begin{Bmatrix} \rho u \\ \rho vu \\ \rho vv + p \\ \rho vw \\ \rho h v \end{Bmatrix} ; \quad F_z(U) = \begin{Bmatrix} \rho u \\ \rho wu \\ \rho wv \\ \rho ww + p \\ \rho h w \end{Bmatrix} \tag{7.54}$$

Before discussing the discontinuous finite element formulation, let us consider some of the basic properties of the 3-D Euler equations.

7.2.2 Basic Properties of the Split 3-D Euler Equations

The basic properties of a split Euler equation system are similar to those seen in 1-D Euler equations. The x-split, 3-D Euler equations have the following form:

$$\frac{\partial U}{\partial t} + \frac{\partial F_x(U)}{\partial x} = 0 \tag{7.55}$$

with the Riemann boundary conditions,

$$U(x,0) = \begin{cases} U_L & \text{if } x < 0 \\ U_R & \text{if } x > 0 \end{cases} \tag{7.56}$$

The Jacobian matrix $\mathbf{A}$ for the problem is given by

$$\mathbf{A} = \frac{\partial F}{\partial U} = \begin{bmatrix} 0 & 1 & 0 & 0 & 0 \\ \tilde{\gamma}h - u^2 - a^2 & (3-\gamma)u & -\tilde{\gamma}v & -\tilde{\gamma}w & \tilde{\gamma} \\ -uv & v & u & 0 & 0 \\ -uw & w & 0 & u & 0 \\ \frac{1}{2}u[(\gamma-3)h - a^2] & h - \tilde{\gamma}u^2 & -\tilde{\gamma}uv & -\tilde{\gamma}uw & \gamma u \end{bmatrix} \tag{7.57}$$

where

$$h = \frac{1}{2}\mathbf{u}\cdot\mathbf{u} + \frac{a^2}{\gamma-1} \; ; \; \tilde{\gamma} = \gamma-1 \; ; \; \mathbf{u} = (u,v,w)^T \tag{7.58}$$

The x-split system is hyperbolic in nature and has the real eigenvalues,

$$\lambda_1 = u-a; \; \lambda_2 = \lambda_3 = \lambda_4 = u; \; \lambda_5 = u+a \tag{7.59}$$

The matrix of the corresponding right eigenvectors is given by

$$\mathbf{R} = \begin{bmatrix} 1 & 1 & 0 & 0 & 1 \\ u-a & u & 0 & 0 & u+a \\ v & v & 1 & 0 & v \\ w & w & 0 & 1 & w \\ h-ua & \frac{1}{2}\mathbf{u}\cdot\mathbf{u} & v & w & h+ua \end{bmatrix} \tag{7.60a}$$

and its inverse is given by

$$\mathbf{R}^{-1} = \frac{\gamma-1}{2a^2} \begin{bmatrix} h+a(u-a)/\tilde{\gamma} & -u-(a/\tilde{\gamma}) & -v & -w & 1 \\ (4a^2/\tilde{\gamma})-2h & 2u & 2v & 2w & -2 \\ -2va^2/\tilde{\gamma} & 0 & 2a^2/\tilde{\gamma} & 0 & 0 \\ -2wa^2/\tilde{\gamma} & 0 & 0 & 2a^2/\tilde{\gamma} & 0 \\ h-a(u+a)/\tilde{\gamma} & -u+(a/\tilde{\gamma}) & -v & -w & 1 \end{bmatrix} \tag{7.60b}$$

These quantities are useful for constructing Roe's approximate solver for the Riemann problem. The characteristics and structure of the solution of the x-split Euler equations are very similar to the 1-D Euler equations and are plotted in Figure 7.11.

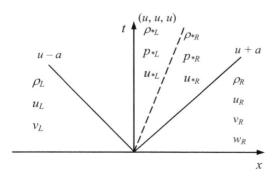

Figure 7.11. Structure of the solution of the x-split 3-D Riemann problem

7.2.3 Discontinuous Finite Element Formulation

To develop a discontinuous finite element formulation, the computational domain is first discretized into a tessellation of elements. Over a typical element, say, element j, Equation 7.53 is integrated with respect to a testing function vector W. Upon integration by parts, we obtain the weak statement of the problem,

$$\int_{\Omega_j} W \frac{\partial U}{\partial t} dV + \oint_{\partial\Omega_j} W \cdot \mathbf{F}(U) \cdot \mathbf{n} \, dS - \int_{\Omega_j} \nabla W : \mathbf{F}(U) \, dV = 0 \quad (7.61)$$

We now replace the function fluxes at the element boundaries by numerical fluxes and (W, U) by the approximations (W_h, U_h) to complete the discontinuous finite element formulation,

$$\int_{\Omega_j} W_h \frac{\partial U_h}{\partial t} dV + \oint_{\partial\Omega_j} W_h \hat{H}(U_h) \, dS - \int_{\Omega_j} \nabla W_h \cdot F(U_h) \, dV = 0 \quad (7.62)$$

where $\hat{H}$ are the numerical fluxes. Various numerical fluxes suitable for the solution of the inviscid Burgers' equation were listed in Chapter 5. These numerical fluxes are equally applicable for the multidimensional calculations with relevant changes.

If, for example, the Roe flux function is chosen,

$$\hat{H}(U_h^-, U_h^+) = \tfrac{1}{2}[\mathbf{F}(U_h^+) + \mathbf{F}(U_h^-)] \cdot \mathbf{n} - \tfrac{1}{2}|\mathbf{A}|(U_h^+ - U_h^-)$$

$$= \mathbf{F}(U_h^-) \cdot \mathbf{n} + \mathbf{A}^-(U_h^+ - U_h^-) \quad (7.63)$$

and the Galerkin approximation is used, then the integral equation for the discontinuous formulation becomes

$$\int_{\Omega_j} \phi_i \frac{\partial U_h}{\partial t} dV + \oint_{\partial\Omega_j} \phi_i \mathbf{F}(U_h^-) \cdot \mathbf{n} \, dS - \oint_{\partial\Omega_j} \phi_i \mathbf{A}^- U_h^- \, dS$$

$$- \int_{\Omega_j} \nabla \phi_i \cdot \mathbf{F}(U_h) \, dV = - \oint_{\partial\Omega_j} \phi_i \mathbf{A}^- U_h^+ \, dS \quad (7.64)$$

In the above two equations, ϕ_i is the shape function and $\mathbf{A}$ is the $\mathbf{n}$-split Jacobian matrix, evaluated using the Roe averages. The other terms are defined as follows:

$$|\mathbf{A}| = \mathbf{L}|\mathbf{\Lambda}|\mathbf{R}, \quad \mathbf{A}^{-1} = \mathbf{L}\mathbf{\Lambda}^-\mathbf{R} \quad (7.65)$$

$$|\mathbf{\Lambda}| = diag(|\lambda_k|), \quad \mathbf{\Lambda}^- = diag(\min(\lambda_k, 0)) \quad (7.66)$$

with $\mathbf{L}$ and $\mathbf{R}$ being the matrices of the left and right eigenvectors of $\mathbf{A}$.

For flows that involve discontinuities, slope or flux limiters must be used to suppress the oscillations as shown above. These limiters are the extension of the 1-D counterparts discussed in Chapter 5. Construction of these limiters for an unstructured mesh in a multidimensional domain is not trivial [12].

With these considerations taken into account, the solutions are approximated using the local polynomial basis functions. Then, one obtains a system of ordinary differential equations,

$$\frac{d\mathbf{U}_{(j)}}{dt} = L(\mathbf{U}) \tag{7.67}$$

where L is the operator and $U_{(j)}$ is the vector of unknown variables defined at the nodal points of element j.

Equation 7.67 can be integrated using the Runge–Kutta integration scheme. The TVD scheme can also be used, provided that appropriate limiters are used. This integration procedure and also the computational process are very similar to the 1-D case and thus require no further elaboration.

Bassi and Rebay [13] were among the first to propose the discontinuous finite element scheme, and applied it to study the compressible flow around a cylinder. Their results show that accurate solutions can be obtained on a relatively coarse mesh using a higher order representation of the unknowns and of the geometry of the domain boundary. They further show that no limiting procedure is needed if the solution is sufficiently smooth. They caution, however, that the discontinuous finite element method requires a higher order approximation of curved boundaries, if accurate numerical results are to be computed. They recommend that for curved boundaries, elements of the order of $m \geq 2$ be used. Some of their results are given in Figure 7.12.

7.3 Multidimensional Compressible Viscous Flows

The compressible viscous flows are governed by the Navier–Stokes equations, which may be written in conservation form,

$$\partial_t U + \nabla \cdot \mathbf{F}_c(U) - \nabla \cdot \mathbf{F}_v(U, \nabla U) = 0 \tag{7.68a}$$

where U and $\mathbf{F}_c$ are the vector, and the convective Euler flux tensor, defined by Equation 7.54. The viscous flux tensor is given by

$$F_{vx}(U, \nabla U) = \begin{bmatrix} 0 \\ \tau_{xx} \\ \tau_{xy} \\ \tau_{xz} \\ u\tau_{xx} + v\tau_{xy} + w\tau_{xz} - q_x \end{bmatrix} \tag{7.68b}$$

$$
F_{vy}(U,\nabla U) =
\begin{bmatrix}
0 \\
\tau_{yx} \\
\tau_{yy} \\
\tau_{yz} \\
u\tau_{yx} + v\tau_{yy} + w\tau_{yz} - q_y
\end{bmatrix}
\qquad (7.68c)
$$

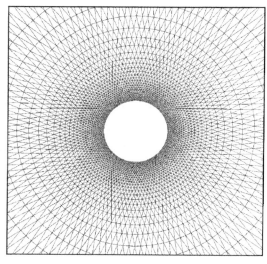

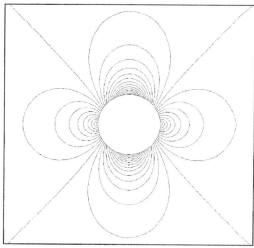

Figure 7.12. Computed results for 2-D compressible flows passing around a cylinder, using the discontinuous finite element method: (a) meshes (128 × 32) used for calculations and (b) March contour [13]

$$F_{vz}(U,\nabla U) = \begin{bmatrix} 0 \\ \tau_{zx} \\ \tau_{zy} \\ \tau_{zz} \\ u\tau_{zx} + v\tau_{zy} + w\tau_{zz} - q_z \end{bmatrix}$$

(7.68d)

$$\tau_{xx} = \frac{4}{3}\mu\frac{\partial u}{\partial x} - \frac{2}{3}\mu\left(\frac{\partial v}{\partial y} + \frac{\partial w}{\partial z}\right) + \mu_b\nabla\cdot\mathbf{u}$$

(7.68e)

$$\tau_{yy} = \frac{4}{3}\mu\frac{\partial v}{\partial y} - \frac{2}{3}\mu\left(\frac{\partial u}{\partial x} + \frac{\partial w}{\partial z}\right) + \mu_b\nabla\cdot\mathbf{u}$$

(7.68f)

$$\tau_{zz} = \frac{4}{3}\mu\frac{\partial w}{\partial z} - \frac{2}{3}\mu\left(\frac{\partial u}{\partial x} + \frac{\partial v}{\partial y}\right) + \mu_b\nabla\cdot\mathbf{u}$$

(7.68g)

$$\tau_{xy} = \tau_{xy} = \mu\left(\frac{\partial u}{\partial y} + \frac{\partial v}{\partial x}\right)$$

(7.68h)

$$\tau_{yz} = \tau_{zy} = \mu\left(\frac{\partial v}{\partial z} + \frac{\partial w}{\partial y}\right)$$

(7.68i)

$$\tau_{zx} = \tau_{xz} = \mu\left(\frac{\partial u}{\partial z} + \frac{\partial w}{\partial x}\right)$$

(7.68j)

$$\mathbf{q} = -\kappa\nabla T$$

(7.68k)

where μ_b is the bulk viscosity.

The vector $\mathbf{F}_v(U,\nabla U)$ is a function of ∇U, which leads to second order derivatives when the viscous fluxes are evaluated. The second order derivatives are not accommodated directly in a weak formulation using a discontinuous finite element space. Thus, as for the incompressible flows, the auxiliary variables are introduced to split the second order derivative terms into a system of first order partial differential equations,

$$\mathbf{S} - \nabla U = 0$$

(7.69a)

$$\partial_t U + \nabla\cdot\mathbf{F}_c(U) - \nabla\cdot\mathbf{F}_v(U,\mathbf{S}) = 0$$

(7.69b)

Following the procedure given for the inviscid compressible flow calculations, the discontinuous finite element formation starts with the integration of the governing equations with respect to a testing function W over element j. Upon substituting numerical fluxes for the function fluxes at the element boundaries, and replacing the exact solution with approximate solutions taken from the finite element broken space, we have the integral formulation of the Navier–Stokes equations,

$$\int_{\Omega_j} W \mathbf{S}_h d\Omega - \oint_{\partial\Omega_j} W \hat{U} dS + \int_{\Omega_j} \nabla W U_h d\Omega = 0 \qquad (7.70\text{a})$$

$$\frac{d}{dt}\int_{\Omega_j} W U_h d\Omega + \oint_{\partial\Omega_j} W \hat{H}(U_h) d\sigma - \int_{\Omega_j} \nabla W \cdot \mathbf{F}(U_h) d\Omega$$

$$+ \oint_{\partial\Omega_j} W \hat{H}_v(U_h, \mathbf{S}_h) d\sigma - \int_{\Omega_j} \nabla W \cdot \mathbf{F}_v(U_h, \mathbf{S}_h) d\Omega = 0 \qquad (7.70\text{b})$$

For this problem, three different types of numerical fluxes are required. First, the convective fluxes $\hat{H}$ may be selected as either the Godunov fluxes or the Roe approximate fluxes or other qualified numerical fluxes. With these fluxes, the discontinuity at the element boundaries is treated as an **n**-split Riemann problem, **n** being the boudnary normal of element j. We saw in the last section how the Roe flux functions are incorporated into the discontinuous formulation. Second, the viscous fluxes $\hat{H}_v$ need to be constructed. The theoretical basis for the construction of these viscous fluxes was given in Chapter 4. Thus, in principle, the fluxes listed in Table 4.2 can be used. For the simplest case, a central approximation, which gives a suboptimal convergence rate, was chosen first by Bassi and Rebay [14]. This same idea can also be applied to the U fluxes, $\hat{U}$. If these choices are made, we then have the following numerical fluxes for the discontinuous finite element formulation of the Navier–Stokes equations:

(i) Convective fluxes,

$$\hat{H}(U_h^-, U_h^+) = \tfrac{1}{2}[\mathbf{F}(U_h^+) + \mathbf{F}(U_h^-)] \cdot \mathbf{n} - \tfrac{1}{2}|\mathbf{A}|(U_h^+ - U_h^-)$$

$$= \mathbf{F}(U_h^-) \cdot \mathbf{n} + \mathbf{A}^-(U_h^+ - U_h^-) \qquad (7.71\text{a})$$

(ii) Viscous fluxes

$$\hat{H}_v(U_h^-, \mathbf{S}_h^-, U_h^+, \mathbf{S}_h^-; \mathbf{n}) = \tfrac{1}{2}[\mathbf{F}(U_h^-, \mathbf{S}_h^-) + \mathbf{F}(U_h^+, \mathbf{S}_h^+)] \cdot \mathbf{n} \qquad (7.71\text{b})$$

(iii) U fluxes

$$\hat{U}(U_h^-, U_h^+; \mathbf{n}) = \tfrac{1}{2}(U_h^- + U_h^+)\mathbf{n} \qquad (7.71\text{c})$$

With these flux expressions substituted into Equation 7.69, followed by numerical integration, one has the following ordinary differential equations,

$$\frac{d\mathbf{U}_{(j)}}{dt} = L(\mathbf{U}) \tag{7.72}$$

where L is the matrix operator and $\mathbf{U}_{(j)}$ is the vector of unknown variables defined at the nodal points of element j.

The system can be integrated using the Runge–Kutta time integrator as discussed in Section 7.1.6.

Bassi and Rebay [14] applied the above discontinuous finite element formulation to solve the Navier–Stokes equations for compressible flows in a 2-D domain. They used the Godunov fluxes, instead of Roe's approximate fluxes, for the convective fluxes, and simple central schemes for U fluxes and viscous fluxes. Some of their results obtained using the constant, linear, quadratic and cubic approximations are given in Figure 7.13.

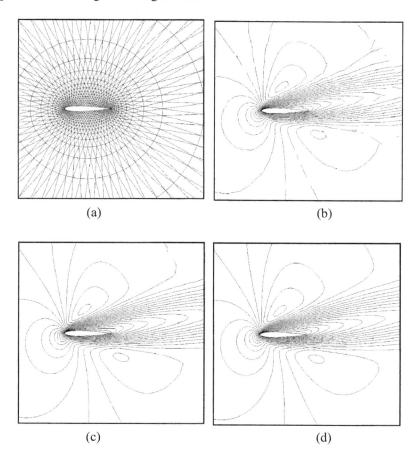

Figure 7.13. Numerical results of compressible viscous flows over an airfoil computed using the discontinuous finite element method: (a) finite element mesh, (b) linear element approximation, (c) quadratic approximation and (d) cubic approximation [14]

Baumann and Oden [15] present a discontinuous Galerkin finite element method technique to obtain a compact, higher order accuracy and stable solver. The method involves a weak imposition of continuity conditions on the state variables and on inviscid and diffusive fluxes across inter-element and domain boundaries. Auxiliary variables are not needed in numerical fluxes constructed by them. Dolejsi [16] recently presented a discontinuous finite element formulation, which employs the internal and boundary penalty terms to provide numerical stability. These methods are applied to the compressible viscous flow calculations.

7.4 ALE Formulation

The arbitrary Lagrangian–Eulerian (ALE) formulation is developed as a moving grid treatment of discontinuities in compressible flow fields. In this Section, we discuss some basics of the formulation, and the discontinuous Galerkin solution of the ALE equations for compressible fluid flow problems. The theoretical background of the ALE description of continuum mechanics problems can be found in the work of Huge and others [17–20], answering the need for a general framework within which flow–structure interactions can be effectively modeled.

7.4.1 ALE Kinematic Description

We consider, as shown in Figure 7.14, three different coordinate systems to describe the motion of a particle.

First, the motion of particle p, initially located at $\mathbf{X}$, is described by $\chi(\mathbf{X}, t)$. At t, its position is $\mathbf{x} = \chi(\mathbf{X}, t)$. If we fix our focus at $\mathbf{x}$, then the particle occupying $\mathbf{x}$ can be considered coming from its position at $\hat{\mathbf{x}}$, and follows the motion described by $\phi(\hat{\mathbf{x}}, t)$. Likewise, the particle occupying $\hat{\mathbf{x}}$ may be considered coming from that located at $\mathbf{X}$, and follows the motion of $\psi(\mathbf{X}, t)$. Thus, the particle motion may be considered coming from $\mathbf{X}$ via $\chi(\mathbf{X}, t)$, or via a combined motion of $\psi(\mathbf{X}, t)$ and $\phi(\mathbf{X}, t)$. Different views of motion can be made, depending upon which coordinates or descriptions are used. All these descriptions, however, refer to the same global inertia frame with its origin at O (see Figure 7.14).

Perhaps, what is important for dynamics is the velocity and acceleration, which are related to the motion. Thus, a time derivative of $\chi(\mathbf{X}, t)$ with $\mathbf{X}$ fixed,

$$\mathbf{u} = \left.\frac{\partial \mathbf{x}}{\partial t}\right|_{\mathbf{X}} = \left.\frac{\partial \chi(\mathbf{X}, t)}{\partial t}\right|_{\mathbf{X}} = \mathbf{u}(\mathbf{X}, t)$$

is the velocity of the particle observed by a person standing still at O. Here the particle is identified, with its initial position at $\mathbf{X}$. However, its current position is at $\mathbf{x} = \chi(\mathbf{X}, t)$ at t, and $\mathbf{x}_1 = \chi(\mathbf{X}, t_1)$ at t_1, etc. In this view, the eyes of the person are fixed on the particle as it moves through space. This is the so-called Lagrangian description, as discussed above, where $\mathbf{X}$ and t are independent variables to describe the motion.

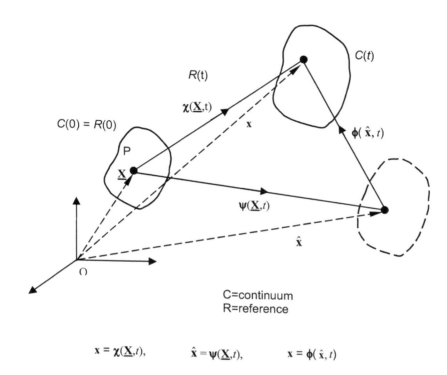

$$x = \chi(\underline{X},t), \qquad \hat{x} = \psi(\underline{X},t), \qquad x = \phi(\hat{x}, t)$$

Figure 7.14. Various configurations used for the ALE description.

If, on the other hand, a probe is placed at x all the time, then it will pick up the velocity of the particle when the particle passes x at t. Later, it will pick up the velocity of another particle that occupies x at $t_1 > t$. The same person, standing still at O, with his eyes focused at position x, will observe a velocity at x and at t as $u(x,t)$,

$$u = \left.\frac{\partial x}{\partial t}\right|_{\underline{X}} = \left.\frac{\partial \chi(\underline{X},t)}{\partial t}\right|_{\underline{X}} = \left.\frac{\partial \chi(\chi^{-1}(x,t),t)}{\partial t}\right|_{\underline{X}} = u(x,t) \tag{7.73}$$

and at different t_1, he observes another velocity $u(x,t_1)$ at x. This velocity $u(x,t_1)$ is no longer the velocity associated with the particle originally at $\underline{X}$. In fact, in this description, the observer cannot remember, and does not care, who passes through x at any time. We take t and x as the independent variables. This is the Eulerian description.

If the person standing still at O focuses his eyes at position $\hat{x}$, he will observe a velocity at $\hat{x}$ and at t as $w(\hat{x},t)$, which is calculated by taking the time derivative of ψ with $\underline{X}$ fixed,

$$\mathbf{w} = \frac{\partial \mathbf{\psi}(\mathbf{X},t)}{\partial t}\bigg|_{\mathbf{X}} = \mathbf{w}(\hat{\mathbf{x}},t) \tag{7.74}$$

and at different t_1, he observes another velocity $\mathbf{w}(\hat{\mathbf{x}},t_1)$ at $\hat{\mathbf{x}}$.

If, now, the person changes his position and stands at $\hat{\mathbf{x}}$ then with his eyes focused at $\mathbf{x}$, he will observe a velocity at $\mathbf{x}$ and at t as $\mathbf{v}(\hat{\mathbf{x}},t)$,

$$\mathbf{v} = \frac{\partial \mathbf{\phi}(\hat{\mathbf{x}},t)}{\partial t}\bigg|_{\hat{\mathbf{x}}} = \mathbf{v}(\hat{\mathbf{x}},t) \tag{7.75}$$

These velocities observed above are related. To derive a relation, we note that

$$\mathbf{x} = \mathbf{\chi}(\mathbf{X},t) = \mathbf{\phi}(\hat{\mathbf{x}},t) \tag{7.76}$$

Taking the time derivative and utilizing the relation $\hat{\mathbf{x}} = \mathbf{\psi}(\mathbf{X},t)$, we have by the chain rule of differentiation,

$$\frac{\partial \mathbf{\chi}}{\partial t}\bigg|_{\mathbf{X}} = \frac{\partial \mathbf{\phi}(\hat{\mathbf{x}}(\mathbf{X},t),t)}{\partial t}\bigg|_{\mathbf{X}} = \frac{\partial \mathbf{\phi}(\hat{\mathbf{x}},t)}{\partial t}\bigg|_{\hat{\mathbf{x}}} + \frac{\partial \mathbf{\phi}(\hat{\mathbf{x}},t)}{\partial \mathbf{\psi}}\frac{\partial \mathbf{\psi}(\mathbf{X},t)}{\partial t}\bigg|_{\mathbf{X}} \tag{7.77}$$

or

$$\mathbf{u}(\mathbf{x},t) = \mathbf{v}(\hat{\mathbf{x}},t) + \frac{\partial \mathbf{x}}{\partial \hat{\mathbf{x}}}\mathbf{w}(\hat{\mathbf{x}},t) = \mathbf{u}(\mathbf{x}(\hat{\mathbf{x}}),t) = \mathbf{u}(\hat{\mathbf{x}},t) \text{ or } u_i = v_i + \frac{\partial x_i}{\partial \hat{x}_j}w_j \tag{7.78}$$

Here, $\partial x_i/\partial \hat{x}_j$ represents the scale change of the particle when it passes through the referential frame $\hat{\mathbf{x}}$, and is often called the deformation tensor.

We may also write

$$\frac{\partial x_i}{\partial \hat{x}_j}w_j = u_i - v_i \tag{7.79}$$

Let us now consider the acceleration. By definition, the material time derivative is the acceleration,

$$\frac{\partial \mathbf{u}(\mathbf{X},t)}{\partial t}\bigg|_{\mathbf{X}} = \frac{\partial \mathbf{u}(\mathbf{x},t)}{\partial t}\bigg|_{\mathbf{x}} + \frac{\partial \mathbf{u}(\mathbf{x},t)}{\partial \mathbf{x}}\frac{\partial \mathbf{x}}{\partial t}\bigg|_{\mathbf{X}}$$

$$= \frac{\partial \mathbf{u}(\mathbf{x},t)}{\partial t} + \mathbf{u}(\mathbf{x},t)\cdot\nabla\mathbf{u}(\mathbf{x},t) = \frac{D\mathbf{u}(\mathbf{x},t)}{Dt} \tag{7.80}$$

where use has been made of the chain rule and the relation $\mathbf{x} = \mathbf{\chi}(\mathbf{X},t)$.

Again the left is the acceleration of the particle identified by $\underline{\mathbf{X}}$, and the observer's eyes are fixed on the particle. The two terms on the right represent the local time derivative (time rate reading of a probe fixed at $\mathbf{x}$) and the effect of neighboring points when the observer's eyes are focused on the position $\mathbf{x}$.

We also have the acceleration for the referential frame,

$$\frac{\partial \mathbf{u}(\underline{\mathbf{X}},t)}{\partial t}\bigg|_{\underline{\mathbf{X}}} = \frac{\partial \mathbf{u}(\hat{\mathbf{x}},t)}{\partial t}\bigg|_{\hat{\mathbf{x}}} + \frac{\partial \mathbf{u}(\hat{\mathbf{x}},t)}{\partial \hat{\mathbf{x}}} \frac{\partial \hat{\mathbf{x}}}{\partial t}\bigg|_{\underline{\mathbf{X}}} = \frac{\partial \mathbf{u}(\hat{\mathbf{x}},t)}{\partial t}\bigg|_{\hat{\mathbf{x}}} + \mathbf{w}(\hat{\mathbf{x}},t)\cdot\hat{\nabla}\mathbf{u}(\hat{\mathbf{x}},t)$$

or

$$\frac{\partial \mathbf{u}(\underline{\mathbf{X}},t)}{\partial t}\bigg|_{\underline{\mathbf{X}}} = \frac{\partial \mathbf{u}(\hat{\mathbf{x}},t)}{\partial t}\bigg|_{\hat{\mathbf{x}}} + \frac{\partial \mathbf{u}(\hat{\mathbf{x}},t)}{\partial \mathbf{x}} \frac{\partial \mathbf{x}}{\partial \hat{\mathbf{x}}} \frac{\partial \hat{\mathbf{x}}}{\partial t}\bigg|_{\underline{\mathbf{X}}}$$

$$= \frac{\partial \mathbf{u}(\hat{\mathbf{x}},t)}{\partial t}\bigg|_{\hat{\mathbf{x}}} + \frac{\partial \mathbf{x}}{\partial \hat{\mathbf{x}}}\mathbf{w}(\hat{\mathbf{x}},t)\cdot\nabla\mathbf{u}(\hat{\mathbf{x}},t)$$

$$= \frac{\partial \mathbf{u}(\hat{\mathbf{x}},t)}{\partial t}\bigg|_{\hat{\mathbf{x}}} + (\mathbf{u}(\hat{\mathbf{x}},t) - \mathbf{v}(\hat{\mathbf{x}},t))\cdot\nabla\mathbf{u}(\hat{\mathbf{x}},t)$$

$$= \frac{\partial \mathbf{u}(\hat{\mathbf{x}},t)}{\partial t}\bigg|_{\hat{\mathbf{x}}} + (\mathbf{u}(\mathbf{x},t) - \mathbf{v}(\mathbf{x},t))\cdot\nabla\mathbf{u}(\mathbf{x},t) \qquad (7.81)$$

where $\hat{\nabla} = \partial/\partial\hat{\mathbf{x}}$. Note that the velocity $\mathbf{u}$ is the same quality and represents the velocity of the particle. Thus, we have

$$\frac{\partial \mathbf{u}(\underline{\mathbf{X}},t)}{\partial t}\bigg|_{\underline{\mathbf{X}}} = \frac{\partial \mathbf{u}(\mathbf{x},t)}{\partial t}\bigg|_{\hat{\mathbf{x}}} + (\mathbf{u}(\mathbf{x},t) - \mathbf{v}(\mathbf{x},t))\cdot\nabla\mathbf{u}(\mathbf{x},t) = \frac{D\mathbf{u}(\mathbf{x},t)}{Dt} \qquad (7.82)$$

In general, for a physical property g, we have the following relation:

$$\frac{Dg}{Dt} = \frac{\partial g(\underline{\mathbf{X}},t)}{\partial t}\bigg|_{\underline{\mathbf{X}}} \qquad \text{(Lagrangian)} \qquad (7.83a)$$

$$= \frac{\partial g(\mathbf{x},t)}{\partial t}\bigg|_{\mathbf{x}} + \mathbf{u}(\mathbf{x},t)\cdot\nabla g(\mathbf{x},t) \qquad \text{(Eulerian)} \qquad (7.83b)$$

$$= \frac{\partial g(\mathbf{x},t)}{\partial t}\bigg|_{\hat{\mathbf{x}}} + \mathbf{w}(\mathbf{x},t)\cdot\hat{\nabla}g(\mathbf{x},t) \qquad \text{(referential)} \qquad (7.83c)$$

$$= \frac{\partial g(\mathbf{x},t)}{\partial t}\bigg|_{\hat{\mathbf{x}}} + (\mathbf{u}(\mathbf{x},t) - \mathbf{v}(\mathbf{x},t))\cdot\nabla g(\mathbf{x},t) \qquad \text{(mixed)} \qquad (7.83d)$$

Clearly, $\mathbf{v} = 0$ corresponds to the Eulerian description, $\mathbf{v} = 0$ to the Lagrangian, and $\mathbf{v} = \mathbf{u} =$ to the referential. Any mixed description can also be written. These relations will be useful for deriving conservation laws in the ALE description.

7.4.2 Conservation of Mass

We are familiar with the equation of mass conservation in fluids, which is expressed in spatial coordinates as

$$0 = \frac{\partial \rho(\mathbf{x},t)}{\partial t} + \nabla \cdot \left(\mathbf{u}(\mathbf{x},t)\rho(\mathbf{x},t) \right)$$

$$= \frac{\partial \rho(\mathbf{x},t)}{\partial t} + \rho(\mathbf{x},t)\nabla \cdot \mathbf{u}(\mathbf{x},t) + \mathbf{u}(\mathbf{x},t) \cdot \nabla \rho(\mathbf{x},t)$$

$$= \frac{D\rho(\mathbf{x},t)}{Dt} + \rho(\mathbf{x},t)\nabla \cdot \mathbf{u}(\mathbf{x},t) \tag{7.84}$$

For incompressible fluids in particular, we have

$$\nabla \cdot \mathbf{u}(\mathbf{x},t) = 0 \tag{7.85}$$

where ∇ is the vector differential operator with respect to $\mathbf{x}$. Let us now consider mass conservation in the Lagrangian description. We have then for two different configurations, Ω_0 at $t = 0$ and Ω at $t = t$,

$$dm = \rho \, d\Omega(\mathbf{x}) = \rho_o \, d\Omega_0(\mathbf{X}) \tag{7.86}$$

Since $d\Omega = J d\Omega_0$, we have $\rho J = \rho_0$ and J is the determinant of the Jacobian, $J = |\partial \mathbf{x}/\partial \mathbf{X}|$. We can choose an initial configuration in which $\rho_0 =$ constant. Then we have the following relation:

$$\left. \frac{\partial (J\rho)}{\partial t} \right|_{\underline{\mathbf{X}}} = \frac{d}{dt}(J\rho) = 0 \tag{7.87}$$

where d/dt refers to the time derivative in the Lagrangian frame with $\underline{\mathbf{X}}$ fixed. Now carrying out the operations, we have

$$\frac{d}{dt}(J\rho) = \rho \frac{dJ}{dt} + J \frac{d\rho}{dt} = \rho J \nabla \cdot \mathbf{u} + J \frac{\partial \rho}{\partial t} + J\mathbf{u} \cdot \nabla \rho = 0 \tag{7.88}$$

or

$$\rho \nabla \cdot \mathbf{u} + \frac{\partial \rho}{\partial t} + \mathbf{u} \cdot \nabla \rho = \frac{D\rho}{Dt} + \rho \nabla \cdot \mathbf{u} = \frac{\partial \rho}{\partial t} + \nabla \cdot (\rho \mathbf{u}) = 0 \tag{7.89}$$

where ∇ is expressed in terms of spatial coordinates $\mathbf{x}$. This is the same as the spatial description.

The change of volume is then given by the following expression:

$$\frac{d}{dt}(d\Omega_0) = \frac{d}{dt}(Jd\Omega) = \dot{J}\,d\Omega = \frac{\dot{J}}{J}d\Omega_0 \tag{7.90}$$

with the time derivative of the Jacobian calculated by

$$\dot{J} = \frac{\Delta\det(\mathbf{F})}{\Delta t} = \det(\mathbf{F})\frac{\mathrm{tr}(\mathbf{F}^{-1}\Delta\mathbf{F})}{\Delta t} = \det(\mathbf{F})\mathrm{tr}(\mathbf{F}^{-1}\dot{\mathbf{F}})$$

$$= J\,\mathrm{tr}(\mathbf{F}^{-1}\mathbf{F}) = J\,\mathrm{tr}(\mathbf{I}) = J\,\mathrm{tr}(\mathbf{D}) = J\nabla\cdot\mathbf{u}$$

Note that we have used the following relations: $\mathrm{tr}(\mathbf{F}^{-1}\mathbf{F}) = \mathbf{F}^{-1}:\mathbf{F}=\mathbf{I}\mathbf{F}:\ \mathbf{F}^{-1} = \mathrm{tr}(\mathbf{I})$, $\mathbf{F} = \partial\mathbf{x}/\partial\mathbf{X}$, $\mathbf{D} = 0.5(\nabla\mathbf{u}+(\nabla\mathbf{u})^T)$ and $J = \det(\mathbf{F})$, where $\mathbf{I}=\nabla\mathbf{u}$ is the velocity gradient tensor.

The equation for the conservation of mass in the ALE description is obtained by substituting the relation in Equation 7.83d into Equation 7.84,

$$\left.\frac{\partial\rho(\mathbf{x},t)}{\partial t}\right|_{\hat{\mathbf{x}}} + (\mathbf{u}(\mathbf{x},t) - \mathbf{v}(\mathbf{x},t))\cdot\nabla\rho(\mathbf{x},t) + \rho\nabla\cdot\mathbf{u} = 0 \tag{7.91}$$

7.4.3 Conservation of Momentum

By definition, Newton's law applies only to the inertial frame. For the case under consideration, it means that it applies only to the observed acceleration by the person standing still at O with his eyes fixed on the particle. More specifically, Newton's law is written as

$$\rho(\underline{\mathbf{X}},t)\frac{d\mathbf{u}(\underline{\mathbf{X}},t)}{dt} = f^B(\mathbf{X},t) + f^S(\mathbf{X},t) \tag{7.92}$$

where superscript B and S represent the body and surface forces acting on the particle identified by $\underline{\mathbf{X}}$. This is the Lagrangian description, and the law of conservation of momentum or Newton's second law. It is noted here that the independent variables here are $\underline{\mathbf{X}}$ and t.

If we now write the above conservation law in terms of $\mathbf{x}$ and t, then we have

$$\rho(\mathbf{x},t)\left(\left.\frac{\partial\mathbf{u}(\mathbf{x},t)}{\partial t}\right|_{\mathbf{x}} + \mathbf{u}(\mathbf{x},t)\cdot\nabla\mathbf{u}(\mathbf{x},t)\right) = f^B(\mathbf{x},t) + f^S(\mathbf{x},t) \tag{7.93}$$

which is the Eulerian description. Note that subscript $\mathbf{x}$ here refers to the Eulerian coordinates.

If we express the conservation of momentum in the mixed frame, we then have

$$\rho(\mathbf{x},t)\frac{\partial \mathbf{u}(\mathbf{x},t)}{\partial t}\bigg|_{\dot{\mathbf{x}}} + \rho(\mathbf{x},t)(\mathbf{u}(\mathbf{x},t)-\mathbf{v}(\mathbf{x},t))\cdot\nabla\mathbf{u}(\mathbf{x},t)=f^B(\mathbf{x},t)+f^S(\mathbf{x},t)$$

(7.94)

which is sometimes called the mixed description.

If the conservation of momentum is written in the referential frame, then we have

$$J(\mathbf{x},\hat{\mathbf{x}})\rho(\hat{\mathbf{x}},t)\left(\frac{\partial \mathbf{u}(\hat{\mathbf{x}},t)}{\partial t}\bigg|_{\hat{\mathbf{x}}} + \mathbf{F}^{-1}\cdot(\mathbf{u}(\hat{\mathbf{x}},t)-\mathbf{v}(\hat{\mathbf{x}},t))\cdot\hat{\nabla}\mathbf{u}(\hat{\mathbf{x}},t)\right)$$

$$= f^B(\hat{\mathbf{x}},t)+f^S(\hat{\mathbf{x}},t)$$

(7.95)

where we have the Jacobian J and the deformation gradient tensor $\mathbf{F}$ defined as follows:

$$F_{ij}=\partial x_i/\partial\hat{x}_j \text{ or } J(\mathbf{x},\hat{\mathbf{x}})=|F_{ij}|=|\partial x_i/\partial\hat{x}_j|$$

(7.96)

and also use has been made of the following relation:

$$\rho(\hat{\mathbf{x}},t)d\Omega(\hat{\mathbf{x}})=\rho(\mathbf{x},t)d\Omega(\mathbf{x}) \text{ or}$$

$$\rho(\mathbf{x},t)=\rho(\hat{\mathbf{x}},t)\frac{d\Omega(\hat{\mathbf{x}})}{d\Omega(\mathbf{x})}=\rho(\hat{\mathbf{x}},t)J(\mathbf{x},\hat{\mathbf{x}})$$

(7.97)

Equation 7.95 is called the referential description.

7.4.4 Conservation of Energy

The same approach presented above can be employed to obtain the equation for the conservation of energy. Let us consider the Eulerian description of energy conservation,

$$\rho\frac{\partial E}{\partial t}+\rho\mathbf{u}\cdot\nabla E=\nabla\cdot(\mathbf{u}\cdot\boldsymbol{\sigma})+\mathbf{u}\cdot\mathbf{f}+\nabla\cdot(\kappa\nabla)$$

(7.98)

Once again, with the relations in Equation 7.83 substituted into the above equation, we have the ALE description for the conservation of energy,

$$\rho\frac{\partial E}{\partial t}\bigg|_{\hat{\mathbf{x}}} + \rho(\mathbf{u}-\mathbf{v})\cdot\nabla E=\nabla\cdot(\mathbf{u}\cdot\boldsymbol{\sigma})+\mathbf{u}\cdot\mathbf{f}+\nabla\cdot(\kappa\nabla T)$$

(7.99)

7.4.5 Summary of ALE Equations

Donea *et al.* [20] showed that all of the above descriptions (Lagrangian, Eulerian, mixed and referential) can be generalized as follows:

$$\frac{d}{dt}\left(\rho\tilde{J}\right) = \tilde{J}\frac{\partial}{\partial x_j}\left(\rho(w_j - u_j)\right) \tag{7.100a}$$

$$\frac{d}{dt}\left(\rho u_i \tilde{J}\right) = \tilde{J}\frac{\partial}{\partial x_j}\left(\rho v_i (w_j - u_j)\right) + \tilde{J}\left(\rho b_i + \frac{\partial \sigma_{ij}}{\partial x_j}\right) \tag{7.100b}$$

$$\frac{d}{dt}\left(\rho e \tilde{J}\right) = \tilde{J}\frac{\partial}{\partial x_j}\left(\rho e(w_j - u_j)\right) + \tilde{J}\left(\rho u_i b_i + \frac{\partial (\sigma_{ij} u_i)}{\partial x_j}\right) \tag{7.100c}$$

where $e = 0.5 u_i^2 + E$ and E is the internal energy. Also, the right hand side is the time derivative in the Lagrangian description, and the right hand side refers to the derivative in the Eulerian description (see Equation 7.83b). The Jacobian is given by

$$\tilde{J} = \left|\frac{\partial \hat{\mathbf{x}}}{\partial \underline{\mathbf{X}}}\right| \tag{7.100d}$$

7.4.6 Constitutive Relations

Constitutive relations must be invariant under changes of reference frames. This means that the quantity remains the same under arbitrary rigid body rotation, which is referred to as material-frame-independence. This constraint has important implications in solid mechanics as it involves the specification of the material's behavior as a function of stress rate. Often the Jaumann rates, or similar types of stress rates, need to be used. For fluids, the constitutive relations involve only stresses, which are frame-indifferent. Thus the same constitutive relations for the Eulerian frames should be applicable to the ALE description [21].

7.4.7 ALE Description of Compressible Flows

With the above results, the governing equations for compressible flows in an ALE framework are written as follows [22, 23]:

$$\partial_t U + \nabla_{\underline{\mathbf{X}}} \cdot \mathbf{F}_c(U) - \nabla_{\underline{\mathbf{X}}} \cdot \mathbf{F}_v(U, \nabla\mathbf{u}) = 0 \tag{7.101a}$$

where the vector of variables, and convective and diffusive fluxes, are given by the following expressions:

$$U = J \begin{Bmatrix} \rho \\ \rho u_1 \\ \rho u_2 \\ \rho u_3 \\ \rho e \end{Bmatrix}; \quad F_{c,i}(U) = \eta_{ji} \begin{Bmatrix} \rho(u_j - V_j^g) \\ \rho u_1(u_j - V_j^g) + p\delta_{j1} \\ \rho u_2(u_j - V_j^g) + p\delta_{j2} \\ \rho u_3(u_j - V_j^g) + p\delta_{j3} \\ \rho e(u_j - V_j^g) + pu_j \end{Bmatrix};$$

$$F_{v,i}(U, \nabla \mathbf{u}) = \eta_{ji} \begin{Bmatrix} 0 \\ \sigma_{j1} \\ \sigma_{j2} \\ \sigma_{j3} \\ u_1\sigma_{i1} + u_2\sigma_{i2} + u_3\sigma_{i3} - q_i \end{Bmatrix} \tag{7.101b}$$

$$J_{ij} = \partial x_i / \partial X_j; \quad \eta_{ij} = J J_{ij}^{-1}; \quad \frac{\partial \eta_{ij}}{\partial X_i} = 0; \quad J = \det \mathbf{J} \tag{7.101c}$$

where $\mathbf{u} = (u_1, u_2, u_3) = (u, v, w)$ and $\mathbf{V}^g = (V_1^g, V_2^g, V_3^g)$ is the velocity of the grid movement. It is seen that the grid movement affects the convection terms only, and the diffusion fluxes are the same as before, thanks to the constitutive relation.

We further note that the above formulation is general. Compared with Equation 7.83d, we have $v = V^g$. When the grid velocity $\mathbf{V}^g$ is set to zero ($\mathbf{V}^g = \mathbf{0}$) the Eulerian description of the Navier–Stokes equations are recovered. Furthermore, if grid velocity is set to the local velocity $\mathbf{u}$, we have Lagrangian fluid equations where material interfaces are exactly resolved.

7.4.8 Discontinuous Finite Element Formulation

The discontinuous finite element formulation is also very similar to that already given in Section 7.4, aside from the factor η_{ij}. The computational procedures are thus very similar. An important computational issue of solving the Navier–Stokes equations in an ALE frame is the determination of the grid velocity. A common approach to update the grid velocity is to solve the following diffusion equation:

$$\nabla \cdot (\alpha(x)\nabla \mathbf{V}^g) = 0 \tag{7.102}$$

with $\alpha(\mathbf{x})$ being a parameter, and Dirichlet conditions on both the moving wall boundary and on the outer boundaries of the computational domain.

The above equation may be solved like the classical elliptic equations. An alternative approach is proposed by Lomtev [22], who used concepts from graph theory to update the grid movement; and thus, no matrix inversion is needed, with the additional advantage of minimizing the grid distortion.

The use of the ALE formulations for compressible flow calculations, with and
vithout shock wave formations, and some of the computational details are given in
3].

:ises

`onsider the nonlinear pure convection (inviscid Burgers') equation,

$$\frac{\partial u}{\partial t} + u \frac{\partial u}{\partial x} = 0, \quad \in \Omega, \ t > 0$$

$$\iota,0) = -x^2, \quad x \in \Omega, \quad t > 0$$

\araracteristics, sketch the characteristics diagrams and find the

\ simplest detonation model (or Semenov's model), which is
'mplifying the compressible flow equations and has the

$$\iota, C, T), \quad t > 0$$

ıs.
Let u and ρ
linearized,

$$\iota r(\rho_0, C, T), \ t > 0$$

$$t > 0$$

bove equations.
yields

table reactant, C is the concentration,
rate, Q is the heat generation of the
erature. For this particular model,

'he Arrhenius equation,

ve 1-D Euler equations
o use different orders of

'ld one equation for the

$$\frac{\partial T}{\partial t} = K(T_b - T)\exp\left(\frac{E}{RT}\right); \ T(0) = T_0$$

where T_b is a constant given by

$$T_b = \frac{Q}{C_p \rho_0 (C_p - R)} C(0) + T(0)$$

Solve the equation to obtain T(t) and plot and discuss the evolution of th
temperature for the detonation process.

3. Consider the 1-D Euler equations for isentropic compressible flows. Sh
    that the flows are governed by the following equations:

$$\frac{\partial U}{\partial t} + \frac{\partial F(U)}{\partial x} = 0$$

$$U = \begin{pmatrix} \rho \\ \rho u \end{pmatrix}, \ F(U) = \begin{pmatrix} \rho u^2 \\ \rho u^2 + p \end{pmatrix}$$

$$p = p(\rho) = C\rho^\gamma$$

Study the characteristics of the above partial differential equatio

4. Consider the small perturbations u and ρ' to a motionless gas. I
    $= \rho' + \rho_0$, where ρ_0 is a constant density value. Show that whe
    the isothermal Euler equations reduce to

$$\frac{\partial \rho}{\partial t} + \rho_0 \frac{\partial u}{\partial x} = 0, \ \frac{\partial u}{\partial t} + \frac{a^2}{\rho_0}\frac{\partial \rho}{\partial x} = 0$$

Analyze the characteristics of the Jacobian matrix of the a
Show also that the combination of the above two equations

$$\frac{\partial^2 \rho}{\partial t^2} + a^2 \frac{\partial^2 \rho}{\partial x^2} = 0$$

where a is the speed of sound,

$$a^2 = \frac{\partial p(\rho_0)}{\partial \rho}$$

5. Develop a discontinuous finite element code to sol
    for compressible flows. The code should allow us

approximation in space and in time and also different types of flux approximations.

6. Consider the 1-D Euler equations for compressible flows through a diverging channel,

$$\frac{\partial U}{\partial t} + \frac{\partial F}{\partial x} - H = 0$$

where

$$U = \begin{bmatrix} \rho \\ \rho u \\ \rho E \end{bmatrix}; \quad F = A \begin{bmatrix} \rho u \\ \rho u^2 + p \\ (\rho E + p)u \end{bmatrix}; \quad H = \frac{dA}{dx} \begin{bmatrix} 0 \\ p \\ 0 \end{bmatrix}$$

The channel is 3.3 m long, with its cross section given by

$$A(x) = 1{,}400 + 0.347\tanh(0.8x - 4) \text{ m}^2$$

Inlet: $M = 1.5$, $p = 47880$ Pa, $\rho = 1.22145$ kg/m^3
$\rho u = 429.2101$ kg/m^{2}s, $\rho E = 19909.39$ J/m^3

Exit: either supersonic or subsonic.

Solve with the discontinuous finite element code developed in Problem 5 using (1) Lax–Friedrichs explicit, (2) Godunov explicit, and (3) *TVD*. Employ constant, linear, quadratic and cubic approximations. Compare and discuss your results.

7. Solve the Euler equations for the two-dimensional domain shown below.

$$\frac{\partial U}{\partial t} + \frac{\partial F_i}{\partial x_i} = 0, \quad U = \begin{bmatrix} \rho \\ \rho V_i \\ \rho E \end{bmatrix}, \quad F_i = \begin{bmatrix} \rho V_i \\ \rho_i V_j + p\delta_{ij} \\ \rho E V_i + p V_i \end{bmatrix}$$

Inlet: $M=2$, $\gamma = 1.4$, $T = 288.33$ K

$$a = \sqrt{\gamma RT} = 340.4616 \text{ m/s}, \quad \rho = 1.225571 \text{ kg/m}^3$$

$$u = 680.9232 \text{ m/s}, \quad v = 0, \quad p = 10331.22 \text{ kg/m}^2$$

$$\rho E = \frac{p}{\gamma - 1} + \frac{1}{2}\rho\left(u^2 + v^2\right) = 54800.37 \text{ kg/m}^2$$

Initial conditions: Use inlet conditions as initial conditions for all nodes. Boundary conditions for the problem: Supersonic inlet. Supersonic exit. Slip wall conditions.

Develop a discontinuous finite element code to solve the above equations. Use constant, linear and quadratic approximations on a triangular mesh.

8.   For the 2-D problem shown in Problem 4, the Navier–Stokes equation has the form given below. Develop programs to solve the Navier–Stokes system of equations for Problem 4 using the discontinuous finite element method. Repeat these programs for a geometry in 3D with the depth of x_3 direction given as 1 in the figure.

$$\frac{\partial \mathbf{U}}{\partial t} + \frac{\partial \mathbf{F}_i}{\partial x_i} + \frac{\partial \mathbf{G}_i}{\partial x_i} = 0$$

$$\mathbf{U} = \begin{bmatrix} \rho \\ \rho V_i \\ \rho E \end{bmatrix}; \quad \mathbf{F}_i = \begin{bmatrix} \rho V_i \\ \rho_i V_j + p\delta_{ij} \\ \rho E V_i + p V_i \end{bmatrix}; \quad \mathbf{G}_i = \begin{bmatrix} 0 \\ -\tau_{ij} \\ -\tau_{ij}V_j + q_i \end{bmatrix}$$

9.   Formulate the shock tube problem in an ALE configuration and develop a discontinuous finite element code to solve the Riemann problem for the shock tube using the ALE Euler equations.

References

[1]   Anderson JD. Modern Compressible Flow with Historical Perspective. New York: McGraw-Hill, 1982.

[2]   Shivamoggi BK. Theoretical Fluid Dynamics, 2nd Ed. New York: John Wiley, 1998.

[3]   Toro EF. Riemann Solvers and Numerical Methods for Fluid Dynamics: a Practical Introduction. 2nd Ed. New York: Springer, 1997.

[4]   Gottlieb JJ, Groth CPT. Assessment of Riemann Solvers for Unsteady One-dimensional Inviscid Flows of Perfect Gases. J. Comput. Phys. 1988; 78:437–458.

[5]   Roe PL. Approximate Riemann Solvers, Parameter Vectors, and Difference Schemes. J. Comput. Phys. 1981; 43:357–372.

[6]   Einfeldt B, Muntz CD, Roe PL, Sjogreen B. On Godunov-type Methods near Low Densities. J. Comput. Phys. 1991; 92:273–295.

[7]   Godunov SK. Finite Difference Method for Numerical Computation of Discontinous Solution of the Equations of Fluid Dynamics. Matematicheskii Sbornik 1959; 47: 271–295, translated from Russian by I. Bohachevsky.

[8]   Tannehill JC, Anderson DA, Pletcher RH. Computational Fluid Mechanics and Heat Transfer. New York: Taylor & Francis, 1997.

[9]   LeVeque RJ. Numerical Methods for Conservation Laws. Basel: Birkhauser Verlag, 1990.

[10]   Cockburn B. Discontinuous Galerkin Methods. School of Mathematics, University of Minnesota, 2003: 1–25.

[11]   Cockburn B. Discontinuous Galerkin Methods for Convection Dominated Problems. NATO Lecture Notes. School of Mathematics, University of Minnesota, 2001.

[12]   Cockburn B, Hou S, and Shu CW. TVB Runge–Kutta Local Projection Discontinuous Galerkin Finite Element Method for Conservation Laws IV: the Multidimensional Case. Math. Comput. 1990; 54: pp.545–581.

[13]   Bassi F, Rebay S. High-Order Accurate Discontinuous Finite Element Method for the Numerical Solution of the Compressible Navier–Stokes Equations. J. Comput. Phys. 1997; 138: 251–285.

[14]   Bassi F, Rebay S. A High-Order Accurate Discontinuous Finite Element Solution of the 2D Euler Equations. J. Comput. Phys. 1997; 131: 267–279.

[15]   Baumann CE and Oden, JT. A Discontinuous Hp Finite Element Method for the Euler and Navier–Stokes Equations. Int. J. Numer. Meth. Fluids. 1999; 31(1): 79–95.

[16]   Dolejsi V. On the Discontinuous Galerkin Method for the Numerical Solution of the Navier–Stokes Equations. Int. J. Numer. Meth. Fluids 2004; 45(6): 1083–1106.

[17]   Hughes TJR, Liu WK, Zimmerman TK, Lagrangian–Eulerian Finite Element Formulation for Incompressible Viscous Flows. Comput. Meth. Appl. Mech. Eng. 1981; **29**: 329–349.

[18]   Nomura T, Hughes TJR. An Arbitrary Lagrangian–Eulerian Finite Element Method for Interaction of Fluid and a Rigid Body. Comput. Meth. Appl. Mech. Eng. 1992; 95: 115–138.

[19]   Hirt CW, Amsden AA, Cook HK. An Arbitrary Lagrangian–Eulerian Computing Method for All Flow Speeds. J. Comput. Phys. 1974; 14: 227–253.

[20]   Donea J, Giuliani S, Halleux JP. An Arbitrary Lagrangian–Eulerian Finite Element Method for Transient Dynamic Fluid–Structure Interactions. Comput. Meth. Appl. Mech. Eng. 1982; 33: 689–723.

[21]   Malvern LE. Introduction to the Mechanics of a Continuous Medium. Englewood Cliffs: Prentice Hall, 1969.

[22]   Lomtev I, Kirby RM, Karniadakis GE. A Discontinuous Galerkin ALE Method for Compressible Viscous Flows in Moving Domains. J. Comput. Phys. 1999; 155: 128–159.

[23]   Kershaw DS, Prasad MK, Shaw MJ, Milovich JL. 3D Unstructured Mesh ALE Hydrodynamics with the Upwind Discontinuous Finite Element Method. Comput. Meth. Appl. Mech. Eng. 1998; 158: 81–166.

8

External Radiative Heat Transfer

Thermal transfer by radiation is an important heat transfer mechanism in thermal systems. Unlike heat conduction and convection, thermal radiation does not require direct contact between the heat transfer parties. Rather, it is a result of the electromagnetic energy radiated from a body at a temperature above absolute zero. In general, thermal radiation problems are classified into two categories. The first category is concerned with the thermal radiation exchange between surfaces, which involves no intervening media. The second category deals with the thermal radiation transfer through absorbing and scattering media. In the heat transfer literature, the former is often referred to as *external radiation*, while the latter as *internal radiation*. The mathematical formulations are different for these two categories, and therefore, naturally lead to the use of different computational approaches.

Thermal radiation problems can be difficult to solve, and thus there are only limited cases where simple solutions are possible. For most practical applications, in which heat flows and temperatures need to be found, numerical solutions are often needed, and may require considerable computational effort. Many methods have been reported in the literature and an extensive review of the methods available today is documented in Modest [1] and Siegal and Howell [2]. Here we focus on the numerical algorithms that have been developed on the basis of the discontinuous Galerkin finite element formulation for the solution of thermal radiation problems. Since the external and internal thermal radiation problems are described using different mathematical equations, it is convenient to discuss the subject in two consecutive chapters.

In this chapter, we consider the solution of the external radiation exchange problems by the discontinuous Galerkin finite element method. We start with the basic concept of external radiation between surfaces, and develop an integral representation of the surface energy exchange in an enclosure, on the basis of thermal energy balance. The use of the discontinuous Galerkin method for the solution of the integral equation is then discussed for 2-D, 2-D axisymmetric and 3-D geometries.

Perhaps the most important part of algorithm development for the numerical solution of external radiative heat transfer is an accurate estimation of kernel functions, which would require the detection of a third party blockage in complex geometric arrangements. The calculations involving blockage detection can be difficult, cumbersome, and time consuming. Various ideas have been developed in the past [3–7] but a systematic description of the algorithms for geometries of all dimensions appears to be lacking. Undoubtedly, the algorithms should have some common features. However, the enclosures of different dimensions can have very different geometric complexities that warrant different numerical treatments. A systematic description of these shadowing algorithms is provided in this chapter. This allows for a better appreciation of these common and different features, and thus helps to develop more efficient algorithms. The detection algorithms described in this chapter combine the best ideas published in the literature. These techniques make use of the organized data structure and the advanced computer graphics schemes used for hidden line removal [3–11]. These algorithms could be considered to be the most efficient schemes to date for the numerical solution of external radiation problems.

The numerical solution of mixed heat transfer problems involving conduction, convection and radiation is discussed. A solution strategy is presented, which is based on a combination of the discontinuous Galerkin method and the conventional finite element method. Of course, the solution of these mixed mode heat transfer problems can be obtained using the discontinuous Galerkin method alone. However, for engineering applications, and for multiphysics model development, a combined approach can be more beneficial if it enables the use of the best properties of each of the methods, or if it is built within an already existing software framework.

For illustrative purposes, some simple examples are presented and discussed in detail. The chapter ends with several more examples of varying degrees of difficulty, covering 2-D, 2-D axisymmetric and 3-D geometries with and without internal geometric blockages.

8.1 Integral Equation for Surface Radiation Exchanges

If the media are not radiatively participating, the thermal radiation energy will exchange between surfaces, which in many engineering applications define an enclosure, as shown in Figure 8.1. The surface radiative energy transfer depends on the local surface temperature of the enclosure and the properties of the surfaces, but not upon the intervening media, which neither absorb, emit, nor scatter. This condition is often satisfied by a vacuum or a transparent medium. This section discusses the basic concept for surface radiation transfer and the equation governing the heat flux distribution along the surfaces.

8.1.1 Governing Equation

Let us now consider the radiation exchange among surfaces that form an enclosure, as illustrated in Figure 8.1. It is assumed that there are no radiation-absorbing or scattering media in the enclosure. Surface element J emits the thermal energy to the other surfaces of the enclosure, while receiving the energy from these surfaces. The heat flux $q(\mathbf{r}_j)$ is supplied to the surface element J to sustain the radiation heat transfer, and is determined by the heat balance on point j, involving incoming and outgoing radiation energy fluxes. Referring to Figure 8.2, the general expression for the heat flux exchange between two surfaces I and J is given by the following integral [1,2]:

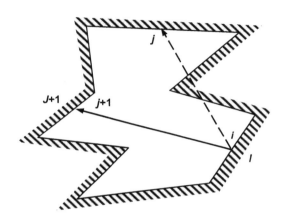

Figure 8.1. Schematic of thermal radiation exchanges among surfaces in an enclosure

$$q(\mathbf{r}_i) = \int_{\lambda=0}^{\infty} \int_{\varphi_i=0}^{\varphi_i=2\pi} \int_{\theta_i=0}^{\theta_i=\pi/2} \varepsilon_{\lambda,i}(\lambda,\varphi_i,\theta_i,\mathbf{r}_i)$$

$$I_{b\lambda,i}(\lambda,\mathbf{r}_i)\cos\theta_i \, \sin\theta_i d\theta_i d\varphi_i d\lambda$$

$$-\sum_{j=1}^{N} \int_{\lambda=0}^{\infty} \int_{A_j} \alpha_{\lambda,i}(\lambda,\varphi_i,\theta_i,\mathbf{r}_i)I_{\lambda,j}(\lambda,\varphi_j,\theta_j,\mathbf{r}_i)\frac{\cos\theta_i\cos\theta_j}{\left|\mathbf{r}_i-\mathbf{r}_j\right|^2}dA_j d\lambda$$

$$(8.1)$$

where the radiation intensity emitted from point j is defined by

$$I_{\lambda,j}(\lambda,\varphi_{r,j},\theta_{r,j},\mathbf{r}_i) = \varepsilon_{\lambda,j}(\lambda,\varphi_{r,j},\theta_{r,j},\mathbf{r}_i)I_{b\lambda,j}(\lambda,\mathbf{r}_i)$$

$$-\sum_{k=1}^{N} \int_{A_k} \rho_{\lambda,j}(\lambda,\varphi_{r,j},\theta_{r,j},\varphi_j,\theta_j,\mathbf{r}_i)I_{\lambda,k}(\lambda,\varphi_k,\theta_k,\mathbf{r}_i)\frac{\cos\theta_j\cos\theta_k}{\left|\mathbf{r}_k-\mathbf{r}_j\right|^2}dA_k$$

$$(8.2)$$

In the above equations, λ is the frequency of the thermal rays, or essentially electromagnetic waves, I is the radiant intensity, ε is the emissivity, ρ is the reflectivity, θ is the angle between the normal of the surface and radiation exchange direction, and φ is the azimuthal angle. Note that in this chapter the radiant intensity I always has subscripts. This should not be confused with the use of I (which does not have a subscript) for surface or surface elements. Also, subscript b refers to the blackbody radiation, and thus by its definition the blackbody radiation intensity is directionally independent. Subscript r refers to the reflected radiation. Also, $\theta_{r,j}$ is the angle between the normal of the differential element centered at point j and the direction of the reflected radiation at point j and $\varphi_{r,j}$ is the azimuthal angle similarly defined. The geometric relations of these quantities are illustrated in Figure 8.2.

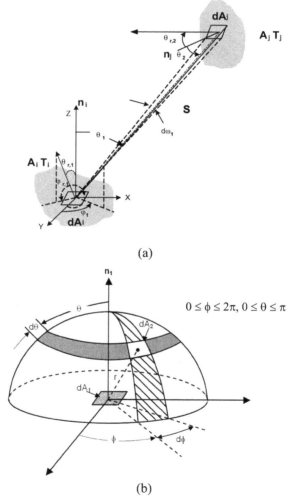

(a)

(b)

Figure 8.2. Schematic representation of the surface energy interchange between two surfaces I and J. (a) Energy exchange between surfaces A_i and A_j, where $\mathbf{S} = |\mathbf{r}_i - \mathbf{r}_j|$ and (b) spherical coordinates showing the angular relations

While the above formulations are general, for most engineering applications, the surfaces of an enclosure can be well approximated as gray, diffuse surfaces [2]. Within the framework of these approximations, the integration over the wavelength λ can be carried out analytically. After these analytical operations and rearrangement, the following boundary integral formulation is obtained for the radiative heat flux $q(\mathbf{r})$ at the surface of an enclosure:

$$q(\mathbf{r}) + \varepsilon(\mathbf{r})E_b(\mathbf{r}) = \varepsilon(\mathbf{r}) \oint K(\mathbf{r},\mathbf{r}') \left[E_b(\mathbf{r}') + \frac{1-\varepsilon(\mathbf{r}')}{\varepsilon(\mathbf{r}')}q(\mathbf{r}') \right] d\Gamma(\mathbf{r}') \quad (8.3)$$

where E_b is the blackbody emissive power and is calculated by the integration of the blackbody radiation intensity over the entire spectrum of wavelengths,

$$E_b(\mathbf{r}) = \int_{\lambda=0}^{\infty} \int_{\varphi=0}^{\varphi=2\pi} \int_{\theta=0}^{\theta=\pi/2} I_{b\lambda}(\lambda,\mathbf{r})\cos\theta\sin\theta\, d\theta\, d\varphi\, d\lambda = \sigma_s T^4(\mathbf{r})$$

$$(8.4)$$

with σ_s being the Stefan–Boltzmann constant. Here, use has been made of the spectral distribution of Planck's blackbody radiation [1],

$$I_{b\lambda}(\lambda,T) = \frac{2\hbar c_0^2}{\lambda^5 \left[\exp(\hbar c_0/\lambda k_B T) - 1\right]} \quad (8.5)$$

where $\hbar$ is the Planck constant, c_0 is the speed of light, and k_B is the Boltzmann constant. Notice that the subscript i on $\mathbf{r}_i$ has been dropped out and $\mathbf{r}_j$ replaced by $\mathbf{r}'$ to simplify the notation. This will remain true hereafter unless indicated otherwise.

8.1.2 Kernel Functions

In Equation 8.3, $K(\mathbf{r},\mathbf{r}')$ is the kernel function for the integral, which for 3-D problems takes the following form [1]:

$$K(\mathbf{r},\mathbf{r}') = -\frac{\mathbf{n}\cdot(\mathbf{r}-\mathbf{r}')\mathbf{n}'\cdot(\mathbf{r}-\mathbf{r}')}{\pi|\mathbf{r}-\mathbf{r}'|^4}\chi = \frac{\cos\theta_r \cos\theta_{r'}}{\pi|\mathbf{r}-\mathbf{r}'|^2}\chi \quad (8.6)$$

where χ assumes a value of one when the surface element I sees the surface element J, as illustrated by the ray connecting i to j; otherwise it is zero if the ray is blocked. Also, θ_r is the angle at point $\mathbf{r}$, and $\theta_{r'}$ at point $\mathbf{r}'$. This is illustrated in Figure 8.2. Thus the parameter χ is a strong function of the geometric configuration, which makes the kernel function highly irregular for a geometrically complex enclosure.

Equation 8.6 is the kernel function for a general 3-D geometry. For 2-D and axisymmetric configurations, the kernel function can be analytically integrated along the z or θ direction. For 2-D geometry, the integration of the kernel function is straightforward,

$$K(\mathbf{r},\mathbf{r}') = -\int_{-\infty}^{\infty}\int_{-\infty}^{\infty} \chi \frac{\mathbf{n}\cdot(\mathbf{r}-\mathbf{r}')\mathbf{n}'\cdot(\mathbf{r}-\mathbf{r}')}{\pi\left|\mathbf{r}-\mathbf{r}'\right|^{4}} dz'dz$$

$$= \frac{\mathbf{n}\cdot(\mathbf{r}-\mathbf{r}')\mathbf{n}'\cdot(\mathbf{r}-\mathbf{r}')}{2\left|\mathbf{r}-\mathbf{r}'\right|^{3}}\chi \tag{8.7}$$

For an axisymmetric configuration, however, integration along the θ direction is much more involved. The derivation is given here for completeness. Referring to Figure 8.3, $\mathbf{n}$ denotes the unit normal of element I at the azimuth angle ϕ' being zero, and $\mathbf{n}'$ refers to the unit normal of element J with any azimuth angle ϕ'. The mathematical expressions for $\mathbf{n}$ and $\mathbf{n}'$ are as follows:

$$\mathbf{n} = (\cos\theta, 0, \sin\theta) \quad \text{and} \quad \mathbf{n}' = (\cos\theta'\cos\phi', \cos\theta'\sin\phi', \sin\theta') \tag{8.8}$$

Substituting these terms into Equation 8.6, the kernel function is rearranged in terms of the azimuth angle ϕ',

$$K(\phi') = -\frac{(c'+d'\cos\phi')(c''+d''\cos\phi')}{\pi(c+d\cos\phi')^{2}}\chi \tag{8.9}$$

where the coefficients are calculated by the following expressions:

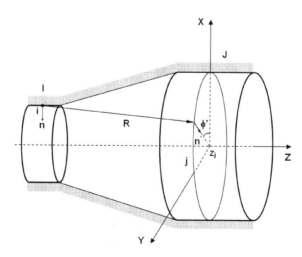

Figure 8.3. Schematic of thermal radiation exchanges between two ring surfaces I and J in a cylindrical enclosure

$$\begin{cases} c = r_i^2 + r_j^2 + z_j^2 \\ d = -2r_i r_j \end{cases} ; \quad \begin{cases} c' = z_j \sin\theta - r_i \cos\theta \\ d' = r_j \cos\theta \end{cases} ; \quad \begin{cases} c'' = z_j \sin\theta' + r_j \cos\theta' \\ d'' = -r_i \cos\theta' \end{cases}$$

Integration with respect to ϕ' and making use of the following relation:

$$\frac{d}{d\phi}\left\{ \tan^{-1}\left(\sqrt{\frac{c-d}{c+d}}\tan\frac{\phi}{2}\right)\right\} = \frac{\sqrt{c^2-d^2}}{2(c+d\cos\phi)} \tag{8.10}$$

one has the final expression for the kernel function,

$$\overline{K}(\phi) = 2\int_0^\phi K(\phi')d\phi' = -\frac{2}{\pi}\left[A\phi + B\tan^{-1}\left(\sqrt{\frac{c-d}{c+d}}\tan\frac{\phi}{2}\right) + C\frac{\sin\phi}{c+d\cos\phi}\right] \tag{8.11}$$

where the coefficients are given by

$$A = \frac{d'd''}{d^2} ; B = 2\left[\frac{(c^2-d^2)(d'f+ed'') + cdef}{d(c^2-d^2)\sqrt{c^2-d^2}}\right]$$

$$C = \frac{def}{d^2-c^2} ; e = \frac{dc'-cd'}{d} ; f = \frac{dc''-cd''}{d}$$

Further integration of Equation 8.11 requires knowledge of the geometric configuration, which is discussed in Section 8.3.2.

8.2 Discontinuous Galerkin Finite Element Formulation

We now derive the discontinuous Galerkin finite element formulation for the surface radiative energy exchanges in an enclosure. By the Galerkin method, the global residuals are forced to zero by use of the orthogonality condition, thereby minimizing the error that could arise from the integral stated in Equation 8.3. More importantly, the double integral enables the shadowing elements to be detected that could be missed by the direct application of the traditional boundary element method [3,11]. Thus, Equation 8.3 is integrated once again over the entire surface, with shape functions ψ_i used as the weighting functions,

$$\oint\left\{ q(\mathbf{r}) + \varepsilon(\mathbf{r})E_b(\mathbf{r})\right.$$

$$-\varepsilon(\mathbf{r}) \oint K(\mathbf{r},\mathbf{r}') \left[E_b(\mathbf{r}') + \frac{1-\varepsilon(\mathbf{r}')}{\varepsilon(\mathbf{r}')} q(\mathbf{r}') \right] d\Gamma(\mathbf{r}') \psi_i \, d\Gamma(\mathbf{r}) = 0 \quad (8.12)$$

Equation 8.12 involves integration over the surface only, and thus the surface element method can be applied naturally to discretize the domain and obtain the solution [12]. Both E_b and q can be interpolated over each of the boundary elements by use of the shape functions discussed in Chapter 3,

$$q(\mathbf{r}) = \sum_{l=1}^{N_e} \psi_l q_l \; ; \; E_b(\mathbf{r}) = \sum_{l=1}^{N_e} \psi_l E_{bl} \quad (8.13)$$

Now, with the boundary discretized and the shape functions chosen as described above, Equation 8.12 becomes

$$\sum_{k=1}^{N} \int_{BE_k} \left\{ \sum_{j=1}^{N_e} \psi_j \, q_j + \varepsilon(\mathbf{r}) E_b(\mathbf{r}) - \varepsilon(\mathbf{r}) \sum_{l=1}^{N} \right.$$

$$\times \int_{BE_l} \left[E_b(\mathbf{r}') + \frac{1-\varepsilon(\mathbf{r}')}{\varepsilon(\mathbf{r}')} \sum_{j=1}^{N_e} \psi_j q_j \right] d\Gamma(\mathbf{r}') \bigg\} K(\mathbf{r},\mathbf{r}') \psi_i \, d\Gamma(\mathbf{r}) = 0 \quad (8.14)$$

Following standard boundary element procedure, the above equation may be written in matrix notation,

$$\mathbf{A}\underline{q} = \mathbf{B}\underline{E}_b \quad (8.15)$$

where an underscore beneath the boldfaced letter denotes the vector, and the elements of matrices $\mathbf{A}$ and $\mathbf{B}$ are calculated by the following expressions:

$$A_{ij} = \int_{BE_k} \psi_i \, \psi_j \, d\Gamma(\mathbf{r})$$

$$- \int_{BE_k} \varepsilon(\mathbf{r}) \sum_{l=1}^{N} \int_{BE_l} \frac{1-\varepsilon(\mathbf{r}')}{\varepsilon(\mathbf{r}')} \psi_j \, K(\mathbf{r},\mathbf{r}') d\Gamma(\mathbf{r}') \psi_i \, d\Gamma(\mathbf{r}) \quad (8.16)$$

$$B_{ij} = - \int_{BE_k} \varepsilon(\mathbf{r}) \psi_i \psi_j \, d\Gamma(\mathbf{r})$$

$$+ \int_{BE_k} \varepsilon(\mathbf{r}) \sum_{l=1}^{N} \int_{BE_l} \psi_j \, K(\mathbf{r},\mathbf{r}') d\Gamma(\mathbf{r}') \; \psi_i \, d\Gamma(\mathbf{r}) \quad (8.17)$$

Depending on the boundary conditions, Equation 8.15 can be rearranged into a standard matrix form,

$$\mathbf{KU} = \mathbf{F} \tag{8.18}$$

where $\mathbf{K}$ is the global matrix, $\mathbf{U}$ the vector containing unknowns, and $\mathbf{F}$ the force vector. The solution of the above equation can be solved using the standard LU decomposition method.

Several points are worth noting. First, Equation 8.12 permits the use of both continuous and discontinuous shape functions. In some of the literature, Equation 8.12 is referred to as the Galerkin boundary integral formulation, as the field variable is defined on the surface only. In the discontinuous finite element sense, cross-boundary continuity is not required as a result of the use of the kernel function. Second, for this particular formulation, the final matrix is not local to an element only. The kernel function links all the surface elements together; consequently, the global matrix involves contributions from all surface elements, although discontinuous elements are used. This is different from the discontinuous finite element formulations discussed in previous chapters. Third, the kernel function is not based on the diffusion theory, and thus it does not necessarily satisfy the diffusion with a point source. In fact, the kernel function describes the straight line path relation between two points on two surface elements, which is much different from commonly known diffusion behavior. Therefore, if there exists a blockage between the two surfaces, then the thermal ray emitted from one surface will not be able to reach the other surface. In the case of diffusion, however, the field variables would bend around or diffuse through the blockage to reach the other surface. Because of these characteristics, the kernel function can be discontinuous, and thus special treatments must be applied to ensure correct detection of these internal blockages in complex geometries, in order to obtain an accurate evaluation of the kernel functions.

8.3 Shadowing Algorithms

The evaluation of kernel functions for surface radiative energy exchange calculations can be carried out in a fairly straightforward manner, if the internal blockages are not present, and the surface of the enclosure is convex everywhere [13, 14]. We will illustrate this point in Example 8.1 below. When a blockage occurs between the surface elements either by the internal structures or by the non-convexness of the enclosure surface, an accurate evaluation of the kernel functions requires a tedious procedure to detect these geometric blockages [3–11]. While the idea is to find out if a line emitted from a point on a surface element is intercepted by any objects in the enclosure before it reaches its designated point on another surface element, a general, efficient computational procedure – namely the shadowing algorithm – for this type of calculation requires careful geometric and floating point considerations. Perhaps the most important aspect of surface radiation exchange computations is the design and implementation of a shadowing algorithm that allows an efficient and effective detection of any geometric obstructions by third parties embedded in a configuration (see Figures 8.1 and 8.3). This is also the most difficult, time consuming, and error-prone part of a computational scheme for thermal radiation exchange calculations, without which these computations can be carried out routinely using essentially any integration algorithms available [12–14].

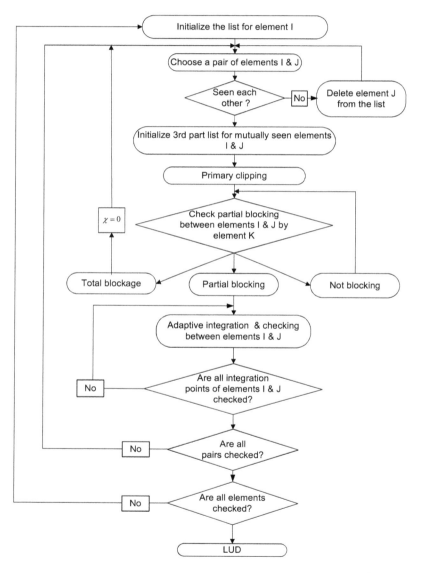

Figure 8.4. Outline of the shadowing algorithm for detection of third body blockages during external radiation calculations

Two mechanisms can prevent a radiation ray from reaching its designated destination, and must be identified in the algorithm. In the first case, the surface element does not see at all the destination point the ray is supposed to reach. This is essentially the case of self-blocking, which occurs with a curved or ring element. In the second case, the ray is blocked by a third party element, due to a geometric configuration, which is coined *third party blocking*. In both cases, the kernel function is zero. The blockage of a ray can be further differentiated into the categories of partial blocking and total blocking, for which special treatments are required. While kernel functions can be evaluated by integration with final checking alone for every pair of elements, this brute force approach can be prohibitively expensive, especially for large-scale calculations involving complex 3-D geometries, and thus should be avoided whenever possible. An efficient algorithm should be able to eliminate all those elements that are not needed, and to perform only on those elements that are absolutely necessary for the computationally intensive integration with blockage detection. The shadowing algorithms developed based on these ideas for the detection of blockages in 2-D, axisymmetric and 3-D configurations are described below. These algorithms all involve data structure creation, sorting, primary and secondary clipping, and adaptive integration, but differ in the details of geometric treatments. The outline of the shadowing detection and adaptive integration algorithm is illustrated in Figure 8.4. Although the general procedures are the same for the geometric configurations of all dimensions, the details are sufficiently different that the detection algorithm development for the 2-D, axisymmetric and 3-D geometries is discussed in separate sections.

8.3.1 Shadowing Algorithm for 2-D Geometry

The shadowing scheme for a 2-D geometry is considered first, because it is the most intuitive and easy to envision. Before starting the search and sort procedure, a list of elements actively engaged in thermal radiation with element I, from which radiation emits, is created and initialized. The data structure creation follows a similar procedure to that given in Bastian and Li [15], and should be applicable to the problems of all dimensions. This list is updated as the procedure proceeds. The first step tests the signs of the product of the surface normals of the two elements (e.g., elements I and J) with the vector connecting the two elements (i.e., $\mathbf{R}$ in Figure 8.5). This is equivalent to testing the sign of $\mathbf{n}\cdot(\mathbf{r} - \mathbf{r}')\,\mathbf{n}'\cdot(\mathbf{r} - \mathbf{r}')$. If the sign of $\mathbf{n}\cdot(\mathbf{r} - \mathbf{r}')\,\mathbf{n}'\cdot(\mathbf{r} - \mathbf{r}')$ is positive, the two elements can see each other. If the sign of $\mathbf{n}\cdot(\mathbf{r} - \mathbf{r}')\,\mathbf{n}'\cdot(\mathbf{r} - \mathbf{r}')$ is zero or negative, the two elements are considered to be unable to see each other. Based on this sign test convention, elements I and J can see each other, whereas elements I and J' cannot, as shown in Figure 8.5. After this test, the elements that are unseen by element I are discarded from the list. The kernel function for them is set to zero, and no further considerations are given to them in terms of thermal radiation exchange with element I.

For those elements remaining in the list, a test is conducted to determine the blockage. The procedure involves additional three steps, the basic idea of which is illustrated in Figure 8.6. Again, before the test is started, another list of third party elements is created and initialized for each pair of mutually seen surface elements, determined as described in the previous paragraph. The algorithm here consists of the coarse screening and the detailed checking. First, a rectangular primary window is set up using the maximum and minimum coordinates of the pair of mutually seen elements, i.e., elements I and J (see Figure 8.6). For this purpose, the standard clip algorithm routinely used in computer graphics for clipping objects, proves to be extremely effective, is thus directly applied [16–18]. The elements lying outside the window are deleted from the list of blocking elements. This check will throw out a majority of unblocking third party shadowing elements from the list of blocking elements. Those screened through this check are further clipped out and discarded from the list of blocking elements, if they lie outside the irregular window defined by 1'2'12. The algorithms used to clip out the elements out of the irregular window, such as 1'2'12, are more involved and computationally intensive. However, the basic procedure is the same as for clipping against a triangle, which is performed during the integration. Thus, the elements are discarded from the list if they lie outside the window 1'2'12. Those lying partially or completely inside the window are further checked for blocking while integration is performed.

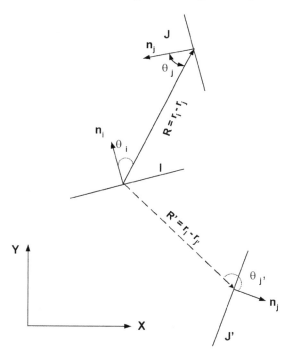

Figure 8.5. Geometric relations between mutually seen and unseen elements: radiation emitted from point i can reach point j but is unable to reach point j' because of the self-blockage by the element J'

For those elements remaining in the list, a test is conducted to determine the blockage. The procedure involves an additional three steps, the basic idea of which is illustrated in Figure 8.6. Again, before the test is started, another list of third party elements is created and initialized for each pair of mutually seen surface elements, determined as described in the previous paragraph. The algorithm here consists of the coarse screening and the detailed checking. First, a rectangular primary window is set up using the maximum and minimum coordinates of the pair of mutually seen elements, i.e., elements I and J (see Figure 8.6). For this purpose, the standard clip algorithm routinely used in computer graphics for clipping objects, proves to be extremely effective, is thus directly applied [16–18]. The elements lying outside the window are deleted from the list of blocking elements. This check will throw out a majority of unblocking third party shadowing elements from the list of blocking elements. Those screened through this check are further clipped out and discarded from the list of blocking elements, if they lie outside the irregular window defined by 1',2',1,2. The algorithms used to clip out the elements out of the irregular window, such as 1',2',1,2, are more involved and computationally intensive. However, the basic procedure is the same as for clipping against a triangle, which is performed during the integration. Thus, the elements are discarded from the list if they lie outside the window 1',2',1,2. And those lying partially or completely inside the window are further checked for blocking while integration is performed.

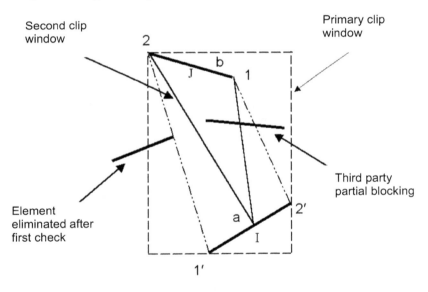

Figure 8.6. Primary and secondary clip windows for screening out third party shadowing elements for blocking the radiation exchange between elements I and J

Numerical integration is performed to compute A_{ij} and B_{ij} for each pair of mutually seen elements I and J, while checking for blockage by each remaining element in the list of partial blocking elements attached to the pair. Some of these partially blocking elements may or may not block every ray between the two elements. To determine if a third party element actually blocks a ray between elements I and J, the third party element is first checked against a triangle, formed by an integration point on element I and element J. The detailed algorithm for clipping against the triangle is given in Figure 8.7, where three different scenarios are depicted. For case 8.7a, the third party element K lies outside the triangle, and thus integration along element J from point i is not affected. As a result, numerical integration is carried out for element J and point i without further checking. For case 8.7b, the third party completely blocks a ray from point i to any points on element J. Thus, the kernel function is set to zero and no integration is carried out for element J. For case 8.7c, which is the most common, the third party element is partially blocking, and therefore every ray from point i to any integration point on element J needs to be further checked (see the next paragraph for detail). To determine to which case the third party element K belongs, the following rule based on the geometric considerations is applied. The nodes of the triangle formed from point i and element J are numbered anti-clockwise, which allows one to determine the surface normal, $\mathbf{n}_s$, of the triangle as positive, when pointing out of the paper. Two vectors are created by connecting one point of element K, say a' (see Figure 8.7c), to two end points of one side of the triangle, e.g., $\vec{l}_1$ and $\vec{l}_2$. The cross product of the two vectors, $\vec{l}_2 \times \vec{l}_1$, is then dot with the surface normal $\mathbf{n}_s$ of the triangle. If the dot product is positive, then the element K is inside the triangle and belongs to the category of Figure 8.7c. If both $\mathbf{n}_s \cdot \vec{l}_2 \times \vec{l}_1$ and $\mathbf{n}_s \cdot \vec{l}_4 \times \vec{l}_3$ are negative, and $\vec{l}_i$ $(i = 1, \ldots, 4)$ is formed from the same side of the triangle, then the third party element K belongs to the category of Figure 8.7a. If both $\mathbf{n}_s \cdot \vec{l}_2 \times \vec{l}_1$ and $\mathbf{n}_s \cdot \vec{l}_4 \times \vec{l}_3$ are positive and $\vec{l}_i$ $(i = 1, \ldots, 4)$ is formed from two different sides of the triangle, then the element K belongs to the category of Figure 8.7c, where a third party element totally blocks the point to element J.

For the case illustrated in Figure 8.7c, an adaptive integration algorithm is applied, while the ray connecting point i to an integration point on element J is checked for blockage, as illustrated in Figure 8.8. In this case, the interception point P is calculated by simultaneously solving the two linear equations describing lines a'b' and ij. The blockage occurs if P lies on line a'b' or element K; otherwise line ij is not blocked. In applying the adaptive integration, two successive numerical integrations, with twice as many integration points in the current step as those in the preceding one, are employed. The error between the two successive integrations is checked and the calculation is considered converged, when it is smaller than a preset value. Our experience indicates that a preset value of 0.001 as a relative error with respect to the diagonal term yields a reasonably fast convergence with sufficient accuracy.

At this point, it may be constructive to revisit the treatment of the third party blockage when a single surface integration is applied. The drawback associated with the traditional boundary element algorithm is the possible miss in detecting the third party shadowing in some special cases. This point is illustrated in Figure 8.9. Because the numerical integration is carried out between the nodal points (1' or 2') and element J, and point a is not included in integration (see Equation 8.3), the shadowing effect of element L, a third party element that actually partially blocks the view between elements I and J, is not accounted for, thereby resulting in numerical errors. The Galerkin boundary element method, however, requires the double integration of elements I and J, which in turn requires the use of internal points on either elements, and thus will be able to detect the existence of element L.

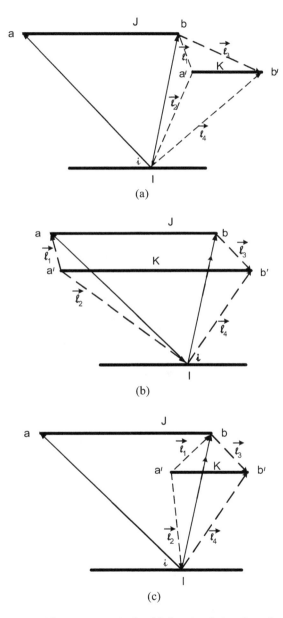

Figure 8.7. Three geometric arrangements for third party obstruction of radiation exchange between point i and element J in relation to the testing triangle: (a) no obstruction, (b) total obstruction and (c) partial obstruction

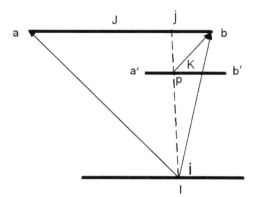

Figure 8.8. Detailed testing of blockage of radiation between points i and j by a third party element K

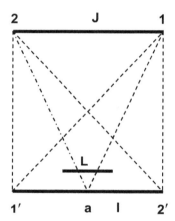

Figure 8.9. Comparison of strategies for detecting third party shadowing by the traditional and Galerkin boundary element methods. The traditional method uses the triangle formed by $1'12$, where the Galerkin used triangle $1'12$ and other triangles such as $a12$

8.3.2 Shadowing Algorithm for Axisymmetric Configurations

The geometric considerations for an axisymmetric configuration can be quite different from the simple 2-D case, although there are some similarities. This is because the self-blocking must be checked, even if a linear surface element is used. This is shown in Figure 8.3.

As in the 2-D case, the shadowing algorithm starts with the creation of a list of elements actively engaged in thermal radiation with element I, from which radiation emits. This list is updated as the procedure proceeds. The first step tests the signs of the product of the surface normals of the two elements (e.g., elements I and J) with the vector connecting the two elements, i.e., $\mathbf{R}$ in Figure 8.3. This is equivalent to testing the sign of $(\mathbf{n} \cdot \mathbf{R})\,(\mathbf{n'} \cdot \mathbf{R})$. If the sign is negative, the two elements can see each other. Otherwise, the two elements are considered as not seeing each other. In axisymmetric cases, the following rule is very useful and can be easily proved. Elements I and J are not mutually seen, if element I cannot see both sections of element J at $\phi' = 0$ and $\phi' = \pi$, as shown in Figure 8.10. In other words, elements I and J are unseen by each other, if both $(\mathbf{n}\cdot\mathbf{R}_p)$ $(\mathbf{n}_p\cdot\mathbf{R}_p)$ and $(\mathbf{n}\cdot\mathbf{R}_b)$ $(\mathbf{n}_b\cdot\mathbf{R}_b)$ are positive or zero. After this test, the elements that are unseen by element I are discarded from the list. The kernel function for them is set to zero and no further considerations are given to them in terms of thermal radiation exchange with element I.

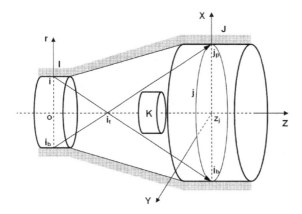

Figure 8.10. Geometric relations between mutually seen and unseen elements: radiation emitted from point i can partially reach element L, but is unable to reach element J

For those remaining in the list, further testing is performed to determine the blockage. The procedure is identical to that used for 2-D calculations, except that the detection for axisymmetric geometry uses the window as shown in Figure 8.11. This check eliminates a majority of unblocking third party shadowing elements from the list of blocking elements. Those screened through this check are further clipped out and discarded from the list of blocking elements, if they lie inside the two triangles defined by $\Delta ii_b i_t$ and $\Delta i_t j_b j_p$, as illustrated in Figure 8.12. Based on this criterion, element K does not block elements I and J that see each other. Thus, those lying partially or completely outside these two triangles are further checked for blocking while integration is performed.

The blocking region on element J by the third party element K is determined by the azimuthal angle ranging from ϕ_1' to ϕ_2'. This region is calculated analytically in this study. Due to the fact that the geometry is symmetric with respect to the plane $\phi' = 0$, the blocking region on element J is confined from $\phi' = 0$ to $\phi' = \pi$. The surface of the third party element K shown in Figure 8.13 can be expressed as

$$x^2 + y^2 - (kt + 1)^2 R_I^2 = 0 \tag{8.19}$$

where the parameters are defined *as* $t = z/z_j$, $R_I = (R_a Z_b - R_b Z_a)/(Z_b - Z_a)$, $k = (R_J - R_I)/R_I$, and $R_J = (R_b - R_a)(Z_J - Z_a)/(Z_b - Z_a) + R_a$. The normal vector on the surface of element K is obtained by the gradient of Equation 8.19,

$$\mathbf{n}_k = \{x, y, -(kt+1)R_I^2 k / z_j\} \tag{8.20}$$

The ray emitted from element I can be described as the following equations with the parameter t:

$$x = a + (b\cos\phi' - a)t \; ; \quad y = bt\sin\phi' \; ; \quad z = z_j t \tag{8.21}$$

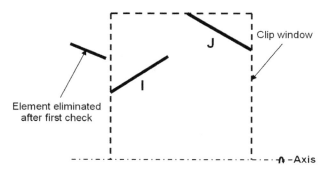

Figure 8.11. Clip windows used for primary checking of shadowing elements for axisymmetric configurations

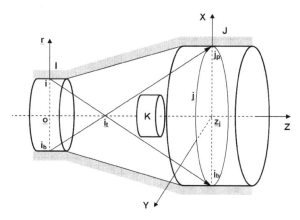

Figure 8.12. Two triangles formed by $\mathbf{ii}_b\mathbf{i}_t$ and $\mathbf{i}_t\mathbf{j}_b\mathbf{j}_p$ provide un-shadowing zones between point *i* and contour **j**

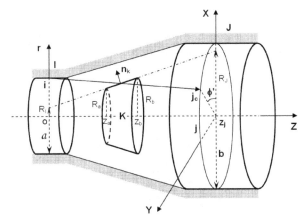

Figure 8.13. Detailed geometrical description between point i and contour $\mathbf{j}$ by a third party element K

Therefore, the normal vector of element K can be further expressed in terms of the parameter t and the azimuth angle ϕ',

$$\mathbf{n}_k = \{a + (b\cos\phi'-a)t,\ bt\sin\varphi',\ -(kt+1)R_I^2 k/z_j\} \tag{8.22}$$

The criteria for which the ray $\overrightarrow{\mathbf{ij}}_e$ is tangent to the surface of element K must satisfy the relation, $\mathbf{n}_k \cdot \overrightarrow{\mathbf{ij}}_e = 0$. This gives

$$\alpha\beta\cos\varphi' = \frac{(\alpha^2 + \beta^2 - k^2)t - (\alpha^2 + k)}{2t - 1} \tag{8.23}$$

where $\alpha = a/R_I$ and $\beta = b/R_I$.

The solution of the above equation, ϕ_t', represents the minimum angle that does not block the ray $\overrightarrow{\mathbf{ij}}_e$ reaching the contour $\mathbf{j}$ on element J at $z = z_j$. The intersection of $\overrightarrow{\mathbf{ij}}_e$ and the surface of element K can be determined by the solution of Equations 8.19 and 8.21, i.e.,

$$(\alpha + (\beta\cos\phi'-\alpha)t)^2 + \beta^2 t^2 \sin^2\phi' = (kt+1)^2 \tag{8.24}$$

The location, where the ray $\overrightarrow{\mathbf{ij}}_e$ is tangent to the surface of element K, is calculated by solving for t and is written as follows:

$$t = \frac{(1-\alpha^2) \pm \sqrt{(1-\alpha^2)(1-\beta^2 + k(2+k))}}{\beta^2 - \alpha^2 - k(2+k)} \tag{8.25}$$

If the solution of Equation 8.23 exists but the parameter t falls out of the range $z_a/z_j < t < z_b/z_j$, the ray $\overrightarrow{\mathbf{ij_c}}$ may intersect the end disks at the top or the bottom of the third party element K. The intersections are defined by two azimuthal angles,

$$\cos\varphi_a' = \frac{(\alpha^2 + \beta^2)t^2 - 2\alpha^2 t + (\alpha^2 - \xi_a^{\,2})}{2\alpha\beta t(1-t)} \tag{8.26}$$

where $t = z_a/z_j$ and $\xi_a = R_a/R_I$; and

$$\cos\varphi_b' = \frac{(\alpha^2 + \beta^2)t^2 - 2\alpha^2 t + (\alpha^2 - \xi_b^{\,2})}{2\alpha\beta t(1-t)} \tag{8.27}$$

where $t = z_b/z_j$ and $\xi_b = R_b/R_I$.

To effectively determine the shadowing regions on element J, blocked by the third party element K, the following algorithm is developed. In the algorithm, every third party element blocking the ray from the radiating point i and the contour $\mathbf{j}$ is treated as a hollow cone. Three angles defined in Equations 8.23, 8.26 and 8.27 are used for determining any possible blocking regions.

(1)   for any third party element K

Initiate $\phi_1 = \phi_2 = 0$;

If ϕ_a' exists, then $\phi_1 = \phi_a'$;
if $R_{z=z_b} > R_b$, then $\phi_2 = \pi$;

If ϕ_b' exists, then $\phi_2 = \phi_b'$;
if $R_{z=z_a} > R_a$, then $\phi_1 = \pi$;

If both ϕ_a' and ϕ_b' exist, then $\phi_1 = \min(\phi_a', \phi_b')$;
if $R_{z=z_a} > R_a$ and $R_{z=z_b} > R_b$, then $\phi_2 = \pi$;

$\phi_{\min,K} = \min(\phi_1, \phi_2)$;
$\phi_{\max,K} = \max(\phi_1, \phi_2)$;

If ϕ_t' exists and $z_a/z_j < t < z_b/z_j$, then $\phi_{\min,K} = \phi_t'$.
If $\phi_{\max,K} = 0$, then $\phi_{\max,K} = \pi$

$K=K+1$

Go to (1) to check next possible third party element.

The process will continue until all possible third party elements are checked. The final blocking region is determined by $\cup_{K=1}^{N_k}(\phi_{min,K},\phi_{max,K})$, where N_K denotes the total number of third party elements. Therefore, the integration in Equation 8.11 should be evaluated in the regions $\{(0,\pi)-\cup_{K=1}^{N_k}(\phi_{min,K},\phi_{max,K})\}$.

8.3.3 Shadowing Algorithm for 3-D Geometry

The shadowing algorithm for a 3-D configuration also follows the outline given in Figure 8.7. As in the 2-D case, the algorithm starts with sorting by signs. Prior to the four-step procedure, a list of elements actively engaged in thermal radiation exchange with element I from which radiation emits is created and initialized. This list, labeled as the main list, may be taken as the entire set of surface elements if there exists no prior information on inter-element relations, which would be the case with $I = 1$. The list initiated may be shorter, starting with $I \geq 2$, because information obtained from the previous testing can be used to preclude some elements from the list to speed up the computation. For example, if elements 1 and 2 are not seen by each other, then element 1 would be excluded from the list for element 2 when it is created. Once it is created, the list is updated as the searching/sorting procedure proceeds. As the first step, elements are sorted by testing whether two elements can see each other at all. To do that, we consider the signs of the dot product of the surface normals of the two elements (e.g., elements I and J) with the vector connecting the two elements (i.e., $\mathbf{R}_{ij}$ in Figure 8.14). This is equivalent to testing the signs of $\mathbf{n}\cdot(\mathbf{r}-\mathbf{r'})$ and $\mathbf{n'}\cdot(\mathbf{r}-\mathbf{r'})$. The thermal rays may reach each other between the two elements, or the two elements can see each other, if $\mathbf{n}\cdot(\mathbf{r}-\mathbf{r'})$ and $\mathbf{n'}\cdot(\mathbf{r}-\mathbf{r'})$ have different signs. Otherwise, the thermal rays emitted from either of the elements cannot reach the other, or the two elements cannot see each other. Based on this sign test convention, elements I and J can see each other (see Figure 8.2), whereas elements I and J' cannot, as shown in Figure 8.5. After this test, the elements that are not seen by element I are discarded from the main list. The kernel function related to them is set to zero and no further considerations are given for them in the procedures ensuing.

The primary clipping is now performed on the elements remaining in the main list, which are considered mutually seen at this moment, to determine the foreign elements for the potential third party blockage of surface radiation exchanges between an element pair of element I and one other element (say element J) selected from the main list. The basic idea of the primary clipping is illustrated in Figure 8.15. Again, before the test is started, another list of third party elements, or the blocking list, is created and initialized for each pair of mutually seen surface elements in the list. This third party or blocking list is a subset of the main list excluding, but attached to, the pair of mutually seen elements, i.e., elements I and J. To begin, a primary window of brick shape is set up using the maximum and minimum coordinates of the element pair. The elements in the blocking list are checked against the primary window. For this purpose, the standard 3-D clip algorithm routinely used in computer graphics for clipping objects, proved to be extremely effective, is thus directly applied [18]. The elements lying outside the window cannot possibly block the thermal rays traveling between elements I and J, and thus are deleted from the blocking list. This check will delete a majority of unblocking third party shadowing elements from the blocking list. For example, as shown in Figure 8.15, this procedure will drop out element L from further consideration but still keeps element K, which is considered a potential candidate for a third-party blockage. Depending on the relative geometric positions of mutual elements I and J, a good portion of the elements can be eliminated, thereby reducing the computational burden for the most tedious ray-tracing checking and adaptive integration.

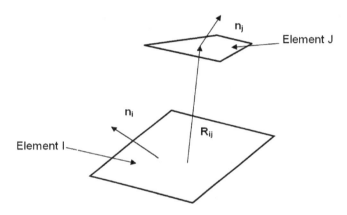

Figure 8.14. Schematic illustration of two 3-D surface elements unseen by each other

The next step involves the secondary clipping, which eliminates the elements not blocking any ray from element I to element J from the active list. This is done as follows. A pyramid is formed by selecting a corner point, say point i, of element I and connecting it to all the corners of element J. A remaining third party element, or active element in the blocking list, is now checked against the pyramid. The detailed algorithm for clipping against the pyramid is given in Figure 8.16, where three different scenarios are depicted. For case 8.16a, the third party element K lies outside the pyramid. For case 8.16b, the third party totally blocks a ray from point i to any points on element J. For cases 8.16c and 8.16d, which are the most common scenarios, the third party element is partially blocking, and therefore every ray from point i to any integration point on element J needs to be further checked; that is, the final checking is required during detailed integration.

To determine which of the three cases the third party element K belongs to, the following rule, based on the geometric considerations, is applied. A pyramid is formed by four lines, which all originate from point i, and each connect to a different vertex of element J. Each of these four lines may pass through the plane defined by element K. One of these lines, marked by ij, is illustrated in Figure 8.16c. By the similarity rules of plane triangles Δipq and Δicr, we have the following relation:

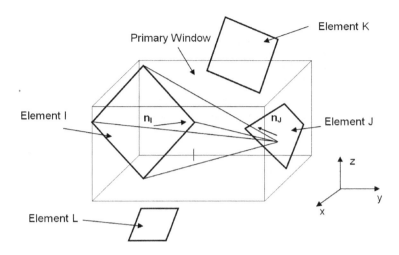

Figure 8.15. Primary testing window against which unblocking elements are eliminated and potentially blocking elements are retained

$$\mathbf{m} = (\mathbf{r}_c - \mathbf{r}_i); \qquad \mathbf{r}_p = \mathbf{r}_i + \mathbf{m}t \; ; \quad t = (\mathbf{r}_j - \mathbf{r}_i) \cdot \mathbf{n}_J / (\mathbf{r}_c - \mathbf{r}_i) \cdot \mathbf{n}_J \quad (8.28)$$

where p is the interception point of line ij with element J, and $\mathbf{n}_J$ is the outnormal of element J. Note that $pq = |(\mathbf{r}_j - \mathbf{r}_i) \cdot \mathbf{n}_J|$, which is the distance between point i and plane J. Thus, if $0 < t < 1$ or $t < 0$, then element K lies outside the pyramid, that is, either above element J or below point i. The element is then eliminated, and a new check for a different element is started. The third party element, or element K, blocks the ray from point i to element J if the following two conditions hold for any one of the four lines connecting point i to the four corners of element J:

$$t > 1 \quad \text{and} \quad p \text{ is inside element } J.$$

To test if p is inside element J, the algorithm as illustrated in Figure 8.17 is employed. The nodes (1,2,3,4) of element J is set up clockwise, and two vectors are created by connecting the interception point p to two nodes of one side of element J, e.g., $p1$ and $p2$. The cross product of the two vectors, $p1 \times p2$, is then dotted with the surface normal $\mathbf{n}_J$ of element J. If the dot product is positive, then the point j (or p) is inside the element J. If the points inside are less than the number of corners of element J, then element K is partially blocking, as in case 8.16c. If all corner points of element J are inside the shadowing area, then element K is totally blocking (see Figure 8.16b). If, during checking, a totally blocking element (or case 8.16b) is found, then the kernel function is set to zero, and a new round of checking starts with a new Gaussian point on element I. If there are partially blocking elements but no totally blocking elements, then final checking is required.

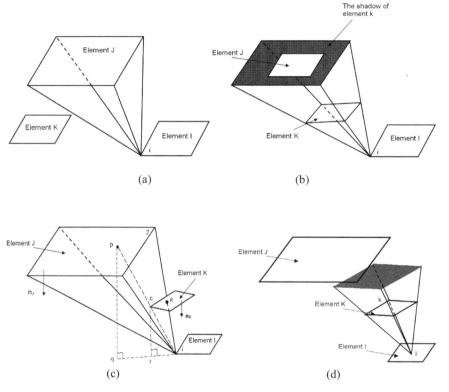

Figure 8.16. Four different scenarios of third party element blockage of thermal rays emitted from point i to element J: (a) no third party blocking, (b) total third party blocking, (c) and (d) partial third party blocking

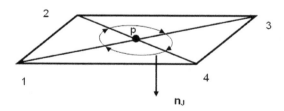

Figure 8.17. Procedure for testing if p is inside element J; n_J is the normal of the element

The above procedure, however, cannot positively determine cases illustrated in Figures 8.16b and 8.16d, which show that element J may be either totally or partially blocked, even if all interception points lie outside element K. Thus for all those elements whose intersection points are all outside, but $0 < t < 1$, additional checking is required. This check involves calculating the intersection point k with element K by line ij. Point k can be calculated by simply exchanging the corresponding points of element J with element K, or

$$\mathbf{m} = (\mathbf{r}_j - \mathbf{r}_i); \quad \mathbf{r}_k = \mathbf{r}_i + \mathbf{m}t_1; \quad t_1 = (\mathbf{r}_c - \mathbf{r}_i) \cdot \mathbf{n}_K / (\mathbf{r}_j - \mathbf{r}_i) \cdot \mathbf{n}_K \quad (8.29)$$

Thus, element K blocks the view from point i to element J, if $0 < t < 1$ and k is inside element K. Of course, the same procedure sketched in Figure 8.17 can be used to determine if k is inside element K.

It is noted that the purpose of this secondary check is to establish the active list of partial blocking elements and can be expensive since all the corners of the elements must be checked. Hence if either p or k is considered blocking during checking, then element K remains in the active list for final integration and a new check is started for the next element waiting to be checked. After this secondary checking procedure, the active list contains only those elements that will most likely block the view between elements I and J. Whether or not they indeed block, will be checked during final integration, which is described below.

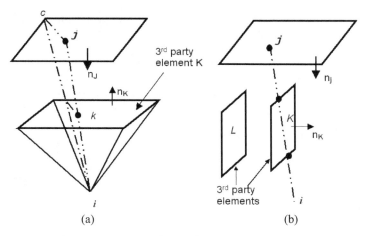

(a)   (b)

Figure 8.18. Procedures used to determine the blockage of the thermal ray emitted from i to j by element K. (a) Element K is not parallel ($\mathbf{m} \cdot \mathbf{n}_K \neq 0$) to the thermal ray ij and k is the interception point between the thermal ray and element K. (b) third party elements are parallel ($\mathbf{m} \cdot \mathbf{n}_K = 0$) to the thermal ray ij, and element K blocks thermal ray ij, but element L does not

After the secondary clipping of partial blocking elements as described above, the actual integration over elements I and J is now performed, along with the detailed final check of the third party element blockage of the thermal ray originating from an integration point on element I to another on element J (see Figure 8.18). In this study, the Gaussian integration quadrature is used. Other integration rules can also be applied. If the blocking list is empty, or no active element is in the list, the integration is carried out without further checking.

Otherwise, final checking of blockage is performed for those elements active in the blocking list, when integration is taken over element J. In this case, each line (or light ray emitting from integration point i on element I, and ending at integration point j on element J) is checked against all partially blocking elements active in the blocking list, determined from the secondary clipping. If the ray connecting two integration points of respective elements I and J is blocked by the third party element K, the value of integration is set to zero. If not, this integration is calculated. The idea of determining various scenarios of third-party blocking during final numerical integration is illustrated in Figure 8.18. Here a line is constructed between points i and j, and the interception point k of the line with the third-party element K is calculated by substituting relevant parameters into Equation 8.29. The interception point k is then tested using the same procedure as illustrated in Figure 8.17 to see if it lies inside element K.

When the line ij is parallel (i.e., $\mathbf{m}\cdot\mathbf{n}_K = 0$) to the third party element (see Figure 8.18b), Equation 8.29 is no longer applicable. In this case, care is taken to ensure that the line passes through the element, and the interception points between the plane edges and the line are calculated. If one of these interception points is between points i and j, then the third party element blocks. If none of these points lie on the line between points i and j, then the third party element does not block.

In order to obtain an accurate value of integration, an adaptive integration is applied. Two successive numerical integrations, with twice as many integration points in the current step as those in the preceding one, are employed in the present study. Other ways of refinement are also possible; for instance, two successive orders of numerical integration may be used instead. The error between the two successive integrations is checked, and the calculation is considered converged, if the error is smaller than a preset value. Our experience indicates that a preset value of 0.001 as a relative error, with respect to the diagonal term, yields a reasonably fast convergence with sufficient accuracy.

The Galerkin formulation requires surface integration over both elements I and J. This mandates that the above search and integration algorithm is carried out for integration of element J with every integration point i on element I and then integration of element I with every integration point j on element J.

The above procedure continues until the initial list associated with element I is exhausted. Then a new surface element is selected and a list of candidate elements is created. The three-step searching and integration computational process is followed. This is repeated until every surface element is calculated.

It is noted here that there are similarities between the secondary clipping and the final check for blockage during adaptive integration, though the latter is computationally more involved. Thus some procedures described in the secondary clipping are applicable to the final check. Additional checks as shown in Figure 8.18 are also employed in this final check stage, to ensure all blockages are detected accurately. Since both procedures can be very time consuming, the secondary clipping helps to save time if the integration points per element exceed the number of corners of the element. For relatively simple geometries and relatively few elements, few integration points are needed and thus the secondary clipping may not necessarily speed up the calculations significantly. For complex geometries, partial blocking occurs very frequently and often the number of integration points needed to accurately integrate the kernel functions is significantly larger than the number of corners. For these cases, the secondary clipping provides a useful means to shorten the list to be checked during integration, and thus allows savings in computing time.

8.4 Coupling with Other Heat Transfer Calculations

In most engineering applications, surface radiation often is not the only heat transfer mechanism, and other phenomena also coexist, such as heat conduction, fluid convection, phase changes, etc. Thus it is required to couple the discontinuous Galerkin surface method described above with other methods to solve the problems involving mixed heat transfer mechanisms. Here we consider the procedure for coupling the discontinuous and convectional finite elements for the analysis. A similar idea can also be applied for the coupling of surface calculations with other domain methods. Below, a coupling problem is considered for a 3-D thermal and fluid flow system, in which the surface radiation, heat conduction, Marongani convection, and buoyancy driven flow are all present. Figure 8.19 shows the finite element mesh used to model the furnace. The chamber of the furnace has a cylindrical roof and a protruded blockage. The roof is fixed at a higher temperature, and by surface radiation the liquid pool below is heated up. The outside of the chamber is at room temperature and the system loses heat to the environment through radiation, which follows the Stefan–Boltzmann law. Because the free surface of the metal is exposed to the radiation, the surface temperature is not uniform. As a result, the Marangoni convection and buoyancy flow arise due to the surface tension and gravity effects. The melt flow and heat transfer are governed by the Navier–Stokes equations and the energy balance equations within the solid wall, the melt, and the surroundings. The governing equations for the fluid flow and heat transfer phenomena are given as follows:

$$\nabla \cdot \mathbf{u} = 0 \tag{8.30}$$

$$\rho \frac{\partial \mathbf{u}}{\partial t} + \rho \mathbf{u} \cdot \nabla \mathbf{u} = -\nabla p + \nabla \cdot \mu \left(\nabla \mathbf{u} + (\nabla \mathbf{u})^T \right) - \rho \beta \mathbf{g} \left(T - T_{ref} \right) \tag{8.31}$$

$$\rho C_p \frac{\partial T}{\partial t} + \rho C_p \mathbf{u} \cdot \nabla T = \nabla \cdot k \nabla T \tag{8.32}$$

where $\mathbf{u}$ is the velocity, ρ is the density, T is the temperature, μ is the viscosity, β is the thermal expansion coefficient, $\mathbf{g}$ is the gravity, C_p is the specific heat, T_{ref} is the reference temperature, and k is the thermal conductivity. The above equations, along with Equation 8.3, describe the surface radiation exchanges in the enclosure formed by the liquid surface, and the surfaces of the furnace facing the liquid. The boundary conditions for the problem are:

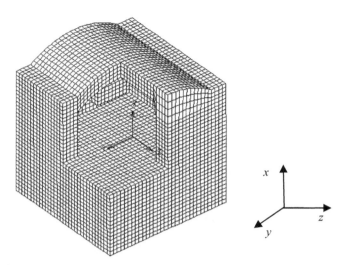

Figure 8.19. Schematic of an industrial furnace, where surface radiation, conduction, and convection coexist.

$$k\mathbf{n} \cdot \nabla T = -\varepsilon\sigma(T^4 - T_\infty^4) \quad \in \partial\Omega_1 \tag{8.33}$$

$$\mathbf{t} \cdot \tau \cdot \mathbf{n} = \frac{d\gamma}{dT}\mathbf{t} \cdot \nabla T \quad \in \partial\Omega_2 \tag{8.34}$$

$$\mathbf{u} \cdot \mathbf{n} = 0 \quad \in \partial\Omega_2 \tag{8.35}$$

$$T = T_H \quad \in \partial\Omega_3 \tag{8.36}$$

where $\partial\Omega_1$ is the outer surface of the furnace, $\partial\Omega_2$ is the top liquid surface, $\partial\Omega_3$ is the top inner surface, and γ is the surface tension, which may be a function of surface temperature. Note that Equation 8.34 represents a shear stress balance along the interface, which causes a fluid motion when the surface temperature is not uniform. This type of flow is often referred to as the Marangoni flow, or surface tension driven flow. The finite element discretization for the solution of convection and conduction heat transfer leads to the following matrix equations [19–23]:

$$\begin{bmatrix} \mathbf{A}(\mathbf{U}) + \mathbf{K} + \frac{1}{\varepsilon_0}\mathbf{EM}_p^{-1}\mathbf{E}^T & \mathbf{B}_T \\ \mathbf{0} & \mathbf{D}_T(\mathbf{U}) + \mathbf{L}_T \end{bmatrix} \begin{bmatrix} \mathbf{U} \\ \mathbf{T} \end{bmatrix} = \begin{bmatrix} \mathbf{F} \\ \mathbf{G}_T \end{bmatrix} \tag{8.37}$$

where ε_0 is the penalty number. This matrix equation is then solved together with the matrix, which describes the surface radiation exchange between the liquid surface and surfaces of the furnace above it, in order to obtain information on fluid flow and temperature distribution in the system. The elements of the matrices are calculated using the following expressions [19–23]:

$$\mathbf{L}_T = \int_\Omega \kappa \nabla \psi \cdot \nabla \psi^T dV \; ; \; \mathbf{B}_T = \int_\Omega Gr_T (\mathbf{g}\psi\psi^T)dV - \int_{\partial\Omega} \frac{\partial \gamma}{\partial T} \mathbf{t} \cdot \nabla T dS$$

$$\mathbf{E}_i = \int_\Omega \hat{\imath} \cdot \nabla \psi \psi^T dV \; ; \quad \mathbf{G}_T = -\int_{\partial\Omega} q_T \psi \, dS \; ; \quad \mathbf{M}_p = \int_\Omega \psi \psi^T dV$$

$$\mathbf{A(U)} = \int_\Omega \rho \psi \, \mathbf{u} \cdot \nabla \psi^T dV \; ; \; \mathbf{D}_T \mathbf{(U)} = \int_\Omega \rho C_p \psi \mathbf{n} \cdot \nabla \psi^T dV$$

$$\mathbf{K}_{ij} = \left(\int_\Omega \nabla \psi \cdot \nabla \psi^T dV \right)\delta_{ij} + \int_\Omega (\hat{\imath} \cdot \nabla \psi)(\hat{\jmath} \cdot \nabla \psi^T) \, dV$$

where ψ is the shape function and $\hat{\imath}, \hat{\jmath}, \hat{k} = (\hat{x}, \hat{y}, \hat{z})$. The coupling of surface radiation calculations with the finite element calculations is represented by the term $\mathbf{G}_T$. Here $\mathbf{q}_T$ for the liquid is calculated by the external radiation calculations. The numerical algorithm integrating the external radiation and finite element calculations can be developed with either direct coupling or iterative coupling. These two types of coupling strategies are described below. The merits and drawbacks of both direct and iterative coupling schemes, as applied to mixed mode heat transfer calculations, are discussed along with Example 8.5 in Section 8.5.

8.4.1 Direct Coupling

While there are many different ways to couple the boundary and finite element methods, the simplest and most natural way is to make use of the physical constraints for the flux and field variables along the common boundaries [24–26]. This is the approach taken here. To facilitate computation, the direct matrix inversion procedure is replaced by one LU decomposition procedure and $n+1$ times of back-substitutions. To do that, Equation 8.15 is re-written as follows:

$$\mathbf{A}_{LU}\mathbf{q}_I = \mathbf{A}_{LU}\left(\mathbf{q}_{I,1} + \mathbf{q}_{I,2} + \cdots + \mathbf{q}_{I,n-1} + \mathbf{q}_{I,n}\right)$$

$$= \mathbf{B}_1 E_{b,1} + \mathbf{B}_2 E_{b,2} + \cdots + \mathbf{B}_n E_{b,n} \tag{8.38}$$

where $\mathbf{A}_{LU}$ stores the LU-decomposed matrix of $\mathbf{A}$, $\mathbf{B}_i$ the ith column of the matrices $\mathbf{B}$, and E_{bi} the ith element of the vector $\mathbf{E}$. With $\mathbf{A}_{LU}$, one back-substitution is needed to solve for $\mathbf{q}'_{I,i}$ from the following equation:

$$\mathbf{q}'_{I,i} = \mathbf{A}_{LU}\mathbf{B}_i \qquad\qquad (8.39)$$

The result is multiplied by $E_{b,i} = (\sigma_s T_i^3)T_i$ to give $\mathbf{q}_{I,i} = [\mathbf{q}'_{I,i}(\sigma_s T_i^3)]T_i$. Applying this procedure to each column of $\mathbf{B}$ and summing up the results, one has the following expression:

$$\mathbf{q}_{\mathbf{I}} = \sum_{i=1}^{n} \mathbf{q}'_{I,i}(\sigma_s T_i^3) T_i \qquad\qquad (8.40)$$

The advantage of this approach is that $\mathbf{q}'_{I,i}$ is constant if ε_i is constant, and thus needs to be calculated only once. Also, in deriving Equation 8.40, we have assumed the temperature at the node points does not experience a jump. With Equation 8.40 substituted into Equation 8.37, the final expression is obtained for the vector potential distribution within the finite element region,

$$\overline{\mathbf{K}}\,\overline{\mathbf{U}} = \overline{\mathbf{F}} \qquad\qquad (8.41)$$

where $\overline{\mathbf{K}}$ is the resultant stiffness matrix, $\overline{\mathbf{U}}$ is the unknown potential vector in the finite element region, and $\overline{\mathbf{F}}$ is the modified force vector that represents the effects of the applied heating source.

8.4.2 Iterative Coupling

The direct coupling procedure described above can be difficult to apply and may also be slow for large scale problems. For these cases, an iterative coupling procedure may be employed instead. The iterative procedure starts with an initially guessed wall temperature distribution to calculate the radiant boundary heat flux distribution using the discontinuous Galerkin method. The calculated heat flux values are then input back into the algorithm for the calculation of fluid flow and temperature distributions in the system. The wall temperature distributions are then updated and the radiant boundary heat flux distribution is recalculated. This recalculated heat flux is then used to update the fluid flow and temperature field and so on and so forth. The procedure continues until the criterion set for convergence is met.

In practice, both direct and iterative procedures have been applied. While it is difficult to draw a precise guideline for which of these coupling procedures should be used for what conditions, a rule of thumb is that the direct method can be more effective for 2-D, or small to moderate size problems, whereas the iterative procedure outperforms the direct coupling for large scale problems. In addition, the iterative solver does not require the temperature at nodal points to be the same, and thus the usual jump condition for a discontinuous formulation needs not be modified. A more thorough discussion on the subject is given at the end of this chapter through a numerical example, i.e., Example 8.5.

8.5 Numerical Examples

The numerical examples selected here are intended to illustrate the external radiation computational procedures and the performance of the discontinuous Galerkin boundary element method as applied to the external radiation calculations. We start with a 2-D problem of simple geometry without internal obstruction, and detail the basic computational procedure for the solution of the problem using a few elements for discretization. This same problem is then solved using the discontinuous Galerkin boundary element method with a refined discretization for a more detailed description of heat flux distribution along the walls. The criterion is also given to check the accuracy of the numerical computations for external radiation problems. The problems of more complex geometries, involving internal extrusions that block the radiative energy exchange between surface elements, are then considered. These complex problems are not examined in full detail, but are sufficiently complete to demonstrate the capability of the algorithms described. As the last example, a coupled problem involving conduction, convection, and external radiation phenomena is discussed.

Example 8.1. As a first example, we consider surface radiation in a 2-D cavity. The cavity is 3 m × 3 m in cross section, with side wall and roof at 1700 K and 1400 K, respectively. The radiant heat fluxes along the four walls need to be determined when the bottom wall is at 600 K. All the surfaces are taken to be gray and diffuse and have an emissivity of 0.5. The problem is illustrated in Figure 8.1e.

Solution. This problem is solved both analytically and numerically, using the discontinuous Galerkin method described. In the analytical approach, a few elements are used for the sake of demonstrating the computational procedure. The numerical approach uses a mesh of considerably more elements for a refined description of the heat flux distribution. For both approaches, the criterion for checking the accuracy of the solutions is discussed.

Analytical Approach. This problem may be solved using radiosity and emissive power as variables, which are given in introductory heat transfer books [25, 26]. Here we solve the problem analytically from the discontinuous Galerkin solution procedure described above. This will allow us to illustrate the computational procedure in full detail. For this purpose, the cavity is discretized into four elements, with each wall treated as an element. Further we use the constant element approximation, which means that the temperature and heat flux are approximated using a box function over each element. Thus, Equations 8.16 and 8.17 are simplified as,

$$A_{ij} = L_i \delta_{ij} - (1 - \varepsilon_i) F_{ij} L_i \qquad (8.1e)$$

$$B_{ij} = -\varepsilon_i \delta_{ij} L_i + \varepsilon_i F_{ij} L_i \qquad (8.2e)$$

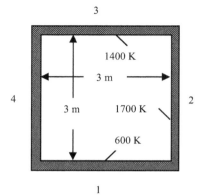

Figure 8.1e. Surface radiation exchange in a simple 2-D cavity

where the subscript i and j refer to the ith and jth elements, respectively, and F_{ij} is a geometric factor calculated by the following expression:

$$F_{ij} = \frac{1}{L_i} \int_{L_i} \int_{L_j} K(\mathbf{r},\mathbf{r}') d\Gamma(\mathbf{r}') \, d\Gamma(\mathbf{r}) \tag{8.3e}$$

Since $K(\mathbf{r},\mathbf{r}')$ is a function of both geometric location and the product of the cosines between the normals of the two participating walls and the vector connecting two points on the two surface elements, its evaluation has to take into account the relative geometric orientations of the two walls. For two adjacent walls, say, elements 1 and 2, which are perpendicular each other, the factor F_{12} is calculated as follows:

$$F_{12} = \frac{1}{A_1} \int_{A} \int_{A_2} \frac{\cos\theta_i \cos\theta_j}{2|\mathbf{r}_i - \mathbf{r}_j|} dA_1 dA_2 \tag{8.4e}$$

where we have used cosine instead of vectors for convenience. From Figure 8.2e, we have the following geometric relations:

$$\mathbf{s}_{ij} = |\mathbf{r}_i - \mathbf{r}_j| = \sqrt{(3 - x_i)^2 + y_i^2} \tag{8.5e}$$

$$\cos\theta_i = \frac{y_j}{|\mathbf{r}_i - \mathbf{r}_j|} = \frac{y_j}{\sqrt{(3 - x_i)^2 + y_i^2}} \tag{8.6e}$$

$$\cos\theta_j = \frac{(3 - x_i)}{|\mathbf{r}_i - \mathbf{r}_j|} = \frac{(3 - x_i)}{\sqrt{(3 - x_i)^2 + y_i^2}} \tag{8.7e}$$

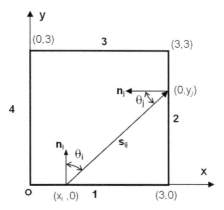

Figure 8.2e. Illustration of the integration procedure for geometric factor calculations

Substituting these relations into Equation 8.4e, F_{12} is calculated analytically,

$F_{12}=0.293$

The same procedure may be applied to other pairs of participating elements. Without repeating the numerical details, we give the results for these F_{ij} factors,

$$F_{13} = F_{31} = F_{24} = F_{42} = 0.414$$

$$F_{21} = F_{23} = F_{32} = F_{41} = F_{14} = F_{34} = F_{43} = F_{12} = 0.293$$

$$F_{11} = F_{22} = F_{33} = F_{44} = 0.0$$

The last relation should be obvious from the geometric relation shown in Figure 8.2e. Substituting these factors, and the sides of the elements, into Equations 8.1e and 8.2e, we can calculate the coefficients A_{ij} and B_{ij}. For example, with $\varepsilon_1 = \varepsilon_2 = \varepsilon_3 = 0.5$, A_{12} and B_{12} are calculated with the results,

$$A_{12} = -(1 - \varepsilon_1)F_{12}L_1 = -0.5 \times 0.293 \times 3 = -0.4395$$

$$B_{12} = \varepsilon_1 F_{12} L_1 = 0.5 \times 0.293 \times 3 = 0.4395$$

Note here that both A_{12} and B_{12} have a length unit (m). Furthermore, the temperatures of all the walls are known, which allow us to calculate the blackbody emissive power at each wall,

$$E_{b1} = \sigma T_1^4 = (5.67 \times 10^{-8})(600)^4 = 7.348 \ \text{kW/m}^2$$

$$E_{b2} = \sigma T_2^4 = (5.67 \times 10^{-8})(1700)^4 = 473.6 \text{ kW/m}^2$$

$$E_{b3} = \sigma T_3^4 = (5.67 \times 10^{-8})(1400)^4 = 217.8 \text{ kW/m}^2$$

$$E_{b4} = \sigma T_2^4 = (5.67 \times 10^{-8})(1700)^4 = 473.6 \text{ kW/m}^2$$

With the coefficients A_{ij} and B_{ij} calculated as described above and substituted into Equation 8.15, we have the following matrix equation:

$$\begin{pmatrix} 3.0000 & -0.4395 & -0.6210 & -0.4395 \\ -0.4395 & 3.0000 & -0.4395 & -0.6210 \\ -0.6210 & -0.4395 & 3.0000 & -0.4395 \\ -0.4395 & -0.6210 & -0.4395 & 3.0000 \end{pmatrix} \begin{pmatrix} q_1 \\ q_2 \\ q_3 \\ q_4 \end{pmatrix}$$

$$= \begin{pmatrix} -1.5000 & 0.4395 & 0.6210 & 0.4395 \\ 0.4395 & -1.5000 & 0.4395 & 0.6210 \\ 0.6210 & 0.4395 & -1.5000 & 0.4395 \\ 0.4395 & 0.6210 & 0.4395 & -1.5000 \end{pmatrix} \begin{pmatrix} 7.348 \\ 473.6 \\ 217.8 \\ 473.6 \end{pmatrix} = \begin{pmatrix} 540.5262 \\ -317.3419 \\ 94.1575 \\ -317.3419 \end{pmatrix}$$

Inverting the above matrix using Gaussian elimination, or any other matrix methods, we obtain the heat fluxes along the four walls,

$$\begin{pmatrix} q_1 \\ q_2 \\ q_3 \\ q_4 \end{pmatrix} = \begin{pmatrix} 159.040 \\ -97.4039 \\ 35.7678 \\ -97.4039 \end{pmatrix}$$

Here the heat flux is in kW/m^2. As a check, the total heat flow balance in the cavity is calculated,

$$\sum_{i=1}^{4} \dot{Q}_i = \sum_{i=1}^{4} q_i L_i \cong 10^{-10} \text{ W}$$

This shows that the total radiative energy in the cavity is conserved, which confirms that the calculated results are correct.

Discontinuous Galerkin Approach. Figure 8.3e compares the heat flux distributions along the surface of a simple 2-D cavity, which are calculated using the analytical method and the Galerkin boundary element scheme with 32 constant elements or 32 linear elements. The linear boundary element mesh is also shown, and the constant elements are defined such that the quantities such as heat flux and emissive powers are evaluated at the center of the element. The numerical calculations used 5 integration points, which seems to be a reasonable choice, resulting in an error of less than 0.001% in identity at worst. This means that all radiation leaving point $\mathbf{r}$ must be intercepted by the enclosure surfaces,

$$\oint K(\mathbf{r},\mathbf{r}')\,d\Gamma(\mathbf{r}') = 1 \tag{8.8e}$$

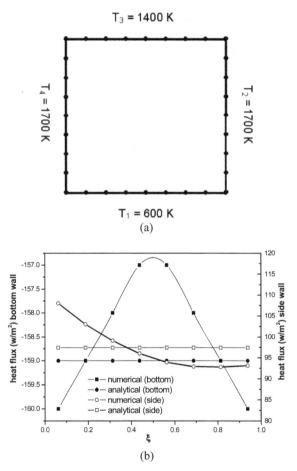

Figure 8.3e. Discontinuous Galerkin solution of external radiation energy exchange using a refined mesh: (a) mesh discretization with dots representing the nodal points, and (b) comparison of the numerical results with the analytic solution for boundary heat flux distributions. Here ξ is the non-dimensional length of side: $\xi = x/L$ (or y/L)

For this simple problem, one integration point gives an error of about 3–5%. Larger errors occur near the corner. Inspection of these results shows that the numerical results using both types of elements are in good agreement with the analytical solutions. We note here also that the analytical solutions used the view factor, and assumed a constant heat flux along one side. As such, the analytical solutions should be viewed as an approximation, or a reference, with which numerical solutions are compared. A further check of the overall heat balance gives a relative error of 0.02% for constant element. Here, the error being measured is the difference between the analytical solution and the average of the numerical heat fluxes along one wall, divided by the analytical value. The negative sign for the heat flux indicates that the heat flows into the surface or out of the square enclosure. These results are consistent with the physical processes. For instance, the fluxes are higher near edges of the bottom wall and lower at the center, which is attributed to the fact that the edges are influenced more by the side walls at higher temperatures. This is seen in the computed results by both linear and constant elements. The results from the linear elements, however, show discrepancies at the corner where the heat flux is physically discontinuous, because of an abrupt change in curvature from one side to the adjacent one.

Example 8.2. Develop a discontinuous Galerkin boundary element algorithm for the numerical solution of surface radiation heat transfer in a 2-D cavity with geometric obstruction.

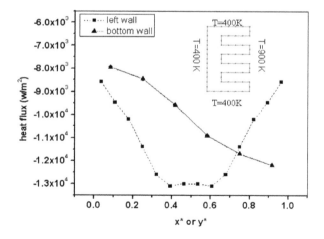

Figure 8.4e. Boundary heat flux distribution in a 2-D enclosure with complex geometric blockages. The insert shows the 2-D geometric configuration used for calculations. Significant blockage occurs between surface elements of the cavity: $\varepsilon = 0.5$, $x^* = x/3$, $y^* = y/7$, width of the bottom = 3 and height of the block = 7

Solution. In this example, a 2-D cavity with various geometric obstructions is considered. The problem definition and the discretization for numerical computations are given in Figure 8.4e. The discontinuous Galerkin solution used 64 constant elements and adaptive integration algorithms were invoked to treat the kernel function for the elements that are partially shadowed by the third party elements. The normalization factor is also checked for this computation and the error is less than 0.5%. The calculated results are consistent with thermal radiation heat transfer principles and the temperature distribution along the left sidewall shows a perfect symmetric profile [1, 2].

Example 8.3. Implement the discontinuous Galerkin boundary element algorithm described in Section 8.3.2 for the numerical solution of radiative heat transfer in a cylinder with internal blockage as shown in Figure 8.5e(a).

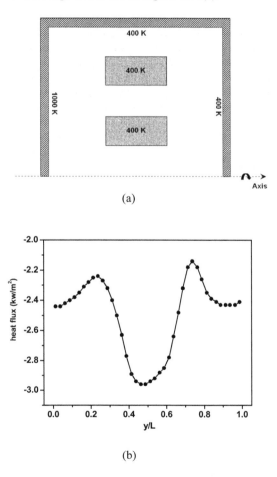

(a)

(b)

Figure 8.5e. A cylindrical cavity with two internal concentric cylinders that result in partial blocking of radiation between some surfaces of the enclosure: (a) geometric arrangement and (b) heat flux distribution along the right side wall of the outer cylinder. Parameters used for calculations: $\varepsilon = 0.5$ for all walls and other conditions are shown in the figure

Solution. As in the 2-D case, the numerical algorithm for an axisymmetric geometry is also checked by analytically calculating the geometric factors and the heat fluxes for an axisymmetric cavity without internal blockages. The procedure is very similar to that given in Example 8.1. It is important that this checking be carried out after the blockage detection algorithm has also been implemented. As expected, for the case without internal blockage, the detection scheme should signal a full view between any two surface elements in the cavity.

After this checking, the discontinuous Galerkin algorithm then can be used to predict the surface radiation exchange in more complex configurations involving geometric blockages. One of these calculations for a hollow cylindrical geometry with two cylindrical rings lying inside the cylindrical cavity is shown in Figure 8.5e(a). For the calculations, 80 constant elements are used, which are determined following the usual procedure for mesh independence check, and adaptive integration algorithms were invoked to treat the kernel function for the elements that are partially shadowed by the third party elements. The calculated results are shown in Figure 8.5e(b), which are consistent with thermal radiation heat transfer principles. The identity condition is checked as well and the worst error is less than 0.2%. For the axisymmetric problems, the identity condition is slightly different from 2-D and 3-D cases, and is given by the following expression:

$$\oint r'(\mathbf{r},\mathbf{r}')K(\mathbf{r},\mathbf{r}')d\Gamma(\mathbf{r}') = 1 \tag{8.9e}$$

The heat flux distribution along the right side wall shows the clear shadowing effects provided by the two internal hollow cylinders. The thermal radiation from the left surface at a higher temperature impinges upon the right surface. In the region where the internal blockages occur, a lower heat flux is calculated. In the region corresponding to the gap between the two internal blockages, the heat flux is high, which is attributed to the fact that the thermal radiation from the left surface is not blocked.

Example 8.4. Consider the radiative heat transfer in a 3-D cavity with blockages and calculate the heat fluxes at the walls of the enclosure using the algorithm discussed in Section 8.3.3.

Solution. This example considers the surface radiation exchange in a rather complex 3-D closure with several internal blocks of various heights. The boundary element meshes and boundary conditions used for computations are given in Figure 8.6e(a). The computed results are selectively plotted in Figures 8.6e(b) and (c). To check the calculations, the identity condition is also calculated and the error is less than 2%. Detailed analysis shows that a major portion of the error comes from the integration between two surface elements that are very close together. The error can be reduced with a further refinement of mesh sizes. It is worth noting that in some calculations reported in the view factor calculations, an error of as high as 70% in identity condition was reported, when much less complex blockages are in place [5].

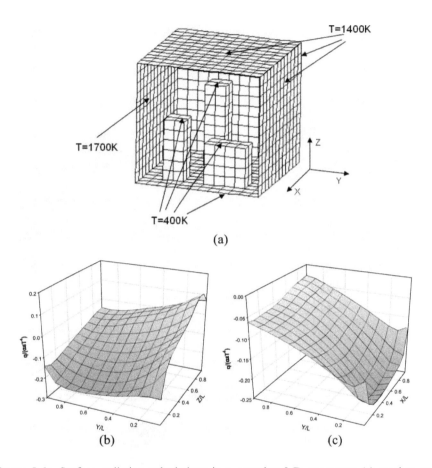

Figure 8.6e. Surface radiation calculations in a complex 3-D geometry: (a) meshes and thermal conditions used for calculations, (b) heat flux on the left lateral surface of the enclosure and (c) heat flux on the top surface of the enclosure. The dimensions of the enclosure is 1.2 m × 1.2 m × 1.2 m, and the temperature used to normalize the heat flux is 1400 K. Emissivity of all surfaces is 0.5, and side surface temperatures of the internal obstructions vary from 400 K to 1700 K. The front surface is removed for the purpose of illustration

Example 8.5. Calculate the temperature distribution and fluid flow in a furnace as shown in Figure 8.19.

Solution. This example considers calculations involving combined heat transfer modes are presented. The radiation algorithm described in Section 8.3.3 has been successfully integrated with a finite element code, following the procedure given in Section 8.4. The finite element code is capable of performing steady state and transient fluid flow and heat transfer calculations in 2-D, 3-D and axisymmetry geometries [20–22]. The computed results for an industrial processing system are shown in Figure 8.19, where the coupling of conduction, convection, and radiation is considered.

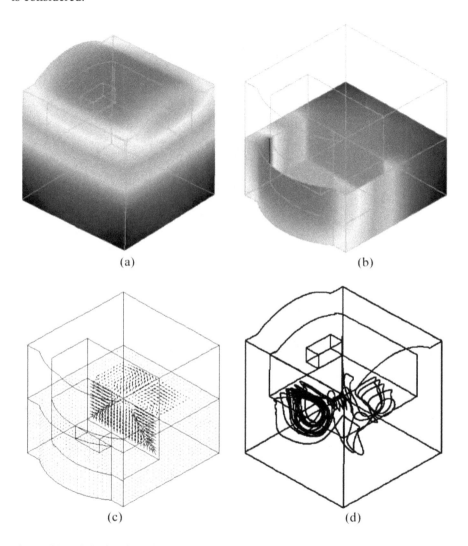

(a)   (b)

(c)   (d)

igure 8.7e. Calculated results of mixed mode heat transfer in an industrial furnace using the coupled Galerkin boundary/finite element method: (a) temperature distribution in the furnace, (b) body-cut view of temperature distribution, (c) body-cut view of velocity distribution in the liquid pool – maximum velocity is 1.9 mm/s and (d) particle trajectory plot. Parameters for calculations: liquid pool is 0.15 m × 0.3 m × 0.3 m and filled with Ga melt. The top surface temperature is 340 K and the environment is at 295 K. The emissivity of all surfaces is 0.5 and the thermal conductivity and specific heat of the furnace walls are 10 W/m K and 130 kJ/kg K, respectively. The gravity is in the opposite x direction (see Figure 8.19). The maximum temperature is 340 K and the minimum temperature is 328 K

The calculations used 25984 8–node brick finite element elements, which are determined to be the "optimal" mesh for the simulation after a grid independence check [20]. There, a total of 2400 constant boundary elements was used for surface radiation calculations. The nonlinear solution is obtained using the successive substitution method. Because E_b is proportional to the fourth power of local surface temperatures, a relaxation parameter of 0.1, which is common in mixed mode calculations, was used to obtain converged results. The convergence was achieved in 21 iterations and residuals are progressively smaller. For this problem, the criterion for nonlinear convergence is set, such that the relative error is less than 1×10^{-4}.

The calculated results are plotted in Figure 8.7e where the overall temperature distribution, the particle tracing, and the velocity fields in a few cutting planes are given. The metal surface is heated up by the radiation from the cylindrical roof at a higher temperature. As a result, the temperature at the middle of the surface is lower than that at the sides of the surface. This temperature field creates a surface force field such that a higher pulling force exists at the middle of the surface, which pulls the fluid particles on the surface from the side towards the center of the surface. Because of the mass conservation and also buoyancy forces, a recirculating flow pattern develops, which is clearly revealed in the cutting plane representations.

As shown in this example, the coupling of surface radiation calculations and the finite element calcualtions for engineering applications can be achieved either directly or iteratively. Numerical experience indicates that the coupling may be as tedious as it gets, or as easy as one would like, depending on the strategies to be implemented. The direct coupling involves incorporating the boundary element matrix $\mathbf{A}^{-1}\mathbf{B}$ into the finite element global matrix by treating the entire boundary integral as one macro-finite element. The implementation of this direct procedure can be very cumbersome and will result in a significant change in the finite element global matrix structure (see Section 8.5.1). For linear problems with a moderate boundary size, this approach is favored, in that the increase in the bandwidth of the global matrix is relatively small, and the results can be obtained directly without iteration between surface and domain calculations. On the other hand, for highly nonlinear problems with a large boundary element size, the direct coupling greatly increases the finite element global matrix bandwidth, and the iterative solution is thus more effective and also easy to implement. For cases falling in between, experience is the key to obtaining faster solutions.

Exercises

1. A 10 m by 30 m rectangular cavity has an emissivity of 0.6 for the two longer walls, placed at the top and bottom, and 0.4 for the shorter ones, placed on two sides. The wall temperatures are uniform at 1000°C for the top and bottom walls, 300°C for the left side wall and 700°C for the right side wall. Using both the analytical and numerical methods, determine the net heat transfer along the four walls and compare the analytic and numerical results.

2. A cylinder 40 cm in diameter and 40 cm high has the bottom disk surface maintained at 950 K with $\varepsilon = 0.75$. The vertical cylindrical surface is perfectly insulated. The top disk has a 20-cm-diameter hole in the center with a surface temperature of 650 K and $\varepsilon = 0.5$. The top surface is a blackbody wall at 400 K. Calculate the heat loss, emissive power, and irradiation for all three surfaces to the top surface, and the temperature of the vertical cylindrical surface.

3. A room is represented by the following 3-D enclosure of a rectangular prism, where the ceiling is 10 m × 6 m with an emissivity of 0.8 and is kept at a constant temperature of 42°C by an embedded electric heater. Heaters are also used to maintain the floor at 50°C, which has an emissivity of 0.9. The right wall, 10 m × 4 m in size, has an emissivity of 0.7 and reaches a temperature of 10°C during a cold, winter day. The front wall, 6 m × 4 m in dimension, and also other walls, are all well insulated, with an emissivity of 0.65. Calculate the net radiation heat transfer from each surface, using one element for each wall and a constant element approximation. Calculate the heat flux distribution on the walls using the following mesh discretization consisting of a 1 m × 1 m element with constant element approximation. Compare the calculated total heat flux for two different discretizations.

4. A 1 m diameter cylinder, 1 m long, is maintained at 1000 K and has an emissivity of 0.65. Another cylinder, 2 m in diameter and 1 m long, encloses the first cylinder and is perfectly insulated. Both cylinders are placed in a large room maintained at 300 K. Calculate the heat lost by the inner cylinder.

References

[1]   Modest MF. Radiative Heat Transfer. New York: McGraw-Hill, 1993.
[2]   Siegel R, Howell JR. Thermal Radiation Heat Transfer, 3rd Ed. Washington, D.C.: Hemisphere Publishing Company, 1992.
[3]   Blobner J, Bialecki RA, Kuhn G. Boundary Element Solution of Coupled Heat Conduction-Radiation Problems in the Presence of Shadow Zones. Numer. Heat Transf. Part B. 2001; 39: 451–478.
[4]   Miyahara S, Kobayashi S. Numerical Calculation of View Factors for an Axially Symmetrical Geometry. Numer. Heat Transf. Part B. 1995; 28: 437–453.
[5]   Dupret F, Nicodeme P, Ryckmans Y, Wouters P, Crochet MJ. Global Modeling of Heat Transfer in Crystal Growth Furnaces. Int. J. Heat Mass Transf. 1990; 33(9): 849–1871.
[6]   Emery AF, Johansson O, Lobo M, Abrous A. A Comparative Study of Methods for Computing the Diffuse Radiation View Factors for Complex Structures. ASME J. Heat Transf. 1991; 113(12): 4132–422.
[7]   Li BQ, Cui X, Song SP. The Galerkin Boundary Element Solution of Surface Radiation Problems. Int. J. Eng. Anal. with Bound. Elem. 2004; 28: 881–892.
[8]   Song SP, Li BQ. Boundary Integral Solution of Thermal Radiation Exchanges in Axisymmetric Furnaces. Numer. Heat Transf. Part B. 2003; 44: 489–507.

[9] Cui X, Li BQ. A Parallel Galerkin Boundary Element Method for Surface Radiation and Mixed Heat Transfer Calculations in Complex 3-D Geometries. Int. J. Numer. Meth. Eng. 2004; 61: 2020–2044.

[10] Cui X, Song SP, Li BQ. A Galerkin Boundary Element Method for Thermal Radiation Problems. ASME Winter Annual Meeting, 2002; Paper #: IMECE2002–33900.

[11] Bialecki RA. Solving Heat Radiation Problems Using the Boundary Element Method. Boston, MA: Computational Mechanics Publications, 1993.

[12] Brebbia C, Telles J, Wrobel L. Boundary Element Techniques. New York: Springer, 1984.

[13] Kuppurao S, Derby J. Finite Element Formulations For Accurate Calculation of Radiant Heat Transfer in Diffuse-Gray Enclosures. Numer. Heat Transf. Part B. 1993; 24: 431–454.

[14] Minkowycz W, Haji-Sheikh A. The Sparrow Galerkin Solution of Radiation Exchange and Transition to Finite Elements. Int. J. Heat Mass Transf. 1999; 42: 1353–1362.

[15] Bastian M, Li BQ. An Efficient Automatic Mesh Generator for Quadrilateral Elements Implemented Using C++. Finite Elem. in Anal. Design 2003; 39: 905–930.

[16] Rogers D. Procedural Elements for Computer Graphics. New York: McGraw-Hill, 1998.

[17] Foley JD. Computer Graphics: Principles and Practice. Reading MA: Addison-Wesley Publishing Company, 1997.

[18] Zhang X. Data Visualization and Mesh Generation for Finite Element Simulation of Fluid Flow and Heat Transfer. MS. Thesis. Pullman: Washington State University, 2001.

[19] Song SP, Li BQ. Finite Element Computation of Shapes and Marangoni Convection in Droplets Levitated by Electric Fields. Int. J. Comp. Fluid Dyn. 2001; 15: 293–308.

[20] Huo Y, Li BQ. Three-Dimensional Marangoni Convection in Electrostatically Levitated Droplets Under Microgravity. Int. J. Heat Mass Transf. 2004; 47: 3533–3547.

[21] Li K, Li BQ. Effects of Magnetic Fields on g-Jitter Induced Convection and Solute Striation During Space Processing of Single Crystals. Int. J. Heat Mass Transf. 2003; 46: 4799–4811.

[22] Li K, Li BQ, Numerical Analyses of g-Jitter Induced Convection and Solidification in Magnetic Fields, J. Thermophys. Heat Transf. 2003; 17(2): 199–209.

[23] Li K, Li BQ. A 3-D Model for g-Jitter Driven Flows and Solidification in Magnetic Fields, J. Thermophys. Heat Transf. 2003; 17(4): 498–508.

[24] Song SP, Li BQ. Coupled Boundary/Finite Element Computation of Magnetically and Electrostatically Levitated Droplets. Int. J. Numer. Meth. Eng. 1999; 44(8): 1055–1077.

[25] Song SP, Li BQ, Khodadadi JA. Coupled Boundary/Finite Element Solution of Magnetothermal Problems, Int. J. Numer. Methods Heat Fluid Flow 1998; 8(3): 321–349.

[26] Song SP, Li BQ. Coupled Boundary/Finite Element Analysis of Magnetic Levitation Processes: Free Surface Deformation and Thermal Phenomena. ASME J. Heat Transf. 1998; 120: 492–504.

[27] Holman JP. Heat Transfer, 9th Ed. New York: McGraw-Hill, 2002.

[28] Incorpera F, DeWitt C. Introduction to Heat Transfer. New York: John Wiley, 2002.

9

Radiative Transfer in Participating Media

In the last chapter, surface (or external) radiation exchange between the walls of an enclosure was considered. An important assumption associated with external radiative heat transfer is that the enclosure either bounds a vacuum or is filled with non-participating media. A medium is considered participating in radiative thermal transfer if it absorbs, emits, or scatters a thermal ray as it travels through the medium. Radiative heat transfer in a participating medium is also referred to as internal radiation, which occurs in many engineering thermal systems. An example of internal radiation is a high temperature combustion gas mixture, which is known to absorb, emit, and scatter the thermal energy. Another example is the semitransparent melt from which optical single crystals are pulled out. For the problems of radiative transfer in a participating medium, the absorption, emission and scattering effects must be considered in order to provide an accurate estimate of thermal energy transfer in the system.

Thermal radiation in participating media is governed by the radiative transfer equation, which describes the energy balance along a thermal ray. Owing to the importance of internal radiation transfer in thermal engineering applications, many numerical techniques have been developed to predict the phenomena and to assist in thermal designs involving radiative heat transfer. The widely used numerical algorithms include the finite difference, discrete ordinates, Monte Carlo, zonal method, and finite element methods as well as other approximation methods such as exponential kernel approximation, direct numerical integration, reduction of the integral order, and the YLX method. These methods have been documented in detail in two recent monographs on radiation heat transfer [1, 2].

This chapter discusses the discontinuous Galerkin finite element method for the solution of thermal radiative transfer problems involving participating media. The application of the discontinuous method to the solution of the internal radiative heat transfer problems has attracted attention only recently, and the literature is still expanding. It starts with the differential-integral equation governing the transfer of the radiation intensity and the boundary conditions required for the solution. Two popular approximation methods for the solution of the radiative transfer equation are also presented, which will be used to compare with the numerical solutions. The general discontinuous Galerkin formulation for the solution of the radiative transfer equation is then presented. This is followed by a discussion of detailed numerical procedures. Analytic expressions are given for 1-D, 2-D and 3-D elemental calculations, whenever possible, so that they can be used to develop an efficient computer code. These expressions are derived for linear elements, which are the most frequently used elements in practical applications. Higher order elements in general would require numerical integration, for which the general expressions are also provided.

For the understanding of the discontinuous Galerkin procedure for numerical solution, examples are useful. A simple 1-D example, for which both analytic and approximate solutions are available, is considered first, and analyzed in detail. This simple example is chosen to illustrate the very basic steps to develop a discontinuous Galerkin solution procedure. Other more complex examples are given for both 2-D and 3-D geometries and for heat transfer problems involving conduction, convection and internal radiation. The chapter ends with an example of practical significance to show how the discontinuous Galerkin method can be combined with other numerical methods to develop numerical models for thermal system design applications.

9.1 Governing Equation and Boundary Conditions

Radiative transfer in a participating medium is described by the radiant intensity, and is affected by the interaction between the traveling thermal rays and the medium, which includes emission, absorption and scattering. The governing equation and the boundary conditions are derived based on the local optical or thermal balance.

9.1.1 Radiative Transfer Equation

The radiative transfer equation governs the distribution of the radiant intensity $I(\mathbf{r},\mathbf{s})$, sometimes called radiation intensity, which is a function of both coordinates $\mathbf{r}$ and direction $\mathbf{s}$ [3]. The radiant intensity is defined as radiative energy flow per unit solid angle, and unit area normal to the thermal rays. The transfer equation is derived from the local conservation of radiative energy, as shown in Figure 9.1, and has the following general form [1, 2]:

$$\frac{1}{c}\frac{\partial I_\lambda(\mathbf{r},\mathbf{s},t)}{\partial t} + \frac{\partial I_\lambda(\mathbf{r},\mathbf{s},t)}{\partial s} = -\beta_\lambda(\mathbf{r})I_\lambda(\mathbf{r},\mathbf{s},t) + \kappa_\lambda(\mathbf{r})I_{b\lambda}(\mathbf{r},t)$$

$$+ \frac{\sigma_\lambda(\mathbf{r})}{4\pi}\int_{4\pi} I_\lambda(\mathbf{r},\mathbf{s}',t)\Phi(\mathbf{s},\mathbf{s}')d\Omega' \qquad (9.1)$$

where $\beta_\lambda(\mathbf{r}) = \kappa_\lambda(\mathbf{r}) + \sigma_\lambda(\mathbf{r})$ is the extinction coefficient, $\kappa_\lambda(\mathbf{r})$ is the absorption coefficient, $\mathbf{s} = \sin\theta\cos\phi\,\hat{i} + \sin\theta\sin\phi\,\hat{j} + \cos\theta\,\hat{k}$, $\Omega(\mathbf{s})$ is the solid angle associated with $\mathbf{s}$, $\sigma_\lambda(\mathbf{r})$ is the scattering coefficient, and $d\Omega = \sin\theta\,d\theta\,d\phi$ is the differential solid angle. The phase function $\Phi(\mathbf{s},\mathbf{s}')$ satisfies the following condition:

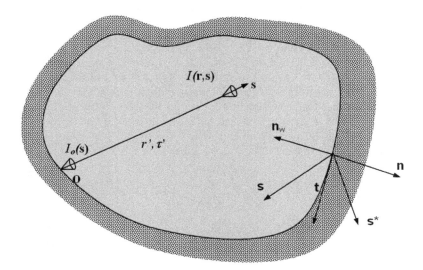

Figure 9.1. Schematic representation of internal radiation heat transfer and symmetry boundary condition

$$\frac{1}{4\pi}\int_{4\pi}\Phi(\mathbf{s},\mathbf{s}')d\Omega'=1 \qquad (9.2)$$

In Equation 9.1, $c = ds/dt$ is the speed with which radiation intensity travels. All the quantities are a function of location in space, time and wave numbers. The intensity and the phase function are also dependent upon directions $\mathbf{s}$ and $\mathbf{s}'$. For many engineering applications, thermal radiation reaches equilibrium far faster than other heat transfer mechanisms and thus a quasi-steady state approximation is often used. This allows us to drop out the transient term. To facilitate discussion, it is further assumed that all quantities are spectral independent, although the numerical algorithms discussed later in this chapter apply equally to the case where these quantities are spectral dependent as well. With these approximations taken into account, Equation 9.1 is simplified as

$$\frac{\partial I(\mathbf{r},\mathbf{s})}{\partial s}=-\beta(\mathbf{r})I(\mathbf{r},\mathbf{s})+\kappa(\mathbf{r})I_b(\mathbf{r})+\frac{\sigma(\mathbf{r})}{4\pi}\int_{4\pi}I(\mathbf{r},\mathbf{s}')\Phi(\mathbf{s},\mathbf{s}')d\Omega' \qquad (9.3)$$

This equation is a first order integral-differential equation for the radiant intensity, and is of the hyperbolic type. In Equation 9.3, the first term on the left measures the change in $I(\mathbf{s}, \mathbf{r})$ over a differential distance in the $\mathbf{s}$ direction, the first term on the right represents the loss to the medium due to absorption or scattering, the second is the local emission, and the third represents the contribution to the intensity in the $\mathbf{s}$ direction that results from the scattering of intensity in other directions. The equation needs to be solved for I in each given direction s at every position.

9.1.2 Boundary Conditions

Like any other boundary value problems, the solution of the radiative transfer equation requires the knowledge of intensity distribution on the boundary surrounding the participating medium. For an opaque surface emitting and reflecting diffusively, the exiting intensity is independent of direction. At a point r on a surface facing the participating medium, the thermal balance leads to the following equation for intensity I (see also Figure 9.1):

$$I(\mathbf{r},\mathbf{s}) = \varepsilon(\mathbf{r})I_b(\mathbf{r}) + \frac{1-\varepsilon(\mathbf{r})}{\pi} \int_{\mathbf{s}'\cdot\mathbf{n}_w<0} I(\mathbf{r},\mathbf{s}') |\, \mathbf{s}'\cdot\mathbf{n}_w\,|\, d\Omega' \qquad (9.4)$$

where $\mathbf{n}$ is the surface normal pointing into the medium and $\mathbf{s}'$ is the direction of irradiation (i.e., incoming radiative heat flux). Here, the first term on the right represents the emission from the surface, and the second term on the right represents the reflected portion of the incoming thermal energy. In the case of a black surface, $\varepsilon(\mathbf{r})=1$ and the last term disappears. The boundary condition is then simplified to

$$I(\mathbf{r},\mathbf{s}) = I_b(\mathbf{r}) \qquad (9.5)$$

In thermal radiation, an opaque surface is defined as the surface of a medium with transmitivity being zero, which means that radiation cannot penetrate it. For an opaque surface that emits diffusively but reflects specularly, the intensity leaving the surface has two contributions: one from diffusive emission and the diffusive part of reflection, and the other from the specular part of reflection. Thus, the thermal balance gives the following expression for the outgoing intensity:

$$I(\mathbf{r},\mathbf{s}) = \varepsilon(\mathbf{r})I_b(\mathbf{r}) + \frac{\rho^d(\mathbf{r})}{\pi} \int_{\mathbf{s}'\cdot\mathbf{n}_w<0} I(\mathbf{r},\mathbf{s}') |\, \mathbf{s}'\cdot\mathbf{n}_w\,|\, d\Omega' + \rho^s(\mathbf{r})I(\mathbf{r},\mathbf{s}_s)$$

$$(9.6)$$

where $\mathbf{s}_s$ is the specular direction, defined as the direction from which a light beam must hit the surface in order to travel in the direction of s after a specular reflection. Also, ρ^d and ρ^s are, respectively, the diffusive and specular reflectivities of the surface. This direction can be determined using the kinematic relation between the incident and the reflected waves [2].

For an opaque surface with arbitrary surface properties, the reflectivity is a function of location and direction. Thus, we have for the outgoing intensity,

$$I(\mathbf{r},\mathbf{s}) = \varepsilon(\mathbf{r})I_b(\mathbf{r}) + \int_{\mathbf{s}'\cdot\mathbf{n}_w<0} \rho''(\mathbf{r},\mathbf{s}',\mathbf{s})I(\mathbf{r},\mathbf{s}') |\, \mathbf{s}'\cdot\mathbf{n}_w\,|\, d\Omega' \qquad (9.7)$$

with $\rho'' = \rho/\pi$. For a semitransparent surface, external radiation may propagate into the enclosure and an effective emissivity may then be used in the above equations. If the bounding surface is totally transparent or is an opening, then the emission from the boundary does not exist, or $\varepsilon = 0$.

When systems possess certain symmetry, the symmetry boundary conditions can be applied, which take the following form:

$$\begin{cases} I(\mathbf{r},\mathbf{s}) = I(\mathbf{r},\mathbf{s}^*) \\ \mathbf{n}\cdot\mathbf{s} = -\mathbf{n}\cdot\mathbf{s}^* \\ \mathbf{s}\times\mathbf{s}^*\cdot\mathbf{n} = 0 \end{cases} \tag{9.8}$$

where the $\mathbf{s}^*$ is the symmetric radiation direction of $\mathbf{s}$, in respect to the tangent of the boundary, with both $\mathbf{s}$ and $\mathbf{s}^*$ lying on the plane of $\mathbf{t}$–$\mathbf{n}$ (see Figure 9.1).

With the radiation intensity distribution known, various quantities of interest to radiation heat transfer can be calculated. Two of them are the heat fluxes and their derivatives, which are determined by the following expressions for gray media:

$$q_i(\mathbf{r}) = \int_{4\pi} I(\mathbf{r},\mathbf{s})\mathbf{s}\cdot\hat{i}\,d\Omega \tag{9.9}$$

$$\nabla\cdot\mathbf{q}(\mathbf{r}) = \kappa(4\sigma_s T^4(\mathbf{r}) - \int_{4\pi} I(\mathbf{r},\mathbf{s})\,d\Omega) \tag{9.10}$$

Note that Equation 9.10 often appears as a source term in the thermal energy conservation equation.

Another quantity that is also important for internal radiation calculations is the incident radiation $G(\mathbf{r})$,

$$G(\mathbf{r}) = \int_{4\pi} I(\mathbf{r},\mathbf{s})\,d\Omega \tag{9.11}$$

where G is the irradiation or incident radiation onto a surface.

9.2 Approximation Methods

The integro-differential equation describes the radiative intensity that depends on five variables: three space coordinates ($\mathbf{r}$) and two direction coordinates ($\mathbf{s}$). The solution of the problem represents a challenging task. Before we present the discontinuous finite element formulation, it is constructive to discuss two popular approximation methods: the discrete ordinate method and the spherical-harmonics method. These methods are very often used to obtain the solution of the radiative transfer equation, and will be used later in this chapter to check the numerical solutions from the discontinuous finite element computations.

9.2.1 The Discrete Ordinate Method

The discrete ordinate method was first proposed by Chandrasekhar [3] to study the stellar and atmospheric radiation phenomena. The method was later extended to study general radiative heat transfer problems [4, 5]. The basic idea of the discrete ordinates is that the integrals over directions are replaced by numerical quadratures of discrete different directions; that is

$$\int_{4\pi} I(\mathbf{r},\mathbf{s})\,d\Omega = \sum_{i=1}^{n} w_i I(\mathbf{r},\mathbf{s}_i) \tag{9.12}$$

where the w_i values are the quadrature weights associated with the directions $\mathbf{s}_i$, $i = 1, 2, 3, ..., n$. As a result, the integro-differential equation (i.e., Equation 9.3) is approximated by a set of n first order partial differential equations,

$$\frac{\partial I(\mathbf{r},\mathbf{s}_j)}{\partial s} = -\beta(\mathbf{r})I(\mathbf{r},\mathbf{s}_j) + \kappa(\mathbf{r})I_b(\mathbf{r}) + \frac{\sigma(\mathbf{r})}{4\pi}\sum_{i=1}^{n} w_i I(\mathbf{r},\mathbf{s}_i)\Phi(\mathbf{s}_j,\mathbf{s}_i),$$

$$j = 1, 2, ..., n \tag{9.13}$$

The boundary condition can be integrated following the same procedure. For an opaque, diffuse surface,

$$I(\mathbf{r},\mathbf{s}_j) = \varepsilon(\mathbf{r})I_b(\mathbf{r}) + \frac{1-\varepsilon(\mathbf{r})}{\pi}\sum_{\mathbf{s}_i \cdot \mathbf{n}_w < 0} w_i I(\mathbf{r},\mathbf{s}_i)\big|\, \mathbf{s}_i \cdot \mathbf{n}_w\,\big|, \mathbf{s}_j \cdot \mathbf{n}_w > 0 \tag{9.14}$$

To carry out the numerical computation, a thermal ray is released from a point on the enclosure surface, $\mathbf{s}_j \cdot \mathbf{n}_w > 0$, and is allowed to travel along the direction of $\mathbf{s}_i$ until it strikes another point, $\mathbf{s}_j \cdot \mathbf{n}_w < 0$, to be absorbed or reflected. Here $\mathbf{n}_w$ is the surface normal pointing inward to the enclosure. The n equations can be solved using the standard numerical or analytical methods.

9.2.2 The Spherical Harmonics Method

The spherical harmonics method seeks the solution of $I(\mathbf{r}, \mathbf{s})$ by transforming the equation of radiative transfer into a set of simultaneous partial differential equations through eliminating the direction dependence. By this method, the radiative intensity at location $\mathbf{r}$ within the medium is treated as a scalar function on the surface of a sphere of unit radius surrounding the point $\mathbf{r}$. As such, the intensity is expressed in terms of spherical harmonics,

$$I(\mathbf{r},\mathbf{s}) = \sum_{l=0}^{\infty}\sum_{m=-l}^{l} I_l^m(\mathbf{r})Y_l^m(\mathbf{s}) \tag{9.15}$$

where $I_l^m(\mathbf{r})$ depends on $\mathbf{r}$ only and $Y_l^m(\mathbf{s})$ is the spherical harmonics defined by

$$Y_l^m(\mathbf{s}) = (-1)^{(m+|m|)/2} \left[\frac{(l-|m|)!}{(l+|m|)!} \right]^{1/2} e^{jm\varphi} P_l^{|m|}(\cos\theta) \tag{9.16a}$$

and satisfies the following eigenvalue differential equation:

$$\frac{1}{\sin\theta} \frac{\partial}{\partial\varphi} \left(\sin\theta \frac{\partial Y_l^m(\theta,\varphi)}{\partial\varphi} \right) + \frac{1}{\sin^2\theta} \frac{\partial^2 Y_l^m(\theta,\varphi)}{\partial^2\varphi} - l(l+1)Y_l^m(\theta,\varphi) = 0$$

$$\tag{9.16b}$$

Here θ and φ are the polar and azimuthal angles of the direction unit vector $\mathbf{s}$, and $P_l^m(\cos\theta)$ is the associated Legendre polynomials. To obtain the solution, Equation 9.15 is substituted into the integro-differential equation, and then integrated over the solid angle of 4π with respect to the spherical harmonics as weighting functions. The use of the orthogonal property of the spherical harmonics then results in a set of $(N+1)^2$ partial differential equations, where N is the highest order retained for l. Ou and Liou [6] gave the general expression for the intensity calculations for constant properties,

$$\left(\frac{\partial}{\partial\tau_x} + i\frac{\partial}{\partial\tau_y} \right) X_l^m + \left(\frac{\partial}{\partial\tau_x} + i\frac{\partial}{\partial\tau_y} \right) Y_l^m + \frac{\partial}{\partial\tau_z} Z_l^m + \left(1 - \frac{\omega A_l}{2l+1} \right) I_l^m$$

$$= (1-\omega)I_b\delta_{0l}, \qquad l \le N, \quad -N \le m \le N \tag{9.17a}$$

where

$$X_l^m = -\frac{\sqrt{(l+m+1)(l+m+2)}}{2(2l+3)} I_{l+1}^{m+1} + \frac{\sqrt{(l-m+1)(l-m+2)}}{2(2l-1)} I_{l-1}^{m+1} \tag{9.17b}$$

$$Y_l^m = +\frac{\sqrt{(l-m+1)(l-m+2)}}{2(2l+3)} I_{l+1}^{m-1} - \frac{\sqrt{(l+m-1)(l+m)}}{2(2l-1)} I_{l-1}^{m-1} \tag{9.17c}$$

$$Z_l^m = -\frac{\sqrt{(l-m+1)(l+m+1)}}{2l+3} I_{l+1}^m + \frac{\sqrt{(l-m)(l+m)}}{2(2l-1)} I_{l-1}^m \tag{9.17d}$$

where $\omega = \sigma/\beta$ is the single scattering albedo. In practice, the approximations with $N>3$ are rarely used. Most applications use $N=1$ or the P_1 approximation.

For an isotropic medium, the governing equation for the P_1 approximation may be expressed in terms of incident radiation G. With the lengthy derivations relegated to the textbooks by Modest [1], and Siegel and Howell [2], only the governing equations are given below,

$$\nabla^2 G = 3\kappa^2 (1 - A\omega)(1 - \omega)\left(G - 4\pi I_b\right) \tag{9.18}$$

and for an opaque, diffuse surface, the boundary condition becomes

$$-\frac{2-\varepsilon}{\varepsilon}\frac{2}{3-A\omega}\mathbf{n}\cdot\nabla G + G = 4\pi I_{bw} \tag{9.19}$$

The heat flux and its derivatives in the domain can then be calculated from Equations 9.9 and 9.10. The physical domain of participating medium radiation may be bounded or unbounded. In the case of a medium surrounded by opaque surfaces, the boundary condition is the Marshak type and the value of the constant A is equal to 2. If the model is a truncation of an infinite domain, the boundary condition is of the Mark type and $A = 3^{1/2}$ [1, 2].

Below we consider an example of radiation in a 1-D slab with participating medium. The problem is solved analytically using different methds presented above and these analytic expressions will be used to check the numerical calculations using the discontinuous finite element method.

Example 9.1. Raidation in a 1-D slab filled with an absorbing but non-scattering medium is perhaps the simplest configuration for radiative transfer study (see Figure 9.1e). Here, steady state radiation from a diffuse, gray source wall (1) across a non-scattering medium goes to a sink wall (2). Conduction and convection are neglected. The walls have prescribed temperatures T_1 and T_2, and are black ($\varepsilon_1 = \varepsilon_2 = 1$). The medium is assumed to be adiabatic with an absorption coefficient of κ. The spacing between the plates is L, and the optical depth τ is based on the coordinate normal to the plates, z. Derive analytic expressions for the radiative heat transfer.

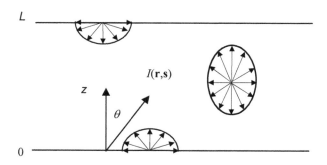

Figure 9.1e. Radiative transfer in a 1-D slab

Solution. We will use three different analytical techniques to solve this same problem. These analytic solutions will be compared with the discontinuous finite element solutions discussed in the next example.

Exact (analytic) solution. Because scattering is not present, $\beta = \kappa$. The different equation for the radiation intensity I, Equation 9.3, becomes

$$\frac{dI}{ds} = -\kappa I + \kappa I_b \tag{9.1e}$$

From the theory of linear differential equations, the general solution for this first order differential equation is,

$$I = I(0)\exp\left(-\int_0^s \kappa ds'\right) + \exp\left(\int_0^s \kappa ds'\right)\int I_b \exp\left(-\int_0^{s''} \kappa ds'\right)ds'' \tag{9.2e}$$

To obtain the exact form for this problem, $I(0)$ needs to be specified. To do that, we select the following coordinates for the convenience of expressing the solution (see also Figure 9.1.1e):

$$\tau = \int_0^z \kappa dz', \quad z = s\mu, \quad \mu = \cos\theta$$

$$s = z/\mu, \quad \kappa ds = \kappa dz/\mu = d\tau/\mu \text{ and } \tau = \int_0^z \kappa dz'/\mu = \int_0^s \kappa ds'/\mu$$

The radiation intensity I is a function of s; and, from the above, is a function of both position τ and direction μ (or θ). Note that for the problem to be one-dimensional, the intensity is constant along the azimuthal direction everywhere in space (see Figure 9.2e). For convenience, we may express the intensity in the following way:

$$I(\tau,\mu) = \begin{cases} I_+ & \mu > 0 \\ I_- & \mu < 0 \end{cases} \tag{9.3e}$$

then at the upper and lower walls (both being diffuse),

$$I_+(\tau = 0) = q_+/\pi, \quad I_-(\tau = \tau_L) = q_-/\pi \tag{9.4e}$$

where q denotes the heat flux supplied to the wall, and has only the z component for this problem. Here subscript $+$ is used to indicate intensity in the forward hemisphere $(\mu > 0)$, or from the lower surface to the upper surface, and subscript $-$ indicates intensity in the backward hemisphere $(\mu < 0)$, or from the upper surface to the lower surface.

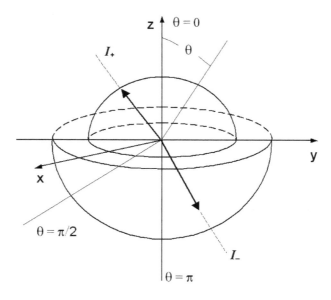

Figure 9.2e. Coordinate system used for deriving analytic solutions

If s is replaced by τ/μ, then the exact solution can be obtained as follows:

$$I_+(\tau,\mu) = \frac{q_+}{\pi}\exp\left[-\frac{\tau}{\mu}\right] + \int_0^\tau I_b(\tau')\exp\left[-\frac{\tau'-\tau}{\mu}\right]\frac{d\tau'}{\mu} \qquad (9.5e)$$

$$I_-(\tau,\mu) = \frac{q_-}{\pi}\exp\left[\frac{\tau_L-\tau}{\mu}\right] - \int_\tau^{\tau_L} I_b(\tau')\exp\left[-\frac{\tau'-\tau}{\mu}\right]\frac{d\tau'}{\mu} \qquad (9.6e)$$

The first terms on the right-hand side of these two equations represent the contribution to the intensity at location τ in direction μ from the wall. The q/π terms represent the (diffuse) intensity leaving the walls, and the exponential terms are the transmissivities along the path. The second terms on the right-hand side represent the contribution to the intensity at location τ in direction μ from emission along the path between location τ and the wall. The hemispherical fluxes at location τ in the slab may be obtained by integrating the intensity field (see Equation 9.9), using the azimuthal symmetry ($d\Omega = -2\pi\,d\mu$)

$$q_+ = 2\pi\int_0^1 I_+\mu d\mu\,; \quad q_- = 2\pi\int_0^{-1} I_-\mu d\mu \qquad (9.7e)$$

Substituting Equations 9.5e and 9.6e into the above relations gives

$$q_+ = 2 \int_0^1 \left\{ q_{+,1} \exp\left(-\frac{\tau}{\mu}\right) + \int_0^\tau E_b(\tau') \exp\left(\frac{\tau'-\tau}{\mu}\right) \frac{1}{\mu} d\tau' \right\} \mu \, d\mu \qquad (9.8e)$$

$$q_- = 2 \int_0^{-1} \left\{ q_{-,2} \exp\left[\frac{\tau_L - \tau}{\mu}\right] - \int_\tau^{\tau_L} E_b(\tau') \exp\left[\frac{\tau'-\tau}{\mu}\right] \frac{1}{\mu} d\tau' \right\} \mu \, d\mu$$

$$\qquad (9.9e)$$

The net radiative flux at location τ then is obtained as follows:

$$q = \int_{4\pi} I(\tau,\theta)\cos\theta \, d\Omega = \int_0^{2\pi} d\varphi \int_0^\pi I(\tau,\theta)\cos\theta \sin\theta \, d\theta$$

$$= 2\pi \int_{-1}^{+1} I(\tau,\theta)\mu \, d\mu = q_+ - q_-$$

$$= 2 \left\{ q_{+,1} E_3(\tau) + \int_0^\tau E_b(\tau') E_2(\tau - \tau') d\tau' \right.$$

$$\left. -q_{-,2} E_3(\tau_L - \tau) - \int_\tau^{\tau_L} E_b(\tau') E_2(\tau'-\tau) d\tau' \right\} \qquad (9.10e)$$

where the exponential integral functions E_n have been used. These functions are defined as

$$E_n(\tau) = \int_0^1 \mu^{n-2} \exp(-\tau/\mu) d\mu, \quad n > 0 \qquad (9.11e)$$

Its properties and recursive relations are given in Modest [1], along with the values tabulated for $n = 1, \ldots, 4$.

Two-flux (S_2) solution. Here we consider the solution of the problem using the discrete ordinates method, which is an approximate method. Since the medium does not scatter, the discrete ordinate equation simplifies (see Equation 9.13),

$$\frac{dI(z,s_i)}{ds} = -\kappa I(z,s_i) + \kappa I_b(z), \ i = 1, \ldots, n \qquad (9.12e)$$

where n is the number of discrete ordinates. The equations can be written in terms of the optical depth, and τ may be used instead of s,

$$\mu_i \frac{dI_{+,i}}{d\tau} + I_{+,i} = I_b, \ -\mu_i \frac{dI_{-,i}}{d\tau} + I_{-,i} = I_b \qquad (9.13e)$$

where the same notations as in the exact solution are used here. Since I_b is constant, these equations may be integrated right away, leading to

$$I_{+,i} = I_b + C_+ e^{-\tau/\mu_i}, \quad I_{+,i} = I_b + C_- e^{\tau/\mu_i} \tag{9.14e}$$

The integration constants C_+ and C_- may be found from the boundary conditions (Equation 9.4e) as

$$\tau = 0: \quad I_{+,i} = I_{w1} = I_b + C_+, \quad \text{or} \quad C_+ = I_{w1} - I_b;$$

$$\tau = \tau_L: \quad I_{-,i} = I_{w2} = I_b + C_- e^{\tau/\mu_i}, \quad \text{or} \quad C_- = (I_{w2} - I_b)e^{-\tau_L/\mu_i} \tag{9.15e}$$

where $I_{w1} = I_+(\tau=0)$ and $I_{w2} = I_-(\tau=\tau_L)$. Thus, we have the results,

$$I_{+,i} = I_b + (I_{w1} - I_b)e^{-\tau/\mu_i}; \quad I_{-,i} = I_b + (I_{w2} - I_b)e^{-(\tau_L-\tau)/\mu_i} \tag{9.16e}$$

$$q = \sum_{i=1}^{n/2} w_i' \mu_i (I_+ - I_-)$$

$$= \sum_{i=1}^{n/2} w_i' \mu_i \left((I_{w1} - I_b)e^{-\tau/\mu_i} - (I_{w2} - I_b) - e^{-(\tau_L-\tau)/\mu_i} \right) \tag{9.17e}$$

So far, the derivation is general up to the nth order. For the non-symmetric S_2 approximation, we have $n = 2$, $w_1' = 2\pi$ and $\mu_1 = 0.5$. The flux is then calculated,

$$q = \sum_{i=1}^{n/2} w_i' \mu_i (I_+ - I_-) = \pi \left((I_{w1} - I_b)e^{-2\tau} - (I_{w2} - I_b) - e^{-2(\tau_L-\tau)} \right) \tag{9.18e}$$

P-1 Solution. In terms of the optical depth τ, the governing equations for the P_1 approximation can be written as

$$\frac{1}{3}\frac{dI_1^0}{d\tau} + I_0^0 = I_b; \quad \frac{dI_0^0}{d\tau} + I_1^0 = 0 \tag{9.19e}$$

Combining the two equations above, we have the second order differential equation for I_0^0,

$$\frac{d^2 I_0^0}{d\tau^2} - 3I_0^0 = -3I_b \tag{9.20e}$$

The general solution to the problem is given by

$$I_0^0(\tau) = I_b + C_1 e^{-\sqrt{3}\tau} + C_2 e^{\sqrt{3}\tau} \; ; \; I_1^0(\tau) = \sqrt{3}\,C_1 e^{-\sqrt{3}\tau} - \sqrt{3}\,C_2 e^{\sqrt{3}\tau} \quad (9.21e)$$

where C_1 and C_2 are two integration constants to be determined. The other term may now be calculated using either equation in Equation 9.20e.

By the P_1 approximation, the radiant intensity $I(\tau,\mu)$ is constructed in the following way:

$$I(\tau,\mu) = I_0^0(\tau) + I_1^0(\tau)Y_1^0 = I_0^0(\tau) + I_1^0(\tau)\mu$$

At the two boundaries, the boundary condition is satisfied in an integral sense [2],

$$\int_0^1 I(0,\mu)\mu d\mu = 0 \text{ and } \int_{-1}^0 I(\tau_L,\mu)\mu d\mu = 0 \qquad (9.22e)$$

which, after $I(\tau,\mu)$ substituted, and integration is completed, yields the following boundary conditions:

$$I_0^0(0) + \tfrac{2}{3}I_1^0(0) = I_{w1} \; ; \; I_0^0(\tau_L) - \tfrac{2}{3}I_1^0(\tau_L) = I_{w2} \qquad (9.23e)$$

These two boundary conditions then allow us to determine the integration constants C_1 and C_2,

$$C_1 = \frac{b_4(I_{w1} - I_b) - b_2(I_{w2} - I_b)}{b_1 b_4 - b_2 b_3} \; ; \qquad C_2 = \frac{b_1(I_{w2} - I_b) - b_3(I_{w1} - I_b)}{b_1 b_4 - b_2 b_3}$$

$$(9.24e)$$

$$b_1 = 1 + 2/\sqrt{3} \; ; \; b_2 = 1 - 2/\sqrt{3} \; ; \; b_3 = b_2 e^{-\sqrt{3}\tau_b} \; ; \; b_4 = b_1 e^{\sqrt{3}\tau_b} \qquad (9.25e)$$

The heat flux at location τ is then calculated by integrating the radiative intensity over the entire solid angle (see also Equation 9.9),

$$q(\tau) = 2\pi \int_{-1}^{+1} I(\tau,\mu)\mu d\mu = 2\pi \int_{-1}^{+1} (I_0^0(\tau) + I_1^0(\tau)\mu)\mu d\mu = \frac{4\pi}{3}I_1^0(\tau)$$

$$(9.26e)$$

Note that the solution for q may also be obtained by directly solving for the incident radiation G and then q.

9.3 Discontinuous Finite Element Formulation

We discuss below the application of the discontinuous finite element method to the solution of radiative heat transfer problems. The method was first used by Reed and Hill [7] to solve a neutron transport equation. An advantage of this method is that no inter-element continuity is enforced, and thus approximation functions are from a finite element broken space, which means that at the same geometric point, the field variable may be considered discontinuous (see Figure 9.2) [8].

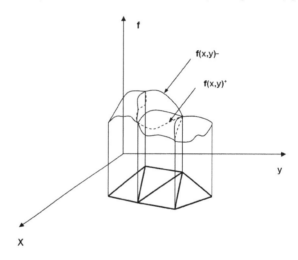

Figure 9.2. Finite element broken space

To apply the method, the domain is first discretized into a collection of finite elements. In this study, unstructured meshes are used, with triangular elements for 2-D problems and tetrahedral elements for 3-D problems. Let us take a 2-D problem to illustrate the integral formulation. Specifically, we consider the ith element in a 2-D mesh, as shown in Figure 9.3, and integrate Equation 9.3 over the element with respect to a weighting function $v(\Omega,\mathbf{r})$,

$$\int_{\Delta\Omega_l}\int_{A_e} v(\Omega,\mathbf{r})\mathbf{s}\cdot\nabla I dA d\Omega$$

$$= \int_{\Delta\Omega_l}\int_{A_e} v(\Omega,\mathbf{r})(-\beta(\mathbf{r})I(\mathbf{r},\mathbf{s})+S(\mathbf{r},\mathbf{s}))dA d\Omega \qquad (9.20)$$

where A_e is the area of the element under consideration (i.e., the ith element), $\mathbf{r} = x\hat{i} + y\hat{j}$ and $S(\mathbf{r}, \mathbf{s})$ is the source function defined by

$$S(\mathbf{r},\mathbf{s}) = \kappa(\mathbf{r})I_b(\mathbf{r}) + \frac{\sigma(\mathbf{r})}{4\pi}\int_{4\pi} I(\mathbf{r},\mathbf{s}')\Phi(\mathbf{s},\mathbf{s}')d\Omega' \qquad (9.21)$$

In the above equations, we have also used the definition, $\mathbf{s} \cdot \nabla I(\mathbf{r,s}) = \partial I(\mathbf{r,s})/\partial s$.

Applying integration by parts once to Equation 9.20, we have the following expression:

$$\int_{\Delta\Omega_l} \int_{V_e} -I\mathbf{s} \cdot \nabla v(\mathbf{r},\Omega)\, dV d\Omega + \int_{\Delta\Omega_l} \int_{\Gamma} v(\mathbf{r},\Omega) I^+ \, \mathbf{n} \cdot \mathbf{s}\, d\Gamma d\Omega$$

$$= \int_{\Delta\Omega_l} \int_{V_e} (-\beta(\mathbf{r}) I(\mathbf{r,s}) + S(\mathbf{r,s})) v(\mathbf{r},\Omega)\, dV d\Omega \qquad (9.22)$$

where Γ is the boundary of the element and superscript $+$ means taking the value outside the element boundary and where $\mathbf{n}$ is the outnormal of the element boundary. In deriving the above equation, we have used the divergence theorem,

$$\mathbf{s} \cdot \int_{\Gamma} \phi I \mathbf{n}\, dA - \mathbf{s} \cdot \int_{V_e} I \nabla \phi\, dV = \mathbf{s} \cdot \int_{V_e} \phi \nabla I\, dV \qquad (9.23)$$

to convert the domain integral into the boundary integral. Note that in selecting the values of I on the boundary, we have chosen those lying just outside the element under consideration. This choice is made in order to be consistent with the upwinding scheme.

We now integrate by parts once again and also use the divergence theorem to convert the volume integral into the surface integral. We then have the following integral formulation for the radiant intensity I:

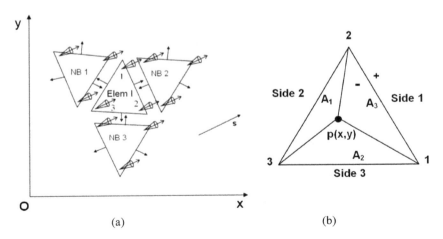

(a)   (b)

Figure 9.3. Illustration of the discontinuous finite element formulation for 2-D internal radiation transfer in absorbing and emitting media using unstructured triangular meshes: (a) element i, its boundary normals, and its neighboring elements, and (b) local node number and side number of a typical triangular element (or ith element)

$$\int_{\Delta\Omega_l}\int_{V_e} v(\mathbf{r},\Omega)\mathbf{s}\cdot\nabla I dVd\Omega + \int_{\Delta\Omega_l}\int_{\Gamma} v(\mathbf{r},\Omega)[I]\mathbf{n}\cdot\mathbf{s}\ d\Gamma d\Omega$$

$$= \int_{\Delta\Omega_l}\int_{V_e}(-\beta(\mathbf{r})I(\mathbf{r},\mathbf{s})+S(\mathbf{r},\mathbf{s}))v(\mathbf{r},\Omega)dVd\Omega \qquad (9.24)$$

Note that in the conventional finite element formulation, the terms on the element boundary disappear when they are combined with neighboring elements or $[I] = 0$. In the discontinuous finite element formulation, however, these terms do not cancel each other when elements are assembled. Instead, the following limiting values are used:

$$I_j^+ = \lim_{\mathbf{r}_j\to\Gamma^+} I(\mathbf{r}_j) \quad \text{and} \quad I_j^- = \lim_{\mathbf{r}_j\to\Gamma^-} I(\mathbf{r}_j) \qquad (9.25)$$

where the superscripts $+$ and $-$ denote the front side and back side of the normal vector, respectively. By this convention, the values denoted by superscript "$-$" are *inside* the element and those by "$+$" are *outside* the element (see Figure 9.3b). This definition is slightly different from the one-D case [4] but the essential idea is the same.

The above treatment assumes that the two values I_j^+ and I_j^- across the element boundaries are not the same, and these jumps are often denoted by the following expression:

$$[I]_j = I_j^+ - I_j^- \qquad (9.26)$$

These jumps may also be modeled by the generic numerical fluxes that are single-valued at the boundaries and are a function of field values across the inter-element boundaries [8–10]. For the problems under consideration, the simplest and most effective treatment of the jump condition is by using the upwinding procedur there, which in the discontinuous finite element literature is sometimes referred to as the inflow boundary value,

$$[I]_j = \begin{cases} [I]_j & \text{if} & \mathbf{n}\cdot\mathbf{s} < 0 \\ 0 & \text{if} & \mathbf{n}\cdot\mathbf{s} > 0 \end{cases} \qquad (9.27)$$

Appropriate interpolation functions now may be chosen from the finite element broken space that does not demand continuity across the inter-element boundaries [8–10]. A natural choice of shape functions for internal radiation applications is made by taking a step function for the solid angle and a polynomial function for the spatial variation, $v(\Omega,\mathbf{r}) = \psi(\Delta\Omega_l)\phi(\mathbf{r})$. Here $\psi(\Delta\Omega_l)$ is the step function of the solid angle differential centered at Ω_l, and $\phi(\mathbf{r})$ is the shape function of the spatial coordinates. Substituting this testing function into the integral expression and re-arranging, one has the following relation:

$$\int_{\Delta\Omega_l} \mathbf{s} \cdot \int_{V_e} \phi\,\nabla I(\mathbf{r},\mathbf{s})\,dVd\Omega + \int_{\Delta\Omega_l}\int_{\Gamma} \phi[I]\mathbf{n} \cdot \mathbf{s}\,d\Gamma d\Omega$$

$$= \int_{\Delta\Omega_l}\int_{V_e} [-\beta(\mathbf{r})\phi I(\mathbf{r},\mathbf{s}) + \phi S(\mathbf{r},\mathbf{s})]\,dVd\Omega \qquad (9.28)$$

which is the final form of the integral presentation of the radiative transfer equation. It is noted that Equation 9.28 reduces to the finite volume formulation if a constant shape function $\phi(\mathbf{r})$ is used, and to the finite element formulation, across the element interface, when $[I] = 0$ is enforced. From this perspective, Equation 9.28 represents a general integral formulation for all these integral-based methods.

Following the standard procedure for elemental calculations, Equation 9.28 can be readily calculated, once the shape functions are specified. Assembling all these discretized terms together for the element, the final results can be expressed in terms of the following matrix form:

$$\mathbf{KU} = \mathbf{F} \qquad (9.29)$$

where **U**, as usual, contains the unknown intensity vector and the matrix elements are summarized as follows:

$$K_{ij} = \int_{V_e} \phi_i\nabla\phi_j \cdot \int_{\Delta\Omega_l} \mathbf{s}\,d\Omega\,dV + \int_{V_e} \phi_i\phi_j\beta\,dV \int_{\Delta\Omega_l} d\Omega$$

$$+ \sum_{k=1}^{NS} \max\!\left(0, -\int_{\Delta\Omega_l} \mathbf{s}\cdot\mathbf{n}_k d\Omega\right)\int_{\Gamma_k} \phi_i\phi_j\,d\Gamma \qquad (9.30)$$

$$F_i = \int_{V_e} \phi_i S(\mathbf{r},\mathbf{s})\,dV \int_{\Delta\Omega_l} d\Omega$$

$$+ \sum_{k=1}^{NS} \max\!\left(0, -\int_{\Delta\Omega_l} \mathbf{s}\cdot\mathbf{n}_k d\Omega\right)\int_{\Gamma_k} \phi_i\phi_j I_{NB,j}\,d\Gamma \qquad (9.31)$$

with NS being the number of boundaries associated with the ith element.

For those elements associated with a boundary element, the boundary condition is imposed as follows if the boundary is gray:

$$I^{+}(\mathbf{r},\mathbf{s}_l) = \varepsilon(\mathbf{r})I_b(\mathbf{r}) + \frac{1-\varepsilon(\mathbf{r})}{\pi} \sum_{j=0,\,\mathbf{s}_j\cdot\mathbf{n}>0}^{N_\Omega} I^{-}(\mathbf{r},\mathbf{s}_j\,') \left|\,\mathbf{s}_j\,'\cdot\mathbf{n}\,\right| \Delta\Omega'_j \qquad (9.32)$$

where N_Ω is the number of discretized solid angles.

The following equation is used for the symmetry boundary condition:

$$I^{+}(\mathbf{r},\mathbf{s}_l) = I^{-}(\mathbf{r},\mathbf{s}_l^{*}) \qquad (9.33)$$

Here $\mathbf{s}^*$ is the symmetric direction of $\mathbf{s}$ with respect to the boundary. Equation 9.29 can be obtained for each element and its neighbors, and the calculations are then performed element by element. Thus, with Equation 9.29, the calculation for the ith element starts with selecting a direction and continues element by element until the entire domain and all directions are covered. Because of the boundary conditions, iterative procedures are required. Experience suggests that the successive substitution method seems to work well for these types of problems.

A few points are worthy of noting. First, if the jump condition $[I]$ is set to zero in Equation 9.28, which means that the inter-element continuity is enforced, then the conventional finite element formulation is recovered. Second, if the zeroth order polynomial is chosen as the spatial interpolation function, then we have the common finite volume formulation. Thus, in this sense, the finite volume method is a subclass of the discontinuous finite element method, and uses the lowest order approximation to the field variables.

Before numerical implementation is discussed for multidimensional problems, we first consider two examples of internal radiation in a 1-D slab, which is filled with participating media. These examples are used here to illustrate in detail the discontinuous finite element procedure for the solution of internal radiation problems with and without a scatttering medium. Numerical results are obtained using both linear and constant element approximations, the latter being the same as the finite volume approach.

Example 9.2. Solve the 1-D radiation problem defined in Example 9.1 using the discontinuous finite element method. Show the details of numerical implementation using three linear elements. Based on that, develop a discontinuous code and use the code to solve the radiative heat transfer problem using 20 linear elements.

Solution. This is perhaps the simplest system for which the discontinuous computational procedures can be illustrated in detail. For this purpose, a mesh consisting of three equal-sized linear elements is used, which is shown in Figure 9.3e, where E_j and N_i represent global element j and node i.

In the discontinuous approximation, two values are associated with a node, which we denote by I^+ and I^-. For consistency, the value inside the element is I^- and that outside is I^+, as shown in Figure 9.2e(b).

Now, let us consider the jth element denoted by E_j, $\Delta V_j \in [z_j, z_{j+1}]$. The discontinuous Galerkin formulation (i.e., Equation 9.28) gives

$$\int_{\Delta\Omega_l}\int_{z_j}^{z_{j+1}} \phi_i \mathbf{s}\cdot\nabla I dz d\Omega + \int_{\Delta\Omega_l}\left(\phi_i\left(\mathbf{n}_j\cdot\mathbf{s}\right)[I]\right)\Big|_{z_j}^{z_{j+1}} d\Omega$$

$$= \int_{\Delta\Omega_l}\int_{z_j}^{z_{j+1}} \phi_i\left(-\beta(\mathbf{r})I(\mathbf{r},\mathbf{s})+S(\mathbf{r},\mathbf{s})\right)dz d\Omega \qquad (9.27\text{e})$$

with $\mathbf{n}_j$ being the outward normal of element j and subscript i refers to the local node number.

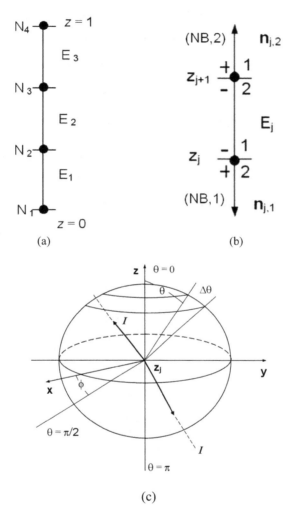

Figure 9.3e. Discretization of a 1-D problem. (a) Element and node numbers and (b) jump condition across the element boundaries: 1 and 2 corresponding to " – " are the local node numbers of element j, $(NB,1)$ and $(NB,2)$ are the neighboring elements adjacent to 1 and 2, and $\mathbf{n}_{j,1}$ and $\mathbf{n}_{j,2}$ are outward normals of element j. Also, 1 and 2 corresponding to " + " are local node numbers of elements $(NB,2)$ and $(NB,1)$. Note that node 1 of element j and node 2 of neighboring element $(NB,1)$ have the same coordinate z_j, but have different values of I, i.e., I^- and I^+. (c) Control angle used for the calculations.

First, the angular discretization is considered. For the 1-D problem, the radiation intensity is independent of φ, which means $\varphi_1 = 0$, and $\varphi_2 = 2\pi$, whence the integration over the lth discretized solid angle $\Delta\Omega_l$ can be carried out analytically,

$$\Delta\Omega_l = \int_{\Delta\Omega_l} d\Omega = \int_{\Delta\Omega_l} \sin\theta d\theta d\varphi$$

$$= (\varphi_2 - \varphi_1)(\cos\theta_1 - \cos\theta_2) = 2\pi(\cos\theta_1 - \cos\theta_2) \tag{9.28e}$$

$$\overline{s}_l = \int_{\Delta\Omega_l} s\, d\Omega$$

$$= 0\hat{i} + 0\hat{j} - 0.5\pi(\cos 2\theta_1 - \cos 2\theta_2)\,\hat{k} = \Delta\omega_l \hat{k} \tag{9.29e}$$

Next we consider the spatial discretization. For a 1-D linear element, the shape function has the following properties:

$$\phi_1(\xi) = 0.5(1-\xi)\,;\ \phi_2(\xi) = 0.5(1+\xi)$$

$$I_l = \phi_1 I_{1,l} + \phi_2 I_{2,l}\,;\ z = \phi_1 z_1 + \phi_2 z_2$$

$$dz = d\phi_1 z_1 + d\phi_2 z_2 = 0.5(z_2 - z_1)d\xi = 0.5\ell_{\,j} d\xi$$

$$\frac{d\phi_1}{dz} = \frac{d\phi_1}{d\xi}\frac{d\xi}{dz} = \frac{-0.5}{0.5\ell_{\,j}} = -\frac{1}{\ell_{\,j}}\,;\ \frac{\partial\phi_2}{\partial z} = \frac{1}{\ell_{\,j}}$$

$$\nabla I_l = \frac{\partial I_{h,l}}{\partial z}\hat{k} = \hat{k}\frac{\partial\phi_1}{\partial z} I_{1,l} + \hat{k}\frac{\partial\phi_2}{\partial z} I_{2,l}$$

where the subscript l denotes the lth direction, and $\ell_j = (z_2 - z_1)_j$ is the length of element j. Letting $i = 1, 2$ in Equation 9.27e and using the result from Equation 2.5e, we have

$$\int_{\Delta\Omega_l}\int_{z_j}^{z_{j+1}} \begin{bmatrix} \phi_1 \\ \phi_2 \end{bmatrix}\left[\frac{\partial\phi_1}{\partial z}\hat{k}I_{1,l} + \frac{\partial\phi_2}{\partial z}\hat{k}I_{2,l}\right]\cdot \mathbf{s}\,dz\,d\Omega$$

$$= \hat{k}\cdot\int_{\Delta\Omega_l} \mathbf{s}d\Omega \int_{z_j}^{z_{j+1}} \begin{bmatrix} \phi_1 \\ \phi_2 \end{bmatrix}\left[\frac{\partial\phi_1}{\partial z}, \frac{\partial\phi_2}{\partial z}\right]dz\begin{bmatrix} I_{1,l} \\ I_{2,l} \end{bmatrix}$$

$$= \Delta\omega_l \int_{-1}^{1}\begin{bmatrix} 0.5(1-\xi) \\ 0.5(1+\xi) \end{bmatrix}\left(-\frac{1}{\ell_j}, \frac{1}{\ell_j}\right)0.5\ell_{\,j}d\xi\begin{bmatrix} I_{1,l} \\ I_{2,l} \end{bmatrix}$$

$$= \frac{\Delta\omega_l}{2}\begin{bmatrix} -1 & 1 \\ -1 & 1 \end{bmatrix}\begin{bmatrix} I_{1,l} \\ I_{2,l} \end{bmatrix} \tag{9.30e}$$

for the first term on the left side of the equation, and

$$\int_{\Delta\Omega_l}\int_{z_j}^{z_{j+1}} \beta\begin{pmatrix}\phi_1\\\phi_2\end{pmatrix}(\phi_1,\phi_2)\,dzd\Omega$$

$$=\frac{\beta\Delta\Omega_l\ell_j}{8}\int_{-1}^{1}\begin{pmatrix}1-\xi\\1+\xi\end{pmatrix}(1-\xi,\ 1+\xi)\,d\xi$$

$$=\frac{\beta\Delta\Omega_l\ell_j}{8}\begin{pmatrix}-\frac{1}{3}(1-\xi)^3,\ \left(\xi-\frac{1}{3}\xi^3\right)\\\left(\xi-\frac{1}{3}\xi^3\right),\ \frac{1}{3}(1+\xi)^3\end{pmatrix}\Bigg|_{-1}^{1}=\frac{\beta\Delta\Omega_l\ell_j}{6}\begin{pmatrix}2,\ 1\\1,\ 2\end{pmatrix} \qquad (9.31e)$$

for the matrix coefficients of the first term on the right side of the equation. With $S(\mathbf{r},\mathbf{s}_l)$ being a constant, the source term can be calculated analytically,

$$\int_{\Delta\Omega_l}\int_{z_j}^{z_{j+1}}\begin{pmatrix}\phi_1\\\phi_2\end{pmatrix}S\,dzd\Omega=\frac{\Delta\Omega_l\ell_j S}{4}\int_{-1}^{1}\begin{pmatrix}1-\xi\\1+\xi\end{pmatrix}d\xi=\frac{\Delta\Omega_l\ell_j S}{2} \qquad (9.32e)$$

The element boundary terms represent jump conditions across the element boundary and require careful treatment. For element j, the shape function has the following values at the two boundaries:

$$\phi_1(z_{j+1})=\phi_1(\xi=1)=0;\ \ \phi_1(z_j)=\phi_1(\xi=-1)=1$$

$$\phi_2(z_{j+1})=\phi_2(\xi=1)=1;\ \ \phi_2(z_j)=\phi_2(\xi=-1)=0$$

$$[I]_1=I_1^+-I_1^-=I_{2,NB_1}-I_{1,E_j};\ \ [I]_2=I_2^+-I_2^-=I_{1,NB_2}-I_{2,E_j}$$

$$[I]_1=\begin{bmatrix}1&0\\0&0\end{bmatrix}\begin{bmatrix}[I]_1\\{[I]_2}\end{bmatrix}=\begin{bmatrix}1&0\\0&0\end{bmatrix}\begin{bmatrix}I_1^+-I_1^-\\I_2^+-I_2^-\end{bmatrix}$$

$$=-\begin{bmatrix}1&0\\0&0\end{bmatrix}\begin{bmatrix}I_1\\I_2\end{bmatrix}_{E_j}+\begin{bmatrix}1&0\\0&0\end{bmatrix}\begin{bmatrix}I_2\\0\end{bmatrix}_{(NB,1)}$$

$$[I]_2=\begin{bmatrix}0&0\\0&1\end{bmatrix}\begin{bmatrix}[I]_1\\{[I]_2}\end{bmatrix}=\begin{bmatrix}0&0\\0&1\end{bmatrix}\begin{bmatrix}I_1^+-I_1^-\\I_2^+-I_2^-\end{bmatrix}$$

$$=-\begin{bmatrix}0&0\\0&1\end{bmatrix}\begin{bmatrix}I_1\\I_2\end{bmatrix}_{E_j}+\begin{bmatrix}0&0\\0&1\end{bmatrix}\begin{bmatrix}0\\I_1\end{bmatrix}_{(NB,2)}$$

This allows the boundary integration to be carried out analytically, whence we have the following results:

$$\int_{\Delta\Omega_l}(\phi_1(\mathbf{n}_j\cdot\mathbf{s})[I])_{z_j}^{z_{j+1}}\,d\Omega=[I_l]_2\int_{\Delta\Omega_l}\phi_1(z_{j+1})(\mathbf{n}_{j,2}\cdot\mathbf{s})d\Omega$$

$$+[I_l]_1 \int_{\Delta\Omega_l} \phi_1(z_j)(\mathbf{n}_{j,1} \cdot \mathbf{s})d\Omega = \mathbf{n}_{j,1} \cdot \overline{\mathbf{s}}_l[I_l]_1 \tag{9.33e}$$

$$\int_{\Delta\Omega_l} \left(\phi_2 (\mathbf{n}_j \cdot \mathbf{s})[I]\right) \Big|_{z_j}^{z_{j+1}} d\Omega = [I_l]_2 \int_{\Delta\Omega_l} \phi_2(z_{j+1})(\mathbf{n}_{j,2} \cdot \mathbf{s})d\Omega$$

$$+[I_l]_1 \int_{\Delta\Omega_l} \phi_1(z_j)(\mathbf{n}_{j,1} \cdot \mathbf{s})d\Omega = \mathbf{n}_{j,2} \cdot \overline{\mathbf{s}}_l[I_l]_2 \tag{9.34e}$$

The above two equations can be written in matrix terms,

$$(\mathbf{n}_{j,i} \cdot \overline{\mathbf{s}})[I]_i = -(-\mathbf{n}_{j,i} \cdot \overline{\mathbf{s}}) \begin{bmatrix} \delta_{i1} & 0 \\ 0 & \delta_{i2} \end{bmatrix} \begin{bmatrix} [I]_1 \\ [I]_2 \end{bmatrix}$$

$$= -(-\mathbf{n}_{j,1} \cdot \overline{\mathbf{s}}) \begin{bmatrix} \delta_{i1} & 0 \\ 0 & \delta_{i2} \end{bmatrix} \begin{bmatrix} I_1^+ - I_1^- \\ I_2^+ - I_2^- \end{bmatrix} \tag{9.35e}$$

where δ_{ik} is the delta function, i.e., $\delta_{ik} = 1$ if $i = k$ and $\delta_{ik} = 0$ if $i \neq k$. The application of the upwinding condition (i.e., Equation 9.27) yields the following results:

$$-(-\mathbf{n}_{j,1} \cdot \overline{\mathbf{s}}_l) \begin{bmatrix} 1 & 0 \\ 0 & 0 \end{bmatrix} \begin{bmatrix} I_1^+ - I_1^- \\ I_2^+ - I_2^- \end{bmatrix}$$

$$= \max(0, -\mathbf{n}_{j,1} \cdot \overline{\mathbf{s}}_l) \begin{bmatrix} 1 & 0 \\ 0 & 0 \end{bmatrix} \begin{bmatrix} I_1 \\ I_2 \end{bmatrix} - \max(0, -\mathbf{n}_{j,1} \cdot \overline{\mathbf{s}}_l) \begin{bmatrix} 1 & 0 \\ 0 & 0 \end{bmatrix} \begin{bmatrix} I_2 \\ 0 \end{bmatrix}_{(NB,1)}$$

$$-(-\mathbf{n}_{j,2} \cdot \overline{\mathbf{s}}_l) \begin{bmatrix} 0 & 0 \\ 0 & 1 \end{bmatrix} \begin{bmatrix} I_1^+ - I_1^- \\ I_2^+ - I_2^- \end{bmatrix} \tag{9.36e}$$

$$= \max(0, -\mathbf{n}_{j,2} \cdot \overline{\mathbf{s}}_l) \begin{bmatrix} 0 & 0 \\ 0 & 1 \end{bmatrix} \begin{bmatrix} I_1 \\ I_2 \end{bmatrix} - \max(0, -\mathbf{n}_{j,2} \cdot \overline{\mathbf{s}}_l) \begin{bmatrix} 0 & 0 \\ 0 & 1 \end{bmatrix} \begin{bmatrix} 0 \\ I_1 \end{bmatrix}_{(NB,2)}$$

$$\tag{9.37e}$$

Note that to simplify the notation, the subscript E_j on the first term of the right hand side has been dropped. Equations 9.30e–9.32e and 9.36e–9.37e are substituted into Equation 9.27e to give

$$\left\{ \frac{\Delta\omega_l}{2} \begin{pmatrix} -1, 1 \\ -1, 1 \end{pmatrix} + \frac{\beta\Delta\Omega_l \, \ell_j}{6} \begin{pmatrix} 2, 1 \\ 1, 2 \end{pmatrix} + \max(0, -\mathbf{n}_{j,1} \cdot \overline{\mathbf{s}}_l) \begin{pmatrix} 1 & 0 \\ 0 & 0 \end{pmatrix} \right.$$

$$\left. + \max(0, -\mathbf{n}_{j,2} \cdot \overline{\mathbf{s}}_l) \begin{pmatrix} 0 & 0 \\ 0 & 1 \end{pmatrix} \right\} \begin{bmatrix} I_{1,l} \\ I_{2,l} \end{bmatrix}$$

$$= \frac{\Delta\Omega_l \ell_j S}{2} + \max(0, -\mathbf{n}_{j,1} \cdot \overline{\mathbf{s}}_l) \begin{pmatrix} 1 & 0 \\ 0 & 0 \end{pmatrix} \begin{bmatrix} I_{2,l} \\ 0 \end{bmatrix}_{(NB,1)}$$

$$+ \max(0, -\mathbf{n}_{j,2} \cdot \overline{\mathbf{s}}_l) \begin{pmatrix} 0 & 0 \\ 0 & 1 \end{pmatrix} \begin{bmatrix} 0 \\ I_{1,l} \end{bmatrix}_{(NB,2)} \qquad (9.38e)$$

The above analytical expression can be derived for any elements and any solid angle discretization. A computer program can be readily developed to calculate these quantities efficiently.

As a check, we consider the simple three equally sized element mesh, with the size $\ell_j = 1/3$ m. The absorption coefficient is 1.0 m^{-1}, the scattering coefficient is 0, and the temperature of the medium is $T = 100$ K. The angular space is discretized into 1×2, which denotes the azimuthal angle is parted into 1 angle, and the polar angle is parted into 2 angles. Therefore, there are two control angles in this problem, and the variables associated with the control angle $l = 1, 2$ can be calculated with the following results:

$$\Delta\Omega_1 = 2\pi(\cos\theta_1 - \cos\theta_2) = 2\pi(\cos 0 - \cos(\pi/2)) = 2\pi$$

$$\Delta\Omega_2 = 2\pi(\cos\theta_1 - \cos\theta_2) = 2\pi(\cos\pi/2 - \cos\pi) = 2\pi$$

$$\overline{\mathbf{s}}_1 = \Delta\omega_1 \hat{k} = 0.5\pi(\cos 2\theta_1 - \cos 2\theta_2) \hat{k} = \pi\, \hat{k}$$

$$\overline{\mathbf{s}}_2 = \Delta\omega_2 \hat{k} = 0.5\pi(\cos 2\theta_1 - \cos 2\theta_2) \hat{k} = -\pi\, \hat{k}$$

$$\mathbf{n}_{j,1} \cdot \overline{\mathbf{s}}_1 = -\hat{k} \cdot \pi\hat{k} = -\pi \quad \mathbf{n}_{j,2} \cdot \overline{\mathbf{s}}_1 = \hat{k} \cdot \pi\hat{k} = \pi$$

$$\mathbf{n}_{j,1} \cdot \overline{\mathbf{s}}_2 = -\hat{k} \cdot (-\pi\hat{k}) = \pi \quad \mathbf{n}_{j,2} \cdot \overline{\mathbf{s}}_2 = \hat{k} \cdot (-\pi\hat{k}) = -\pi$$

$$\max(0, -\mathbf{n}_{j,1} \cdot \overline{\mathbf{s}}_1) = \max(0, -(-\pi)) = \max(0, \pi) = \pi$$

$$\max(0, -\mathbf{n}_{j,2} \cdot \overline{\mathbf{s}}_2) = \max(0, \pi) = \pi$$

$$\max(0, -\mathbf{n}_{j,2} \cdot \overline{\mathbf{s}}_1) = \max(0, -\pi) = 0$$

$$\max(0, -\mathbf{n}_{j,1} \cdot \overline{\mathbf{s}}_2) = \max(0, -\pi) = 0$$

The source of the element can be calculated by

$$S = kI_b = 1.0 \times 5.67 \times 10^{-8} \times (100)^4 = 1.805$$

For element 1 as shown in Figure 9.1e, we have for $l = 1$

$$\left\{ \frac{\pi}{2} \begin{pmatrix} -1, 1 \\ -1, 1 \end{pmatrix} + \frac{1 \times 2\pi}{6 \times 3} \begin{pmatrix} 2, 1 \\ 1, 2 \end{pmatrix} + \pi \begin{pmatrix} 1 & 0 \\ 0 & 0 \end{pmatrix} \right\} \begin{bmatrix} I_{1,1} \\ I_{2,1} \end{bmatrix}$$

$$= \frac{2\pi S}{2 \times 3} + \pi \begin{pmatrix} 1 & 0 \\ 0 & 0 \end{pmatrix} \begin{bmatrix} I_{2,1} \\ 0 \end{bmatrix}_{(NB,1)} \tag{9.39e}$$

or

$$\begin{bmatrix} 2.269 & 1.920 \\ -1.222 & 2.269 \end{bmatrix} \begin{bmatrix} I_{1,1} \\ I_{2,1} \end{bmatrix} = \begin{bmatrix} 1.890 \\ 1.890 \end{bmatrix} + \begin{pmatrix} \pi & 0 \\ 0 & 0 \end{pmatrix} \begin{bmatrix} I_{2,1} \\ 0 \end{bmatrix}_{(NB,1)} \tag{9.40e}$$

Similarly, for $l = 2$,

$$\begin{bmatrix} 2.269 & 1.920 \\ -1.222 & 2.269 \end{bmatrix} \begin{bmatrix} I_{1,2} \\ I_{2,2} \end{bmatrix} = \begin{bmatrix} 1.890 \\ 1.890 \end{bmatrix} + \begin{pmatrix} 0 & 0 \\ 0 & \pi \end{pmatrix} \begin{bmatrix} 0 \\ I_{1,2} \end{bmatrix}_{(NB,2)} \tag{9.41e}$$

We can now apply the above formulae to obtain solutions element by element. The procedure is iterative, starting with the radiant intensity I initially set to zero everywhere and sweeping from one side of the boundary to the other. Let us start with element 1 for $l = 1$, $(I_{2,1})_{(NB,1)} = I_{w1,1} = 0$, $I_{w1,1}$ being the value of the domain boundary. Consequently, Equation 9.40e is solved with the result for element 1,

$$\begin{bmatrix} I_{1,1} \\ I_{2,1} \end{bmatrix}^{elem1} = \begin{bmatrix} 0.8804 \\ 8.8040 \end{bmatrix} \tag{9.42e}$$

We now proceed to calculate the values for element 2. Noticing that $(I_{2,1})_{(NB,1)} = (I_{2,1})^{elem1} = 8.804$ for element 2, we have again from Equation 9.40e

$$\begin{bmatrix} I_{1,1} \\ I_{2,1} \end{bmatrix}^{elem\ 2} = \begin{bmatrix} 0.9255 \\ 1.3313 \end{bmatrix} \tag{9.43e}$$

Similarly, for element 3, $(I_{2,1})_{(NB,1)} = (I_{2,1})^{elem\ 2} = 1.3313$. We thus obtain the solution for element 3 as follows:

$$\begin{bmatrix} I_{1,1} \\ I_{2,1} \end{bmatrix}^{elem\ 3} = \begin{bmatrix} 1.3544 \\ 1.5623 \end{bmatrix} \tag{9.44e}$$

The same procedure can be applied with Equation 9.41e except that $(I_{1,2})_{(NB,1)}$ should be used instead. After all elements are calculated, the intensity at a global node is obtained by simple averaging. For example, the first node is, with element 1 only,

$$I_1 = I_1^{elem1} \tag{9.45e}$$

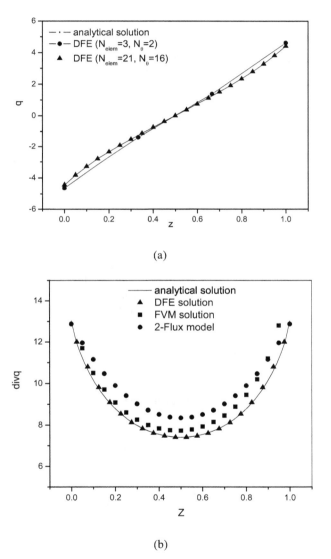

(a)

(b)

Figure 9.4e. Comparison of the numerical and analytic solutions for 1-D radiation transfer problem: (a) boundary heat flux distribution and (b) distribution of divergence of the heat flux

For the second node, the intensity is an averaged value,

$$I_2 = 0.5(I_2^{elem1} + I_1^{elem2}) \tag{9.46e}$$

The intensities of other nodes of the mesh can be calculated by the same method. With this, the relative error is calculated and compared with the convergence criterion,

$$\left| \max \left(\sum_{j=1}^{N_\Omega} I_{i,j}^{k+1} \Delta\Omega_j - \sum_{j=1}^{N_\Omega} I_{i,j}^{k} \Delta\Omega_j \right) \middle/ \sum_{j=1}^{N_\Omega} I_{i,j}^{k+1} \Delta\Omega_j \right| < 1.0 \times 10^{-6} \tag{9.47e}$$

Here for the present case, the subscript k denotes the kth iteration, and the convergence criterion is set to 1.0×10^{-6}. After the convergence is reached, the heat flux in the medium is calculated by Equation 9.9, and for the simple 3-element mesh and 2-solid-angle angular space discretization, the numerically calculated heat flux is compared with the analytical solution as in Figure 9.4e(a).

With the above data, a computer code is developed and the results using 20 (N = 21) linear elements are also shown in Figure 9.4e. Clearly, excellent agreement exists between the numerical and analytic solutions, suggesting that the discontinuous finite element method is useful for this type of problem. Even a 3-element mesh gives a reasonably good trend. It is noted, however, that the two-flux model gives an averaged value in the half sphere, and thus it lies in between the intensities at different directions. The computed results for the distribution divergence obtained from the two-flux model and the discontinuous finite element (DFE) method are also plotted in Figure 9.4e(b), along with the solutions obtained from analytical techniques. Examination of these results illustrates that the DFE results with linear elements match well with the analytic solutions. For comparison, results from a piecewise constant approximation are also shown and denoted by the finite volume method (FVM).

Example 9.3. Solve the same problem of internal radiation in a 1-D slab but with a scattering medium.

Solution. One of the important phenomena in radiative transfer processes is scattering, which changes the local energy balance. This problem illustrates the effects of scattering medium on internal radiation. To describe the scattering effect, the source term is used such that the radiation intensity in one direction at a certain point is affected by the intensities in all directions at the point. The scattering effect can be rather easily handled by the discontinuous method as a source term $S(\mathbf{r}, \mathbf{s}_l)$, and normally requires an iterative procedure. Some calculated results obtained using the discontinuous formulation for a 1-D radiation slab, filled with isotropic scattering media of different scattering paramters, are shown in Figure 9.5e.

Inspection of these results shows that for a medium with a larger scattering coefficient, the scattering effect on the radiation intensity increases as the intensity is further away from the boundary at which it originates, and the largest effect occurs when the intensity reaches the other boundary. In fact, for the intensity at $\theta=0$, the value of the intensity is reduced by 40% when it reaches the upper boundary with the scattering coefficient $\sigma=1$. The effect at the lower boundary at which $I(\theta=0)$ originates, however, is rather small. As a result of scattering, the heat flux (absolute value) is smaller near the walls; however, the distribution is anti-symmetric, as expected. Figure 9.5e(b) compares the results of the boundary fluxes calculated using the DFE method for different scattering coefficients.

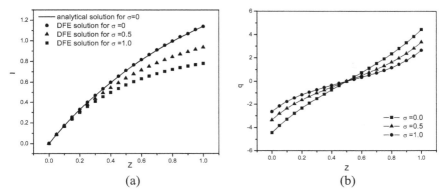

(a)   (b)

Figure 9.5e. Effect of scattering on radiative transfer calculated using the discontinuous finite element method: (a) intensity distribution and (b) wall flux distribution

9.4 Numerical Implementation

The above 1-D examples have illustrated some basic procedures involved in the use of the discontinuous finite element method for the solution of internal radiation problems. In this section, numerical details are given to form the matrix equations for the discontinuous Galerkin finite element formulation. We discuss these procedures for the 2-D and 3-D calculations. The 1-D case will be given in the example section. The 2-D axisymmetric case requires special treatment, which we relegate to Section 9.6 for discussion.

9.4.1 2-D Calculations

Let us consider again the ith element and its neighbors as shown in Figure 9.3a. For the sake of discussion, the inter-element boundaries are plotted separately. The nodal values of the variable are defined within the element. Since discontinuity is allowed across the element boundaries, the common geometric node does not have the same field variable value. This is an essential difference between the conventional and the discontinuous finite element formulations.

For a 2-D linear triangular element, the shape functions may be written in terms of natural coordinates,

$$
\begin{bmatrix} \phi_1 \\ \phi_2 \\ \phi_3 \end{bmatrix} = \begin{bmatrix} \lambda_1 \\ \lambda_2 \\ \lambda_3 \end{bmatrix} \tag{9.34}
$$

Here λ_j ($j = 1, 2, 3$) is defined by the area ratio, $\lambda_j = A_j/A_e$, where A_e is the area of the element and A_j the sub-triangular area formed from two vertices and point p inside the element (see Figure 9.3b). With relevant global coordinates substituted in λ_j, the shape functions take the following form:

$$
\begin{bmatrix} \phi_1 \\ \phi_2 \\ \phi_3 \end{bmatrix} = \frac{1}{2A_e} \begin{bmatrix} A_{23} & -X_{23} & Y_{23} \\ A_{31} & -X_{31} & Y_{31} \\ A_{12} & -X_{12} & Y_{12} \end{bmatrix} \begin{bmatrix} 1 \\ x \\ y \end{bmatrix} \tag{9.35}
$$

where the definition of elements in the matrix of Equation 9.35 is given by the following equations:

$$
A_{ij} = \begin{vmatrix} x_i & x_j \\ y_i & y_j \end{vmatrix}, \quad X_{ij} = \begin{vmatrix} 1 & 1 \\ y_i & y_j \end{vmatrix}, \quad Y_{ij} = \begin{vmatrix} 1 & 1 \\ x_i & x_j \end{vmatrix} \tag{9.36}
$$

The radiation intensity inside the element is interpolated using the above shape functions,

$$
I(x, y; \mathbf{s}) = I_1(\mathbf{s})\phi_1(x, y) + I_2(\mathbf{s})\phi_2(x, y) + I_3(\mathbf{s})\phi_3(x, y) \tag{9.37}
$$

Substituting the above equation into Equation 9.28, and noting that the intensity with control angles is constant due to a step function approximation, one has the following expression:

$$
\int_{\Delta\Omega_l} \mathbf{s} \cdot \int_{A_e} \begin{bmatrix} \phi_1 \\ \phi_2 \\ \phi_3 \end{bmatrix} [\nabla\phi_1, \nabla\phi_2, \nabla\phi_3] \begin{bmatrix} I_1 \\ I_2 \\ I_3 \end{bmatrix} dAd\Omega
$$

$$
+ \int_{\Delta\Omega_l} \int_{\Gamma} \begin{bmatrix} \phi_1 \\ \phi_2 \\ \phi_3 \end{bmatrix} [\phi_1 \quad \phi_2 \quad \phi_3] \begin{bmatrix} [I]_1 \\ [I]_2 \\ [I]_3 \end{bmatrix} (\mathbf{n} \cdot \mathbf{s}) d\Gamma d\Omega
$$

$$
= \int_{\Delta\Omega_l} \int_{A_e} \left\{ -\beta(\mathbf{r}) \begin{bmatrix} \phi_1 \\ \phi_2 \\ \phi_3 \end{bmatrix} [\phi_1, \phi_2, \phi_3] \begin{bmatrix} I_1 \\ I_2 \\ I_3 \end{bmatrix} + \begin{bmatrix} \phi_1 \\ \phi_2 \\ \phi_3 \end{bmatrix} S(\mathbf{r}, \mathbf{s}) \right\} dAd\Omega \tag{9.38}
$$

For a 2-D triangular element, the above integration can be carried out analytically. Taking the derivative of the shape functions,

$$
\begin{bmatrix} \nabla \phi_1 \\ \nabla \phi_2 \\ \nabla \phi_3 \end{bmatrix} = \frac{1}{2A_e} \begin{bmatrix} -X_{23} & Y_{23} \\ -X_{31} & Y_{31} \\ -X_{12} & Y_{12} \end{bmatrix} \begin{bmatrix} \hat{i} \\ \hat{j} \end{bmatrix}
\tag{9.39}
$$

and noticing that $\nabla \phi_i$ ($i = 1, 2, 3$) is independent of the area integral, one has

$$
\int_{\Delta\Omega} \int_{A_e} \begin{bmatrix} \phi_1 \\ \phi_2 \\ \phi_3 \end{bmatrix} [\mathbf{s} \cdot \nabla \phi_1, \mathbf{s} \cdot \nabla \phi_2, \mathbf{s} \cdot \nabla \phi_3] dA d\Omega
$$

$$
= \int_{A_e} \begin{bmatrix} \phi_1 \\ \phi_2 \\ \phi_3 \end{bmatrix} dA \int_{\Delta\Omega} [\mathbf{s} \cdot \nabla \phi_1, \mathbf{s} \cdot \nabla \phi_2, \mathbf{s} \cdot \nabla \phi_3] d\Omega = \frac{A_e}{3} D
\tag{9.40}
$$

where D is a matrix and $\bar{\mathbf{s}}$ is calculated by

$$
D = \begin{bmatrix} \bar{\mathbf{s}}_l \cdot \nabla \phi_1 & \bar{\mathbf{s}}_l \cdot \nabla \phi_2 & \bar{\mathbf{s}}_l \cdot \nabla \phi_3 \\ \bar{\mathbf{s}}_l \cdot \nabla \phi_1 & \bar{\mathbf{s}}_l \cdot \nabla \phi_2 & \bar{\mathbf{s}}_l \cdot \nabla \phi_3 \\ \bar{\mathbf{s}}_l \cdot \nabla \phi_1 & \bar{\mathbf{s}}_l \cdot \nabla \phi_2 & \bar{\mathbf{s}}_l \cdot \nabla \phi_3 \end{bmatrix}
\tag{9.41a}
$$

$$
\bar{\mathbf{s}}_l = \int_{\Delta\Omega_l} \mathbf{s}\, d\Omega = [0.5(\theta_2 - \theta_1) - 0.25(\sin 2\theta_2 - \sin 2\theta_1)]
$$

$$
\times [(\sin\varphi_2 - \sin\varphi_1)\hat{i} - (\cos\varphi_2 - \cos\varphi_1)\hat{j}]
$$

$$
- 0.25(\cos 2\theta_2 - \cos 2\theta_1)(\varphi_2 - \varphi_1)\hat{k}
\tag{9.41b}
$$

The second integral in Equation 9.38 represents the jump condition (or numerical fluxes) across the boundary of the ith element and its neighbor (see Figure 9.3a). For a linear triangular element, it is split into three terms, one for each side of the element,

$$
\int_{\Delta\Omega_l} \int_\Gamma \begin{bmatrix} \phi_1 \\ \phi_2 \\ \phi_3 \end{bmatrix} [\phi_1, \phi_2, \phi_3] \begin{bmatrix} [I]_1 \\ [I]_2 \\ [I]_3 \end{bmatrix} (\mathbf{n} \cdot \mathbf{s}) d\Gamma d\Omega
$$

$$
= \int_{\Gamma_1} \begin{bmatrix} \phi_1 \\ \phi_2 \\ \phi_3 \end{bmatrix} [\phi_1, \phi_2, \phi_3] \begin{bmatrix} [I]_1 \\ [I]_2 \\ [I]_3 \end{bmatrix} d\Gamma \int_{\Delta\Omega_l} (\mathbf{n}_1 \cdot \mathbf{s}) d\Omega
$$

$$+ \int_{\Gamma_3} \begin{bmatrix} \phi_1 \\ \phi_2 \\ \phi_3 \end{bmatrix} [\phi_1, \phi_2, \phi_3] \begin{bmatrix} [I]_1 \\ [I]_2 \\ [I]_3 \end{bmatrix} d\Gamma \int_{\Delta\Omega_l} (\mathbf{n}_3 \cdot \mathbf{s}) d\Omega \qquad (9.42)$$

The line integration associated with the element can be carried out analytically. For side 1, $\phi_3 = 0$; and $\phi_1 + \phi_2 = 1$, we therefore have the following matrix expression:

$$\int_{\Gamma_1} \begin{bmatrix} \phi_1 \\ \phi_2 \\ \phi_3 \end{bmatrix} [\phi_1 \quad \phi_2 \quad \phi_3] d\Gamma_1 = \int_{\Gamma_1} \begin{bmatrix} \phi_1 \\ \phi_2 \\ 0 \end{bmatrix} [\phi_1 \quad \phi_2 \quad 0] d\Gamma_1$$

$$= \int_0^1 \begin{bmatrix} \phi_1\phi_1 & \phi_1(1-\phi_1) & 0 \\ (1-\phi_1)\phi_1 & (1-\phi_1)(1-\phi_1) & 0 \\ 0 & 0 & 0 \end{bmatrix} L_1 d\phi_1 = \frac{L_1}{6} \begin{bmatrix} 2 & 1 & 0 \\ 1 & 2 & 0 \\ 0 & 0 & 0 \end{bmatrix} \qquad (9.43)$$

where L_1 is the length of side 1 (see Figure 9.3b). The term involving the solid angle integration can also be treated analytically,

$$-\mathbf{n}_1 \cdot \bar{\mathbf{s}}_l = \int_{\Delta\Omega_l} (-\mathbf{n}_1 \cdot \mathbf{s}) d\Omega$$

$$= [0.5(\theta_2 - \theta_1) - 0.25(\sin 2\theta_2 - \sin 2\theta_1)]$$

$$\times [(\sin \varphi_1 - \sin \varphi_2) n_{1x} - (\cos \varphi_1 - \cos \varphi_2) n_{1y}]$$

$$-0.25(\cos 2\theta_1 - \cos 2\theta_2)(\varphi_2 - \varphi_1) n_{1z} \qquad (9.44)$$

Note that for a 2-D problem, $n_z = 0$. In the DFE treatment, the jump terms have to be selected depending on the sign of $-\mathbf{n} \cdot \bar{\mathbf{s}}_l$. This is different from the conventional finite element formulation in which the across-element continuity is enforced, and the inter-element boundary terms cancel each other. One treatment that works effectively with linear elements is the upwinding scheme. By this scheme, one has

$$-\mathbf{n}_1 \cdot \bar{\mathbf{s}}_l \begin{bmatrix} [I]_1 \\ [I]_2 \\ [I]_3 \end{bmatrix} = \max(0, -\mathbf{n}_1 \cdot \bar{\mathbf{s}}_l) \begin{bmatrix} I_1 \\ I_2 \\ I_3 \end{bmatrix}^{elem\ i} - \max(0, -\mathbf{n}_1 \cdot \bar{\mathbf{s}}_l) \begin{bmatrix} I_1 \\ I_2 \\ I_3 \end{bmatrix}_{(NB,1)} \qquad (9.45)$$

By the same token, the calculations for the other two sides can also be performed analytically. By definition, we have $(I_j)_{(NB,1)} = (I_j^+)^{elem\ i}$ (see Figure 9.3b). Thus the calculated results for these two sides may be summarized below for convenience,

$$\int_{\Gamma_j} \begin{bmatrix} \phi_1 \\ \phi_2 \\ \phi_3 \end{bmatrix} [\phi_1, \phi_2, \phi_3] \begin{bmatrix} [I]_1 \\ [I]_2 \\ [I]_3 \end{bmatrix} d\Gamma \int_{\Delta\Omega_l} (\mathbf{n}_j \cdot \mathbf{s}) d\Omega$$

$$= \frac{L_j}{6} CofB_{(4-j,4-j)} \max(0, -\mathbf{n}_j \cdot \bar{\mathbf{s}}_l) \left\{ \begin{bmatrix} I_1 \\ I_2 \\ I_3 \end{bmatrix}^{elem\ i} - \begin{bmatrix} I_1 \\ I_2 \\ I_3 \end{bmatrix}_{(NB,j)} \right\} \qquad (9.46a)$$

where Cof B_{kl} is a matrix obtained by setting to zero the elements in the kth row and lth column of matrix B, which is defined by

$$B = \begin{bmatrix} 2 & 1 & 1 \\ 1 & 2 & 1 \\ 1 & 1 & 2 \end{bmatrix} \qquad (9.46b)$$

The first term on the right hand side of Equation 9.38 represents the attenuation of radiation intensity due to extinction. It may be calculated analytically when $\beta(\mathbf{r})$ is a constant with the result,

$$\int_{\Delta\Omega_l} d\Omega \int_{A_e} \begin{bmatrix} \phi_1 \\ \phi_2 \\ \phi_3 \end{bmatrix} [\phi_1, \phi_2, \phi_3] \beta(\mathbf{r}) dA$$

$$= \beta \int_{\Delta\Omega_l} d\Omega \int_{A_e} \begin{bmatrix} \phi_1\phi_1 & \phi_1\phi_2 & \phi_1\phi_3 \\ \phi_2\phi_1 & \phi_2\phi_2 & \phi_2\phi_3 \\ \phi_3\phi_1 & \phi_3\phi_2 & \phi_3\phi_3 \end{bmatrix} dA$$

$$= \frac{\beta A_e \Delta\Omega_l}{12} \begin{bmatrix} 2 & 1 & 1 \\ 1 & 2 & 1 \\ 1 & 1 & 2 \end{bmatrix} \qquad (9.47)$$

If all these discretized terms are assembled together, the equation for the element can be written in terms of the following matrix form:

$$\mathbf{KU} = \mathbf{F} \qquad (9.48)$$

where the expressions for the matrix elements are summarized as follows:

$$K_{ij} = \frac{A_e}{3} D_{ij} + \frac{\beta A_e \Delta\Omega_l}{12} B_{ij} + \sum_{k=1}^{NS} \frac{L_k}{6} [CofB_{(4-k,4-k)}]_{ij} \max(0, -\mathbf{n}_k \cdot \bar{\mathbf{s}}_l)$$

$$(9.49a)$$

$$F_i = \int_{A_e} \phi_i S dA \int_{\Delta\Omega_l} d\Omega$$

$$+ \sum_{k=1}^{NS} \frac{L_k}{6} \max(0, -\mathbf{n}_k \cdot \overline{\mathbf{s}}_l) \sum_{j=1}^{N_e} [CofB_{(4-k,4-k)}]_{ij} I_{(NB,k),j} \qquad (9.49b)$$

with NS being the number of boundaries associated with the ith element and N_e the number of the nodes of the element (see also Equation (9.62) below). The integration involving $S(\mathbf{r},\mathbf{s})$ is discussed in Section 9.5.3.

The calculations will start with those associated with a boundary element, where the boundary condition is imposed for a gray boundary as follows:

$$I^+(\mathbf{r},\mathbf{s}_l) = \varepsilon(\mathbf{r})I_b(\mathbf{r}) + \frac{1-\varepsilon(\mathbf{r})}{\pi} \sum_{j=0, \mathbf{s}_j \cdot \mathbf{n} < 0}^{N_\Omega} I^-(\mathbf{r},\mathbf{s}_j{}') |\mathbf{s}_j \cdot \mathbf{n}| \Delta\Omega'_j \qquad (9.50)$$

and the following equation is for the symmetry boundary condition,

$$I^+(\mathbf{r},\mathbf{s}_l) = I^-(\mathbf{r},\mathbf{s}_l^*) \qquad (9.51)$$

Here $\mathbf{s}^*$ is the symmetric direction of $\mathbf{s}$, with respect to the boundary, and can be calculated by Equation 9.8. Note that Equation 9.28 can be obtained for each element and its neighbors, and the calculations are then performed element by element. Thus, with Equation 9.28, the calculation for the ith element starts with selecting a direction and continues element by element until the entire domain and all directions are covered. Because of the boundary conditions, iterative procedures are required. The successive substitution method seems to work well for this type of problem. In the above, it is assumed that the medium is not scattering and thus the scattering term is set to zero. When the scattering term is known, the source term can be readily calculated using the Gaussian integration and included in the force vector $\{f\}$.

Let us now illustrate the above procedure through a numerical example of radiation in a 2-D cavity.

Example 9.4. We consider a 2-D problem of internal radiative heat transfer, which is schematically illustrated in Figure 9.6e(a). The cavity is filled with an absorbing and non-scattering medium. The absorptivity of the medium is $\kappa = 1.0$. The discontinuous finite element procedure discussed above is used to solve the problem.

Solution. Since the medium is non-scattering, $\sigma = 0$. At $x = 0$ and $x = 1$ the boundaries are black cold walls, that is, the emissivity and temperature of the wall are one and 0 respectively. Symmetry boundary conditions are applied at $y = 0$ and $y = 1$. The temperature is assumed to vary from the left to the right and the variation is described by the function, $T(x,y) = 100(1+0.75\sin(2\pi x))$. Since the symmetry boundary condition is applied at the top and bottom walls, and the temperature varies only with the x coordinate, the analytic solution for the problem can be obtained by integrating the radiative transfer equation as shown in Example 9.1,

$$q(x) = 2\pi \int_0^x I_b(x')E_2(x-x')\,dx' - 2\pi \int_x^1 I_b(x')E_2(x'-x)\,dx' \qquad (9.48e)$$

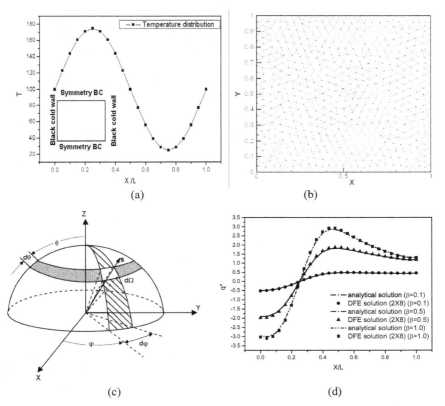

(a)

(b)

(c)

(d)

Figure 9.6e. Internal radiation in a 2-D square: (a) prescribed temperature distribution and boundary conditions, (b) unstructured triangular mesh, (c) schematic of angular space discretization, and (d) the heat flux q_x distribution along x/L obtained by the discontinuous finite element (DFE) method and the analytical solutions for three extinction coefficients $\beta = 0.1$, $\beta = 0.5$ and $\beta = 1.0$

This problem is solved using the DFE method in a 2-D unstructured triangular mesh consisting of 1142 elements, as shown in Figure 9.6e(b). The unstructured mesh is generated using the front advancing technique. The angular space discretization is 2×8; that is, the angular space is divided into 2 in the azimuthal direction (θ) and 8 in the polar direction (φ), which is shown in Figure 9.6e(c). The distribution of the non-dimensionalized radiative heat flux $q* = q_x/(100^4\sigma_s)$ at boundary $y = 0.0$ is computed for various conditions, and is plotted in Figure 9.6e(d), along with the analytical solutions. Apparently, excellent agreement is obtained between the analytical and the discontinuous finite element solutions for various $\beta = \kappa$ parameters.

9.4.2. 3-D Calculations

The above procedures are applicable to 3-D calculations. The element arrangement for 3-D is shown in Figure 9.4a. For a 3-D element, the area integral above is replaced by a volume integral, and the boundary line integral above by a surface integral, respectively. For a linear tetrahedral element, the integrations can be carried out analytically.

In an analogy to a 2-D triangular element, the shape function for a tetrahedral element has the following form when written in the global coordinate system:

$$
\begin{bmatrix} \phi_1 \\ \phi_2 \\ \phi_3 \\ \phi_4 \end{bmatrix} = \frac{1}{6V_e} \begin{bmatrix} V_{234} & -X_{234} & -Y_{234} & -Z_{234} \\ -V_{341} & X_{341} & Y_{341} & Z_{341} \\ V_{412} & -X_{412} & -Y_{412} & -Z_{412} \\ -V_{123} & X_{123} & Y_{123} & Z_{123} \end{bmatrix} \begin{bmatrix} 1 \\ x \\ y \\ z \end{bmatrix}
\tag{9.52}
$$

where the definition of elements in Equation 9.52 is as follows:

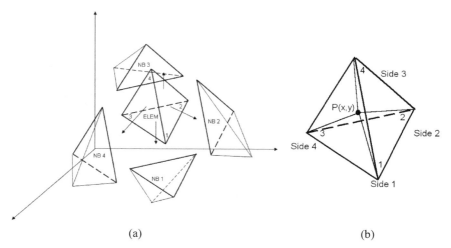

| (a) | (b) |

Figure 9.4. Illustration of the discontinuous finite element formulation for 3-D internal radiation transfer in absorbing and emitting media using unstructured tetrahedral meshes: (a) element i, its surface normals, and its neighboring elements, and (b) local node number and side number of a typical tetrahedral element (or ith element)

$$V_{ijk} = \begin{vmatrix} x_i & x_j & x_k \\ y_i & y_j & y_k \\ z_i & z_j & z_k \end{vmatrix} ; X_{ijk} = \begin{vmatrix} 1 & 1 & 1 \\ y_i & y_j & y_k \\ z_i & z_j & z_k \end{vmatrix} ;$$

$$Y_{ijk} = \begin{vmatrix} 1 & 1 & 1 \\ z_i & z_i & z_k \\ x_j & x_j & x_k \end{vmatrix} ; Z_{ijk} = \begin{vmatrix} 1 & 1 & 1 \\ x_i & x_j & x_k \\ y_i & y_j & y_k \end{vmatrix} \tag{9.53}$$

Here the lower case x_j, y_j and z_j denote the coordinates x, y, and z of the jth node of the tetrahedron under consideration (see Figure 9.4b).

The radiation intensity within a tetrahedron is interpolated by

$$I(x,y,z;\mathbf{s}) = \sum_{i=1}^{4} \phi_i I_i(\mathbf{s}) = \phi_1 I_1(\mathbf{s}) + \phi_2 I_2(\mathbf{s}) + \phi_3 I_3(\mathbf{s}) + \phi_4 I_4(\mathbf{s}) \tag{9.54}$$

Substituting the above expression into Equation 9.28 yields

$$\int_{\Delta\Omega_l} \mathbf{s}d\Omega \cdot \int_{V_e} \begin{bmatrix} \phi_1 \\ \phi_2 \\ \phi_3 \\ \phi_4 \end{bmatrix} [\nabla\phi_1, \nabla\phi_2, \nabla\phi_3, \nabla\phi_4] \begin{bmatrix} I_1 \\ I_2 \\ I_3 \\ I_4 \end{bmatrix} dV$$

$$+ \sum_{i=1}^{4} \int_{\Delta\Omega} \int_{\Gamma} \begin{bmatrix} \phi_1 \\ \phi_2 \\ \phi_3 \\ \phi_4 \end{bmatrix} [\phi_1,\phi_2,\phi_3,\phi_4] \begin{bmatrix} [I]_1 \\ [I]_2 \\ [I]_3 \\ [I]_4 \end{bmatrix} (\mathbf{n}_i \cdot \mathbf{s}) d\Gamma d\Omega$$

$$= \int_{\Delta\Omega_l} \int_{V_e} \left\{ -\beta(\mathbf{r}) \begin{bmatrix} \phi_1 \\ \phi_2 \\ \phi_3 \\ \phi_4 \end{bmatrix} [\phi_1 \ \phi_2 \ \phi_3 \ \phi_4] \begin{bmatrix} I_1 \\ I_2 \\ I_3 \\ I_4 \end{bmatrix} + \begin{bmatrix} \phi_1 \\ \phi_2 \\ \phi_3 \\ \phi_4 \end{bmatrix} S(\mathbf{r},\mathbf{s}) \right\} dV d\Omega$$

$$\tag{9.55}$$

Once again, the derivative of the shape functions can be obtained analytically with the following result:

$$\begin{bmatrix} \nabla\phi_1 \\ \nabla\phi_2 \\ \nabla\phi_3 \\ \nabla\phi_4 \end{bmatrix} = \frac{1}{J} \begin{bmatrix} -X_{234} & -Y_{234} & -Z_{234} \\ X_{341} & Y_{341} & Z_{341} \\ -X_{412} & -Y_{412} & -Z_{412} \\ X_{123} & Y_{123} & Z_{123} \end{bmatrix} \begin{bmatrix} \hat{i} \\ \hat{j} \\ \hat{k} \end{bmatrix} \tag{9.56}$$

where $J = 6V_e$ is the Jacobian of the tetrahedral element.

This will allow us to analytically integrate the volume terms in Equation 9.28. Following the same procedure as for the 2-D calculations, we have the following result for the 3-D tetrahedral elements:

$$\int_{\Delta\Omega_l} \int_{V_e} \begin{bmatrix} \phi_1 \\ \phi_2 \\ \phi_3 \\ \phi_4 \end{bmatrix} [\mathbf{s}\cdot\nabla\phi_1, \mathbf{s}\cdot\nabla\phi_2, \mathbf{s}\cdot\nabla\phi_3, \mathbf{s}\cdot\nabla\phi_4]\, dA\, d\Omega$$

$$= \int_{V_e} \begin{bmatrix} \phi_1 \\ \phi_2 \\ \phi_3 \\ \phi_4 \end{bmatrix} dV \int_{\Delta\Omega_l} [\mathbf{s}\cdot\nabla\phi_1, \mathbf{s}\cdot\nabla\phi_2, \mathbf{s}\cdot\nabla\phi_3, \mathbf{s}\cdot\nabla\phi_4]\, d\Omega$$

$$= \frac{V_e}{4} \begin{bmatrix} \bar{\mathbf{s}}_l\cdot\nabla\phi_1 & \bar{\mathbf{s}}_l\cdot\nabla\phi_2 & \bar{\mathbf{s}}_l\cdot\nabla\phi_3 & \bar{\mathbf{s}}_l\cdot\nabla\phi_4 \\ \bar{\mathbf{s}}_l\cdot\nabla\phi_1 & \bar{\mathbf{s}}_l\cdot\nabla\phi_2 & \bar{\mathbf{s}}_l\cdot\nabla\phi_3 & \bar{\mathbf{s}}_l\cdot\nabla\phi_4 \\ \bar{\mathbf{s}}_l\cdot\nabla\phi_1 & \bar{\mathbf{s}}_l\cdot\nabla\phi_2 & \bar{\mathbf{s}}_l\cdot\nabla\phi_3 & \bar{\mathbf{s}}_l\cdot\nabla\phi_4 \\ \bar{\mathbf{s}}_l\cdot\nabla\phi_1 & \bar{\mathbf{s}}_l\cdot\nabla\phi_2 & \bar{\mathbf{s}}_l\cdot\nabla\phi_3 & \bar{\mathbf{s}}_l\cdot\nabla\phi_4 \end{bmatrix} \qquad (9.57)$$

where $\bar{\mathbf{s}}$ is the same as given by Equation 9.44. Note that, for linear elements, $\nabla\phi_i$ $(i = 1, 2, 3, 4)$ is constant and is defined solely by the nodal coordinates of the tetrahedral element and thus can be taken outside the volume integral. The consideration for the solid angles and sign of $-\mathbf{n}_j\mathbf{s}$ $(j = 1, 2, 3, 4)$ is also the same as for the 2-D case discussed above, except that $\mathbf{n}$ now refers to the outnormal of the boundary surfaces of the tetrahedral element and has three components. After some algebraic manipulations, one has the final result for the surface integral along the element boundary:

$$\int_{\Gamma_j} \begin{bmatrix} \phi_1 \\ \phi_2 \\ \phi_3 \\ \phi_4 \end{bmatrix} [\phi_1, \phi_2, \phi_3, \phi_4] \begin{bmatrix} [I]_1 \\ [I]_2 \\ [I]_3 \\ [I]_4 \end{bmatrix} d\Gamma \int_{\Delta\Omega_l} (\mathbf{n}_j\cdot\mathbf{s})\, d\Omega$$

$$= \frac{A_j}{12} \text{CofC}_{(5-j,5-j)} \max(0, -\mathbf{n}_j\cdot\bar{\mathbf{s}}_l) \left\{ \begin{bmatrix} I_1 \\ I_2 \\ I_3 \\ I_4 \end{bmatrix}^{elem\ i} - \begin{bmatrix} I_1 \\ I_2 \\ I_3 \\ I_4 \end{bmatrix}_{(NB,j)} \right\} \qquad (9.58)$$

where A_j is the jth face of the element and $\text{CofC}_{(k,l)}$ is a matrix obtained by setting to zero the elements in the kth row and lth column of the matrix C, which is defined by

$$C = \begin{bmatrix} 2 & 1 & 1 & 1 \\ 1 & 2 & 1 & 1 \\ 1 & 1 & 2 & 1 \\ 1 & 1 & 1 & 2 \end{bmatrix} \tag{9.59}$$

Also, the absorption term can be integrated analytically if β is a constant,

$$\int_{\Delta\Omega_l} \int_{V_e} \beta \begin{bmatrix} \phi_1 \\ \phi_2 \\ \phi_3 \\ \phi_4 \end{bmatrix} [\phi_1 \quad \phi_2 \quad \phi_3 \quad \phi_4] dV d\Omega$$

$$= \frac{\beta V_e \Delta\Omega_l}{20} \begin{bmatrix} 2 & 1 & 1 & 1 \\ 1 & 2 & 1 & 1 \\ 1 & 1 & 2 & 1 \\ 1 & 1 & 1 & 2 \end{bmatrix} = \frac{\beta V_e \Delta\Omega_l}{20} C \tag{9.60}$$

Again, the above equations for 3-D calculations can be summarized in the same matrix form as given by Equations 9.48 and 9.49.

9.4.3 Integration of the Source Term

The $S(\mathbf{r},\mathbf{s})$ term has two contributions; one describes the emitting effect, and the other, the scattering effect. The integration of these two terms is now considered.

9.4.3.1 The Emitting Contribution
The emitting term is related to the temperature of the medium. For a gray medium, one has

$$I_b(\mathbf{r}) = \sigma_s T^4(\mathbf{r}) / \pi \tag{9.61}$$

We can approximate the temperature at point $\mathbf{r}$ and then calculate $I_b(\mathbf{r})$ using Equation 9.61. Alternatively, we can directly interpolate for $I_b(\mathbf{r})$ using the shape function and the nodal intensity values. If the later is taken, then one has the following expression:

$$I_b(\mathbf{r}) = \sum_{i=1}^{N_e} \phi_i(\mathbf{r}) I_{b,i} \tag{9.62}$$

with N_e being the number of nodes of the element under consideration.

Therefore, the integration of the emitting term is calculated by numerical quadratures,

$$F_{i,sr} = \int_{V_e} \int_{\Delta\Omega_l} \phi_i I_b(\mathbf{r}) d\Omega dV = \Delta\Omega_l \sum_{j=1}^{N_e} \int_{V_e} \kappa(\mathbf{r})\phi_i(\mathbf{r})\phi_j(\mathbf{r}) I_{b,j} dV$$

$$= \Delta\Omega_l \sum_{m=1}^{N_g} \sum_{j=1}^{N_e} \kappa(\mathbf{r}_m) \phi_i(\mathbf{r}_m)\phi_j(\mathbf{r}_m) w_m I_{b,j} \mid J(\mathbf{r}_m) \mid \qquad (9.63)$$

where N_g is the number of integration points, w_m the integration weights, and $|J(\mathbf{r}_m)|$ the Jacobian. If the absorption coefficient κ is a constant, then the integration can be calculated analytically,

$$f_{i,sr} = \frac{\kappa A_e \Delta\Omega_l}{12} \sum_{j=1}^{N_e} B_{ij} I_{b,j} \text{ for a 2-D triangle} \qquad (9.64a)$$

$$f_{i,sr} = \frac{\kappa V_e \Delta\Omega_l}{20} \sum_{j=1}^{N_e} C_{ij} I_{b,j} \text{ for a 3-D tetrahedron} \qquad (9.64b)$$

where matrices B and C are given by Equations 9.46b and 9.59, respectively.

9.4.3.2 The Scattering Contribution
It is known that particles present in a medium will scatter the radiative intensity traveling in one direction into all other directions. Likewise, the radiation in other directions may also be scattered by the particles into the given direction in a scattering medium. Scattering effects are usually classified into isotropic scattering and anisotropic scattering. The former scatters energy to all other directions with the same energy distribution, whereas anisotropic scattering redirects radiation energy in different directions with varying energy distributions. The isotropic scattering function is simple and easy to calculate, that is,

$$\Phi(\mathbf{s},\mathbf{s'}) = 1 \qquad (9.65)$$

Anisotropic scattering is more complex and certainly needs more computing time since the scattering function is directionally dependent. There are two different models being used for anisotropic scattering functions: forward scattering and backward scattering. Forward scattering means more energy is scattered into the forward direction than the backward direction. Backward scattering means just the opposite, that more energy is scattered into the backward direction. The scattering functions, either forward or backward, may be described by the following generic expression:

$$\Phi(s,s') = \sum_{j=1}^{N_\Phi} c_j P_j(\cos\varpi) \qquad (9.66)$$

where N_Φ is the number of the terms used to represent Φ and ϖ is calculated by

$$\varpi = \cos\theta\cos\theta' + (1 - \cos^2\theta)(1 - \cos^2\theta')^{1/2}\cos(\varphi' - \varphi) \tag{9.67}$$

and P_j is the Legendre polynomial, which has the following properties:

$$P_0(x) = 1, \quad P_1(x) = x, \quad P_{n+1}(x) = \frac{2n+1}{n+1}xP_n(x) - \frac{n}{n+1}P_{n-1}(x) \tag{9.68}$$

For the calculations given here, the values of coefficient c_j in Equation 9.66 are taken from the work of Kim and Lee [11], where the coefficients of the polynomial for different models are obtained by slightly modifying the Mie coefficients [11, 12].

Equation 9.66 describes the dependence of the scattering function on the directions for anisotropic scattering phenomena. In calculations, the angular space is discretized into a finite number of control angles. While the scattering function at the axle direction of a control angle may be used as the average scattering function, a better approach is to average the scattering function over each control angle using the following expression [13]:

$$\overline{\Phi}(\mathbf{s},\mathbf{s}') = \frac{\displaystyle\int_{\Delta\Omega'}\int_{\Delta\Omega}\Phi(\mathbf{s},\mathbf{s}')d\Omega d\Omega'}{\displaystyle\int_{\Delta\Omega'}\int_{\Delta\Omega}d\Omega d\Omega'} \tag{9.69}$$

The above computational procedure, and scattering functions, can be readily incorporated into the discontinuous finite element formulation as a source term. The integration of the source term may be made using the integration quadrature rules,

$$f_{i,sc} = \int_{V_e}\frac{\sigma(\mathbf{r})}{4\pi}\int_{\Delta\Omega_l}\sum_{j=1}^{N_\Omega}\int_{\Delta\Omega'_j}\phi_i(\mathbf{r})I(\mathbf{r},\mathbf{s}'_j)\Phi(\mathbf{s},\mathbf{s}')d\Omega'd\Omega dV$$

$$= \sum_{m=1}^{N_g}\frac{\sigma(\mathbf{r}_m)}{4\pi}w_m\phi_i(\mathbf{r}_m)\,|\,J(\mathbf{r}_m)\,|\,\sum_{j=1}^{N_\Omega}I(\mathbf{r}_m,\mathbf{s}'_j)\overline{\Phi}(\mathbf{s}_l,\mathbf{s}'_j)\Delta\Omega_l\Delta\Omega'_j \tag{9.70}$$

where subscript i refers to the node number local to the element, N_Ω is the number of discretized angles, and $\mathbf{r}_m = (x_m, y_m, z_m)$. The force, calculated as described above, is then added to the ith node of the element.

Example 9.5. Employ the above numerical procedure to solve a 3-D cube filled with an absorbing, emitting and scattering medium.

Solution. This is perhaps the most general problem of internal radiation in a 3-D geometry. When the medium is scatterting, the radiation of a given direction is redirected into all other directions. Radiation in other directions may also be scattered into the direction under consideration in a scattering medium. The scattering effect is included as part of the source term for radiation. Scattering effects are usually classified into two categories: isotropic scattering and anisotropic scattering. The former scatters energy to all other directions with the same energy distribution, whereas the latter scatters radiation energy to different directions with varying energy distributions.

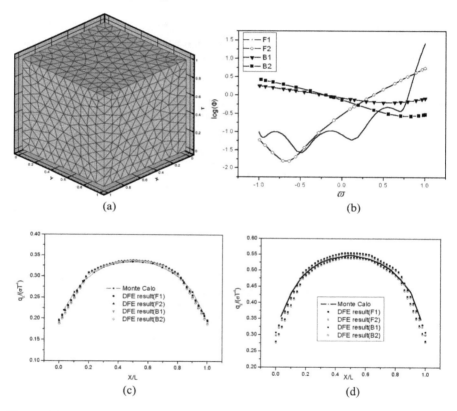

(a)   (b)   (c)   (d)

Figure 9.7e. Comparison of heat flux distributions computed using the DFE, FVM and Monte Carlo methods for radiative heat transfer in a cube filled with anisotropic scattering media. (a) Mesh distribution. (b) Scattering function distributions used for computations. (c) Heat flux q_z^* distributions along x/L at $y/L = 0.5$ and $z/L = 0.5$ on the top surface of the cube with scattering albedo $\omega = \sigma/\beta = 0.5$ and $\beta = 1.0$. (d) Heat flux q_z^* distributions along x/L at $y/L = 0.5$ and $z/L = 1$ on the top surface of the cube with scattering albedo $\omega = \sigma/\beta = 0.5$ and $\beta = 2.0$

The calculated results for anisotropic scattering are given in Figure 9.7e. The present calculations are also compared with those obtained using the Monte Carlo method. The comparison between the DFE results and those reported in Jendoubi and Lee [5] is gratifying for all these cases, suggesting that the DFE method is useful for the radiative heat transfer calculations.

9.5 Radiation in Systems of Axisymmetry

In some cylindrical systems, such as combustion chambers, boilers, gas turbines and optical crystal growth furnaces, an axisymmetric approximation often may be made to predict the thermal performance. An important implication of this approximation is that the axisymmetric and periodic conditions associated with these systems can be applied, and thus fully three-dimensional calculations may be replaced by the corresponding two-dimensional calculations, thereby resulting in savings in both computational cost and storage requirement.

9.5.1 Governing Equation in Cylindrical Coordinates

For the systems of axisymmetry, the cylindrical coordinate system is more convenient to use. For radiative heat transfer in the systems, the temperature and radiative properties vary only in the r and z directions, but not in the azimuthal direction φ_c. Here, φ_c is the azimuthal angle in the cylindrical coordinate system and independent of azimuthal direction angle φ. With the coordinate system shown in Figure 9.5, the radiative transfer equation can be written as

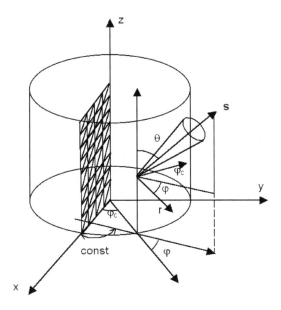

Figure 9.5. Schematic illustration of an internal radiation problem in cylindrical coordinates

$$\mathbf{s} \cdot \nabla I(\mathbf{r}, \mathbf{s}) = \sin\theta \cos\varphi \frac{\partial I(r,z;\mathbf{s})}{\partial r} - \frac{\sin\theta \sin\varphi}{r} \frac{\partial I(r,z;\mathbf{s})}{\partial \varphi} + \cos\theta \frac{\partial I(r,z;\mathbf{s})}{\partial z}$$

$$= -\beta(\mathbf{r})I(\mathbf{r},\mathbf{s}) + \kappa(\mathbf{r})I_b(\mathbf{r}) + \frac{\sigma(\mathbf{r})}{4\pi} \int_{4\pi} I(\mathbf{r},\mathbf{s}')\Phi(\mathbf{s},\mathbf{s}')d\Omega' \qquad (9.71)$$

where φ is the azimuthal direction of radiation intensity. From the geometric relation shown in Figure 9.5, it is clear that $\varphi_c + \varphi = $ constant along $\mathbf{s}$. It is important to note here that for an axisymmetric problem, the radiation intensity also depends on the polar (i.e., φ) direction. However, as shown below, the axisymmetry and periodic conditions intrinsic with an axisymmetric problem may be used to map the quantities at any φ using the data at $\varphi = 0$, thereby making the calculations possible over a 2-D mesh.

Let us consider a typical element, that is, the ith triangular element and its neighbors, as shown in Figure 9.3a. Remember that this 2-D triangular element generates a corresponding 3-D element by rotating around the z axis at a prescribed angle. For the sake of discussion, the inter-element boundaries are plotted separately. The nodal values of the variable are defined within the element to conform to the rule of selecting interpolation functions from the finite element broken space. Because discontinuity is allowed across the element boundaries in the discontinuous formulation, the common geometric node does not have the same field variable value. This is a crucial difference between the conventional and the discontinuous finite element formulations.

For a 2-D linear triangular element located on the r–z plane, the shape functions may be constructed as follows:

$$\begin{bmatrix} \phi_1 \\ \phi_2 \\ \phi_3 \end{bmatrix} = \frac{1}{2A_e} \begin{bmatrix} M_{23} & -Z_{23} & R_{23} \\ M_{31} & -Z_{31} & R_{31} \\ M_{12} & -Z_{12} & R_{12} \end{bmatrix} \begin{bmatrix} 1 \\ z \\ r \end{bmatrix} \qquad (9.72)$$

where A_e is the area of the triangular element on the r–z plane, and the definition of elements in the matrix of Equation 9.72 is as follows:

$$M_{ij} = \begin{vmatrix} z_i & z_j \\ r_i & r_j \end{vmatrix}, \quad R_{ij} = \begin{vmatrix} 1 & 1 \\ r_i & r_j \end{vmatrix}, \quad Z_{ij} = \begin{vmatrix} 1 & 1 \\ z_i & z_j \end{vmatrix} \qquad (9.73)$$

The radiation intensity inside the element is interpolated using the above shape functions,

$$I(z,r;\mathbf{s}) = I_1(\mathbf{s})\phi_1(z,r) + I_2(\mathbf{s})\phi_2(z,r) + I_3(\mathbf{s})\phi_3(z,r) \qquad (9.74)$$

Substituting the above equation into Equation 9.28, and noticing that the intensity with control angles is constant due to a step function approximation, one reaches the following expression:

$$\int_{\Delta\Omega_l}\mathbf{s}\cdot\int_{V_e}\begin{bmatrix}\phi_1\\\phi_2\\\phi_3\end{bmatrix}[\nabla\phi_1,\nabla\phi_2,\nabla\phi_3]\begin{bmatrix}I_1\\I_2\\I_3\end{bmatrix}dVd\Omega$$

$$+\int_{\Delta\Omega_l}\int_{\Gamma_p}\begin{bmatrix}\phi_1\\\phi_2\\\phi_3\end{bmatrix}[\phi_1\;\;\phi_2\;\;\phi_3]\begin{bmatrix}[I]_1\\{[I]_2}\\{[I]_3}\end{bmatrix}(\mathbf{n}\cdot\mathbf{s})d\Gamma d\Omega$$

$$+\int_{\Delta\Omega_l}\int_{\Gamma_\varphi}\begin{bmatrix}\phi_1\\\phi_2\\\phi_3\end{bmatrix}[\phi_1\;\;\phi_2\;\;\phi_3]\begin{bmatrix}[I]_1\\{[I]_2}\\{[I]_3}\end{bmatrix}(\mathbf{n}\cdot\mathbf{s})d\Gamma d\Omega$$

$$=\int_{\Delta\Omega_l}\int_{V_e}\left\{-\beta(\mathbf{r})\begin{bmatrix}\phi_1\\\phi_2\\\phi_3\end{bmatrix}[\phi_1,\;\phi_2,\;\phi_3]\begin{bmatrix}I_1\\I_2\\I_3\end{bmatrix}+\begin{bmatrix}\phi_1\\\phi_2\\\phi_3\end{bmatrix}S(\mathbf{r},\mathbf{s})\right\}dVd\Omega \qquad (9.75)$$

9.5.2 Volume Integration

Let us consider the first term in Equation 9.75, which involves the volume integration over a 3-D element generated by the ith triangular element. Taking the derivative of the shape functions,

$$\begin{bmatrix}\nabla\phi_1\\\nabla\phi_2\\\nabla\phi_3\end{bmatrix}=\frac{1}{2A_e}\begin{bmatrix}-Z_{23}&R_{23}\\-Z_{31}&R_{31}\\-Z_{12}&R_{12}\end{bmatrix}\begin{bmatrix}\hat{k}\\\hat{r}\end{bmatrix} \qquad (9.76)$$

one has the following result:

$$\int_{\Delta\Omega_l}\mathbf{s}\cdot\int_{V_e}\begin{bmatrix}\phi_1\\\phi_2\\\phi_3\end{bmatrix}[\nabla\phi_1,\nabla\phi_2,\nabla\phi_3]\begin{bmatrix}I_1\\I_2\\I_3\end{bmatrix}dVd\Omega$$

$$=\begin{bmatrix}\bar{\mathbf{s}}\cdot\mathbf{A}_{11}&\bar{\mathbf{s}}\cdot\mathbf{A}_{12}&\bar{\mathbf{s}}\cdot\mathbf{A}_{13}\\\bar{\mathbf{s}}\cdot\mathbf{A}_{21}&\bar{\mathbf{s}}\cdot\mathbf{A}_{22}&\bar{\mathbf{s}}\cdot\mathbf{A}_{23}\\\bar{\mathbf{s}}\cdot\mathbf{A}_{31}&\bar{\mathbf{s}}\cdot\mathbf{A}_{32}&\bar{\mathbf{s}}\cdot\mathbf{A}_{33}\end{bmatrix}\begin{bmatrix}I_1\\I_2\\I_3\end{bmatrix} \qquad (9.77)$$

where $\bar{\mathbf{s}}$ is calculated by the expression,

$$\bar{\mathbf{s}}=\int_{\Delta\Omega_l}\mathbf{s}\,d\Omega$$

$$=[0.5(\theta_2-\theta_1)-0.25(\sin 2\theta_2-\sin 2\theta_1)][(\sin\varphi_2-\sin\varphi_1)\hat{i}-(\cos\varphi_2-\cos\varphi_1)\hat{j}]$$

$$-0.25(\cos 2\theta_2 - \cos 2\theta_1)(\varphi_2 - \varphi_1)\hat{k} \qquad (9.78)$$

and $\mathbf{A}_{ij}$ is a vector given by

$$\mathbf{A}_{ij} = \int_{V_e} \phi_i \nabla \phi_j dV = \Delta\varphi_c \sum_{m=1}^{N_g} \phi_i(r_m, z_m)\nabla \phi_j(r_m, z_m) w_m r_m |J(r_m, z_m)|$$

$$(9.79)$$

Here, the integral is evaluated numerically using the Gaussian quadrature. N_g is the number of integration points, and J is the Jacobian.

The first volume integral on the right hand side of Equation 9.75 may be calculated numerically with the result,

$$\int_{\Delta\Omega_l} d\Omega \int_{A_e} \begin{bmatrix} \phi_1 \\ \phi_2 \\ \phi_3 \end{bmatrix} [\phi_1, \phi_2, \phi_3]\beta(\mathbf{r})dV = \begin{bmatrix} B_{11} & B_{12} & B_{13} \\ B_{21} & B_{22} & B_{23} \\ B_{31} & B_{32} & B_{33} \end{bmatrix} \qquad (9.80)$$

where the matrix element B_{ij} is calculated by

$$B_{ij} = \Delta\Omega_l \Delta\varphi_c \sum_{m=1}^{N_g} \phi_i(r_m, z_m)\beta(r_m, z_m)\phi_j(r_m, z_m) w_m r_m |J(r_m, z_m)| \qquad (9.81)$$

The evaluation of the scattering term was discussed in Section 9.4.5. Below we discuss the integration over surfaces in Equation 9.75. As shown below, some of these calculations are simplified considerably with the use of the symmetric and periodic conditions associated with a cylinder.

9.5.3 Surface Integration Over Γ_p

The second integral in Equation 9.75 represents the jump condition (or numerical fluxes) across the boundary of the ith element and its neighbor (see Figure 9.3b). For a linear triangular element, the second integral is split into three terms, one for each side of the element,

$$\int_{\Delta\Omega} \int_{\Gamma_p} \begin{bmatrix} \phi_1 \\ \phi_2 \\ \phi_3 \end{bmatrix} [\phi_1, \phi_2, \phi_3] \begin{bmatrix} [I]_1 \\ [I]_2 \\ [I]_3 \end{bmatrix} (\mathbf{n} \cdot \mathbf{s}) d\Gamma d\Omega$$

$$= \Delta\varphi_c \int_{L_l} \begin{bmatrix} \phi_1 \\ \phi_2 \\ \phi_3 \end{bmatrix} [\phi_1, \phi_2, \phi_3] \begin{bmatrix} [I]_1 \\ [I]_2 \\ [I]_3 \end{bmatrix} d\gamma \int_{\Delta\Omega} (\mathbf{n}_1 \cdot \mathbf{s}) d\Omega$$

$$+\Delta\varphi_c \int_{L_2} \begin{bmatrix} \phi_1 \\ \phi_2 \\ \phi_3 \end{bmatrix} [\phi_1,\phi_2,\phi_3] \begin{bmatrix} [I]_1 \\ [I]_2 \\ [I]_3 \end{bmatrix} d\gamma \int_{\Delta\Omega} (\mathbf{n}_2 \cdot \mathbf{s}) d\Omega$$

$$+\Delta\varphi_c \int_{L_3} \begin{bmatrix} \phi_1 \\ \phi_2 \\ \phi_3 \end{bmatrix} [\phi_1,\phi_2,\phi_3] \begin{bmatrix} [I]_1 \\ [I]_2 \\ [I]_3 \end{bmatrix} d\gamma \int_{\Delta\Omega} (\mathbf{n}_3 \cdot \mathbf{s}) d\Omega \qquad (9.82)$$

where L_1, L_2, and L_3 are the lengths of the corresponding three sides of the triangular element on the symmetry plane (see Figure 9.3b).

The surface integrals can be evaluated numerically. Thus, one has the following result for the surface integral:

$$\int_{\Delta\Omega} \int_{\Gamma_p} \begin{bmatrix} \phi_1 \\ \phi_2 \\ \phi_3 \end{bmatrix} [\phi_1,\phi_2,\phi_3] \begin{bmatrix} [I]_1 \\ [I]_2 \\ [I]_3 \end{bmatrix} (\mathbf{n} \cdot \mathbf{s}) d\Gamma d\Omega$$

$$= \sum_{k=1}^{3} \text{Cof}_{(4-k,4-k)} \begin{bmatrix} C_{11}^k & C_{12}^k & C_{13}^k \\ C_{21}^k & C_{22}^k & C_{23}^k \\ C_{31}^k & C_{32}^k & C_{33}^k \end{bmatrix} \begin{bmatrix} [I]_1 \\ [I]_2 \\ [I]_3 \end{bmatrix} \qquad (9.83)$$

where the elements of matrix $\mathbf{C}^k$ are calculated by the following numerical integration:

$$C_{ij}^k = \Delta\varphi_c \int_{\Delta\Omega} \phi_i \phi_j \, d\gamma \int_{\Delta\Omega} (\mathbf{n}_k \cdot \mathbf{s}) d\Omega$$

$$= \Delta\varphi_c \mathbf{n}_k \cdot \bar{\mathbf{s}} \sum_{m=1}^{N_g} \phi_i(r_m, z_m) \phi_j(r_m, z_m) w_m r_m |J(r_m, z_m)| \qquad (9.84)$$

Also, $\text{Cof}_{(4-k, 4-k)}$ $(k = 1, 2, 3)$ means setting to zero the elements in the $(4-k)$th row and the $(4-k)$th column of the matrix with the index pair $(4-k, 4-k)$ referring to the low double indexes of the matrix element $C_{4-k, 4-k}$. For example, $\text{Cof}_{(4-1, 4-1)} = \text{Cof}_{(3, 3)}$ means forming a new matrix from matrix $\mathbf{C}^1$ by setting the elements in the column and row of C_{33} to zero. Note that the above expression has been written for a curved element. For a linear element, the Jacobian J is a constant, which can be taken out of the summation term.

In the discontinuous finite element treatment, the jump terms at the element boundaries have to be selected depending on the sign of $\mathbf{n}_j \cdot \bar{\mathbf{s}}$. This is different from the continuous finite element formulation in which the across-element continuity is enforced, and the inter-element boundary terms cancel each other out, such that a jump condition does not arise. One treatment of these jump terms that works effectively with linear elements is the upwinding scheme [4, 5]. This scheme is also used here,

$$
(\mathbf{n}_j \cdot \overline{\mathbf{s}})
\begin{bmatrix} [I]_1 \\ [I]_2 \\ [I]_3 \end{bmatrix}
= -(-\mathbf{n}_j \cdot \overline{\mathbf{s}})
\begin{bmatrix} [I]_1 \\ [I]_2 \\ [I]_3 \end{bmatrix}
$$

$$
= \max(0,-\mathbf{n}_j \cdot \overline{\mathbf{s}})
\begin{bmatrix} I_1 \\ I_2 \\ I_3 \end{bmatrix}^{elem\ i}
- \max(0,-\mathbf{n}_j \cdot \overline{\mathbf{s}})
\begin{bmatrix} I_1 \\ I_2 \\ I_3 \end{bmatrix}_{(NB,j)}
\tag{9.85}
$$

Similarly, the calculations for the other two sides can also be performed analytically.

9.5.4 Integration Over Γ_φ

The surface integration over Γ_φ entails the evaluation of the following two terms:

$$
\int_{\Delta\Omega_l} \int_{\Gamma_\varphi}
\begin{bmatrix} \phi_1 \\ \phi_2 \\ \phi_3 \end{bmatrix}
[\phi_1 \quad \phi_2 \quad \phi_3]
\begin{bmatrix} [I]_1 \\ [I]_2 \\ [I]_3 \end{bmatrix}
(\mathbf{n}\cdot\mathbf{s})\,d\Gamma d\Omega
$$

$$
= \int_{\Delta\Omega_l} \int_{\Gamma_t}
\begin{bmatrix} \phi_1 \\ \phi_2 \\ \phi_3 \end{bmatrix}
[\phi_1 \quad \phi_2 \quad \phi_3]
\begin{bmatrix} [I]_1 \\ [I]_2 \\ [I]_3 \end{bmatrix}
(\mathbf{n}_t \cdot \overline{\mathbf{s}})\,d\Gamma d\Omega
$$

$$
+ \int_{\Delta\Omega_l} \int_{\Gamma_b}
\begin{bmatrix} \phi_1 \\ \phi_2 \\ \phi_3 \end{bmatrix}
[\phi_1 \quad \phi_2 \quad \phi_3]
\begin{bmatrix} [I]_1 \\ [I]_2 \\ [I]_3 \end{bmatrix}
(\mathbf{n}_b \cdot \overline{\mathbf{s}})\,d\Gamma d\Omega
\tag{9.86}
$$

where Γ_t and Γ_b denote the top and bottom surfaces, respectively (see Figure 9.6).

For an axisymmetric problem, the surface normals for these two surfaces are calculated by the following expressions:

$$
\mathbf{n}_t = -\sin(\Delta\varphi_c/2)\hat{i} + \cos(\Delta\varphi_c/2)\hat{j}
\tag{9.87}
$$

$$
\mathbf{n}_b = -\sin(\Delta\varphi_c/2)\hat{i} - \cos(\Delta\varphi_c/2)\hat{j}
\tag{9.88}
$$

Examination of the integrals above shows that the surface integrals can be carried out over the triangular element on the r–z plane. Consequently, we have the following expressions:

$$
\int_{\Delta\Omega_l} \int_{\Gamma_\varphi}
\begin{bmatrix} \phi_1 \\ \phi_2 \\ \phi_3 \end{bmatrix}
[\phi_1 \quad \phi_2 \quad \phi_3]
\begin{bmatrix} [I]_1 \\ [I]_2 \\ [I]_3 \end{bmatrix}
(\mathbf{n}\cdot\mathbf{s})\,d\Gamma d\Omega
$$

$$
= \begin{bmatrix} D_{11} & D_{12} & D_{13} \\ D_{21} & D_{22} & D_{23} \\ D_{31} & D_{32} & D_{33} \end{bmatrix} \left\{ (\mathbf{n}_t \cdot \overline{\mathbf{s}}) \begin{bmatrix} [I]_1 \\ [I]_2 \\ [I]_3 \end{bmatrix}_t + (\mathbf{n}_b \cdot \overline{\mathbf{s}}) \begin{bmatrix} [I]_1 \\ [I]_2 \\ [I]_3 \end{bmatrix}_b \right\}
\tag{9.89}
$$

where the matrix element D_{ij} involves a pure 2-D calculation and may be calculated numerically using the standard Gaussian quadrature,

$$
D_{ij} = \int_{\Gamma_{r-z}} \phi_i \phi_j \, d\Gamma = \sum_{m=1}^{N_g} \phi_i(r_m, z_m)\phi_j(r_m, z_m)w_m \left| J(r_m, z_m) \right|
\tag{9.90}
$$

Here Γ_{r-z} means the area on the r–z plane is used. Note that for a linear triangular element, the above expression may also be evaluated analytically, for the purpose of which some of the formulae given in Chapter 3 should be useful.

The jump conditions across the element interface can be treated using the upwinding scheme for both the top and bottom surfaces. The use of this scheme leads to the following expressions:

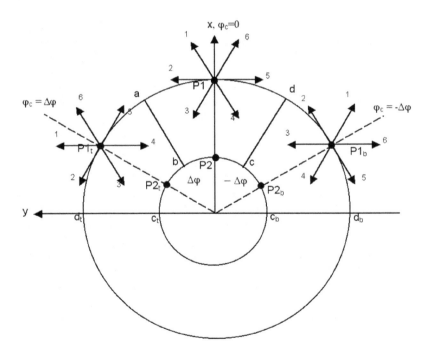

Figure 9.6. Schematic illustration of the mapping of radiation directions for axisymmetric problems

$$(\mathbf{n}_t \cdot \overline{\mathbf{s}}) \begin{bmatrix} [I]_1 \\ [I]_2 \\ [I]_3 \end{bmatrix} = \max(0, -\mathbf{n}_t \cdot \overline{\mathbf{s}}) \begin{bmatrix} I_1 \\ I_2 \\ I_3 \end{bmatrix}^{elem\ i,t} - \max(0, -\mathbf{n}_t \cdot \overline{\mathbf{s}}) \begin{bmatrix} I_1 \\ I_2 \\ I_3 \end{bmatrix}_{NBt} \quad (9.91)$$

$$(\mathbf{n}_b \cdot \overline{\mathbf{s}}) \begin{bmatrix} [I]_1 \\ [I]_2 \\ [I]_3 \end{bmatrix} = \max(0, -\mathbf{n}_b \cdot \overline{\mathbf{s}}) \begin{bmatrix} I_1 \\ I_2 \\ I_3 \end{bmatrix}^{elem\ i,b} - \max(0, -\mathbf{n}_b \cdot \overline{\mathbf{s}}) \begin{bmatrix} I_1 \\ I_2 \\ I_3 \end{bmatrix}_{NBb} \quad (9.92)$$

Here, *NBt* and *NBb* denote neighbor elements for the top and bottom surface, respectively. Note that only one neighboring element exists for the top or bottom surface, as shown in Figure 9.6.

9.5.5 Mapping

An important assumption in our calculations is that the intensities are needed only at the nodal points of the triangular element on the $r-z$ plane. The quantities are stored in memory during calculation. The intensities at any other location in the entire cylinder can be obtained from the intensity values stored at the triangular element through an appropriate mapping procedure. The mapping procedure exploits the symmetry and periodic conditions associated with the axisymmetry of the problem. If the angle of the rotation $\Delta\varphi$ is appropriately selected, then the intensities in the neighboring elements along the φ direction can be mapped from those at nodes *P1*, *P2*, and *P3*, respectively. This procedure allows the 3-D calculations to be performed using the 2-D mesh on the $r-z$ plane only.

The mapping procedure for finite volume analysis of radiative heat transfer was studied by Chai *et al.* [13, 14]. Here, a similar idea is applied to incorporate the mapping into the DFE formulation to facilitate the radiative heat transfer calculations over a 2-D mesh. From Equations 9.91 and 9.92, it is clear that when the upwinding procedure is used, intensity values at the adjacent elements are needed in order to calculate the in-flow contributions from either the top or the bottom surface. Since the intensity is stored at the nodal points of the element on the $r-z$ plane, the intensity field at the neighboring elements may be obtained using the axisymmetry and periodic conditions. As shown in Figure 9.6, the element $a-b-c-d$ is on the middle (or $r-z$) plane, and $a-b-c_t-d_t$ is the element that shares the same boundary with the top surface $a-b-c-d$, while the element $a-b-c_b-d_b$ shares the same boundary with bottom surface $a-b-c-d$. The angle between the lines $P1P2$ and $P1_tP2_t$ and that between the lines $P1P2$ and $P1_bP2_b$, is equal to $\Delta\varphi$ and $-\Delta\varphi$ respectively, because of the angular discretization.

Because the radiation intensity possesses axisymmetry or rotational symmetry, one thus has the following relation between the intensities at the three elements under consideration:

$$I_i(\varphi_c = 0) = I_i(\varphi_c = -\Delta\varphi) = I_i(\varphi_c = \Delta\varphi), i = 1, ..., N_\varphi \quad (9.93)$$

where N_φ is the number of discretized polar angles. The indexes on $i + 1$ and $i - 1$ are cyclic, such that $i + 1 \to 1$ if $i = N_\varphi$, and $i - 1 \to N_\varphi$ if $i = 1$. Thus, all the intensities on the other planes are the same as those on the symmetry (i.e., r–z) plane with an appropriate rotation. To comply with the condition of axisymmetry, the net radiation energy flux across any plane passing through the $r = 0$ axis should be zero. Thus, the following relation among the intensities at $\varphi_c = 0$ is:

$$I_i(\varphi = 0) = I_{N_\varphi - i + 1}(\varphi = 0) \tag{9.94}$$

This condition implies that, to avoid numerical errors, N_φ needs to be an even number, which imposes a constraint on the way the polar angle discretization is made. While this constraint may be a nuisance, it is beneficial in that only half of the radiation directions need be solved, thereby permitting an increasing speed of computation.

With reference to Figure 9.6, the intensity in the direction 1 at $\varphi_c = 0$ is parallel to that in the direction 2 at $\varphi_c = -\Delta\varphi$ and also to that in the direction 6 at $\varphi_c = \Delta\varphi$. This periodic condition should hold true for other corresponding directions as well. This condition, in combination with the axisymmetry condition discussed above, allows us to obtain information on the intensities at the top Γ_t and bottom Γ_b from those saved at the element defined at the r–z plane. To illustrate this, we first consider the intensity I_{t2} of direction 2 at the top surface. The quantity I_{t2} is mapped from the known values at the r–z plane as follows:

$$I_{t2}^{\;-} = I_2(\varphi_c = 0) \text{ and } I_{t2}^{\;+} = I_1(\varphi_c = \Delta\varphi) = I_1(\varphi_c = 0) \tag{9.95}$$

The same relation applies to the intensities at the other directions. The mapping can be applied in a similar fashion to the intensities at the bottom surface with the results,

$$I_{b2}^{\;-} = I_2(\varphi_c = 0) \text{ and } I_{t2}^{\;+} = I_3(\varphi_c = -\Delta\varphi) = I_3(\varphi_c = 0) \tag{9.96}$$

Clearly, the same relation can be used for the other directions as well. Here, to be consistent with our notations, subscript $+$ refers to the outside of the elements.

9.5.6 Treatment of the Emitting and Scattering Term

We may follow the same procedure in Section 9.5.3.1 to calculate the emitting term, taking into account the axisymmetry. Using quadrature rules, the integration can be carried out numerically,

$$F_{i,sr} = \int_{V_e} \int_{\Delta\Omega_l} \phi_i I_b(\mathbf{r}) d\Omega dV = \Delta\Omega_l \sum_{j=1}^{N_e} \int_{V_e} \kappa(\mathbf{r}) \phi_i(\mathbf{r}) \phi_j(\mathbf{r}) I_{b,j} dV$$

$$= \Delta\Omega_l \Delta\varphi_c \sum_{m=1}^{N_g}\sum_{j=1}^{N_e} \kappa(r_m,z_m)\phi_i(r_m,z_m)\phi_j(r_m,z_m)r_m w_m \,|\,J(r_m,z_m)| \quad (9.97)$$

The scattering term in the axisymmetric system may be treated in a very similar fashion to that given in Section 9.5.3.2 for 2-D and 3-D geometries. With the appropriate interpolation, the integration of the source term can be readily carried out over the r–z plane only. If the numerical integration is used, we then have the following expression:

$$F_{i,sc} = \int_{V_e} \frac{\sigma(\mathbf{r})}{4\pi} \int_{\Delta\Omega_l} \sum_{j=1}^{N_\Omega} \int_{\Delta\Omega'_j} \phi_i(\mathbf{r})I(\mathbf{r},\mathbf{s}')\Phi(\mathbf{s},\mathbf{s}')d\Omega'd\Omega dV$$

$$= \Delta\varphi_c \sum_{m=1}^{N_g} \frac{\sigma(r_m,z_m)}{4\pi} w_m \phi_i(r_m,z_m)r_m \,|\,J(r_m,z_m)|$$

$$\times \sum_{j=1}^{N_\Omega} I(\mathbf{r}_m,\mathbf{s}'_j)\bar{\Phi}(\mathbf{s}_l,\mathbf{s}'_j)\Delta\Omega_l\Delta\Omega'_j \quad (9.98)$$

where subscript i refers to the node number local to the element and N_Ω is the number of control solid angles for integration. The force, calculated as described above, is then added to the ith node of the element.

Example 9.6. Use of the above numerical procedure to solve the internal radiation in an irregular 2-D cavity of axisymmetry filled with a cold medium.

Solution. We consider a conical enclosure filled with a cold medium ($T_{ref} = 0.0$). The enclosure is 2 m high with a top radius of 2.1547 m and a bottom radius of 1 m. The top and side walls are black and cold, whereas the bottom wall is black and at $T_w = 100$ K. The extinction coefficient of medium β is 1.0 m^{-1}, but the scattering albedo ω ($\omega = \sigma/\beta$) varies from 0.0 to 1.0. The computations used a mesh consisting of 400 linear triangular elements and an angular discretization of 12×8. With an increased scattering albedo, the boundary heat flux goes up because more energy is scattered to the boundary than is absorbed by the medium. The results calculated using the DFE method agree well with the finite volume results (FVM, [19]) for different scattering albedos, as shown in Figure 9.8e.

9.6 Use of RTE for External Radiation Calculations

Radiation exchange between surfaces is the other important category of radiation, and the solution of the external radiation problems using the discontinuous Galerkin boundary element method was discussed in Chapter 8. Unlike the radiation in the absorbing medium, radiation between surfaces occurs in a vacuum or a non-absorbing medium. The prediction of external thermal energy exchange between surfaces is often complicated by the fact that some applications of surface radiation exchanges involve complex geometric arrangements that are either designed to obstruct radiation exchanges or are an integral part of an overall thermal system design. It was shown in Chapter 8 that a very complex, time consuming part of a numerical algorithm for external radiation heat transfer calculations is to detect these internal blockages between a thermal ray from one surface to another in an enclosure.

Since the radiative transfer equation (RTE) describes the transfer of radiative energy in a medium, it should also be applicable to the special cases where the medium is not participating. Consequently, the RTE should be able to be used to solve the external radiation problems. There are perhaps two important advantages associated with this approach, though it rarely is considered in the literature. The first advantage is that a detailed geometric obstruction present in the enclosure needs no special treatment and the domain needs to be discretized as usual for internal radiation calculations. This eliminates a major headache in developing very precise third party detection algorithms. The second advantage is that if a code is developed for internal radiation calculations, then there is no need to develop another separate code for external radiation calculations. Both external and internal radiation problems can be handled using a unified approach, thereby simplifying the computational procedure. This is particularly important for developing multiphysics models for practical applications.

Here we consider the use of the RTE to solve the external radiation problem on the basis of the discontinuous finite element procedures discussed above. We will further demonstrate this use of RTE through a 2-D example with internal blockage and compare the results with those obtained using the Galerkin discontinuous finite element method discussed in Chapter 8.

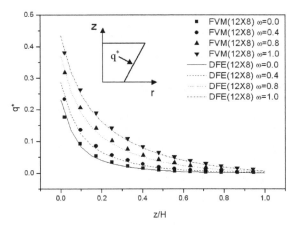

Figure 9.8e. Comparison of computed non-dimensional heat flux along the outer sidewall of the cone filled with absorbing, emitting and scattering media with the results reported in the literature (FVM) for different scattering albedos $\omega = 0.0$, $\omega = 0.4$, $\omega = 0.8$ and $\omega = 1.0$

The radiative transfer equation (RTE), when applied to a non-participating medium, reduces to a very simple form,

$$\frac{\partial I}{\partial s} = 0 \tag{9.99}$$

Here the transient effect is neglected.

Taking a two dimensional domain, we have from Section 9.3 the following discretized form of discontinuous finite element formulation,

$$\bar{s} \cdot \int_{\Delta A_i} \phi_i \nabla \phi_j \, dA \begin{bmatrix} I_1 \\ I_2 \\ I_3 \end{bmatrix} + \sum_{k=1}^{NS} \mathbf{n}_k \bar{s} \int_{\Gamma_k} \phi_i \phi_j \, d\Gamma \begin{bmatrix} [I]_1 \\ [I]_2 \\ [I]_3 \end{bmatrix} = 0 \tag{9.100}$$

This can be further written in matrix form, as has been demonstrated repeatedly in the previous sections.

Thus, the computational methodology developed for the solution of internal radiation in 1-D, 2-D, 3-D and axisymmetric geometries can be directly applied to solve the external radiation problems. This is done by simply setting the relevant properties (i.e., absorptivity, scattering coefficient, scattering functions, and emissivity) of the medium to zero. The properties of the surface of the enclosure are considered in the boundary conditions for internal radiations. We present one numerical example below.

Example 9.7. Consider a 2-D enclosure with an internal blockage as shown in Figure 9.9e(a), along with the temperature boundary conditions. The emissivity of all boundaries is 1.0, and the medium in the enclosure is non-participating. Solve the problem using both the discontinuous Galerkin boundary element method and the discontinuous Galerkin finite element method. Discuss the numerical results.

Solution. Because the blockage exists in this enclosure, the calculation of radiative energy transfer between the surfaces requires the detection of the third party blockage. The algorithm developed in Chapter 8 is applied here. To ensure the numerical accuracy, a total of 320 boundary elements are used to discretize the boundary of the enclosure. On the other hand, the calculation using the RTE approach requires the full discretization of the domain. For this problem, a structured triangular mesh is used, which has 2304 elements, but the boundary is just discretized into 96 boundary elements, and the angular space discretization is $\theta \times \varphi = 2 \times 80$. The calculated heat flux of the bottom surface is calculated and compared with the solution of the boundary element method. The reuslts are shown in Figure 9.9e, where the results of the two methods agree very well, suggesting that the RTE can indeed be applied to solve the external radiation problems with good accuracy. To further compare the speed of the DFE method and the boundary element method, numerical experiments were performed, where both meshes for the boundary element and the discontinuous calculations have 120 boundary elements. Since the geometry is symmetric, the symmetry boundary condition can be easily implemented using the discontinuous Galerkin method. Numerical calculations show that for this particular testing problem, the discontinuous finite element method is faster than the boundary element method to obtain the results of the same accuracy.

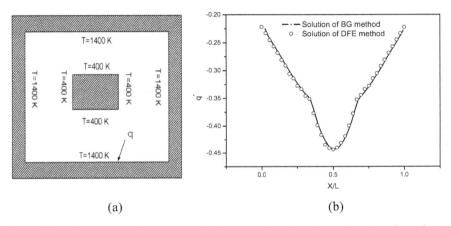

(a)   (b)

Figure 9.9e. Comparison of external radiation transfer calcuations using the discontinous Galerkin boundary and finite element methods: (a) a 2-D cavity with internal blockage, along with the wall temperatures, and the cavity is filled with a non-participating medium and (b) calculated radiation heat flux distribution along the bottom wall

9.7 Coupling of the Discontinuous Method with Other Methods

There have been well-established numerical methods for the solution of a wide range of heat transfer problems such as heat conduction and convection. Thus, it is natural to test the idea of coupling the discontinuous Galerkin method with these well known methods for the mixed heat transfer calculations that involve internal radiation and conduction/convection. This way, the advantages of each method can be fully utilized. To demonstrate such a coupled approach, we again consider a problem of combined conduction and internal radiation in a gray medium. The differential heat balance equation may be readily written,

$$\rho C_p \frac{\partial T}{\partial t} = \nabla \cdot (k\nabla T - \mathbf{q}_r) + q''' \tag{9.101}$$

with the term $\nabla \cdot \mathbf{q}_r$ calculated using the radiative transfer equation as described above. The coupling entails the use of the conventional method for heat conduction and the discontinuous method for the internal radiation calculations. Since the radiation heat flux contribution appears as a divergence term in the source part of the heat balance equation, it opens up two possibilities for coupling the discontinuous and conventional finite elements for the mixed heat transfer calculations. The two coupling approaches are described below.

In the first approach, the heat balance equation is formulated following the same procedure as used in the conventional finite element method. This will lead to a global matrix equation with the nodal temperatures as the unknowns. To incorporate the internal radiation effect, $\nabla \cdot \mathbf{q}_r$ is calculated over a finite element where the internal radiation takes place and then is coupled to the global matrix equation as a source term. This represents a simple and direct approach. In this way, the boundary condition on $\mathbf{n} \cdot \mathbf{q}_c$ is required, which of course must satisfy the total heat flux ($\mathbf{q}_c + \mathbf{q}_r$) balance along the boundary [8–10].

In the second approach, which is often taken by many researchers using the finite volume formulations for temperature calculations, the term $\nabla \cdot \mathbf{q}_r$ is integrated out and $\mathbf{n} \cdot \mathbf{q}_r$ at the element boundaries are used. If this approach is taken for the conventional and discontinuous finite element coupling, then one would have the following expression embedded in the conventional finite element formulation for the heat balance equation:

$$\int_{V_i} \nabla \cdot \mathbf{q}_r \, \phi \, dV = \oint_{\partial V_i} \mathbf{q}_r \cdot \mathbf{n} \phi \, dS - \int_{V_i} \nabla \phi \cdot \mathbf{q}_r \, dV \qquad (9.102)$$

Thus, this approach requires the information on $\mathbf{q}_r$ in the interior of an element and along the domain boundaries. By this approach, a specification of total heat flux ($\mathbf{n} \cdot \mathbf{q}_r$ and $\mathbf{n} \cdot \mathbf{q}_c$) at the boundary term is required, which is more convenient for problems involving different phases [10]. It is noted here that $\nabla \cdot \mathbf{q}_r$ is not calculated using a numerical differentiation of $\mathbf{q}_r$, and thus there is no loss in numerical accuracy if $\nabla \cdot \mathbf{q}_r$ is used. It is noted also that if the shape function ϕ is chosen as a delta function, the volume term on the right hand side vanishes and the formulation reduces to the popular finite volume formulation.

By either of these approaches, the combined heat conduction and radiation calculations require iterative procedures. In a typical iteration process, the temperature distribution is calculated using the conventional finite elements while the internal radiation intensities are calculated by the discontinuous finite elements. The iteration starts with the calculation of temperature without radiative heat transfer. The solution of the intensity distribution, and hence the divergence of heat fluxes, are then calculated using the temperature information. The radiation heat flux divergence is then treated as a heating source and the temperature distribution is updated. This process repeats itself until a convergence on temperature and intensity is obtained.

9.8 Constant Element Approximation

As mentioned at the end of Section 9.4, the discontinuous Galerkin formulation includes the conventional finite element and finite volume formulations. The latter is recovered if one chooses to use the constant element formulation for the spatial approximations. This holds true for radiative transfer problems.

Consider a 2-D case using a triangular mesh. The treatment is identical to other dimensional and/or other mesh cases, including the geometry of axisymmetry. Our starting point is Equation 9.28. If a constant element is applied, then for the element i under consideration, we have

$$\phi_j = 1; \; \nabla\phi_j = 0 \; \text{ and } \; I_j = 1 \; \text{ for } j = 1, 2, 3 \tag{9.103}$$

Thus, the first term in Equation 9.28 vanishes. The second term becomes

$$\sum_{j=1}^{NS} \int_{\Gamma_j} d\Gamma_j \int_{\Delta\Omega_l} (\mathbf{n}_j \cdot \mathbf{s}) \, d\Omega[I] = \sum_{j=1}^{NS} L_j \max(0, -\mathbf{n}_j \cdot \overline{\mathbf{s}}_l)(I_{Elemi} - I_{NBj})$$

$$\tag{9.104}$$

The first and the second terms on the right can also be calculated easily,

$$\int_{A_e} \beta dV \int_{\Delta\Omega_l} d\Omega = \beta A_e \Delta\Omega_l \tag{9.105}$$

$$\int_{A_e} S dV \int_{\Delta\Omega_l} d\Omega = S A_e \Delta\Omega_l \tag{9.106}$$

The above equations can be summarized

$$\left(\sum_{j=1}^{NS} L_j \max(0, -\mathbf{n}_j \cdot \overline{\mathbf{s}}_l) + \beta A_e \Delta\Omega_l \right) I_{Elemi}$$

$$= \sum_{j=1}^{NS} L_j \max(0, -\mathbf{n}_j \cdot \overline{\mathbf{s}}_l) I_{NBj} + S_{Elemi} A_e \Delta\Omega_l \tag{9.107}$$

where S_{Elemi} is the source term evaluated using the information at element i. The source calculation can be done using numerical integration as stated in Section 9.4, with one integration point rule.

Example 9.8. Use both the analytical and discontinuous finite element methods to solve the problem of combined conduction and radiation in a 1-D slab,
Solution. This example is concerned with a combined heat transfer problem that involves heat conduction and internal radiation. The problem is again 1-D with the wall set at different temperatures so that heat conduction is required to predict the temperature distribution. Here the first approach is taken to couple the discontinuous and conventional finite element methods for the combined calculations and the divergence of the radiation heat fluxes is treated as a volumetric source.

We consider a stagnant, gray, non-scattering medium between black, parallel walls. The medium has a constant thermal conductivity k and a constant absorption (or extinction) coefficient $K_a(= K_e)$. The medium is adiabatic (has no energy sources or sinks), and the walls are isothermal at temperatures T_1 and T_2. The exact formulation of the coupled radiative and conductive transfer in this problem is carried out as follows.

Analytic solution. The conductive heat flux normal to the slab at any location z is

$$q_c = -k\frac{dT}{dz} \tag{9.49e}$$

From Equation 9.9 and example 9.1, the radiative flux normal to the slab at any location is

$$q_r = 2\pi \int_{-1}^{1} I\mu d\mu = 2\left\{ q_{+,1}E_3(\tau) - q_{-,2}E_3(\tau_L - \tau) \right.$$

$$\left. + \text{sgn}(\tau - \tau')\int_0^{\tau_L} E_b(\tau')E_2(|\tau - \tau'|)d\tau' \right\} \tag{9.50e}$$

where $\tau = \kappa z$ for κ independent of wavelength and position. The energy balance equation then becomes

$$\frac{dq_r}{dz} + \frac{dq_c}{dz} = 0 \tag{9.51e}$$

Making use of the Leibtnitz rule and introducing the non-dimensional temperature $\tilde{T} = T/T_1$, we obtain a nonlinear, integro-differential equation for the temperature distribution [2],

$$N\frac{d^2\tilde{T}}{d\tau^2} = \tilde{T}^4 - \frac{1}{2}\left\{ E_2(\tau) + \tilde{T}_2^4 E_2(\tau_L - \tau) + \int_0^{\tau_L} \tilde{T}^4(\tau')E_1(|\tau - \tau'|)d\tau' \right\} \tag{9.52e}$$

with boundary conditions $\tilde{T}(0) = 1$ and $\tilde{T}(\tau_L) = \tilde{T}_2$. The parameter N is the conduction-radiation parameter based on T_1,

$$N = \frac{k\kappa}{4\sigma T_1^3} = \frac{4}{3}\frac{k}{k_r} \tag{9.53e}$$

where k_r is the radiative conductivity. The conduction-radiation parameter is a measure of the relative importance of energy transport by radiation and conduction in a gray medium. For $N \to 0$ radiative transfer dominates (radiative equilibrium) and for $N \to \infty$ conduction dominates (non-participating medium).

Integrating Equation 9.52e and evaluating at $\tau = 0$ gives

$$\frac{q}{\sigma T_1^4} = -4N \frac{d\tilde{T}}{d\tau}(\tau = 0) + 1 - 2\tilde{T}_2^4 E_3(\tau_L) - 2 \int_0^{\tau_L} \tilde{T}^4(\tau) E_2(\tau) d\tau \quad (9.54e)$$

This expression can be integrated numerically to obtain the temperature distribution in the slab [1, 2].

Discontinuous Galerkin Solution. The combined continuous/discontinuous finite element method is used, as discussed in Section 9.7. The calculated temperature distributions across the 1-D slab are depicted in Figure 9.10e as a function of radiation numbers. Shown also in the figure are the analytical solutions taken from Modest [1] and Siegal and Howell [2]. Once again, for the entire range of the radiation parameter, the comparison between the analytical and numerical solutions is excellent; validating the combined discontinuous/conventional finite element approach for the combined conduction/radiation problem. It is noted that the coupled thermal system, however, represents a highly nonlinear system, and appropriate relaxation parameters are required to obtain converged results. Our experience shows that selection of these parameters is often dependent on the radiation numbers. For the calculations shown in Figure 9.10e, for example, a relaxation value of 0.04 was used, when the radiation number is 0.001.

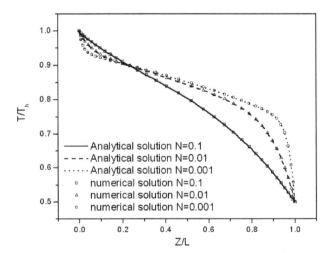

Figure 9.10e. Dependence of temperature distribution across the slab bounded by two black walls upon the radiation number N. Combined heat conduction and internal radiation are considered. The parameters used for calculations: $T(0) = 1$, $T(L) = 0.5$, and $T_h = T_{high}$

Example 9.9. Obtain numerical solution of mixed convection and radiation in a 2-D cavity.

Solution. In this example, natural convection in a simple cavity filled with a participating medium is considered. The two side walls of the cavity are fixed at two different temperatures, which combine with the gravitational forces to drive the melt flow in the cavity. As a result of the melt being thermally absorbing and emitting, internal thermal radiation plays an important role in redistributing the thermal energy. The mathematical equations governing the mixed heat transfer phenomena are given as follows:

$$\nabla \cdot \mathbf{u} = 0 \tag{9.55e}$$

$$\frac{\partial \mathbf{u}}{\partial t} + \mathbf{u} \cdot \nabla \mathbf{u} = -\nabla p + \Pr \nabla^2 \mathbf{u} - \Pr \operatorname{Ra}(T-1) \tag{9.56e}$$

$$\frac{\partial T}{\partial t} + \mathbf{u} \cdot \nabla T = \nabla^2 T - \operatorname{Rc} \nabla \cdot \mathbf{q}_{r,l} \tag{9.57e}$$

where $\Pr = v_l / \alpha_l$, $\operatorname{Ra} = g_0 \beta_{T,l} T_m L^3 / \alpha_l v_l$, and $\operatorname{Rc} = n_l^2 L \sigma T_m^3 / k_l$. The radiative heat transfer in both the melt and solid is described by an integral-differential equation, which becomes, when non-dimensionalized,

$$\frac{1}{\tau} \frac{dI}{ds} = -I(\hat{\mathbf{s}}) + (1 - \varpi) I_b + \frac{\varpi}{4\pi} \int_{4\pi} I(\hat{\mathbf{s}}_i) \Phi(\hat{\mathbf{s}}_i, \hat{\mathbf{s}}) d\Omega_i \tag{9.58e}$$

where $\tau(= \beta L)$ is the optical thickness, β is the extinction coefficient and ϖ is the single scattering albedo.

The divergence of the radiative heat flux in the energy balance equation can be calculated once the radiation intensity distribution is known,

$$\frac{1}{\tau} \nabla \cdot \mathbf{q}_r = 4(1 - \varpi) I_b(\mathbf{r}, \hat{\mathbf{s}}) - \frac{1 - \varpi}{\pi} \int_{4\pi} I(\mathbf{r}, \hat{\mathbf{s}}) d\Omega \tag{9.59e}$$

In the above equations, the following scale factors are used: L for length, α_l / L for velocity, L^2 / α_l for time, T_m for temperature, g_0 for gravity, b_0 for magnetic field intensity, $n_l^2 \sigma T_m^4$ for heat flux, and $n_l^2 \sigma T_m^4 / \pi$ for radiation intensity.

To solve the above equations, appropriate boundary conditions need to be applied. For the system under consideration, the following constraints are applied at the boundaries:

$$\mathbf{u} = 0 \text{ at all boundaries;}$$

$T = T_H$ at $x = 0$; $T = T_C$ at $x = 1$;

$q_c + Rcq_r = 0$ at $y = 0$ and $y = 1$

$$I_w(\mathbf{r}_w, \hat{\mathbf{s}}) = \frac{(1-\varepsilon)}{\pi} \int_{\hat{\mathbf{s}}'\cdot\mathbf{n}>0} I_w(\mathbf{r}_w, \hat{\mathbf{s}}')|\hat{\mathbf{n}} \cdot \hat{\mathbf{s}}'|d\Omega' + \varepsilon(\mathbf{r}_w)I_b(\mathbf{r}_w) \qquad (9.60e)$$

where T_w is the boundary temperature and ε is its emissivity. The unit vector $\hat{\mathbf{n}}$ is the surface normal pointing out of the domain.

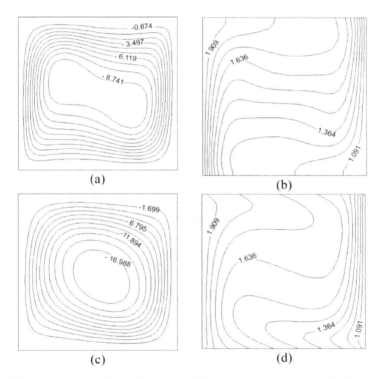

(a)   (b)

(c)   (d)

Figure 9.11e. Internal radiation effects on melt flow and temperature distributions (Ra $= 10^5$, $\tau = 1.0$): (a streamline and b temperature) Rc $= 0$, and (a streamline and b temperature) Rc $= 1$

The solution of the above problem is solved using the coupled discontinuous and conventional Galerkin finite element methods, with the former for the radiative transfer equation and the latter for other equations. The procedure follows the first approach described in Section 9.7. The calculated results are given in Figure 9.11e, which compares with fluid flow and temperature field distributions in the cavity with and without internal radiation considered. Clearly, the internal radiation has a strong effect on flow and heat transfer in this system. When internal radiation is not considered, both flow and temperature distributions are anti-symmetric in the cavity. The flow is nested by two rotating vortices and the thermal boundary layers form along the vertical sidewalls, which is the well known natural convection phenomena in opaque fluids [16–18].

The flow and temperature distributions are sensitive to internal radiation when the medium inside the cavity participates in the energy transfer. Due to the fourth power law of radiation, the bulk temperature of the medium is increased. There are also dramatic changes in temperature distributions near hot and cold walls. The fluid near the hot wall is heated directly by the energy emitted by the high temperature wall, while the fluid near the cold wall has to release the heat to the low temperature wall in order to keep the energy balance within the cavity. Consequently, the temperature distribution is no longer anti-symmetric. The temperature gradient near the cold wall is much higher than that near the hot wall, and the contours of temperature near the boundaries are closer and almost parallel to the vertical walls.

Example 9.10. *A Multiphysics Model for Crystal Growth*
This is a full numerical model for an industrial crystal growth process. This example is included here to show that the discontinuous and continuous finite elements can be seamlessly integrated using the approach discussed above to develop a comprehensive numerical model for thermal processing systems of practical significance.

We consider in particular a process for the single crystal growth from oxide melt. In this process the crucible is heated by Joule heating, which is induced by a set of surrounding coils, and is responsible for melting the material contained in the crucible through the combined mode of heat conduction, convection, and radiation. The whole furnace is placed in a container of much larger size whose wall temperature is controlled at a fixed value. During the growth, a seed crystal is dipped into the melt. With the dynamic control of the thermal environment in the furnace, a crystal grows as it is pulled from the melt. The crystal exchanges thermal radiation with the melt surface, the crucible wall and the inner surfaces of insulation materials, all of which are at different temperatures and exhibit different surface emissivities. In addition, internal radiation takes place within a semi-transparent crystal and oxide melt. The system under consideration assumes axisymmetry. Other assumptions are given in Song *et al.* [17].

The mathematical description of the above problem consists of the Maxwell equations for electromagnetic field distribution, the momentum equation for fluid flow, the energy balance equation for heat conduction and convection, the solid–liquid interface energy balance for the moving boundary, radiative transfer equation for internal radiation, and the surface energy balance equation for external surface exchange between the melt and crystal surface and the furnace walls.

The comprehensive model is developed using a variety of different numerical methods and an iterative procedure is employed to couple these methods together. The electromagnetic field is described by the Maxwell equations, which are solved using a hybrid boundary element and finite element method [20]. The fluid flow, convection and conduction equations are solved using the conventional Galerkin finite element method. The liquid–solid front represents a free surface unknown *a priori* and is solved using the deforming finite elements. The radiative heat transfer equation is solved using the discontinuous Galerkin method as described in this chapter and the external radiation exchange between the melt and crystal surfaces to the surface of the furnace is calculated using the discontinuous Galerkin boundary element method described in Chapter 8. The iterative coupling of the external and internal radiation calculations with the fluid flow calculations follows the iterative procedure described in Chapter 8 and the first approach in Section 9.7, respectively.

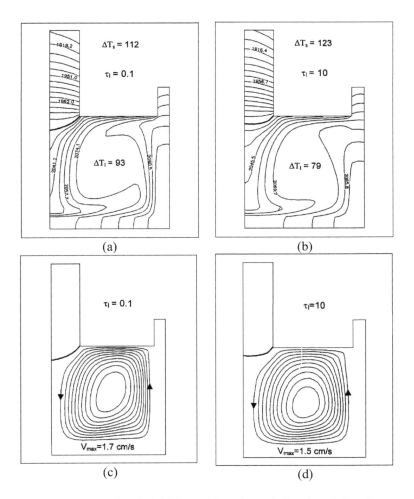

Figure 9.12e. The effect of optical thickness (τ) on the melt flow (a and c) and temperature distribution (b and d) in an optical single crystal growth process

Figure 9.12e shows the results of the effect of the melt optical thickness on the temperature distributions in the melt and the crystal, melt streamline contours, and the solidification shapes. The melt is considered to be totally transparent when its optical thickness is zero, whereas it is opaque when its optical thickness tends to infinity. It is seen that the temperature gradient in the optically thin melt decreases as its optical thickness increases. This is attributed to the fact that the contribution of internal radiation is significant for the optically thin melt, and consequently, more Joule heating is required to maintain the constant radius of the crystal. The higher temperature gradient in the melt drives a stronger melt flow by thermal buoyancy and Marangoni-flow driving forces. As the optical thickness becomes high enough, the melt becomes nearly opaque and the heat conduction and convection dominate over the internal radiation. The interface shape in the optically thin melt tends to be flatter with an increase in the optical thickness of the melt. However, the shape becomes more deeply convex toward the melt, as the optical thickness of the melt continuously increases. In general, the interface shape is strongly dependent upon the temperature distributions in the melt and the crystal. For the case under consideration, the interface exhibits a higher deflection with a larger temperature gradient in the melt. Further calculations show that the melt becomes practically optically opaque when the optical thickness is about 10000. Additional information on model development and simulations can be found in Song *et al.* [17].

Exercises

1. A semi-infinite, absorbing-emitting, non-scattering medium 1 m thick at uniform temperature is in contact with a gray-diffuse wall at $T_w = 3000$ K and with emissivity $\varepsilon_w = 0/75$. The medium is gray, and has a constant absorption coefficient $\kappa = 0.20$ cm^{-1}. Determine the net radiative heat flux at the wall (in W/m^2).

2. Obtain an expression for the temperature distribution $T(y)$ in a stagnant, conducting, absorbing, emitting, isotropic scattering, gray slab of thickness L, with no internal generation in terms of the wall temperatures $T_{1,2}$ and emissivities $\varepsilon_{1,2}$ the thermal conductivity of the medium κ, the extinction coefficient K_e, and L.

3. A stagnant, conducting, absorbing, emitting, scattering, gray medium of thickness $L = 10$ cm is heated on one side with a constant heat flux of 1.588 W/cm^2. The other side is maintained at a constant temperature of 500 K. The effective (constant) thermal conductivity of the medium is 0.02 W /cm K and the effective extinction coefficient for isotropic scattering is 1 cm^{-1}. Both walls have an emissivity of 0.8. Determine the temperature of the heated wall. Compare this result with the limiting results of pure conduction and radiation. Compare the results obtained using the two flux (S$_2$) method, the P$_1$ method, and the discontinuous Galerkin method.

4. Two infinite parallel plates at temperatures T_1 and T_2, having respective emissivities ε_1 and ε_2, and are separated by a distance D. The space

between them is filled with a gray medium having a constant absorption coefficient κ. Show that the temperature distribution in the medium is given by the solution: $(T^4 - T_2^4)/(T_1^4 - T_2^4) = (\varepsilon_1^{-1} - 0.5)/(\varepsilon_1^{-1} + \varepsilon_2^{-1} - 1)$. Use the discontinuous Galerkin method to obtain the temperature distribution and compare with the analytic solution.

5.   A gray gas is contained between two parallel plates, as shown below. The plates both have emissivity $\varepsilon = 0.5$. Plate 1 is held at a temperature $T_1 = 1500$ K, and plate 2 is at $T_2 = 700$ K. The medium between the plates is also gray, non-scattering and has a uniform absorption coefficient of $\kappa = 0.1 \text{m}^{-1}$. The plate geometry is shown below.

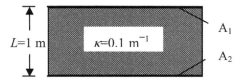

Predict the heat flux between the surfaces (W/m²) and plot the temperature profile $[T^4(\alpha) - T_2^4]/(T_1^4 - T_2^4)$ in the gas with $\alpha = \kappa x$. Solve the problem using the two-flux method.

6.   A rectangular enclosure, infinitely long in directions normal to the cross section shown, has the conditions and properties listed in the figure. The enclosure is filled with an absorbing, emitting, non-scattering gray gas in radiative equilibrium (no heat conduction or convection and no internal sources). Find the heat flux that must be supplied to each surface to maintain the specified temperatures. Develop a discontinuous Galerkin finite element code for the numerical solution of the energy equation including radiative transfer.

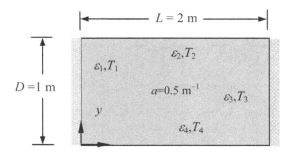

7.   Develop a Galerkin finite element formulation for the two-dimensional analysis of radiative heat transfer with an emitting, absorbing, and conducting gray medium in a two-dimensional rectangular enclosure that is infinitely long in directions normal to the cross section shown. There is no scattering in the medium. The rectangular region has a uniform heating q''' W/m³ throughout its volume. The steady two-dimensional temperature

distribution is to be determined. For simplicity, use a grid having the same increment size in both the x and y directions. All of the boundary walls are black and are at the same temperature T_w.

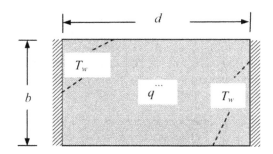

8. A parallel plate channel is heated with a uniform heat flux q along the outside of both of its walls. A semi-transparent absorbing, emitting, heat conducting medium is flowing between the walls with fully developed Poiseuille flow having a parabolic velocity distribution $u(y)$. The medium has absorption coefficient a and thermal conductivity k. Thermal properties are assumed constant. The channel wall interior surfaces are black. Set up a discontinuous Galerkin finite element procedure to determine the distribution of temperature within the medium across the channel.

References

[1]   Modest MF. Radiative Heat Transfer. New York: McGraw-Hill, 1993.
[2]   Siegel R, Howell JR. Thermal Radiation Heat Transfer, 3rd Ed. Washington, D.C.: Hemisphere Publishing Company, 1992.
[3]   Chandrasekhar S. Radiative Transfer. New York: Dover Publications, 1960.
[4]   Cumber PS, Beeri Z. A Parallelization Strategy for the Discrete Transfer Radiation Model. Numer. Heat Transf: Part B. 1998; 34: 401–421.
[5]   Jendoubi S, Lee HS, Discrete Ordinates Solutions for Radiatively Participating Medium in A Cylindrical Enclosure. J. Thermophys. Heat Transf. 1993; 7: 213–19.
[6]   Ou SCS, Liou KN. Generalization of the Spherical Harmonic Method to Radiative Transfer in Multi-Dimensional Space. Int. J. Quant. Spectroscopy Radiative Transf. 1982; 28(4): 271–288.
[7]   Reed WH, Hill TR. Triangular Mesh Methods for the Neutron Transport Equation. Los Alamos Scientific Laboratory Report LA-UR-73-479, Los Alamos, NM, 1973.
[8]   Oden JT, Babuka I, Baumann C. A Discontinuous Hp Finite Element Method for Diffusion Problems. J. Comput. Phys. 1998; 146(2): pp.491–519.
[9]   Cui X, Li BQ. A Discontinuous Finite Element Formulation for Internal Radiation Problems. Numer. Heat Transf. Part B. 2004; 46(3): 223–24.
[10] Cui X, Li BQ. A Discontinuous Finite Element Formulation for Multi-Dimensional Radiative Transfer in Absorbing, Emitting and Scattering Medium. Numer. Heat Transf. Part B. 2004; 46(5): 399–428.
[11] Kim TK, Lee H. Effect of Anisotropic Scattering on Radiative Heat Transfer in Two-Dimensional Rectangular Enclosures. J. Heat Mass Transf. 1998; 31(8): 1711–1721.

[12] Wiscombe WJ. Improved Mie Scattering Algorithms. Appl. Optics 1980; 19(9): 1505–1509.

[13] Chai JC, Patankar SV. Finite-Volume Method for Radiation Heat Transfer. Advances in Numer. Heat Transf. Volume 2, Chapter 4. Washington D.C.: Taylor & Francis, 2000; 109–138.

[14] Chui EH, Raithby GD. Prediction of Radiative Transfer in Cylindrical Enclosures with the Finite Volume method. J. Thermophys. Heat Transf. 1992; 6: 605–611.

[15] Murthy JY, Mathur SR. Radiative Heat Transfer in Axisymmetric Geometries Using an Unstructured Finite-Volume Method. Numer. Heat Transf. Part B. 1998; 33: 397–416.

[16] Shu Y, Li BQ, Lynn KG. Numerical Modeling of Internal Radiation and Solidification in Semitransparent Melts in Magnetic Field. Numer. Heat Transf. Part A. 2004; 45: 1–20.

[17] Song SP, Li BQ, Lynn KG. An Integrated Model for Optical Single Crystal Growth from Oxide Melts. ASME National Heat Transfer, Las Vegas, 2003; Paper#: HT2003–47403.

[18] Tan ZQ, Howell JR. Combined Radiation and Natural Convection in a Two-Dimensional Participating Square Medium. Int. J. Heat Mass Transf. 1992; 34(3): 785–793.

[19] Kim MY, Baek SW. Modeling of Radiative Heat Transfer in an Axisymmetric Cylindrical Enclosure with Participating Medium. J. Quant. Spectrosc. Radiat. Transf. 2005; 90(3-4): 377-388.

[20] Song SP, Li BQ. Coupled Boundary/Finite Element Analysis of Magnetic Levitation Processes: Free Surface Deformation and Thermal Phenomena. ASME J. Heat Transf. 1998; 120: 492–504.

10

Free and Moving Boundary Problems

Many different thermal fluids systems involve moving interfaces or internal boundaries. An important feature these systems often have in common is the presence, in the mathematical model, of an initially unknown (free) boundary or a boundary that moves throughout the analysis, the determination of which is an important part of the solution procedure. Practical examples involving a moving boundary include, but are not limited to, piston-driven flows, extrusion of liquids, bubble and droplet deformation and oscillation, solidification, epitaxial growth of thin films, electrodeposition, glass forming and coating of solid substrates. The effect of these moving interfaces often contributes significantly to the physics of the problems and it is thus essential to solve these problems accurately.

Free and moving boundary problems are challenging owing to the complexity associated with the often severely deformed boundaries and/or broken surfaces, multiple time and length scales, and the nonlinearity resulting from the coupling of the interface dynamics with the dynamics of the material. Only in special cases can problems of this type be solved analytically. An accurate mathematical description of these problems of practical significance usually requires numerical solution. Ideally one would like to track the moving boundary as a sharp front (allowing discontinuities in quantities such as stress and energy across the interface) without smearing the information at the front. Also, one would like to solve the field equations within each region separated by the interfaces with a satisfactory accuracy. If the interfaces become multiply connected, it is desirable to follow the evolution of the interfaces through such topological changes.

Numerous numerical techniques have been developed for solving free and moving boundary problems, because of their great fundamental and practical signficance. The computational approaches in general fall into two main categories: (a) moving grid methods and (b) fixed grid methods. Floryan and Rasmussen [1] presented a survey of these methods. Recent advances in the area are discussed by Scardovelli and Zaleski [2]. A comprehensive review of level set methods and their applications, which have become increasingly popular for free and moving boundary problems, is given by Sethian and Smeraka [3]. Use of the phase field theory to model the moving surface phenomena driven by local curvature and surface energy is reviewed by Anderson and McFadden [4].

This chapter discusses the numerical solution of free surface and moving boundary problems within the discontinuous finite element setting. A description of surface geometry and the differential and integral relations for curved surfaces developed from the theory of differential geometry are presented first. A derivation of the boundary conditions at moving boundaries is given within the framework of thermodynamics and fluid mechanics, which is valuable in appreciating both the physics governing these boundaries and the development of numerical schemes for the solution of these problems. These derivations rarely appear in a single textbook. This is followed by a discussion of both the moving grid and the fixed grid methods for free and moving boundary problems; these methods have been successfully implemented in the framework of other numerical methods such as finite volumes and finite elements. The procedures for incorporating these methods into the discontinuous finite element formulations are given. Recently, the phase field theory has received considerable attention for modeling microstructures and free/moving boundary problems. Unlike the numerical methods, which are developed to enforce the interface boundary conditions, the phase field model is developed based on the microscopic physics that governs the interfacial phenomena. The discontinuous finite element formulation of the phase field model for 1-D, 2-D and 3-D simulation of microstructure evolution in solidification systems is presented. The discontinuous finite element solution of coupled flow, thermal and phase field equations is discussed. Discontinuous algorithms are also presented for numerical analysis of grain misorientation and crystal lattice distortions as the source of the driving force for grain boundary interaction during polycrystalline liquid–solid transformation.

10.1 Free and Moving Boundaries

In many fluid dynamics problems, the computational domain is restricted by a free or moving surface. A problem involving a free and a moving boundary is shown in Figure 10.1. An important feature of this type of problem is that the shape of the surface is unknown *a priori* as it is dependent upon the flow and temperature fields. The solution of these problems demands that the free/moving surface and the flow and thermal fields are determined simultaneously during a computational process. The equations for fluid flow and thermal transport defined in Chapter 1 continue to apply in the case where free/moving boundaries are present. Because the position of the moving boundaries is not known before the solution, it is necessary to impose additional boundary conditions in order to determine the shape of the moving boundaries. While these boundary conditions may vary from problem to problem, the general requirement is that at the moving boundary, the kinematic and mechanical equilibrium conditions must be satisfied for fluid flow, and the energy and species conditions are met if the thermal and concentration fields are involved. The boundary conditions for moving and free surfaces were summarized in Section 1.6. They will be revisited in some detail in Section 10.3.

In the literature, a free surface is referred to as the interface between a gas and a liquid, as illustrated by the boundary between the gas and liquid 2 in Figure 10.1. This designation comes from the large difference in the densities of the gas and the liquid (e.g., the density ratio for water and air is ~1000). A consequence of this large difference in density is that the inertia of the gas phase may be ignored in comparison with that of the liquid. Thus, the liquid flows independently, or freely, with respect to the gas and the free surface is unconstrained to move. The only influence of the gas is the pressure it exerts on the liquid surface.

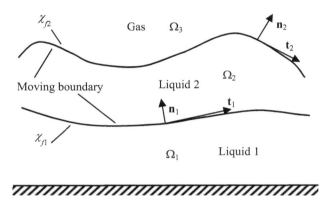

Figure 10.1. A two fluid system involving moving boundaries: a free surface defined by χ_{f2} and an internal moving boundary by χ_{f1}

A moving boundary, on the other hand, is referred to as the interface between phases of comparable densities, such as a phase boundary between a solid and a liquid or an internal moving boundary between two different liquids. In many studies, a free surface is also considered as a moving boundary or vice versa. This view is also taken in this book, because the numerical technique is essentially the same for both free and moving boundaries. A distinction between a free surface and a moving boundary to a large extent is arbitrary. However, they share a common feature, that is, the position of the interface or boundary, either free or moving, is unknown *a priori* and must be part of the solution.

10.2 Basic Relations for a Curved Surface

In this section, the basic relations for a curved surface are discussed. These include both the geometric relations and the differential and integral relations for curved surfaces. These relations are useful in describing the boundary conditions at a moving boundary and in developing numerical algorithms.

10.2.1 Description of a Surface

A moving boundary in essence is a surface in motion. Many different ways may be used to define the geometry of a moving boundary. The most common approach is to define a 3-D surface by two surface parameters as shown in Figure 10.2. By this definition, a surface is traced out by a position vector, **r**,

$$\mathbf{r} = \mathbf{r}(x(\xi,\eta), y(\xi,\eta), z(\xi,\eta)) \tag{10.1}$$

Clearly, a surface is mapped out by the above equation as ξ and η move on the surface.

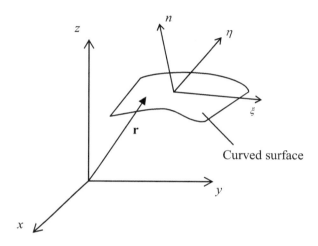

Figure 10.2. Local (ξ, η, n) and global (x, y, z) coordinates for a curved surface

As shown in Figure 10.2, the global Cartesian coordinates (x, y, z) are a function of the local coordinates (ξ, η, n), which are erected on the surface. For the convenience of subsequent discussion, we define the following geometric parameters (see also Section 3.6.2),

$$E = \mathbf{r}_{,\xi} \cdot \mathbf{r}_{,\xi} ; \qquad F = \mathbf{r}_{,\xi} \cdot \mathbf{r}_{,\eta} ; \qquad G = \mathbf{r}_{,\eta} \cdot \mathbf{r}_{,\eta} ;$$

$$L = \mathbf{r}_{\xi\xi} \cdot \mathbf{n}; \qquad M = \mathbf{r}_{\xi\eta} \cdot \mathbf{n}; \qquad N = \mathbf{r}_{\eta\eta} \cdot \mathbf{n} ;$$

$$H_1 = EG - F^2 ; \qquad \hat{\xi} = \mathbf{r}_{,\xi} / \sqrt{E}; \qquad \hat{\eta} = \mathbf{r}_{,\xi} / \sqrt{G} \tag{10.2}$$

where subscript "," denotes the differentiation, e.g., $\mathbf{r}_{,\eta} = \partial \mathbf{r}/\partial \eta$, and the hat "^" on ξ and η means the unit vector along the direction of ξ and η, respectively.

With the above notations, the unit normal vector to the surface is calculated by the following expression,

$$\mathbf{n} = \frac{\mathbf{r}_{,\xi} \times \mathbf{r}_{,\eta}}{|\mathbf{r} \times \mathbf{r}|} = \frac{\mathbf{r}_{,\xi} \times \mathbf{r}_{,\eta}}{\sqrt{EG - F^2}} = \frac{\mathbf{r}_{,\xi} \times \mathbf{r}_{,\eta}}{H_1} \tag{10.3}$$

In a surface coordinate system, the normal vector varies along the surface. The tangential derivatives of the normal vector, which appear in various surface relations, are calculated by

$$\mathbf{n},_{\xi} = \frac{FM - GL}{H_1}\mathbf{r},_{\xi} + \frac{FL - EM}{H_1}\mathbf{r},_{\eta} \qquad (10.4\text{a})$$

$$\mathbf{n},_{\eta} = \frac{FN - GM}{H_1}\mathbf{r},_{\xi} + \frac{FM - EN}{H_1}\mathbf{r},_{\eta} \qquad (10.4\text{b})$$

An important geometric property of the surface is the mean curvature, H, which is calculated by the following expression [5, 6]:

$$2H = -\nabla_s \cdot \mathbf{n} = \frac{EN - 2MF + LG}{EG - F^2} \qquad (10.5)$$

where ∇_s is the surface vector differential operator [5] and its definition is discussed in the next section (see Equation 10.7).

For many applications, a surface may be more conveniently defined as a function of height, $h = f(x, y, t)$. In this definition of a moving boundary, the surface normal and curvature can be readily calculated once the expression for f is known. For convenience, the formulae for the surface normal and the curvature of various common surfaces are listed in Table 10.1. These functions are useful when the moving boundary is single-valued.

In the fluid mechanics literature, a moving boundary is also denoted by a phase-characteristic function $\chi(x, y, z, t)$ with $\chi = 1$ in phase 1 and $\chi = 0$ in phase 2 [2]. The advantage of this generic definition is that it is valid for both single-valued and multiple-valued surfaces.

Example 10.1. Like vectors, we may define a curvature vector $\mathbf{k}$ for a curve as follows:

$$\mathbf{k} = \frac{d\mathbf{s}}{ds} \qquad (10.1\text{e})$$

which points to the origin of the curvature as shown for the curve in Figure 10.1e. This definition is independent of the orientation of the curve. The magnitude of the curvature is calculated by

$$|\kappa| = |\mathbf{k}| = \left|\frac{d\mathbf{s}}{ds}\right| \qquad (10.2\text{e})$$

which is related to the radius of curvature of the curve R by

Table 10.1. List of functions, normals and curvatures for various surfaces

	Function	Normal, $\mathbf{n} =$	Curvature, $2H =$
Cartesian	$y - f(x) = 0$	$\dfrac{(-f_{,z};1)}{\sqrt{1+f_{,z}^2}}$	$\dfrac{f_{,zz}}{[1+f_{,z}^2]^{3/2}}$
	$z - f(x,y) = 0$	$\dfrac{(-f_{,x};-f_{,y};1)}{\sqrt{1+f_{,x}^2+f_{,y}^2}}$	$\dfrac{\left[\begin{array}{c}f_{,xx}(1++f_{,y}^2)+f_{,yy}(1+f_{,x}^2)\\-2f_{,xy}f_{,x}f_{,y}\end{array}\right]}{f[1+f_{,x}^2+f_{,y}^2]^{3/2}}$
Cylindrical	$z - f(r,\theta) = 0$	$(-f_{,r};-f^{-1}f_{,\theta};1)g^{-1}$ $g = \sqrt{f_{,r}^2+(f_{,\theta}/f)^2+1}$	$g^{-1}f_{,r}+f^{-1}(g^{-1}f_{,r})_{,r}$ $+f^{-2}(g^{-1}f_{,\theta})_{,\theta};$
	$z - f(r) = 0$	$\dfrac{(-f_{,r};1)}{\sqrt{f_{,r}^2+1}}$	$\dfrac{f_{,r}}{f\sqrt{f_{,r}^2+1}}+\dfrac{f_{,rr}}{\sqrt{(f_{,r}^2+1)^3}}$
	$r - f(z) = 0$	$\dfrac{(-f_{,z};1)}{\sqrt{1+f_{,z}^2}}$	$\dfrac{ff_{,z}-[1+f_{,z}^2]}{f[1+f_{,z}^2]^{3/2}}$
	$r - f(z,\theta) = 0$	$(1;-f_{,\theta}/f;-f_{,z})d^{-1};$ $d = \sqrt{1+(f_{,\theta}/f)^2+f_{,z}^2}$	$-(fd)^{-1}+f^{-1}\left((fd)^{-1}f_{,\theta}\right)_{,\theta}$ $+\left(d^{-1}f_{,z}\right)_{,z}$
	$r - f(\theta) = 0$	$(1;-f^{-1}f_{,\theta})b^{-1};$ $b = \sqrt{1+(f_{,\theta}/f)^2}$	$\dfrac{f^2+2f_{,\theta}^2}{(bf)^3}-f^{-1}\left(\dfrac{f_{,\theta}}{bf}\right)_{,\theta}$
Spherical	$r - f(\theta,\varphi) = 0$	$(1;-f^{-1}f_{,\theta};-\dfrac{1}{f\sin\theta}f_{,\varphi})e^{-1}$ $e = \sqrt{1+(f_{,\theta}/f)^2+\left(\dfrac{f_{,\varphi}}{f\sin\theta}\right)^2}$	$\dfrac{2}{e}-\dfrac{(e^{-1}f_{,\theta}\sin\theta)_{,\theta}}{f\sin\theta}-\dfrac{(e^{-1}f_{,\varphi})_{,\varphi}}{f\sin^2\theta}$
	$r - f(\theta) = 0$	$(1;-f_{,\theta}/f)a^{-1};$ $a = \sqrt{1+(f_{,\theta}/f)^2}$	$\dfrac{(2f^2+f_{,\theta}^2)\sin\theta}{af^3\sin\theta}-\dfrac{(af_{,\theta}\sin\theta)_{,\theta}}{f^2\sin\theta}$

$$R = \frac{1}{|\kappa|} \qquad\qquad (10.3e)$$

Calculate the surface divergence of the normal vector for the curve shown in Figure 10.1e and discuss the meaning of the signs in terms of curvature vector.

Solution. With the above definitions for the curvature, we may link $\mathbf{k}$ to the outnormal for the curve as shown in Figure 10.1e,

$$\mathbf{k} = \kappa(s)\,\mathbf{n}(s) \qquad\qquad (10.4e)$$

where $\mathbf{n}(s)$ (such as $\mathbf{n}_1(s)$ and $\mathbf{n}_2(s)$ as shown in Figure 10.1e) is the outnormal of the curve at any point on the curve. Its direction is defined by the right-hand rule such that the outnormal points to the right when one walks along the curve in the direction of $\mathbf{s}$ as marked in the figure. This is consistent with the line integral that the left side is the interior of the integral domain when one walks along the $\mathbf{s}$ direction. From Equation 10.4e, it is clear that κ can be either negative or positive, depending on the relative orientation between $\mathbf{k}$ and $\mathbf{n}(s)$. If $\mathbf{k}$ and $\mathbf{n}(s)$ have the same direction, $\kappa = |\mathbf{k}|$; on the other hand, if $\mathbf{k}$ is opposite to $\mathbf{n}(s)$, then $\kappa = -|\mathbf{k}|$.

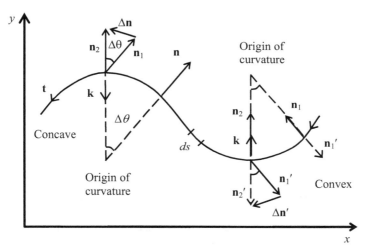

Figure 10.1e. A 2-D curve with local normals and curvature vectors

By definition, the curvature can also be calculated by the surface divergence of the normal vector, $\kappa = -\nabla_s \cdot \mathbf{n}$. For a concave curve, one has from the geometric relation shown in Figure 10.1e,

$$\nabla_s \cdot \mathbf{n} = \mathbf{t} \cdot \frac{d\mathbf{n}}{ds} = \mathbf{t} \cdot \frac{1 \cdot \Delta\theta \cdot \mathbf{t}}{\Delta\theta \cdot R} = \frac{1}{R} = -\kappa \qquad (10.5e)$$

The minus sign of κ here is merely an indication that the curvature vector $\mathbf{k}$ is opposite to $\mathbf{n}(s)$ or $\mathbf{k} = \kappa\mathbf{n}(s) = -(1/R)\mathbf{n}(s)$.

For a convex curve, on the other hand, one can show that

$$\nabla_s \cdot \mathbf{n} = \mathbf{t} \cdot \frac{d\mathbf{n}}{ds} = \hat{s} \cdot \frac{d(-\mathbf{n}')}{ds} = -\mathbf{t} \cdot \frac{1 \cdot \Delta\theta \cdot \mathbf{t}}{\Delta\theta \cdot R} = -\frac{1}{R} = -\kappa \qquad (10.6e)$$

The minus sign of κ in this case represents the fact that κ is positive and $\mathbf{k}$ is in the same direction as $\mathbf{n}(s)$, or $\mathbf{k} = \kappa\mathbf{n}(s) = (1/R)\mathbf{n}(s)$.

From the above two equations, we see that the radius of curvature for a curve can also be calculated by the following expression:

$$\frac{1}{R} = \frac{d\theta}{ds} \tag{10.7e}$$

This relation will be used in Section 10.3.2 for deriving stress balance relations at a moving interface.

By definition, the Gaussian mean curvature for a curve is given by

$$2H = \kappa \tag{10.8e}$$

which may be positive or negative in order to be consistent with κ.

This description should be valid for a 3-D surface, where the surface curvature or Gaussian mean curvature is given by the following relation:

$$2H = \kappa_1 + \kappa_2 = -\nabla_s \cdot \mathbf{n} \tag{10.9e}$$

with κ_1 and κ_2 being the two principal curvatures of the surface.

10.2.2 Differential and Integral Relations for Curved Surfaces

Some differential and integral relations developed in the area of differential geometry for a curved surface are useful for the description of moving boundary problems. These relations are briefly discussed here in vector notations. Detailed derivations of these relations, their tensor representations including the use of the Christoffel symbols and their applications to flows in a thin surface layer can be found in [5–7].

Perhaps a good starting point for the subject is the tangential derivative or surface gradient on a curved surface, which is denoted by $\nabla_s = \mathbf{t} \cdot \nabla$. Its use for general surface element calculations was discussed in Section 3.6.2. The surface gradient is also related to the spatial gradient as follows:

$$\nabla_s = \nabla - \mathbf{n}(\mathbf{n} \cdot \nabla) = (\mathbf{I} - \mathbf{nn}) \cdot \nabla \tag{10.6}$$

where $\mathbf{I}$ is the unit tensor, $I_{ij} = \delta_{ij}$. A proof of this relation is straightforward and is left as an exercise at the end of the chapter. Written in surface coordinates only, the surface vector differential operator becomes

$$\nabla_s = \frac{1}{H_1^2}\mathbf{r}_{,\xi}\left(G\frac{\partial}{\partial\xi} - F\frac{\partial}{\partial\eta}\right) + \frac{1}{H_1^2}\mathbf{r}_{,\eta}\left(-F\frac{\partial}{\partial\xi} + E\frac{\partial}{\partial\eta}\right) \tag{10.7}$$

For a special case where the two parametric curves (ξ, η) on the surface are perpendicular, $F = 0$ and $H_1^2 = EG - F = EG$, whence we have [5]

$$\nabla_s = \frac{1}{E}\mathbf{r}_{,\xi}\frac{\partial}{\partial\xi} + \frac{1}{G}\mathbf{r}_{,\eta}\frac{\partial}{\partial\eta} \tag{10.8}$$

A vector may be represented using the local coordinates (ξ, η, n) erected on a curved surface. Let this vector be $\mathbf{V}$. When expressed in the local coordinate system, $\mathbf{V}$ takes the form of

$$\mathbf{V} = V_\xi \mathbf{r},_\xi + V_\eta \mathbf{r},_\eta + V_n \mathbf{n} \tag{10.9}$$

The surface divergence of this vector can be written as

$$\nabla_s \cdot \mathbf{V} = \nabla_s \cdot (V_\xi \mathbf{r},_\xi) + \nabla_s \cdot (V_\eta \mathbf{r},_\eta) + \nabla_s \cdot (V_n \mathbf{n}) \tag{10.10}$$

Through vector analysis and the use of relations defined in Equation 10.2, it is straightforward to show that

$$\nabla_s \cdot (V_\xi \mathbf{r},_\xi) = \frac{1}{H_1} \frac{\partial}{\partial \xi} (H_1 V_\xi); \quad \nabla_s \cdot (V_\eta \mathbf{r},_\eta) = \frac{1}{H_1} \frac{\partial}{\partial \eta} (H_1 V_\eta) \tag{10.11}$$

The last term in Equation 10.10 has an important geometric implication and its expansion is given in detail here,

$$\nabla_s \cdot (V_n \mathbf{n}) = \frac{1}{H_1^2} \mathbf{r},_\xi \cdot \left(G \frac{\partial (V_n \mathbf{n})}{\partial \xi} - F \frac{\partial (V_n \mathbf{n})}{\partial \eta} \right)$$

$$+ \frac{1}{H_1^2} \mathbf{r},_\eta \cdot \left(-F \frac{\partial (V_n \mathbf{n})}{\partial \xi} + E \frac{\partial (V_n \mathbf{n})}{\partial \eta} \right)$$

$$= \frac{V_n}{H_1^2} \mathbf{r},_\xi \cdot (G\mathbf{n},_\xi - F\mathbf{n},_\eta) + \frac{V_n}{H_1^2} \mathbf{r},_\eta \cdot (E\mathbf{n},_\eta - F\mathbf{n},_\xi) + \mathbf{n} \cdot \nabla_s V_n$$

$$= -V_n \frac{EN - 2FM + GL}{H_1^2} = -V_n 2H \tag{10.12}$$

where use has been made of the relation, $\mathbf{n} \cdot \nabla_s = 0$ and those given in Equation 10.2. From the above equation, we have the vector identity for the surface divergence operator,

$$\nabla_s \cdot (V_n \mathbf{n}) = V_n \nabla_s \cdot \mathbf{n} + \mathbf{n} \cdot \nabla_s V_n \tag{10.13}$$

If we further set $V_n = 1$, then we have the well known expression for surface curvature upon combining Equations 10.12 and 10.13:

$$\nabla_s \cdot \mathbf{n} = -2H \tag{10.14}$$

which is the relation that has been most frequently quoted in the literature on free surface calculations.

Another vector identity for surface gradient is frequently used in surface related calculations,

$$\nabla_s \cdot (\mathbf{U} \times \mathbf{V}) = \mathbf{V} \cdot \nabla_s \times \mathbf{U} - \mathbf{U} \cdot \nabla_s \times \mathbf{V} \qquad (10.15)$$

which is useful for deriving the Stokes theorem on a curved surface [6].

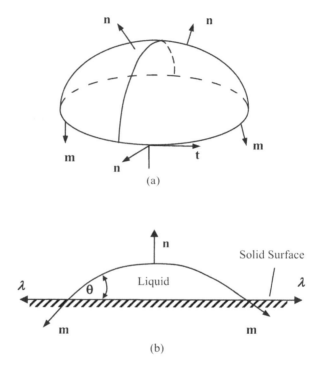

Figure 10.3. Illutration of relations for curved surfaces, $\mathbf{n}$, $\mathbf{m}$ and $\mathbf{t}$ are perpendicular to each other and form a right-hand system: $\mathbf{n} = \mathbf{m} \times \mathbf{t}$: (a) a 3-D surface and (b) a 2-D curve and the definition of contact angle θ

Some integral relations derived for solid geometry have their counterparts in the surface differential geometry. One often-used relation is the surface divergence theorem,

$$\int_S \nabla_s \cdot \mathbf{V} \, dS = \int_{\partial S} \mathbf{m} \cdot \mathbf{V} \, d\Gamma + \int_S (\nabla_s \cdot \mathbf{n})\mathbf{n} \cdot \mathbf{V} \, dS \qquad (10.16)$$

It is important to note that the second term on the right of Equation 10.16 is an extra term arising from a curved surface, which shows the curvature contribution to the continuity. Here $\mathbf{m}$ is a unit vector normal to the curve bounding the surface and is tangential to the surface. In this way, $\mathbf{m}$ is also related to the contact angles. A contact angle θ is defined as the angle between two of the interfaces at the three-phase line of contact, $\cos\theta = \mathbf{m}\cdot\boldsymbol{\lambda}$. The relation between $\mathbf{n}$ and $\mathbf{m}$ is shown in Figure 10.3 for both 2-D and 3-D cases.

For a plane surface, $\mathbf{n}$ is not a function of (ξ, η) and the curvature of the surface is zero or $\nabla_s \cdot \mathbf{n} = -2H = 0$. Therefore, for a plane surface, the following relation is obtained:

$$\int_S \nabla_s \cdot \mathbf{V}\, dS = \int_{\partial S} \mathbf{m} \cdot \mathbf{V}\, d\Gamma \qquad (10.17)$$

which is nothing but the well known Gaussian divergence theorem for a 2-D coordinate system.

Some very useful relations can also be derived from Equation 10.15. If we let $\mathbf{V} = \phi\mathbf{c}$, $\mathbf{c}$ being a constant vector, then $\nabla_s \cdot \mathbf{V} = \phi \nabla_s \cdot \mathbf{V} + \mathbf{c}\cdot\nabla_s\phi$, whence we have the following relation:

$$\int_S \mathbf{c} \cdot \nabla_s\phi\, dS = \int_{\partial S} \phi\mathbf{m} \cdot \mathbf{c}\, d\Gamma + \int_S (\nabla_s \cdot \mathbf{n})\phi\mathbf{n} \cdot \mathbf{c}\, dS \qquad (10.18)$$

This relation may be employed to project the curvature effect onto a plane surface; the technique is sometimes used in a deforming grid method to satisfy the normal stress balance condition along the moving interface [8]. From Equation 10.17 and noticing that $\mathbf{c}$ is an arbitrary constant vector, one can easily deduce the following vector relation:

$$\int_S \nabla_s\phi\, dS = \int_{\partial S} \phi\mathbf{m}\, d\Gamma + \int_S (\nabla_s \cdot \mathbf{n})\phi\mathbf{n}\, dS \qquad (10.19)$$

If ϕ is chosen as a scalar constant, then we can convert the surface integral for a curvature calculation into a line integral along the curve bounding the surface,

$$\int_{\partial S} \mathbf{m}\, d\Gamma = -\int_S (\nabla_s \cdot \mathbf{n})\mathbf{n}\, dS \qquad (10.20)$$

This relation is considered useful for imposing the boundary conditions along an interface for the discontinuous formulation using piecewise constant approximation or finite volume method [9].

The Stokes curl theorem for a curved surface is now considered. By taking $\mathbf{V} = \mathbf{c}$, $\mathbf{c}$ being a constant vector, and substituting it into Equation 10.14, we have the following relation upon the use of the divergence theorem:

$$\mathbf{c} \cdot \int_S \nabla_s \times \mathbf{U}\, dS = \mathbf{c} \cdot \int_{\partial S} \mathbf{m} \times \mathbf{U}\, d\Gamma + \mathbf{c} \cdot \int_S (\nabla \cdot \mathbf{n})\mathbf{n} \times \mathbf{U}\, dS \qquad (10.21)$$

The fact that $\mathbf{c}$ is a constant vector leads immediately to the Stokes theorem for a curved surface,

$$\int_S \nabla_s \times \mathbf{U} dS = \int_{\partial S} \mathbf{m} \times \mathbf{U} d\Gamma + \int_S (\nabla \cdot \mathbf{n}) \mathbf{n} \times \mathbf{U} dS \tag{10.22}$$

Once again, we see that an extra term arises from the effect of curvature. For a plane surface, the second term on the right vanishes and the relation becomes the well known Stokes theorem for a 2-D system.

A choice of $\mathbf{V} = \phi \nabla \psi$ would give us the following differential and integral relations, when the divergence theorem is used,

$$\nabla_s \cdot (\phi \nabla_s \psi) = \nabla_s \phi \cdot \nabla_s \psi + \phi \nabla_s^2 \psi \tag{10.23}$$

$$\int_S (\nabla_s \phi \cdot \nabla_s \psi + \phi \nabla_s^2 \psi) dS$$
$$= \int_{\partial S} \phi \mathbf{m} \cdot \nabla_s \psi \, d\Gamma + \int_S \phi (\nabla_s \cdot \mathbf{n}) \mathbf{n} \cdot \nabla_s \psi \, dS \tag{10.24}$$

Since $\mathbf{n} \cdot \nabla_s = 0$, $\mathbf{n}$ being perpendicular to ∇_s, we have

$$\int_S (\nabla_s \phi \cdot \nabla_s \psi + \phi \nabla_s^2 \psi) dS = \int_{\partial S} \phi \mathbf{m} \cdot \nabla_s \psi \, d\Gamma \tag{10.25}$$

Interchanging ϕ and ψ in Equation 10.24 yields the following relation:

$$\int_S (\nabla_s \phi \cdot \nabla_s \psi + \psi \nabla_s^2 \phi) dS$$
$$= \int_{\partial S} \psi \mathbf{m} \cdot \nabla_s \phi \, d\Gamma + \int_S \psi (\nabla_s \cdot \mathbf{n}) \mathbf{n} \cdot \nabla_s \phi \, dS \tag{10.26}$$

Subtracting Equations 10.26 from 10.24, we have Green's theorem for a curved surface,

$$\int_S (\phi \nabla_s^2 \psi - \psi \nabla_s^2 \phi) dS = \int_{\partial S} \mathbf{m} \cdot (\psi \nabla_s \phi - \phi \nabla_s \psi) \, d\Gamma \tag{10.27}$$

In particular, if ψ is a constant, one has the following simplified relation:

$$\int_S \nabla_s^2 \phi \, dS = \int_{\partial S} \mathbf{m} \cdot \nabla_s \phi \, d\Gamma \tag{10.28}$$

This is again the Green theorem for a 2-D coordinate system, which is known from calculus textbooks.

It is noted that in the above formulae, $\nabla_s \psi$ is a vector tangential to the surface, and this choice causes the curvature term to vanish. If an arbitrary vector $\mathbf{V}$ is chosen to replace $\nabla_s \psi$, then the curvature contributions will need to be included.

10.3 Physical Constraints at a Moving Boundary

We now discuss physical constants imposed on a moving boundary, which lead to the kinematic, stress, and thermal conditions at the boundary. These conditions are required to determine the moving boundaries and the field variables during the solution phase.

10.3.1 Kinematic Conditions at a Moving Boundary

In Chapter 7, the Rankine–Hugoniot condition was derived for a discontinuous boundary caused by a shock wave. This relation is also applicable here for a moving boundary. Let us assume that the interface moves at a velocity $\mathbf{U}$ with its normal component being $U_n = \mathbf{U} \cdot \mathbf{n}$. Here $\mathbf{n}$ is the outnormal of the interface point from phase 1 to phase 2. In the case where there is no phase change, the interface velocity satisfies the following continuity relation:

$$U_n = \mathbf{u}_1 \cdot \mathbf{n} = \mathbf{u}_2 \cdot \mathbf{n} \qquad (10.29)$$

In the case of phase change, say, evaporation from the liquid to vapor phase, there may be a mass flow per unit surface $\dot{m}$ from phase 1 to phase 2. Written in the frame of reference to the moving boundary, we have the relative velocity $u' = \mathbf{u} \cdot \mathbf{n} - U_n$. Applying the mass conservation over a different control volume across the interface, as was done in Chapter 7, we have the following relation,

$$\rho_1 u'_1 = \rho_2 u'_2 = \dot{m} \qquad (10.30)$$

or in an Eulerian frame of reference,

$$\rho_1 (\mathbf{u}_1 \cdot \mathbf{n} - U_n) = \rho_2 (\mathbf{u}_2 \cdot \mathbf{n} - U_n) = \dot{m} \qquad (10.31)$$

In the case of $\dot{m} = 0$, one recovers Equation 10.29 immediately.

The tangential velocity is continuous across a moving boundary due to the no-slip boundary condition.

For a person who sits on and moves along with the interface, he/she sees no rate change of the interface, $d\chi/dt = 0$, with χ being the interface parameter. Relating this Lagrangian description to an Eulerian description, one has

$$\frac{d\chi}{dt} = \frac{\partial \chi}{\partial t} + \mathbf{u} \cdot \nabla \chi = 0 \qquad (10.32)$$

The above equation should be interpreted in a weak form in that the derivatives of the discontinuous function χ are singular or do not exist [2]. In this weak formulation, the equations are interpreted by the spatial integrals of the equations. Equation 10.32 may be further written as

$$\frac{d\overline{\chi}}{dt} = \frac{\partial\overline{\chi}}{\partial t} + \mathbf{u} \cdot \mathbf{n} = 0; \quad \text{with} \quad \overline{\chi} = \frac{\chi}{|\nabla\chi|} \quad \text{and} \quad \mathbf{n} = \frac{\nabla\chi}{|\nabla\chi|} \tag{10.33}$$

Clearly, this equation represents volume evolution and corresponds to the interface motion with velocity $\mathbf{u}$.

10.3.2 Stress Conditions at a Moving Interface

Stresses develop in moving fluids and are a function of surface normal. A stress tensor $\boldsymbol{\sigma}$ has 9 components, or σ_{ij} ($i,j =1,2,3$). Hence, $\boldsymbol{\sigma} \cdot \mathbf{n}$ is the force acting on the surface whose normal is $\mathbf{n}$. This force in general is in a different direction from normal and can be deomposed into tangential and normal components: $\mathbf{t} \cdot \boldsymbol{\sigma} \cdot \mathbf{n}$ and $\mathbf{n} \cdot \boldsymbol{\sigma} \cdot \mathbf{n}$. A fluid element on a moving interface cannot exprerience an infinite acceleration. Consequently, the stresses must be continuous across the interface plus an additional surface force such as surface tension.

For a curved surface, the surface tension contributes to the stress balance to both the normal and tagential directions. This is shown in Figure 10.4a.

To derive the needed relation, we consider a differential surface element as shown in Figure 10.4b. The balance of the normal stress balance across the interface yields the following relation along S_x:

$$-\sigma_{n_1} \cdot dS_x dS_y + \sigma_{n_2} \cdot dS_x dS_y + 2\gamma\, dS_y \sin\left(\frac{d\theta_x}{2}\right) + 2\gamma\, dS_x \sin\left(\frac{d\theta_y}{2}\right) = 0 \tag{10.34}$$

where $\gamma = \gamma(S_x, S_y)$ is the surface tension and in writing the above balance equation, we have combined the contributions in the normal direction,

$$\left(\gamma + \frac{\partial\gamma}{\partial S_x}\frac{dS_x}{2}\right)\sin\frac{d\theta_x}{2} + \left(\gamma - \frac{\partial\gamma}{\partial S_x}\frac{dS_x}{2}\right)\sin\frac{d\theta_x}{2} = 2\gamma\sin\frac{d\theta_x}{2} \tag{10.35}$$

A similar equation is derived for the variation along S_y. Furthermore, if $d\theta$ is small, $\sin(d\theta/2) \approx d\theta/2$, whence we obtain the following equation:

$$-\sigma_{n_1} + \sigma_{n_2} + \gamma\left(\frac{d\theta_x}{dS_x} + \frac{d\theta_y}{dS_y}\right) = 0 \tag{10.36}$$

Making use of the definition of curvature (see Example 10.1), we have the final expression for the normal stress balance across the moving interface,

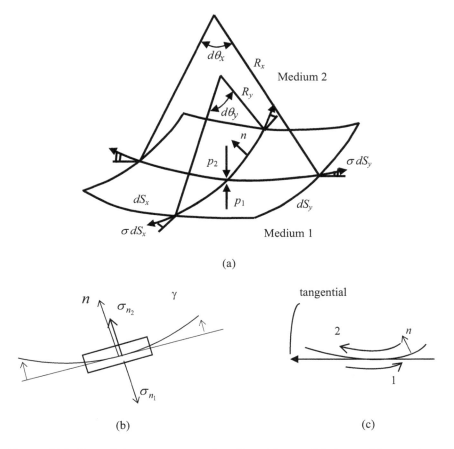

Figure 10.4. Stress balance on a curved surface: (a) overall balance, (b) normal stress balance and (c) tangential stress balance

$$-\sigma_{n_1} + \sigma_{n_2} = -\gamma\left(\frac{1}{R_x} + \frac{1}{R_y}\right)$$ (10.37)

We now consider the balance of the tangential stress in the x direction,

$$(\tau_{xn,2} - \tau_{xn,1})dS_x dS_y + \left(\gamma + \frac{1}{2}dS_x \frac{\partial\gamma}{\partial S_x}\right)dS_y \cos\frac{d\theta_x}{2}$$

$$-\left(\gamma - \frac{1}{2}dS_x \frac{\partial\gamma}{\partial S_x}\right)dS_y \cos\frac{d\theta_x}{2} = 0$$ (10.38)

where τ is the tangential (or shear) stress. Taking the differential quantities to their limit of zero gives the expression required for the tangential stress in the x direction,

$$\tau_{xn,2} - \tau_{xn,1} = -\frac{\partial \gamma}{\partial S_x} \tag{10.39}$$

Similarly, we have the tangential stress balance in the y direction,

$$\tau_{yn,2} - \tau_{yn,1} = -\frac{\partial \gamma}{\partial S_y} \tag{10.40}$$

The above relations for stress balance (i.e., Equations 10.37–10.40) can be summarized and written in vector notation:

$$\boldsymbol{\sigma}_1 \cdot \mathbf{n} - \boldsymbol{\sigma}_2 \cdot \mathbf{n} = \gamma(R_1^{-1} + R_2^{-1})\mathbf{n} + \nabla_s \gamma \tag{10.41a}$$

or in tensor notation,

$$\sigma_{1ij} n_j - \sigma_{2ij} n_j = \gamma(R_1^{-1} + R_2^{-1})n_i + (\gamma_{,i} - n_i n_j \gamma_{,j}) \tag{10.41b}$$

where we have used the relation: $\nabla_s \cdot \gamma = \nabla \gamma - \mathbf{n}(\mathbf{n} \cdot \nabla \gamma)$ and $\mathbf{n}$ is the normal point from medium 1 to medium 2.

If for a free surface, that is, medium 1 is a gas, then $\sigma_{1ij} = -P_a \delta_{ij}$. Thus we have

$$-\boldsymbol{\sigma}_2 \cdot \mathbf{n} = P_a \mathbf{n} + \gamma(R_1^{-1} + R_2^{-1})\mathbf{n} + \nabla_s \cdot \gamma \tag{10.42a}$$

or

$$-\sigma_{2ij} n_j = P_a \delta_{ij} n_j + \gamma(R_1^{-1} + R_2^{-1})n_i + (\gamma_{,i} - n_i n_j \gamma_{,j}) \tag{10.42b}$$

In particular, for a spherical droplet with a constant surface tension and with flow stress neglected,

$$P - P_a = 2\gamma R^{-1} \tag{10.43}$$

which is the well known Laplace–Young relation [10].

10.3.3 Thermal Conditions at a Moving Interface

At a moving boundary, the thermal condition is derived based on the energy balance across the moving interface. Following exactly the same procedure as shown above for the mechanical balance, we obtain the energy balance statement across the moving interphase:

$$q_{1,n} - q_{2,n} = -\rho_1 L_{21} \mathbf{u}_{1i} \cdot \mathbf{n}_{12} \qquad (10.44)$$

where $q_{1,n}$ is the heat flux into the interface from phase 1, $q_{2,n}$ is the heat flux leaving the interface from phase 2, L_{21} is the heat released per unit mass at the boundary when phase 2 is converted to phase 1, $\mathbf{u}_{1i}$ is the velocity of the interface with which phase 1 moves into phase 2 and $\mathbf{n}_{12}$ points from phase 1 to phase 2. Note that $\rho_1 \mathbf{u}_{1i} \cdot \mathbf{n}_{12}$ is the mass of phase 1 produced per unit surface per unit time at the moving boundary.

Written explicitly for a phase change boundary, the above equation becomes

$$\mathbf{n} \cdot k_1 \nabla T_1 - \mathbf{n} \cdot k_2 \nabla T_2 = \rho_1 L_{21} \mathbf{u}_{1i} \cdot \mathbf{n} = -\rho_1 L_{21} \frac{\partial \overline{\chi}_f}{\partial t} \qquad (10.45)$$

where $q_{i,n} = -\mathbf{n} \cdot k_i \nabla T_i$ and $\mathbf{n} = \mathbf{n}_{12}$ have been substituted into Equation 10.44 and Equation 10.33 has been used.

To be in thermal equilibrium, the temperature must be the same at the interface, and hence we have the following relation:

$$T_1 = T_2 = T_i \qquad (10.46)$$

where T_i is the interface temperature, which may depend on the concentration or other surface-related quantities at the interface.

When the interface is a boundary between two different phases, $T_i = T_{ph}$ is the temperature at which phase transformation takes place. Here subscript ph stands for phase. From thermodynamics considerations, T_{ph} is a function of pressure, concentration and curvature of the interface. Although insignificant for most thermal fluids applications, the curvature effect becomes conspicuous for microscale problems, and is responsible for the interfacially driven phenomena such as solidification microstructure formation and spinodal decomposition.

We consider a phase transformation of liquid to solid. The transformation may be denoted by the following reaction:

$$L_l \leftrightarrow S_s \qquad (10.47)$$

where L_l stands for liquid and S_s for solid. An analysis of the Gibbs free energy at the phase interface shows that the phase transformation temperature T_{ph} is given by [11],

$$T_i = T_{ph} = T_{ph}^0 + m_L C_L - 2H\Gamma_G(\theta) - v_n / \mu(\theta) + g_L p \qquad (10.48)$$

where T_{ph}^0 is the phase transformation temperature of a pure material at a flat surface and standard pressure (1 atm). Also, m_L is the slope of the liquidus line in a phase diagram, C_L is the liquid concentration, g_L is the derivative of the temperature with respect to the pressure, and p is the pressure relative to the standard pressure. In addition, H is the curvature, $\Gamma_G = \gamma(\theta)T_{ph}^0/(\rho L)$ is the surface energy coefficient, v_n the interfacial velocity in the normal direction, which is positive with the normal point from solid to liquid for a solid–liquid phase transformation, $\mu(\theta)$ is the interfacial mobility, which is related to the molecular kinetic coefficient, and θ denotes the orientation of the crystal being grown from the liquid. The curvature term represents the classical Gibbs–Thompson effect [11, 35].

For many thermal fluids systems, mass transfer is also involved. The boundary conditions at a moving interface involving species transport can be obtained in a manner similar to Equation 10.44 by considering the species conservation across the moving interface.

Example 10.2. Derive the Gibbs–Thompson relation and the interfacial kinetic effect term in Equation 10.48 for a sphere undergoing solidification. Assume that the properties of the liquid and the solid are the same and neglect the concentration effect.

Solution. We consider the Gibbs–Thompson relation first. From the definition of the Gibbs free energy, we have for the liquid and solid phases,

$$G_l = \underline{H}_l - S_l T \; ; \qquad\qquad G_s = \underline{H}_s - S_s T \qquad\qquad (10.10e)$$

where G stands for the Gibbs free energy, $\underline{H}$ for enthalpy, and S for entropy, and subscripts denote the liquid and the solid respectively.

For solidification on a planar surface, the liquid and the solid are at the thermodynamic equilibrium, and thus $G_l = G_s$. Consequently, we have

$$T_m = \frac{\underline{H}_l - \underline{H}_s}{S_l - S_s} = \frac{h_l - h_s}{s_l - s_s} = \frac{L}{\Delta s} \qquad\qquad (10.11e)$$

where the lower case h and s denote the enthalpy and entropy measured per unit mass.

For solidification on a curved surface, an excessive energy is required to overcome the surface energy (tension) effect to create a near surface. The increase in the Gibbs free energy for the growth of a layer of solid with thickness ΔR upon a spherical solid, due to the presence of the curvature, has to taken into account the surface energy required to growth the layer. Then the total Gibbs free energy change is thus given by

$$\Delta G = G_l - G_s = (L - \Delta s T_i)\rho 4\pi R^2 \Delta R + 8\pi R \Delta R \gamma \qquad\qquad (10.12e)$$

The γ term means the excessive energy required to create the new surface. At the equilibrium, $\Delta G = 0$, whence we have the following result:

$$T_i = T_m - 2\gamma / (R\rho \Delta s) = T_m - 2H\Gamma_G \tag{10.13e}$$

where $\Gamma_G = \gamma/\Delta s = \gamma T_m/(\rho L)$. This is the well known Gibbs–Thompson relation

Let us now consider the kinetic term. This term arises from the non-equilibrium effect at the interface between the solid and liquid. A non-equilibrium condition exists at the interface, because any motion of the interface requires a driving force at the interface, which causes departure from local equilibrium. This driving force is provided by the local kinetic motion of the molecules that go into and out of the liquid phase. The net rate of atomic jumps across the interface measures the growth rate due to the balance between the molecular attachment and detachment. This rate represents the growth of the solid and is calculated by the following relation:

$$v_n = v_{n0}\left(1 - \exp\left(-\frac{\Delta G_M}{R_g T_i}\right)\right) \tag{10.14e}$$

where R_g is the gas constant. The free energy favors the growth of the solid phase for solidification, and is calculated by the following expression,

$$\Delta G_M = \Delta H_M - \Delta S_M T_i = \Delta S_M T_m - \Delta S_M T_i = \Delta S_M \Delta T \tag{10.15e}$$

where subscript M denotes quantities per unit mole and $\Delta T = T_m - T_i$ is the kinetic undercooling. Since the undercooling is small, we may expand the exponential term, and thus we have from Equation 10.14e

$$v_n \approx v_{n0}\left(\frac{\Delta S_M \Delta T_m}{R_g T_m}\right) \tag{10.16e}$$

where use has been made of $T_m \approx T_i$. Rearranging, we have the expression,

$$T_i = T_m - v_n\left(\frac{\Delta S_M}{v_{n0} R_g T_m}\right) = \frac{v_n}{\mu}; \qquad \mu = \left(\frac{v_{n0} R_g T_m}{\Delta S_M}\right) \tag{10.17e}$$

We note that in general, γ and μ are a function of crystal orientations.

10.4 Moving Grids vs. Fixed Grids for Numerical Solutions

We now turn our attention to the numerical algorithms for the solution of moving boundary problems. Both the moving and fixed grid methods have been developed to solve free surface problems [1, 2]. The moving grid method is also called the Lagrangian method, whereas the fixed grid method is referred to as the Eulerian method. There are advantages and limitations associated with these methods.

In moving grid methods, the field equations are solved on a mesh that moves in accordance with the moving boundary, and the interfaces are represented by continuously updated discretization and/or remeshing. In a typical finite element algorithm using the moving grid technique, the moving or free surface shapes are the boundaries of separate flow regions, and the change of these shapes is constantly traced by deforming the elements in each of the regions. With the moving grids, boundary conditions at a moving interface are applied at the exact location of the interface. The moving grid approach provides an accurate account of the shape morphology but is limited to relatively simple shapes.

The fixed grid approach is based on the incorporation of the interface boundary conditions as sources in the momentum and energy equations. For example, in the case of solidification problems, the release of the latent heat can be incorporated into the source for the energy balance equation. In this approach, mesh is not deformed. Different flow regions are modeled by different material properties. There are different Eulerian approaches for moving boundary problems; the difference mainly lies in the way in which the interface is evolved and interpolated. Two popular and relatively straightforward approaches are the volume of fluid (VOF) and the marker in a cell (MAC) methods. One other approach is the level set method, where the interfaces are implicitly defined as the zero level set of a continuous function. This function is updated in order to capture the motion of the interfaces. A major advantage of the fixed grid approach is that it can handle complex geometry of free and moving surfaces with ease. Since the boundary conditions at a moving boundary are included in the governing transport equations, a fixed grid method leads to the smearing of boundary information. With extremely fine grids, however, the fixed grid techniques may also match the accuracy of the Lagrangian grid method.

Both the moving and fixed grid approaches can be incorporated into a discontinuous finite element formulation for moving boundary problems. In the next two sections, we discuss the basic ideas of the moving and fixed grid methods and the use of these methods in a discontinuous finite element setting for the numerical solution of problems involving free surfaces, internal moving interfaces and phase change boundaries.

10.5 Moving Grid Methods

In this approach, a free surface or moving boundary is tracked by constructing a Lagrangian grid that is embedded in and moves with the boundary. Because the grid and fluid move together, the grid is made to explicitly track the free surface or the moving boundary.

The use of moving grids to solve thermal and fluid flow problems is not entirely new. In Section 7.4, the arbitrary Lagrangian–Eulerian (ALE) formulation was introduced for computational compressible fluid dynamics involving shock waves and other discontinuities. The same formulation has also been applied to solve moving boundary problems involving incompressible fluid flows and flow structure interactions [12–14]. For modeling of geometrically complicated moving boundaries, a costly remeshing procedure is required with the Lagrangian approach

[15]. Most moving grid methods are applied to the surfaces of relatively simple geometry. To reduce the cost of remeshing, a simple and yet effective grid-moving strategy is employed, which may be considered as a special implementation of the ALE formulation [16]. It is noted that the moving grid technique provides the most precise means to exactly locate the interface positions, with the boundary conditions precisely applied along these boundaries. This is desirable for the problems where a precise knowledge of the moving boundary shapes is required. One such case is the deformation of a single droplet by electric forces (see Section 12.6). We consider below the moving grid techniques and their application within the framework of the discontinuous fintie element formulations for numerical solution of two types of commonly encountered moving boundary problems: (a) moving boundaries between fluids and (b) interphase boundary between two phases.

10.5.1 Moving Boundaries Between Fluids

We consider a fluid flow system consisting of two immiscible fluids and a gas phase as shown in Figure 10.1. There two moving boundaries exist, with one marking the interface (i.e., free surface) between the liquid and the gas, and the other the interface between the two liquids. Further we assume that the fluids are both incompressible and isothermal. The temperature effects can be readily incorporated if needed. In reference to Figure 10.1, the mathematical description of the problem is given by the following equations:

$$\nabla \cdot \mathbf{u}_j = 0 \quad \in \Omega_1 \bigcup \Omega_2 \tag{10.49}$$

$$\rho \frac{\partial \mathbf{u}_j}{\partial t} + \rho \mathbf{u}_j \cdot \nabla \mathbf{u}_j = -\nabla p_j + \nabla \cdot \boldsymbol{\tau} + \mathbf{F} \quad \in \Omega_1 \bigcup \Omega_2 \tag{10.50}$$

where $j (= 1, 2)$ refers to the two fluids. The solution of above fluid flow equations may be obtained by applying the appropriate boundary conditions discussed in Section 10.3. For the problem being considered, these conditions are expressed by the following equations:

$$\frac{\partial \chi_{f2}}{\partial t} + \mathbf{u}_2 \cdot \nabla \chi_{f2} = 0 \quad \in \Omega_2 \bigcap \Omega_3 \tag{10.51}$$

$$\mathbf{n}_2 \cdot \boldsymbol{\sigma}_2 \cdot \mathbf{n}_2 = -p_{amb} + 2H\gamma_2 \quad \in \Omega_2 \bigcap \Omega_3 \tag{10.52}$$

$$\mathbf{t}_2 \cdot \boldsymbol{\sigma}_2 \cdot \mathbf{n}_2 = \nabla_s \gamma_2 \quad \in \Omega_2 \bigcap \Omega_3 \tag{10.53}$$

$$\frac{\partial \chi_{f1}}{\partial t} + \mathbf{u}_1 \cdot \nabla \chi_{f1} = \frac{\partial \chi_{f1}}{\partial t} + \mathbf{u}_2 \cdot \nabla \chi_{f1} = 0 \quad \in \Omega_1 \bigcap \Omega_2 \tag{10.54}$$

$$\mathbf{n}_1 \cdot \boldsymbol{\sigma}_1 \cdot \mathbf{n}_1 - \mathbf{n}_1 \cdot \boldsymbol{\sigma}_2 \cdot \mathbf{n}_1 = 2H\gamma_{12} \quad \in \Omega_1 \cap \Omega_2 \tag{10.55}$$

$$\mathbf{t}_1 \cdot \boldsymbol{\sigma}_1 \cdot \mathbf{n}_1 - \mathbf{t}_1 \cdot \boldsymbol{\sigma}_2 \cdot \mathbf{n}_1 = \nabla_s \gamma_{12} \quad \in \Omega_1 \cap \Omega_2 \tag{10.56}$$

$$\mathbf{n}_1 \times (\mathbf{u}_1 - \mathbf{u}_2) = \mathbf{0} \quad \in \Omega_1 \cap \Omega_2 \tag{10.57}$$

$$\int_{\Omega_1 + \Omega_2} dV = V_0 \quad \in \Omega_1 \cup \Omega_2 \tag{10.58}$$

where χ_f represents the moving boundary, V_0 is the initial volume of the fluids, and $\boldsymbol{\sigma} = -p\mathbf{n} + \boldsymbol{\tau}$ is the stress tensor. Note that Equation 10.54 combines Equations 10.29 and 10.32.

The constraint of the volume conservation (i.e., Equation 10.58) means that the total mass remains unchanged for a closed system. This constraint is added to the system of equations and usually results in a unique additive pressure constant that satisifes the mass conservation of a constraned system. We note that this constraint applies to the steady state flows. In transient flows (with no net inflow or outflow), the volume is conserved naturally through initial conditions and problem definition.

The basic idea of a moving grid method is such that the nodes (or grid points) on a moving surface move at the same velocity and remain on the moving surface by deforming the computational grids. Since the nodes always track the moving interface, the method is also called the front tracking technique. To avoid mesh distortion, the nodes in certain regions are also allowed to move during calculations. One simple way of deforming the mesh is illustrated in Figure 10.5, where nodes in the mesh are constrained to move in a designated direction.

Since the moving boundaries are unknown *a priori*, the coordinates of the boundaries are constructed from the finite element interpolation basis functions, just as the variables are interpolated using the shape functions,

$$x = \sum_{i=1}^{n_e} x_i \phi_i(s); \quad y = \sum_{i=1}^{n_e} y_i \phi_i(s); \quad z = \sum_{i=1}^{n_e} z_i \phi_i(s) \tag{10.59}$$

where s is the surface coordinates and n_e is the number of nodes per each element lying along the moving boundary. During the calculations, (x_i, y_i, z_i) is solved as part of the final solution and thus is updated at each time step.

In writing Equation 10.50, the time derivative is the Eulerian time derivative: the nodal field values (i.e., $\mathbf{U}$, χ_f) are for nodes fixed in space. With a moving grid algorithm, the nodes are allowed to move so that the geometry of the moving boundary is precisely tracked. Consequently, these nodes are not fixed in the Eulerain frame of reference. Since the nodes are moved in certain designated directions so as to allow the instant positions of these nodes to be monitored, they are not fixed in the Lagrangian frame of reference carried by the fluid particles. Thus, the time derivative in Equation 10.50 must be transformed to time derivatives that follow the moving nodes in the designated directions. This is essentially the mixed Lagrangian–Eulerian description discussed in Section 7.4.

The required transformation between the two time derivatives is given by the following relation:

$$\frac{\delta}{\delta t} = \frac{d}{dt}\Bigg|_{\mathbf{x}_f} = \frac{\partial}{\partial t} + \frac{\partial \mathbf{x}_m}{\partial t}\nabla \tag{10.60}$$

where $\mathbf{x}_m$ is the coordinates of a node point in the moving-node region, ∇ is the spatial derivative with respect to the Eulerian frame of reference and $\partial/\partial t$ is the Eulerian time derivative. This is precisely the same equation obtained by combining Equations 7.80 and 7.81 and identifying $\partial \mathbf{x}_m/\partial t$ as $\mathbf{v}$ in Equation 7.85, which is the velocity observed at a moving node. Consequently, to account for the effect of grid movement, the governing equations need to be modified only by the following substitution:

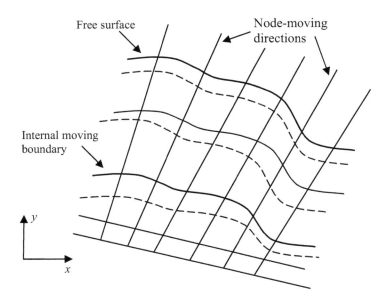

Figure 10.5. Front tracking of moving and free boundaries using the Lagrangian method. The curved lines are grid lines in the moving mesh region and straightlines are non-moving grid lines. Dash lines are grid lines at $t = t$, which are moved to solid curves at $t = t + \Delta t$. Note that all the nodes within the moving grid region, which in this case is defined by the free and the internal moving boundaries, are moved as so to miminize grid distortion. Here, nodes in the moving grid region are constrained to move in designated node-moving directions

$$\frac{\partial}{\partial t} \rightarrow \frac{\delta}{\delta t} - \frac{\partial \mathbf{x}_m}{\partial t} \nabla$$
(10.61)

The treatment of these conditions at the non-moving (or fixed) boundaries is exactly the same as discussed in Chapter 6. The conditions at the moving boundaries, however, need to be added to the discretized momentum equations for these fluids. To incorporate the moving boundary conditions in a discontinuous finite element formulation, we consider a pair of elements that share a common moving surface as shown in Figure 10.6. Let us assume that the boundary is an internal moving boundary between the two liquids. Then, for element j_2, a normal velocity $u_{n,2}$ ($u_{n,2} = \mathbf{u}_{element\ 2} \cdot \mathbf{n}$) and $\mathbf{t}_1 \cdot \mathbf{\sigma}'_1 \cdot \mathbf{n}_1$ are assumed, and $-\mathbf{t}_1 \cdot \mathbf{\sigma}_2 \cdot \mathbf{n}_1 = \nabla_s \gamma_{12} - \mathbf{t}_1 \cdot \mathbf{\sigma}'_1 \cdot \mathbf{n}_1$ is applied along the moving surface boundary. This allows the calculation of the tangential component of the velocity (i.e., $u_{t,2} = \mathbf{t}_1 \cdot \mathbf{u}_2$) and the normal stress component $\mathbf{n}_1 \cdot \mathbf{\sigma}_2 \cdot \mathbf{n}_1$. With these quantities known, $\mathbf{n}_1 \cdot \mathbf{\sigma}_1 \cdot \mathbf{n}_1 = 2H\gamma_{12} + \mathbf{n}_1 \cdot \mathbf{\sigma}_2 \cdot \mathbf{n}_1$ and $\mathbf{t}_1 \cdot \mathbf{u}_2$ are applied, and $\mathbf{t}_1 \cdot \mathbf{\sigma}_1 \cdot \mathbf{n}_1$ and $u_{n,1}$ ($u_{n,1} = \mathbf{u}_{element\ 1} \cdot \mathbf{n}$) are calculated for element j_1. The differences, $|u_{n,1} - u_{n,2}|$ and $|\mathbf{t}_1 \cdot \mathbf{\sigma}_1 \cdot \mathbf{n}_1 - \mathbf{t}_1 \cdot \mathbf{\sigma}'_1 \cdot \mathbf{n}_1|$, are included for convergence check. If convergence is not achieved, $u_{n,2}$ and $\mathbf{t}_1 \cdot \mathbf{\sigma}_1 \cdot \mathbf{n}_1$ are updated and the above calculations for elements j_1 and j_2 will be repeated.

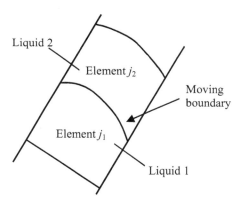

Figure 10.6. Two elements share a common moving boundry

It is noted here that the above scheme represents just one of the ways the interface boundary conditins are incorporated. One could also choose to specify other velocity and stress components to develop a different iterative procedure.

For a free surface problem, we partition the domain such that element j_2 lies in the gas phase. Then $\mathbf{\sigma}_2$ is known and no calculations will be required for element j_2. We can apply the stress conditions, both normal and tagential, on element j_1 only and find all the component velocities at the free surface boundary.

An algorihtm for the discontinuous finite element solution of moving surface problems now may be described as follows. We employ an explicit time scheme to be consistent with the field calculations. Then the free boundary position is calculated by integrating, along the free/moving surface boundaries, Equation 10.51 and one of the relations in Equation 10.54, using the available information on the velocity field at the previous time step. Equations 10.49 and 10.50 for fluid flows are then solved as described in Chapter 6. For elements in the moving node region, Equation 10.61 is used instead to account for the node movement. For the two elements sharing the common moving boundary, the boundary conditions are applied as discussed above for Figure 10.6. The iterative procedure continues until convergence is achieved for this time step. This then is repeated for the next time step.

For steady state calculations, one can follow the above time marching scheme until steady state is reached. Alternatively, one can solve a steady state problem directly by setting transient terms to zero. A basic algorithm that works well for a continuous finite element solution using segragated solver should be applicable here. This algorithm is based on the updating of free surface using the normal velocity condition and is dscribed below. One starts with an initial guess of the free/moving boundary location χ_f, then the flow is calculated by applying the boundary conditions the same way as for the transient algorithm described above. Then the kinematic condition on the normal component of the velocity at the free boundary is used to update the location of the free surface χ_{f2},

$$\mathbf{u}_2 \cdot \mathbf{n}_2 = 0 \quad \in \Omega_2 \cap \Omega_3 \tag{10.62}$$

and the location of the internal free boundary χ_{f1} (see Figure 10.1),

$$\mathbf{u}_1 \cdot \mathbf{n}_1 = 0 \quad \in \Omega_1 \cap \Omega_2 \tag{10.63}$$

The calculations continue with updated χ_f until convergence is achieved.

In the above algorithms, the normal stress component is applied as a stress boundary condition, which involves the calculation of the Gaussian mean curvature. The mean curvature is related to χ_f by the following expression:

$$2H = -\nabla_s \cdot \mathbf{n} \tag{10.64}$$

where $\mathbf{n}$ is calculated using the relations developed in Section 10.2. The calculation for surface gradient appearing in the tangential stress balance can be carried out by following the procedure given in Section 3.6.2.

In the above discussion, the curvature contribution to the stress balance is assumed to be calculated by directly integrating Equation 10.64. This would involve a considerable effort in reconstructing the free surface in order to obtain a desired acuracy. A simplified procedure may be applied using the integral relations obtained from the theory of differential geometry in Section 10.2.2. We consider the stress term arising from the Galerkin formulation of the momentum equation,

$$\int_\Omega \phi_i \nabla \cdot \boldsymbol{\sigma} \, dV = \int_{\partial \Omega_f} \boldsymbol{\sigma} \cdot \mathbf{n} \phi_i \, dS + \int_{\partial \Omega_l} \boldsymbol{\sigma} \cdot \mathbf{n} \phi_i \, dS - \int_\Omega \boldsymbol{\sigma} \cdot \nabla \phi_i \, dV \tag{10.65}$$

where we have applied Green's theorem and $\partial\Omega = \partial\Omega_f \cup \partial\Omega_l$ with subscripts f and l representing the moving and fixed boundaries, respectively. Since the standard procedure can be applied for the fixed boundary term, we single out the free boundary term for consideration. For simplicity, we consider the treatment of the free surface,

$$\int_{\partial\Omega_f} \boldsymbol{\sigma} \cdot \mathbf{n}\,\phi_i dS = \int_{\partial\Omega_f} (-p_{amb}\mathbf{n} + 2H\gamma\,\mathbf{n})\phi_i dS \tag{10.66}$$

The second term involving the curvature is integrated further using the theorem of surface divergence given in Section 10.2.2,

$$\int_{\partial\Omega_f} 2H\gamma\mathbf{n}\,\phi_i dS = -\int_{\partial\Omega_f} (\nabla_s \cdot \mathbf{n})\gamma\mathbf{n}\,\phi_i dS$$

$$= \int_{\partial\Omega_{f,m}} \gamma\mathbf{m}\,\phi_i d\Gamma_m - \int_{\partial\Omega_f} \nabla_s (\gamma\phi_i)\,dS \tag{10.67}$$

where $\mathbf{n}$ and $\mathbf{m}$ are surface normal and surface boundary normal, as shown in Figure 10.3. Note that the term involving $\mathbf{m}$ permits the prescription of contact angles between gas, liquid and solid. Clearly, the above idea directly applies to an internal moving boundary as well, with obvious substitutions.

For steady state calculations, an algorithm based on the normal stress balance may also be applied [17]. In fact, it becomes more efficient if curvature effects are strong. In this case, the function describing a free/moving boundary is first expanded globally over the entire surface and the normal stress balance equation is solved using the collocation method to determine the expansion coefficients for the surface function. For many applications, collocation points are taken to be the Gaussian integration points at a boundary element used for fluid flow calculations. This scheme was considered to converge very fast; the only drawback is the direct estimate of the curvature, which involves a second order derivative. Thus, a higher order approximation is necessary when this approach is taken [17].

10.5.2 Moving Phase Boundaries

In thermal and fluids applications involving phase transitions, the phase boundaries are unknown and need to be determined as part of the solution. These problems are often encountered in the solid-to-liquid and liquid-to-gas transitions. We discuss these types of problems below.

10.5.2.1 Solid–Liquid Phase Transition
For problems involving the solid–liquid transformation, the moving boundary is marked by the inerface between the liquid and solid phases and at the boundary mass transfer may also take place if there are impurity elements in the phases. In reference to Figure 10.7, the governing equations for a solidification problem

decribing the conservation of momentum, energy and species are written as follows:

$$\nabla \cdot \mathbf{u} = 0 \quad \in \Omega_l \tag{10.68}$$

$$\rho \frac{\partial \mathbf{u}}{\partial t} + \rho \mathbf{u} \cdot \nabla \mathbf{u} = -\nabla p + \nabla \cdot \boldsymbol{\tau}$$

$$-\rho \mathbf{g}[\beta_T (T - T_{ref}) + \beta_{C1}(C_1 - C_{ref})] + f \quad \in \Omega_l \tag{10.69}$$

$$\rho_j C_{pj} \frac{\partial T_j}{\partial t} + \rho_j C_{pj} \mathbf{u}_j \cdot \nabla T_j = \nabla \cdot (k_j \nabla T_j) + Q_j \quad \in \Omega_l \bigcup \Omega_s \tag{10.70}$$

$$\rho_j \frac{\partial C_j}{\partial t} + \rho_j \mathbf{u}_j \cdot \nabla C_j = \nabla \cdot (D_j \nabla T_j) + R_j \quad \in \Omega_l \bigcup \Omega_s \tag{10.71}$$

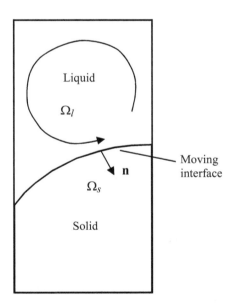

Figure 10.7. Illustration of a moving phase boundary defined by the solid–liquid interface

where C_j ($j =1,2$) denotes the concentrations of the physical phases (liquid, solid) present in the system, f is the body force excluding the gravitational force, and the standard Boussinesq assumption has been used to account for solute and temperature effects. Also, C is the concentration of a foreign element, Q_j and R_j are the generation terms for the energy and species. Although written for a two-phase system, the equations can be readily modified to describe a multiphase system. For this type of problems, the velocity in the solid region is often specified by applications.

The boundary conditions for the equations are standard for fixed boundaries. For a moving boundary, the kinematic conditions, species conservation and mechanical and energy balances must be satisifed. From discussion in Section 10.3, these conditions are written as follows:

$$T_m = T_l = T_s \quad \in \Omega_l \bigcup \Omega_s \tag{10.72}$$

$$k_l \mathbf{n} \cdot \nabla T_l - k_s \mathbf{n} \cdot \nabla T_s = \rho_s L \mathbf{n} \cdot \left(\mathbf{u}_s - \frac{\partial \mathbf{x}_f}{\partial t} \right) \quad \in \Omega_l \bigcup \Omega_s \tag{10.73}$$

$$\rho_l \mathbf{n} \cdot \left(\mathbf{u}_l - \frac{\partial \mathbf{x}_f}{\partial t} \right) - \rho_s \mathbf{n} \cdot \left(\mathbf{u}_s - \frac{\partial \mathbf{x}_f}{\partial t} \right) = 0 \quad \in \Omega_l \bigcup \Omega_s \tag{10.74}$$

$$\mathbf{n} \times (\mathbf{u}_s - \mathbf{u}_l) = 0 \quad \in \Omega_l \bigcup \Omega_s \tag{10.75}$$

$$C_s \mathbf{n} \cdot \left(\mathbf{u}_s - \frac{\partial \mathbf{x}_f}{\partial t} \right) - C_l \frac{\rho_l}{\rho_s} \mathbf{n} \cdot \left(\mathbf{u}_l - \frac{\partial \mathbf{x}_f}{\partial t} \right) = \mathbf{n} \cdot \left(-\frac{\rho_l}{\rho_s} D_l \nabla C_l + D_s \nabla C_s \right)$$
$$\in \Omega_l \bigcup \Omega_s \tag{10.76}$$

$$C_s = \kappa C_l \quad \in \Omega_l \bigcup \Omega_s \tag{10.77}$$

$$T_l = T_s = T_m + \frac{dT}{dC_l} C_l \quad \in \Omega_l \bigcup \Omega_s \tag{10.78}$$

where L is the latent heat released when the liquid is converted to the solid, $\mathbf{n}$ is the normal pointing from the liquid to the solid, $\mathbf{x}_f$ is the coordinates of the interface ($\mathbf{u}_i = \partial \mathbf{x}_f / \partial t$ being the interface velocity at whcih the solid extends into the liquid) and T_m is the melting temperature. Here the pressure and curvature effects on T_m are neglected. Note that Equation 10.73 is obtained by letting $s = 1$, $l = 2$, $L = L_{21}$, $\mathbf{u}_i - \mathbf{u}_s = \partial \mathbf{x}_f / \partial t - \mathbf{u}_s = \mathbf{u}_{1i}$, $q_{j,n} = \mathbf{n} \cdot k_j \nabla T_j$ ($j = l,s$) and $\mathbf{n} = -\mathbf{n}_{12}$ in Equation 10.44.

The deforming grid approach presented above for moving boundary problems can be modified for the numerical solution of a phase change problem. A major modification would be for the presciption of thermal boundary condition on the elements across the phase boundary. We consider this modification for a discontinuous finite element formulation. To accurately track the interface, we apply Equation 10.72 on element j_2 (see Figure 10.6) and solve for $k_l \nabla T_l$. Equation 10.73 is then applied on element j_1 and T_s is determined. Then $|T_s - T_m|$ is used to check the convergence and $T_s - T_m$ is used to determine in which direction to move the boundary. Specific location $\mathbf{x}_f$ is determined by the linear interpolation between the temperatures of the elements and $\partial\mathbf{x}_f/\partial t \approx (\mathbf{x}_f\,(t+\Delta t) - \mathbf{x}_f(t))/(\Delta t)$. With this location determined, the whole procedure is repeated. Numerical experience with continuous finite element formulations shows that this type of approach based on the detection of the isothermal line provides a more accurate front tracking than the algorithm based on the energy balance equation [18, 19].

A discontinuous-based algorithm for the solution of the above problem may be described as follows. At a typical time step, an inital phase boundary is available from the previous time step, and the governing equations for variables ($\mathbf{u}$, $\underline{\tau}$, p, T, $\mathbf{q}$) are solved using the discontinuous formulations given in Chapter 6. For the elements across the moving boundary, the boundary conditions and moving boundary shape are determined as described above. The iteration continues with updated values until all variables are converged. The procedure is then repeated for the next time step.

An important advantage of the above algorithm is that it works for both the steady state and transient simulations. Thus, for a steady state calculation, one simply turns off the transient terms and replaces $\partial\mathbf{x}_f/\partial t$ by a steady state velocity of phase boundary movement [20].

10.5.2.2 Liquid–Vapor Phase Transition

In the case of evaporation from liquid to vapor such as boiling, the algorithm above may be used with some straightforward modifications. If we consider a pure liquid evaporates into its vapor phase, then the fluid flow and heat balance equations apply to the liquid only,

$$\nabla\cdot\mathbf{u} = 0 \quad \in\Omega_l \tag{10.79}$$

$$\rho\frac{\partial\mathbf{u}}{\partial t} + \rho\mathbf{u}\cdot\nabla\mathbf{u} = -\nabla p + \nabla\cdot\boldsymbol{\tau} - \rho\beta_T\mathbf{g}(T - T_{ref}) + \boldsymbol{f} \quad \in\Omega_l \tag{10.80}$$

$$\rho C_p\frac{\partial T}{\partial t} + \rho C_p\mathbf{u}\cdot\nabla T = \nabla\cdot(k\nabla T) + Q \quad \in\Omega_l \tag{10.81}$$

The boundary conditions at the vapor–liquid interface are again derived from the relations given in Section 10.2,

$$\rho\frac{\partial\chi_f}{\partial t} + \rho\mathbf{u}\cdot\mathbf{n} = \dot{m} \quad \in\Omega_l\cap\Omega_g \tag{10.82}$$

$$-k_l\mathbf{n}\cdot\nabla T_l = \rho_l L\mathbf{n}\cdot\left(\mathbf{u} - \frac{\partial\mathbf{x}_f}{\partial t}\right) = \dot{m}L \quad \in\Omega_l\cap\Omega_g \tag{10.83}$$

$$T_l = T_{sat}(P_l) \quad \in\Omega_l\cap\Omega_g \tag{10.84}$$

$$\mathbf{n} \cdot \boldsymbol{\sigma} \cdot \mathbf{n} = -p_{amb} + 2H\gamma \quad \in \Omega_l \cap \Omega_g \tag{10.85}$$

$$\mathbf{t} \cdot \boldsymbol{\sigma} \cdot \mathbf{n} = \frac{d\gamma}{dT} \nabla_s T \quad \in \Omega_l \cap \Omega_g \tag{10.86}$$

$$p_{amb} = p_{sat} \quad \in \Omega_l \cap \Omega_g \tag{10.87}$$

where subscript *sat* stands for saturation, T_{sat} is the saturation temperature at the liquid pressure, $\mathbf{n}$ points from the liquid to the vapor phase, $\dot{m}$ is the mass flux or rate of evaporation per unit area, and L is the latent heat released per unit mass when the vapor is condensed into the liquid. To obtain Equation 10.83, we have substituted $l = 1$, $L = L_{21}$, $\partial \mathbf{x}/\partial t - \mathbf{u}_l = \mathbf{u}_{1i}$, $q_{l,n} = -\mathbf{n} \cdot k_l \nabla T_l$ and $\mathbf{n} = \mathbf{n}_{12}$ in Equation 10.44 and used the definitions of $\dot{m} = \rho(\mathbf{u} - \mathbf{u}_i) \cdot \mathbf{n}$ and $\partial \mathbf{x}/\partial t = \mathbf{u}_i$.

The numerical solution of these types of problems follows the procedure similar to that described above. These deforming grid methods would be useful for modeling the details of the vapor–liquid interface.

A typical algorithm using a discontinuous finite element formulation would be as follows. For a given time step, the calculation starts with an initial or previously determined interface shape. The governing equations are then solved using the discontinuous finite element method discussed in Chapter 6, with Equations 10.83–10.87 applied. Equation 10.83 is used to determine the mass flux $\dot{m}$ and the moving boundary coordinates are determined using Equation 10.84. The calculations iterate until convergence and continue for the next time step.

10.6 Fixed Grid Methods

In the fixed grid methods, the governing equations are solved in the Eulerian frame of reference. The interface mechanical balance conditions are written as a interfacial body force. The governing equations take the following form:

$$\nabla \cdot \mathbf{u}_j = 0 \quad \in \Omega_1 \cup \Omega_2 \tag{10.88}$$

$$\rho_j \frac{\partial \mathbf{u}_j}{\partial t} + \rho_j \mathbf{u}_j \cdot \nabla \mathbf{u}_j = -\nabla p_j + \nabla \cdot \boldsymbol{\tau}_j$$

$$+2H\gamma\delta(\mathbf{x} - \mathbf{x}_f)\mathbf{n} + (\nabla_s \gamma)\delta(\mathbf{x} - \mathbf{x}_f) + \mathbf{F} \quad \in \Omega_1 \cup \Omega_2 \tag{10.89}$$

where $\mathbf{n}$ is the interface normal pointing from fluid 1 to fluid 2, the terms involving γ represent the contributions from the free/moving interfaces and the delta function is defined as

$$\delta(\mathbf{x} - \mathbf{x}_f) = \begin{cases} 1, & \mathbf{x} = \mathbf{x}_f \\ 0, & \mathbf{x} \neq \mathbf{x}_f \end{cases} \tag{10.90}$$

The above equations can be solved using the discontinuous finite element method presented in Chapter 6, once the geometry of moving boundary $\mathbf{x}_f$ is known. All the fixed grid methods involve two computational procedures for a free/moving surface problem: (1) solving the equations described above in an Eulerian mesh and (2) evolving a moving interface based on the flow calculations, using the volume grid, or surface (line) grid or particles. A typical senario of fixed grid methods for free surface calculations is shown in Figure 10.8. The discontinuous framework for the solution of transport equations in an Eulerian mesh was discussed in detail in Chapter 6. We consider below three popular methods for evoloing the moving surface: the volume of fluid (VOF) method, the marker and cell (MAC) method and the level set method.

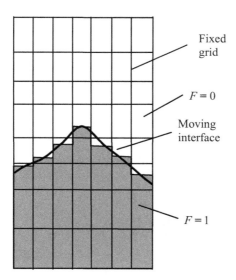

Figure 10.8. Illustration of fixed grid methods for free surface calculations

10.6.1 Volume of Fluid Method

The volume of fluid is based on the idea that the free surface is tracked by the following advection equation [1, 21],

$$\frac{\partial F}{\partial t} + \mathbf{u} \cdot \nabla F = 0 \tag{10.91}$$

where F is the volume of fluid, which is assigned a value of 1 in the liquid and 0 in the gas or the other fluid. Steep gradients in the variable F represent free surface locations.

A general volume of fluid scheme consists of two distinct steps. In the first step, the fluid volume is reconstructed on the basis of the fractional fill states. This reconstruction represents an estimate of the spatial location of the fluid within the mesh. The fluid volume is then advected on the basis of the reconstruction and a given velocity field. This advection itself leads to new fractional fill states. The reconstruction of the fluid volume within an element depends upon its fill state and the fill state of its neighbors. Neighbors are defined here as elements sharing a common side. For instance, for a structured mesh, a quadrilateral element has up to four neighbors sharing its four sides. A brick element has up to six neighbors. For an unstructured mesh, however, the number of the neighboring elements is not the same and is decided by a specific mesh generator.

Slopes and curvatures are computed by using the fluid volume fractions in neighboring cells. The essential element in this process is to remember that the volume fraction should be a step function, i.e., having a value of either one or zero. Information on the volume fractions in neighboring cells can then be used to locate the position of fluid (and its slope and curvature) within a particular cell.

We discuss below the implementation of the VOF method for structured and unstructured meshes.

10.6.1.1 Structured Mesh
An important step in the VOF method is the reconstruction of the free surface. Figure 10.9 illustrates the two ideas used for construction of the interfaces from the volume of fluid data. The SLIC, which stands for Simple Line Interface Calculations, was first proposed by DeBar [22]. The method, as shown in Figure 10.9(a), applies the piecewise constant approximation to model the interface. An abrupt stair stepping interface is obtained, in contrast with the moving grid method, which constantly tracks the interface shape. A more popular method used today is the PLIC method, which stands for the Piecewise Linear Interface Calculations. The method was first introduced by Parker and Youngs [23]. The basic idea of the algorithm is illustrated in Figure 10.9(b). It is noted that across the element boundary, the interface has a jump.

Surface Reconstruction. For simplicty, a two-dimensional (2-D) computational domain with square cells is considered. Other regular cells can be extended rather easily by following the same procedure. We further consider the VOF/PLIC method, by which the interface is constructed by cutting a cell using a straight line defined by the equation,

$$n_x x + n_y y = \alpha \tag{10.92}$$

The determination of the constants (n_x, n_y, α) of Equation 10.92 is in general carried out in two steps: (1) evaluation of the interface normal $\mathbf{n} = (n_x, n_y)$, and (2) determination of the line constant α, so that the fraction of the cell area cut by this line and occupied by the reference phase is equal to F.

If the cut volume and the normal direction $\mathbf{n}$ in a computational cell are known, the constant α is then obtained by a simple integration and by enforcing the volume conservation. This can be done either with a numerical rootfinding technique or directly with analytical formulae describing the relation $\alpha = \alpha\,(F)$. For a 2-D problem, an analytic expression is derived to solve for α [2],

$$F_{ij} = \frac{\alpha^2}{2n_x n_y}\left[1 - H_v(\alpha - n_x h)\left(\frac{\alpha - n_x h}{\alpha}\right)^2 - H_v(\alpha - n_y h)\left(\frac{\alpha - n_y h}{\alpha}\right)^2\right]$$

$$(10.93)$$

Heaviside function defined as

$$H_v(x) = \begin{cases} 1, & \text{if } x > 0 \\ 0, & \text{if } x \le 0 \end{cases}$$

$$(10.94)$$

With the above equations, α can be calculated once the normal vector $\mathbf{n}$ and mesh size h are known. Also, for convenience of description, the double index is used.

In the VOF/PLIC reconstruction the normal vector $\mathbf{n}$ is determined by the volume fraction gradient. A simple approach is the Parker and Youngs method [23], which has gained popularity in finite volume solutions. The method is illustrated using a 3×3 block of square cells shown in Figure 10.10, with $\Delta x = \Delta y = h$. Here, the normal $\mathbf{n}$ is first estimated at the four corners of the central cell (i, j), that is, element ij, with a finite difference formula, for example the x component n_x at the top-right corner is given by

$$n_{x,i+1/2,j+1/2} = (F_{i+1,j+1} + F_{i+1,j} - F_{i,j+1} - F_{i,j})/2h$$

$$(10.95)$$

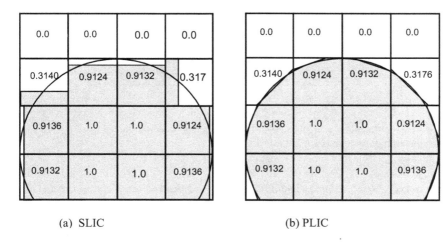

(a) SLIC   (b) PLIC

Figure 10.9. Reconstruction of interfaces for a circle using the SLIC (a) and the PLIC (b) methods. The solid line is a smoothed line connecting the middle points of the reconstructed interfaces in each element. The SLIC method forces the reconstructed interface to align with one of the mesh coordinates. The PLIC method, on the other hand, allows the reconstruction to be tangential to the interface. Numbers indicate the fraction of fluid volume in the element [24].

Similar expressions can be derived for the y component n_y and $\mathbf{n}$ at the other three corners. Then the required cell-centered vector is obtained by averaging the four cell-corner values

$$\mathbf{n}_{x,i,j} = (\mathbf{n}_{i+1/2,j+1/2} + \mathbf{n}_{i+1/2,j-1/2} + \mathbf{n}_{i-1/2,j+1/2} + \mathbf{n}_{i-1/2,j-1/2})/4 \quad (10.96)$$

With $\mathbf{n}$ and α so determined using Equations 10.93 and 10.96, the interface in a cell is constructed as a straight line. The front of the surface is constructed with the above procedure applied to every qualified cell. There are many improvements to the above simple method since it was introduced [2, 9, 24, 25].

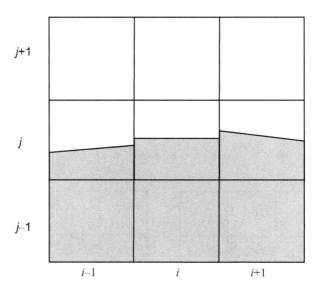

Figure 10.10. Illustration of the VOL/PLIC reconstruction of free surface [1]. Note that across the boundary there is a jump in F

Advection of Fluid Volume. The advection of the fluid volume stems from a mass balance around each element. This is done by integrating the equation for the volume of fluid, i.e., Equation 10.91, with the result,

$$\int_{V_i} \frac{\partial F}{\partial t} dV = -\int_{S} F\mathbf{u} \cdot \mathbf{n} \, dS \quad (10.97)$$

where $\nabla \cdot \mathbf{u} = 0$ is used [21, 24]. Here, V_i is the total volume of element i and the F value under the surface integral is taken at the boundary of the element. To simplify the notation, we use a single index on an element. According to the reconstruction procedure discussed above, the F value has a jump across the element interface (see Figures 10.9 and 10.10). The time rate of change of F_i is approximated by a forward difference over a given time step,

$$\frac{F_i^{n+1} - F_i^n}{\Delta t} = -\int_S F\mathbf{u} \cdot \mathbf{n}\, dS = \frac{1}{V_i}\sum_k Q_{ik} \qquad (10.98)$$

where superscripts n and $n+1$ denote values from the two successive time steps. Obviously, the time step restriction applies here [21]. Also, Q_{ik} is the flow rate of fluid into element i across side k. The net flow rate into element i is simply the sum of Q_{ik} over all sides k. Once the various flow rates Q_{ik} are evaluated at a given time step n, the fluid volume can be advected by determining the new fractional fills F_i^{n+1}. By mass conservation, we have the relation for side k shared by elements i and j,

$$Q_{ik} = -Q_{jk} \qquad (10.99)$$

The treatment of the jump in F at the element boundary during advection is discussed in [21] and a similar idea is presented here. Consider the two elements i and j in Figure 10.11 that are separated by a common edge k. We rewrite Q_{ik} as

$$Q_{ik} = -\int_S f_{ik}\mathbf{u} \cdot \mathbf{n}\, dS ; \qquad\qquad 0 \le f_{ik} \le 1 \qquad (10.100)$$

where f_{ik} is the fraction of side k touched by fluid within element i, and $\mathbf{n}$ is the normal vector pointing outward from element i along side k. The integral in Equation 10.100 is evaluated over the entire area S of side k. The fractional area f_{ik} contains information about the reconstruction of fluid along two elements sharing side k and is determined solely by the two elements. The use of f_{ik} should be obvious. In general, f_{ik} is taken as an averaged value of the two elements sharing side k, if both elements contain fluid, as shown in Figure 10.11(b). For the case where one of the elements is empty, the simple rule states that no fluid would come from the empty element. The empty element, however, can receive fluid from the adjacent element that contains fluid. Figure 10.11 shows the typical situations for the advection of fluid.

The volume fraction obtained by the above VOF advection step needs to be adjusted locally and globally in order to eliminate unphysical partial elements and to satisfy the requirement of global mass conservation [21].

One important correction is in the near wall region. When an element has one side attaching the wall, there may be a certain amount of volume left in the ith element, which makes it practically impossible to empty these near-wall elements. As time advances, the bulk of fluid may leave behind a row of partial elements, forming artificial droplets, rather than empty elements. A practice is to reset a partial element to zero if it is not adjacent to at least one full element. Similarly, a partial element is reset to be full if its immediate neighbors are all full elements to avoid an isolated partial element inside the fluid bulk.

The global balance of the fluid volume is usually not maintained due to the imperfection of the velocity field. Since the continuity equation is expressed in a Galerkin weak form when the discontinuous finite element is applied, a divergence-free condition is not satisfied exactly. The error in the approximation will cause an artificial compressibility of the fluid during the Lagrangian advection step, and introduce local and global imbalance in the fluid volume. The error may accumulate with a time marching scheme and thus it is necessary to make adjustments to ensure the global balance of the fluid volume. One procedure is to adjust the volume fraction of partial elements by using the summation of local imbalances,

$$F_p = F_p + \frac{V_{imb} F_p}{\sum_i F_{pi} V_{pi}} \tag{10.101}$$

where F_p and V_p are respectively the volume fraction and the volume of partial elements and the summation is taken over all partial elements. Also, V_{imb} is the amount of the total volume imbalance, which is the difference between the volume flowing across the external boundary (in–out) and the change of total volume inside the domain. During the process, if the volume fraction of a nearly full element has an unphysical value greater than one, it is reset to one.

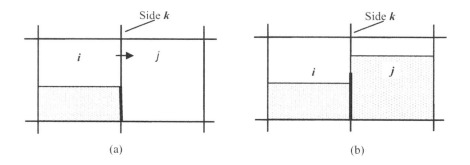

(a)   (b)

Figure 10.11. Schematic of advection of the volume of fluid across the element boundary: (a) advection of fluid into an empty element and (b) advection of fluid between two partially filled elements. The heavy dark vertical line indicates the f_{ik} values to be used for advection

10.6.1.2 Unstructured Mesh

The VOF method can also be incorporated into the computational algorithms using an unstructured mesh, which often is the choice for the discontinuous finite element computations. While the basic concept and computational procedures remain the same as for the structured mesh, the actual implementation differs considerably, particular in the part of geometric considerations, because of different orientations and organization of the mesh used for specific applications [26, 27]. The VOF implementation for an unstructured mesh also involves two steps, which are described below.

Free Surface Reconstruction. For an unstructured mesh, the SLIC method is difficult to apply and algorithms are developed on the basis of the piecewise linear reconstruction. Unlike the structured mesh, the number of the neighboring elements is not necessarily the same for an unstructured mesh. Thus, additional local data and data structure need to be stored for calculations. Consider the case shown in Figure 10.12. The interface is determined from the intersection of the free surface and the the boundaries of the element under consideration, i.e., element i. Using the PLIC method, the intersection line is given by the following relation, which is the same as for the structured mesh

$$g(\mathbf{x}) = \mathbf{n} \cdot \mathbf{x} - c = 0 \qquad (10.102)$$

If the unit normal vector $\mathbf{n}$ is known, the constant c is computed by requiring the volume fraction of the polygon of fluid enclosed by the corresponding line interface to be equal to the given volume fraction for element i.

For an unstructured mesh, the simple geometric relations developed in the last section are not applicable. For the purpose of computing $\mathbf{n} = \nabla F / |\nabla F|$, the method of least squares gradient is useful [26]. By this approach, the volume fraction Taylor series expansions F_{Ts} are formed from the reference element volume fraction to each neighbor of known volume fraction F_k. The sum of the quantities $(F_{Ts} - F_k)^2$ over the list of immediate neighbors is then minimized in the least squares sense. This procedure results in a 2×2 linear system:

$$\mathbf{A}_i \cdot \nabla F_i = \mathbf{b}_i \qquad (10.103)$$

for element i with N_i immediate neighbors. Here, matrix $\mathbf{A}_i$, force vector $\mathbf{b}_i$ and unknown vector ∇F_i are calculated by

$$\mathbf{A}_i = \begin{bmatrix} \sum\limits_{k=1}^{N_i} \dfrac{x_{ik}^2}{d_{ik}} & \sum\limits_{k=1}^{N_i} \dfrac{x_{ik} y_{ik}}{d_{ik}} \\ \sum\limits_{k=1}^{N_i} \dfrac{x_{ik} y_{ik}}{d_{ik}} & \sum\limits_{k=1}^{N_i} \dfrac{y_{ik}^2}{d_{ik}} \end{bmatrix} ; \quad \mathbf{b}_i = \begin{bmatrix} \sum\limits_{k=1}^{N_i} \dfrac{x_{ik} F_{ik}}{d_{ik}} \\ \sum\limits_{k=1}^{N_i} \dfrac{y_{ik} F_{ik}}{d_{ik}} \end{bmatrix} ;$$

$$\nabla F_i = \begin{bmatrix} (\nabla F_i)_x \\ (\nabla F_i)_y \end{bmatrix} \qquad (10.104)$$

with $x_{ik} = x_k^c - x_i^c$, $y_{ik} = y_k^c - y_i^c$, $F_{ik} = F_k - F_i$ and $d_{ik} = (x_{ik}^2 + x_{ik}^2)^{1/2}$. Also, superscript c denotes the value at the centroid of that polygonal element and subscripts i and k refer to the ith and kth elements, respectively.

The solutions of the linear system (Equation 10.103) are obtained with the result,

$$(\nabla F_i)_x = (b_1 A_{22} - b_2 A_{12}) / (A_{11} A_{22} - A_{12} A_{21}) \qquad (10.105a)$$

$$(\nabla F_i)_y = (b_2 A_{11} - b_2 A_{21})/(A_{11} A_{22} - A_{12} A_{21}) \qquad (10.105\text{b})$$

The special case of ($A_{11} A_{22} - A_{12} A_{21}$) ≈ 0 corresponds to the physical condition of an almost constant volume fraction field in the neighborhood of element i. Thus the interface reconstruction procedure is applied only to wet elements ($0 < F < 1$).

The computation of the line constant c in Equation 10.102 cannot be carried out analytically as is done for the structured mesh. If the conservation of fluid volume in the ith element is imposed, the line constant c is the root of the following equation [42, 43],

$$A_i(c) = F_i^{n+1} \qquad (10.106)$$

The root can be obtained numerically by a root finding algorithm. With c being known, the vertices of the reconstructed polygon of fluid delimited inside the ith element by the interface line can be determined. The algorithm for this type of calculation is commonly applied for data visualization and established techniques can be used for this purpose [28–30].

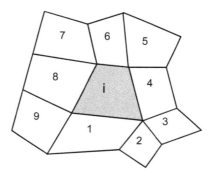

Figure 10.12. Element i and its $N_i(=9)$ neighbors for reconstruction of free surface [26]

VOF Evolution. The scheme presented above for the evolution of fluid volume in a structured mesh is not easily adapted here. Instead, for an unstructured mesh, a Lagrangian–Eulerian advection seems to work well [26, 27]. The idea is to move the fluid portion of an element in a Lagrangian sense, and compute how much of the fluid remains in element i, and how much of it passes into each of its neighboring elements. The algorithm involves four basic steps, as illustrated in Figure 10.13.

To start, the fluid portion inside a non-empty element is used to construct a polygon in that element, which is denoted as element i in Figure 10.13(a). If the element is full, the polygon of fluid coincides with the element. The vertices of this polygon are material points in the fluid flow. Each material point undergoes a Lagrangian displacement (ξ, η) according to the velocity components (u, v),

$$u = \frac{d\xi}{dt}; \qquad v = \frac{d\xi}{dt} \qquad (10.107)$$

where the velocities are calculated by solving the governing equations.

This polygon is then set in a Lagrangian motion and the vertices of this polygon at the end of the next time step are calculated by

$$\mathbf{x}_{p,new} = \mathbf{x}_{p,old} + \mathbf{u}\Delta t \qquad (10.108)$$

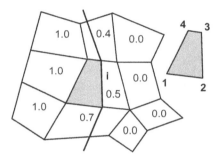

(a) Volume of fluid prior to advection in element i.

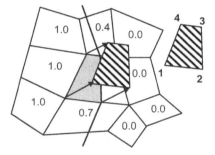

(b) Volume of fluid in element i is advected into other elements.

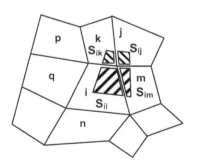

(c) Budget of VOF orginating from element i.

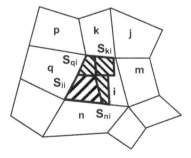

(d) VOF in element i after advection

Figure 10.13. Illustration of the evolution of volume of fluids on an unstructured mesh

This procedure is shown in Figure 10.13(b). The resulting polygon intersects with the mesh and the budget of fluid volume in each of the involved neighboring elements is determined. This process is illustrated in Figure 10.13(c). The next step is to collect all the contributions to the fluid volume from all the adjacent elements and sum them as the final fluid volume in element i, as shown in Figure 10.13(d). This completes the VOF evolution.

As for the structured mesh case, both local and global adjustments are needed to remove the unphysical fluid volumes and to ensure the global mass balance. This is identical to that discussed at the end of Section 10.6.1.1.2.

10.6.2 The Marker-and-cell Method

The marker-and-cell (MAC) method is perhaps the earliest and yet easiest numerical method devised for time-dependent, free-surface flow problems [31, 32]. This scheme is based on a fixed, Eulerian grid, structured or unstructured. The location of fluid within the grid is determined by a set of marker particles that move with the fluid, but otherwise have no volume, mass or other properties. A surface is constructed by the profiles of these particles.

One of the most important atributes of the MAC method is its capability of capturing very complex free surface shapes. Figure 10.14 shows a typical free surface problem that can be solved using the MAC method. Here the complex rolling structure is faithfully represented by the profile of the marker particles. Both volume and surface markers have been used in the literature; the basic idea and algorithm are very similar. Here we outline the surface marker algorithm presented by Chen *et al.* [32] for a structured mesh computation. The same procedure is also applicable to an unstructured mesh.

In the MAC method, evolution of surfaces is computed by moving the markers in accordance with local fluid velocities. Some special treatments are required to define the fluid properties in newly filled grid cells and to cancel values in cells that are emptied.

To advect the marker particles, a simple bilinear interpolation is used to find the velocity inside an element,

$$u(x,y) = \sum_{i=1}^{N_e} \psi_i(\xi(x,y),\eta(x,y))u_i \tag{10.109}$$

where x and y are the coordinates of the marker point and N_e is the number of nodes per element. The marker particles are then advected in a Lagrangian manner using a straightforward first order explicit scheme,

$$x_i^{n+1} = x_i^n + u(x_i^n, y_i^n)\Delta t \quad ; \quad y_i^{n+1} = y_i^n + v(x_i^n, y_i^n)\Delta t \tag{10.110}$$

where superscripts (i.e., n, $n+1$) represent the time step. Once the points have been advected, a parametric representation of the interface is constructed from the particle locations.

For a 2-D curve, a cubic polynomial with continuous first and second order derivatives, *cubic splines*, represents a good choice [32]. The parametric representation is often periodic as the interfaces are mostly self-connected (drops, bubbles, periodic wave trains, . . .). For a 3-D surface, construction of a smoothed surface from the marker particles can be time consuming.

As the interface evolves, the markers drift along the interface following tangential velocities and more markers may be needed if the interface is stretched by the flow. The markers need to be redistributed in order to ensure a homogeneous distribution of points along the interface. This is done at each time step using the interpolating curve $(x(s), y(s))$. As s is an approximation of the arc length, if a redistribution length l is chosen, the new number of markers is $N_{new} = s_N/l$ and the points are redistributed as $(x_i^{new}, y_i^{new} = (x(il), y(il))$. Here l is usually chosen as h, which yields an average number of one marker per computational cell. Decreasing this length does not apparently improve the accuracy and in some cases leads to instabilities [32].

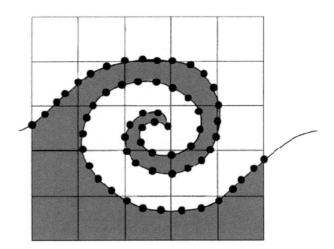

Figure 10.14. Mark-and-cell simulation of roll-up structure of fluid motion

10.6.3 The Level Set Method

The level set method is a computational technique for tracking a propagating interface over time. This method has been used in a variety of aspects of image processing and is now adopted for computational free surface flows [3, 33]. In the level set method, free surface flows are modeled as immiscible gas–liquid two-phase flows. The free surface is identified as a zero level set, i.e. $\phi(\mathbf{x}, t) = 0$, where $\mathbf{x} = (x, y)$ in two dimensions or $\mathbf{x} = (x, y, z)$ in three dimensions. The sharp jumps in density and viscosity at gas–liquid interfaces can cause numerical instabilities if not treated properly. To ease this problem, fluid properties, such as density, viscosity, etc., are smeared over a narrow transition zone around the free surface.

As discussed in Section 10.2, the kinematic boundary condition at a free surface can be interpreted in a Lagrangian way: a particle on the surface always stays on the surface. This condition translates into a constraint for a level set value of a point on the contour with motion $\mathbf{x}(t)$. The value must always be zero on the surface,

$$\phi(\mathbf{x}(t), t) = 0 \tag{10.111}$$

where ϕ is the level set function. By the chain rule, we have the differential equation,

$$\frac{\partial \phi}{\partial t} + \nabla \phi(\mathbf{x}(t), t) \cdot \frac{d\mathbf{x}(t)}{dt} = 0 \tag{10.112}$$

Written in terms of the interfacial velocity, the equation for the advection of the level set function is obtained,

$$\frac{\partial \phi}{\partial t} + \mathbf{u} \cdot \nabla \phi = 0 \tag{10.113}$$

where $\mathbf{u} = (u, v)$ in two dimensions or (u, v, w) in three dimensions is the fluid velocity. Thus, the evolution of the level set function defines the motion of a free surface.

Defining $\mathbf{F}_n$ as the speed in the outward normal direction,

$$\mathbf{F}_n = \frac{d\mathbf{x}(t)}{dt} \cdot \mathbf{n} ; \qquad \mathbf{n} = \frac{\nabla \phi}{|\nabla \phi|}\bigg|_{\phi=0} \tag{10.114}$$

the evolution for ϕ is described by the Hamilton–Jacobi equation,

$$\frac{\partial \phi}{\partial t} + \mathbf{F}_n |\nabla \phi| = 0 \tag{10.115}$$

which shows that ϕ evolves as a signed distance function.

To model the free surface evolution, the level set function ϕ is initially assigned with a signed distance function,

$$\phi = \begin{cases} -d & \text{for} & x \in \Omega_2 \\ 0 & \text{for} & x \in \partial\Omega_2 & \text{(free surface)} \\ d & \text{for} & x \in \Omega_1 \end{cases} \tag{10.116}$$

where d is the absolute normal distance to the free surface. For immiscible incompressible fluids, because of large density and viscosity jumps, particularly the density jump, the properties are often calculated by modeling the surface with a small transition zone defined as $|\phi| \leq \varepsilon$, where ε, the half-thickness of the interface, is typically one or two grid distances. By defining an infinitely differentiable smoothed Heaviside function $H(\phi)$ [3],

$$H_v(\phi) = \begin{cases} 0 & \text{if} \quad \phi < -\varepsilon \\ \frac{1}{2}\left[1 + \frac{\phi}{\varepsilon} + \frac{1}{\pi}\sin\left(\frac{\pi\phi}{\varepsilon}\right)\right] & \text{if} \quad |\phi| \leq \varepsilon \\ 1 & \text{if} \quad \phi > +\varepsilon \end{cases} \qquad (10.117)$$

the density and viscosity are smoothed in such a way that they are $(\rho_1+\rho_2)/2$ and $(\mu_1+\mu_2)/2$ at the surface front ($\phi=0$), respectively, and near the interface,

$$\rho(\phi)=\rho_2+(\rho-\rho_2)H_v(\phi); \qquad \mu(\phi)=\mu_2+(\mu-\mu_2)H_v(\phi) \qquad (10.118)$$

The surface tension is spread over the transition zone as a δ-function-like volume force in the momentum equation [16],

$$2H(\phi)\gamma \delta(\phi)\mathbf{n} + \nabla_s\gamma \delta(\phi) = \mathbf{f}_b \qquad (10.119)$$

where $\mathbf{f}_b$ is the body force, $\mathbf{n}$ is the normal, $2H(\phi)$ is the curvature that is computed in terms of ϕ,

$$2H(\phi) = \nabla \cdot \mathbf{n}\big|_{\phi=0} = \nabla \cdot \frac{\nabla\phi}{|\nabla\phi|}\bigg|_{\phi=0} \qquad (10.120)$$

and the delta function $\delta(\phi)$ is obtained by taking the gradient of the smoothed Heaviside function,

$$\delta(\phi) = \nabla_\phi H(\phi) = \begin{cases} 0 & \text{if} \quad |\phi| > \varepsilon \\ \frac{1}{2\varepsilon}\left[1 + \cos\left(\frac{\pi\phi}{\varepsilon}\right)\right] & \text{if} \quad |\phi| \leq \varepsilon \end{cases} \qquad (10.121)$$

Thus, the kinematic and dynamic boundary conditions at the free surface are automatically embedded in the formulation of the level set method. This is certainly a very useful feature when it comes to numerical implementation.

Since the level function is defined as the distance from the interface, it is necessary that ϕ perserves this feature as evolution continues. This means that ϕ must be a distance function, and satisfies the following condition [3, 35]:

$$|\nabla(\phi)| = 1 \qquad (10.122)$$

By Equation 10.113 (or equivalently Equation 10.115), the level set function will cease to be an exact distance function even after one time step. In moving surface problems, it is possible that steep gradients develop in ϕ, making it difficult to maintain a finite thickness of transition zone. This would cause the numerical errors in computation of the normal and curvature of the moving surface. To avoid these problems, the level set function is reinitialized after evolution over a time step. One widely applied algorithm for reinitializing ϕ to be an exact signed distance function from the moving boundary is to iteratively solve the following equation:

$$\frac{\partial \phi}{\partial \tau} = S(\phi_0)(1 - |\nabla \phi|) \tag{10.123}$$

until it reaches a steady state, at which time the feature of signed distance function is preserved in ϕ,

$$|\nabla \phi| = 1 \qquad \text{at } \tau \to \infty \tag{10.124}$$

Note here that τ is a pseudo-time introduced to satisfy the constraint of ϕ being a signed distance function. Here $\phi_0(\mathbf{x}) = \phi(\mathbf{x},t)$ and $\phi(\mathbf{x},t)$ is calculated using Equation 10.115. Clearly, with Equation 10.123, given a function $\phi_0(\mathbf{x})$ that is not a distance function, one can always evolve it into a function f that is an exact signed distance function satisfying Equation 10.124.

To determine the level set function ϕ from Equation 10.123, the sign function $S(\phi_0)$ needs to be specified. Sussman et al. [35] suggest the use of the following equation for $S(\phi_0)$:

$$S(\phi_0) = \frac{\phi_0}{\sqrt{\phi_0^2 + \varepsilon^2}} \tag{10.125}$$

where parameter ε is taken to be on the order of the grid size.

Equations 10.115 and 10.123 both are of the Hamilton–Jacobi type and are particularly suited for the numerical treatment using the discontinuous finite element schemes [34].

Thus, a typical level set algorithm for modeling the free-surface problems would involve the following computational steps: (1) the field variables are solved using the discontinuous flow solvers with a previously determined interface shape; (2) the function $\mathbf{F}_n$ in Equation 10.117 is then calculated from the field distribution; (3) the level set function $\phi(\mathbf{x},t)$ is initialized, if not yet, as a signed normal distance from the moving boundary; (4) the function $\phi(\mathbf{x},t)$ is evolved using Equation 10.115 and the result $\phi(\mathbf{x},t+\Delta t)$ is denoted as $\phi_0(\mathbf{x})$; (5) with $\phi_0(\mathbf{x})$, Equation 10.123 is continued in τ marching until steady state and the result is denoted as $\phi(\mathbf{x}, \tau=\infty)$; (6) the level set function is reset to $\phi(\mathbf{x}, \tau=\infty)$, i.e., let $\phi(\mathbf{x}, t+\Delta t) = \phi(\mathbf{x}, \tau=\infty)$. Steps 1 to 6 are repeated for the next time step. For this explicit time scheme, a critical time step needs to be observed, that is, the time step needs to satisfy the *CFL* condition.

10.6.4 Fixed Grid Methods for Phase Change Problems

The fixed grid methods presented above for flow calculations may be considered as flow-based moving boundary methods and can be readily modified to predict the interface morphologies for phase change problems. An important aspect of this modification is to estimate the interface velocity. While in principle these methods can be used to solve phase change problems, only the level set method has gained popularity. There are also algorithms developed based on the enthalpy formulation, which is very effective in treating phase change phenomena and has been used in various applications. We discuss below these flow-based methods and enthalpy-based methods.

10.6.4.1 Flow-based Methods
As for the moving grids method, the flow-based fixed grid methods discussed in the previous sections can also be modified to solve the phase change problems. If pure materials are considered, the governing equations for the problems should also include the thermal balance equation,

$$\nabla \cdot \mathbf{u}_j = 0 \quad \in \Omega_1 \bigcup \Omega_2 \tag{10.126}$$

$$\rho \frac{\partial \mathbf{u}_j}{\partial t} + \rho \mathbf{u}_j \cdot \nabla \mathbf{u}_j = -\nabla p_j + \nabla \cdot \boldsymbol{\tau}_j$$

$$+2H\gamma\delta(\mathbf{x} - \mathbf{x}_f)\mathbf{n} + (\nabla_s \gamma)\delta(\mathbf{x} - \mathbf{x}_f) + \mathbf{F} \quad \in \Omega_1 \bigcup \Omega_2 \tag{10.127}$$

$$\rho C_p \frac{\partial T_j}{\partial t} + \rho C_p \mathbf{u}_j \cdot \nabla T_j = -\nabla \cdot \mathbf{q}_j + \rho L \frac{\partial f_s}{\partial t} \delta(\mathbf{x} - \mathbf{x}_f) + Q_j$$

$$\in \Omega_1 \bigcup \Omega_2 \tag{10.128}$$

where $\partial f_s/\partial t$ is the fraction of new phase formed (e.g., f_s is the solid fraction if the solidification is considered), and $\mathbf{F}$ includes the drag forces resulting from phase change and/or solids [36]. Also, the properties of the materials are assumed to be the same for solidification problems. For liquid–gas transition, the equations apply only to the liquid phase.

To calculate the velocity of the phase boundary, the energy balance along the interface is often used,

$$k_1 \mathbf{n} \cdot \nabla T_1 - k_2 \mathbf{n} \cdot \nabla T_2 = \rho_1 L \mathbf{u}_i \cdot \mathbf{n} \tag{10.129}$$

where $\mathbf{n}$ is the normal of the phase boundary pointing from phase 1 to phase 2 and $\mathbf{u}_i$ is the interface velocity with which phase 1 moves into phase 2. At the interphase boundary, the temperature must be the same,

$$T_i = T_{ph} = T_{ph}^0 + m_L C_L + 2H\Gamma_G(\theta) - v_n / \mu(\theta) + g_L p \tag{10.130}$$

where T_i is the interface temperature. Note that Equations 10.129 and 10.130 are the same Equation 10.44 and 10.48, which are reproduced here for convenience.

For phase change problems, the velocity at which the moving interface evolves is often defined by the interfacial energy balance. Consequently, Equation 10.129, is used to calculate the velocity for the evolution of the phase interface. In the case of the level set method, this velocity will be used to estimate **F** also. In addition, the condition set by Equation 10.130 needs to be met. Thus, if the fixed grid methods described in the above sections, VOF, MAC and LSM, are used, some obvisous modifications are necessary, which often result in an iterative procedure within a time step [35, 37, 38].

We take the level set method as an example to illustrate the necessary modifications required to solve this type of problem. The level set method has been used in solving both solidification and vaporization problems within the framework of finite volumes [35, 38]. A typical level set algorithm, when incorporated into a discontinuous finite element formulation, involves 7 steps instead of 6, as discussed in Section 10.6.3. In step 2, Equation 10.129 is used to calculate velocity and hence $\mathbf{F}_n$ in Equation 10.115. Step 7 is added to check if T satisfies Equation 10.130. This is usually done by interpolation to find the temperature at the interface T_i. If Equation 10.130 is satisfied, then the calculation continues with the next time step. If not, it will go back to step 1 and iterate between step 1 through step 7 until Equation 10.130 is satisfied. Then the calculation continues with the next time step.

It is worth noting here that when the properties of the materials cannot be assumed to be the same, such as for the case of liquid–vapor transition occurring in boiling, the governing equations and boundary conditions need to be modified accordingly [37].

10.6.4.2 Enthalpy-based Methods

Useful numerical methods, other than the three flow-based approaches described above, have been developed specifically to solve the phase change problems. These methods have met with success and can be readily incorporated into a discontinuous finite element setting as well. These methods are based on the enthalpy formulation and are very useful for phase change problems involving a transition temperature range, which is often characterized by a mushy zone [11, 36, 39].

By observing that the latent heat, L, corresponds to the isothermal change in the enthalpy, h, for a material at the transition melt temperature, T_m, the following relationship is introduced,

$$h(T) = \int_{T_{ref}}^{T} C_p(T)dT + LH_v(T - T_m) \tag{10.131}$$

where H_v is the Heaviside function. From the above definition, the equivalent (or effective) specific heat, C_{p*}, is then introduced by

$$C_{p*}(T) = \frac{dh(T)}{dT} = C_{p*}(T) + L\delta(T - T_m) \tag{10.132}$$

where δ is the delta function. Through the use of this formulation for the specific heat, the heat flux jump condition is eliminated from the problem. This approach is computationally effective since a two region problem with a jump condition has been reduced to a single region problem with rapidly varying properties. For application in a discontinuous finite element setting, the above equation is replaced by

$$C_{p*}(T) = \frac{dh(T)}{dT} = C_p(T) + L\delta*(T - T_m, \Delta T) \tag{10.133}$$

where δ^* is the delta form function; δ^* has a larger but finite value in the interval ΔT entered about T_m and is zero outside the interval. The interval ΔT is often referred to as the "mushy" zone and corresponds to the difference between the liquidus and solidus temperatures for a material. For pure materials that change phase at a specified temperature, this is an approximation; but it is accurate for non-pure substances that have truly distinct liquidus and solidus temperatures.

With the above definition of the effective heat capacity, the energy balance equation is simplified as

$$\rho C_{p*} \frac{\partial T}{\partial t} + \rho \mathbf{u} \cdot \nabla T = -\nabla \cdot \mathbf{q} + Q \qquad \in \Omega_1 \cup \Omega_2 \tag{10.134}$$

Though the equivalent specific heat model is useful for latent heat effects, caution needs to be exercised with regard to the time integration of this type of phase change model. In general, the transition temperature, ΔT, is small compared to the overall temperature variation. Thus, the time-stepping algorithm must be controlled such that every node undergoing phase change attains a temperature value in the interval bracketed by ΔT. If a nodal point steps over this temperature interval, the latent heat effect is not registered by the node and an incorrect temperature response and energy balance will be predicted. Some approaches have been developed to alleviate the problem and are described below.

One approach is to evaluate the specific heat at a point by computing the slope of the enthalpy–temperature curve based on the temperature at the point. The method performs satisfactorily if at a given time step the integration points in an element may not detect the presence of the solidification front in the element. An alternative approach, which is considered more accurate and convenient, is to compute the required specific heat at a point by the following expression [39, 40]:

$$C_{p*} = \left(\frac{\nabla h \cdot \nabla h}{\nabla T \cdot \nabla T} \right)^{1/2} \tag{10.135}$$

This equation is computed by first determining the enthalpy at the nodes of the element using the provided enthalpy–temperature curve, that is, Equation 10.131. The element shape functions are then used to approximate the enthalpy distribution within the element as well as the temperature distribution in the usual manner. The value of the C_{p*} at the integration point is then computed using the above formula. A drawback of this method is that if the time step is large enough so as to allow the solidification front to pass an element altogether, the proper amount of latent heat will not be released in the element, thereby resulting in a faster than desired temperature drop. Experience indicates that this method works best with linear elements [39].

For transient problems, the change in enthalpy from the value at the previous time step and the change in temperature from the value at the previous time step may be used to construct an effective specific heat model,

$$C_{p*} = \frac{dh}{dT} = \frac{h(T^n) - h(T^{n-1})}{T^n - T^{n-1}} \qquad (10.136)$$

where the superscripts refer to the time step numbers. As this method always detects the passage of the solidification front at an integration point, it performs better than the previous methods on coarser meshes. Equation 10.136 may also be evaluated using the nodal values, leading to yet another approximation. A deficiency of this approach is that even though the correct amount of energy is always released, the rate of release is typically lagged in time.

10.7 Phase Field Modeling of Moving Boundaries

The fixed and moving grid approaches presented above are devised to treat the boundary conditions for a continuum mechanics description of fluid flow and heat transfer involving moving boundaries. The physics governing the microscopic phenomena within the interfacial layer is not considered in these continuum descriptions. Phase field models present a statistical mechanics description of the interfacial phenomena within and near the interphase layer, with allowance for the molecular or microscopic physics, and have emerged as a viable approach to study both free and moving boundary problems [4, 40]. The phase field theory is particularly powerful in modeling the thermal and fluids phenomena driven dominantly by interfacial forces [4]. In what follows, we present the basic ideas of phase field models, their coupling with momentum and energy balance equations for thermal systems and the discontinuous finite element formulation for the phase field models. Numerical results are also given to illustrate the capability of the phase field model to resolve fine structures under various conditions.

10.7.1 Basic Ideas of Phase Field Models

The phase field theory is developed as a result of studying nonlinear critical phenomena during phase transitions in superconducting materials and other physical systems [40]. From the statistical mechanics perspective, particles are described by Langevin's dynamic equation [41, 42], which incorporates the stochastic fluctuating forces, in addition to the systematic forces. The central idea in phase field modeling of phase transitions is the concept of localized statistical averaging, that is, coarse-graining: which means that some of the microscopic degrees of freedom in a given system are integrated out, leaving an effective system (characterized by an effective free energy or Hamiltonian) with fewer degrees of freedom embodied by the coarse-grained order parameter (block magnetization in the Ising spin model, density difference for simple liquids, etc.). We have seen the volume averaged momentum equations for flows in porous media (see Section 1.7). The coarse grain averaging is similar, but is carried out over a molecular ensemble.

In statistical mechanics, all static thermodynamic quantities such as entropy are calculated from the partition function Z, which is related to the free energy F, where $Z = \exp(-F(T)/k_bT)$, and k_b is the Boltzmann constant. By the coarse-grain averaging process, a block is selected and averaging is carried out over a number of molecules in the cell for a microscopic quantity (e.g., magnetic spins in the Ising model) yielding a block quantity, and then summing over those microscopic quantities that give rise to a given block yields a localized partition function Z', which is related to a coarse-grained free energy F, where $Z' = \exp(-F([\phi], T)/k_bT)$. Here ϕ is the phase field parameter, which marks the change from one phase to another. In this way, Z is a sum of all Z' in the system. The square brackets indicate that $F[\phi]$ is a functional of the phase parameter ϕ. The phase field model for phase transition and interfacial phenomena is concerned about the behavior of the phase field parameter ϕ, which is related to the coarse grain free energy. In general, the free energy $F([\phi], T)$ is a functional of the phase field parameter ϕ and it assumes the following form:

$$F([\phi],T) = \int \left(\tfrac{1}{2}\varepsilon^2 \mid \nabla\phi(\mathbf{r}) \mid^2 + f(\phi(\mathbf{r}),T) \right) d\Omega \tag{10.137}$$

where ε is the interfacial gradient thickness parameter and $f(\phi(\mathbf{r}), T)$ is a potential function and is a function (but not functional) of $\phi(\mathbf{r})$ and temperature. Because the phase field model is formulated locally and is able to resolve the microscopic phenomena, it has been adopted to model the interfacial phenomena in thermal and fluids systems that are diffusive in nature [4, 43].

Figure 10.15 shows a typical double wall function associated with phase transition as a function of the phase parameter for a 1-D system. At a temperature above the critical point, that is, the transition point, there exists only one stable phase corresponding to the minimum of the potential function. This is illustrated in Figure 10.15(a). When the temperature drops below the critical point, two phases will co-exist, as marked by the double dip in the potential function distribution, as shown in Figure 10.15(b).

In a time-dependent system, an evolution of the phase parameter $\phi(\mathbf{r}, t)$ represents the evolution of the interface. The phase field model is derived based on the fact that the system evolves towards a state that minimizes the free energy,

$$\tau \frac{\partial\phi(\mathbf{r},t)}{\partial t} = -\frac{\delta F}{\delta\phi} + \eta(\mathbf{r},t) \tag{10.138}$$

where τ is the relaxation time, and $\eta(\mathbf{r}, t)$ represents the stochastic noise that describes the random effects of the environment, and is averaged to zero over the realization of the noise field $\eta(\mathbf{r}, t)$. For most thermal and fluids applications, this term may be set to zero. At the static equilibrium, or the steady state,

$$\frac{\delta F}{\delta \phi} = 0 \qquad\qquad (10.139)$$

The variational derivative in Equation 10.138 can be carried out explicitly with time held constant,

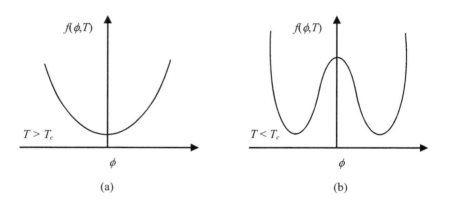

Figure 10.15. Potential energy above (a) and below (b) the cirtical temperature T_c for phase transition

$$
\begin{aligned}
\frac{\delta F}{\delta \phi} &= \frac{\delta}{\delta \phi(\mathbf{r})} \int_V \delta(\tfrac{1}{2}\varepsilon^2 |\nabla \phi(\mathbf{r}')|^2 + f(\phi(\mathbf{r}'), T\phi(\mathbf{r}'))) \, dV(\mathbf{r}') \\
&= \frac{1}{\delta \phi(\mathbf{r})} \int_V \delta\phi(\mathbf{r}') \left(-\nabla \cdot (\varepsilon(\mathbf{r}')^2 \nabla \phi(\mathbf{r}')) + \frac{\partial f(\phi(\mathbf{r}'), T\phi(\mathbf{r}'))}{\partial \phi(\mathbf{r}')} \right) dV(\mathbf{r}') \\
&= -\int_V \delta_f(\mathbf{r}'-\mathbf{r}) \left(-\nabla \cdot (\varepsilon(\mathbf{r}')^2 \nabla \phi(\mathbf{r}')) + \frac{\partial f(\phi(\mathbf{r}'), T\phi(\mathbf{r}'))}{\partial \phi(\mathbf{r}')} \right) dV(\mathbf{r}') \\
&= \int_V \frac{\delta\phi(\mathbf{r}')}{\delta \phi(\mathbf{r})} \left(-\nabla \cdot (\varepsilon(\mathbf{r}')^2 \nabla \phi(\mathbf{r}')) + \frac{\partial f(\phi(\mathbf{r}'), T\phi(\mathbf{r}'))}{\partial \phi(\mathbf{r}')} \right) dV(\mathbf{r}') \\
&= -\nabla \cdot \left(\varepsilon^3 \nabla \phi \right) + \frac{\partial f}{\partial \phi} \qquad\qquad (10.140)
\end{aligned}
$$

where $\delta\phi(\mathbf{r})/\delta\phi(\mathbf{r}')=\delta_f(\mathbf{r}-\mathbf{r}')$, $\delta_f(\mathbf{r}-\mathbf{r}')$ being the delta function. With this substituted into Equation 10.138, we have the following evolution equation for the phase field $\phi(\mathbf{r},t)$:

$$\tau\frac{\partial\phi}{\partial t}=\varepsilon^2\nabla^2\phi-\frac{\partial f}{\partial\phi} \tag{10.141}$$

With f known, this equation can be used to evolve the free surfaces and boundaries. The phase field model has been used in other physical systems and here it is used to describe some of the thermal fluids systems that involve free surface and interfacial boundaries.

Figure 10.16 sketches the distribution of the phase field parameter ϕ as a function of distance (x) across the interface or moving boundary. Far away from the interface $(x = 0)$, the phase field ϕ assumes constant values. Near the interface, there exists a very sharp change in $\phi(x)$. The change of $\phi(x)$ measured by $\partial\phi(x)/\partial x$ shows almost a delta function, and its width represents the interfacial thickness between the two phases.

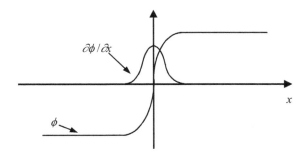

Figure 10.16. 1-D phase field model illustrating the phase transition at $x = 0$ and the derivative of the phase parameter with respect to the x coordinate

It is noted that the phase field behavior shown in Figure 10.16 is very similar to that of the level set function. One should realize that the latter is based on a mathematical description of a signed distance function from the sharp interface, and the signed function needs to be adjusted during each time step. This is in contrast with the phase field model, which has its roots in statistical mechanics and the phase field parameter ϕ evolves continuously based on the physical description of interfacial phenomena.

10.7.2 Governing Equations for Interfacial Phenomena

In the phase model description, the equations describing the interface balance are built into the governing equations and the interface boundaries are evolved using the phase field parameters. The phase field parameter ϕ can be used to evolve a free or moving or phase boundary in thermal fluids systems [4, 44]. With ϕ as a

parameter marking a moving interface, which itself evolves, the Navier–Stokes
equations and the energy balance equations for a moving boundary problem are
written as

$$\nabla \cdot \mathbf{u} = 0 \tag{10.142}$$

$$\phi \left(\rho \frac{\partial \mathbf{u}}{\partial t} + \mathbf{u} \cdot \nabla \mathbf{u} \right) = -\nabla p + \nabla \cdot \boldsymbol{\tau} + \mu_c \nabla \phi + \rho \mathbf{g} + \mathbf{F} \tag{10.143}$$

$$\rho C \frac{\partial T}{\partial t} + \rho C \mathbf{u} \cdot \nabla \mathbf{u} = \nabla k \cdot \nabla T + \rho L \frac{\partial \phi}{\partial t} \tag{10.144}$$

$$\tau \frac{\partial \phi}{\partial t} = \varepsilon^2 \nabla^2 \phi - \frac{\partial f}{\partial \phi} \tag{10.145}$$

where $\boldsymbol{\tau}$ is the share stress, L is the energy release from one fluid to the other and μ_c
is the chemical potential. For the sake of simplicity, subscript j denoting different
phases is dropped and the materials properties are assumed to be a function of ϕ.
The above equations can be solved with the appropriate boundary conditions
imposed [43]. Note that $\mathbf{F}$ contains the drag force result from liquid–solid
transition, which is also a function of the phase field parameter.

In the above equations, both phases across the interface boundaries are
included. The interfacial contributions to the momentum and thermal balance
equations are accounted for through two source terms including interfacial
mechanical and thermal energies. To see that Equation 10.143 yields the interfacial
mechanical force balance equation, we integrate the momentum equation over a
small pillbox across the interface, and obtain the following equation:

$$(-p\mathbf{I} + \boldsymbol{\tau})_1 \cdot \mathbf{n} = (-p\mathbf{I} + \boldsymbol{\tau})_2 \cdot \mathbf{n} + 2\gamma H \mathbf{n} + \nabla_s \gamma \tag{10.146}$$

where subscripts denote different flow regions. This is nothing but the interfacial
mechanical balance equation discussed in Section 10.2.3.

Example 10.3. Derive an expression describing the expansion of a bubble in a fluid
using the phase field model, neglecting the fluid motion.

Solution. With the fluid motion neglected, the growth model can be readily
derived. For this problem, a potential ϕ is the difference between the energy inside
and outside the bubble. The phase field model then becomes

$$\tau \frac{\partial \phi}{\partial t} = \varepsilon^2 \nabla^2 \phi - \frac{\partial f}{\partial \phi} \tag{10.18e}$$

We assume that the order parameter is a function of time and distance from the center of the bubble,

$$\phi(r,t) = \phi(r - R(t)) \tag{10.19e}$$

where $R(t)$ is the radius at time t. Substituting into the phase field model, we obtain

$$-\tau \frac{dR}{dt}\frac{\partial\phi}{\partial r} = -\frac{\partial V_s}{\partial\phi} + \varepsilon^2 \frac{\partial^2\phi}{\partial r^2} + \varepsilon^2 \frac{2}{r}\frac{\partial\phi}{\partial r} - \frac{\partial V_a}{\partial\phi} \tag{10.20e}$$

The first two terms on the right cancel because of the following relation:

$$-\frac{\partial V_s}{\partial\phi} + \varepsilon^2 \frac{\partial^2\phi}{\partial r^2} = 0 \tag{10.21e}$$

Equation 10.20e further is multiplied by $\partial\phi/\partial r$ and integrated from the center of the bubble to infinity,

$$-\tau \frac{dR}{dt}\int_0^\infty \left(\frac{\partial\phi}{\partial r}\right)^2 dr = 2\varepsilon^2 \int_0^\infty \frac{1}{r}\left(\frac{\partial\phi}{\partial r}\right)^2 dr - \int_0^\infty \frac{\partial V_a}{\partial\phi}\frac{\partial\phi}{\partial r} dr \tag{10.22e}$$

Since ϕ changes sharply near the interface, the second term can be approximated to the first order $1/r = (\delta r + R) = 1/R$, whence we have the following differential equation for the rate change of the bubble radius:

$$-\tau \frac{dR}{dt} = \varepsilon^2 \left(\frac{2}{R} - \frac{\Delta V_a}{\gamma}\right) \tag{10.23e}$$

where $\Delta V_a = V(\infty) - V(0)$ and γ is the surface tension,

$$\gamma = \varepsilon^2 \int_0^\infty \left(\frac{\partial\phi}{\partial r}\right)^2 dr \tag{10.24e}$$

Integration of the equation over time with the initial condition $R(t=0) = R_0$ yields the change of the radius as a function of time,

$$\frac{\gamma}{\Delta V_a}(R(t) - R_0) + \frac{2\sigma^2}{\Delta V_a^2}\ln\frac{2\sigma - \Delta V_a R(t)}{2\sigma - \Delta V_a R_0} = \frac{\varepsilon^2}{\tau}t \tag{10.25e}$$

In the case of static equilibrium, the time dependent term in Equation (4.23e) disappears, and we recover the force balance equation at the bubble surface,

$$\frac{2\gamma}{R_e} = \Delta V_a \tag{10.26e}$$

which identifies the potential difference to be the pressure difference inside and outside the bubble. With R_e, the solution for Equation 10.23e can be written as

$$\frac{R(t) - R_0}{R_e} + \ell n \frac{R_e - R(t)}{R_e - R_0} = \frac{2\varepsilon^2}{R_e^2\,\tau}t \tag{10.27e}$$

We have two limiting cases,

$$R(t) = R_0 + \frac{\varepsilon^2 \Delta V_a}{\sigma\,\tau}t \quad \text{if } R_e \to 0 \text{ and } \frac{\varepsilon^2 \Delta V_{ct}}{\sigma\,\tau} = \text{const.} \tag{10.28e}$$

and

$$R^2(t) = R_0^2 - \frac{4\varepsilon^2}{\tau}t \quad \text{if } R_e \to \infty \tag{10.29e}$$

The second limit clearly shows the curvature effect, which is typical for interfacial phenomena, and when it is neglected, the radius expands at a constant speed.

There are certain advantages associated with the use of a phase field model for free surface and moving boundary problems. The model is, in general, easy to construct and non-equilibrium conditions can be imposed by boundary and initial conditions. The interface energy emerges naturally and is linked to microscopic states in the materials. Numerically, a unified equation makes the numerical formulation and computer implementation easy. At present, the phase field model has been primarily used in modeling phase change or phase transition problems, where the internal moving boundaries and inter-phase front morphologies often evolve in a rather complex pattern.

10.7.3 Discontinuous Finite Element Formulation

The discontinuous formulations for the fluid flow and heat transfer problems presented in previous chapters should be directly applied here. The interfacial contribution may be treated as the source/sink terms and incorporated into the force term at the right hand side of the matrix equation. The phase field model is in essence a diffusion model with a nonlinear source term that drives the interface evolution. These equations can be easily discretized and solved using the discontinuous finite element formulation developed for heat conduction problems in Chapter 4. To illustrate the procedure of coupling a phase field model and other mechanical and thermal balance equations, a discontinuous formulation for a phase change problem that involves the evolution of a solid–liquid interface is presented below. We first consider the cases without the fluid flow and other local phenomena. These complications are then included. These results demonstrate the usefulness of the phase field theory for modeling interfacial phenomena driven by local forces.

10.7.4 Phase Field Modeling of Microstructure Evolution

Microstructure evolution during solidification represents a moving boundary problem with very complex internal boundary shapes. In addition, the spatial and temporal resolutions required to resolve these fine structures are so small that intensive computation warrants massive parallel computing for a realistic simulation. For this purpose, the discontinuous finite element method should be a very suitable candidate.

10.7.4.1 Governing Equations
Let us consider a typical solidification problem as illustrated in Figure 10.17, where the liquid solidifies as a result of applied cooling. Here we are concerned with the modeling of the microscale features such as dendritic structures formed during freezing. The governing equation for phase field modeling of the local solidification process is given by the following pair of equations describing the phase field evolution and temperature distribution, with convection neglected:

$$\frac{\partial T}{\partial t} + \frac{1}{S} g'(\phi) \frac{\partial \phi}{\partial t} = \nabla^2 T \tag{10.147}$$

$$\tau(\mathbf{n}) \frac{\partial \phi}{\partial t} = [\phi - \lambda T (1 - \phi^2)](1 - \phi^2) + \nabla \cdot [\mathbf{\Lambda} \cdot \nabla \phi] \tag{10.148}$$

where $g(\phi) = \phi(10 - 15\phi + 6\phi^2)$ and $\mathbf{\Lambda}$ is a tensor of local materials properties and crystallographic orientations of the solid,

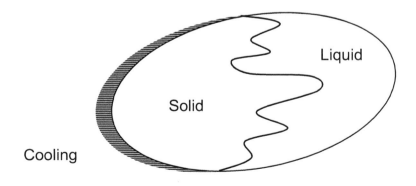

Figure 10.17. Schematic of a solidification problem

$$\Lambda = \begin{bmatrix} \Lambda_{11} & \Lambda_{12} & \Lambda_{13} \\ \Lambda_{21} & \Lambda_{22} & \Lambda_{23} \\ \Lambda_{31} & \Lambda_{32} & \Lambda_{33} \end{bmatrix} \tag{10.149}$$

with its components determined by the following expressions:

$$\Lambda_{ii} = W^2(\mathbf{n})\frac{\partial \phi}{\partial x_i}, \quad i=1,2,3$$

$$\Lambda_{12} = -W(\mathbf{n})\frac{\partial W(\mathbf{n})}{\partial \varphi}\csc^2\theta; \ \Lambda_{13} = W(\mathbf{n})\frac{\partial W(\mathbf{n})}{\partial \theta}\cos\varphi$$

$$\Lambda_{21} = W(\mathbf{n})\frac{\partial W(\mathbf{n})}{\partial \varphi}\csc^2\theta; \ \Lambda_{23} = W(\mathbf{n})\frac{\partial W(\mathbf{n})}{\partial \theta}\sin\varphi$$

$$\Lambda_{31} = -W(\mathbf{n})\frac{\partial W(\mathbf{n})}{\partial \theta}\sin\varphi; \ \Lambda_{32} = -W(\mathbf{n})\frac{\partial W(\mathbf{n})}{\partial \theta}\sin\varphi \tag{10.150}$$

Here $\mathbf{n}$ is the normal of the moving surface and (θ,φ) the Euler angles. The outward normal $\mathbf{n}$ and other parameters are related to the phase field parameter,

$$\mathbf{n} = \frac{\nabla\phi}{|\nabla\phi|}; \ \tau(\mathbf{n}) = \tau_0 a_s^2(\mathbf{n}), \ W(\mathbf{n}) = W_0 a_s(\mathbf{n}) \tag{10.151}$$

where τ_0 and W_0 are two parameters, and a_s is a complex function of the phase field parameter,

$$a_s(\mathbf{n}) = (1-3\varepsilon_4)\left(1 + \frac{4\varepsilon_4}{1-3\varepsilon_4}\frac{(\partial\phi/\partial x)^4 + (\partial\phi/\partial y)^4 + (\partial\phi/\partial z)^4}{|\nabla\phi|^4}\right) \tag{10.152}$$

where ε_4 is the anisotropic parameter. For a 2-D problem, the phase field equation reduces to the well known expression [44],

$$\frac{\bar{\varepsilon}^2}{m}\frac{\partial\phi}{\partial\tau} = \phi(1-\phi)[\phi - \tfrac{1}{2} + 30\bar{\varepsilon}\alpha ST\phi(1-\phi)]$$

$$+ \nabla\cdot(\varepsilon^2\nabla\phi) - \frac{\partial}{\partial x}(\varepsilon\varepsilon'\frac{\partial\phi}{\partial y}) + \frac{\partial}{\partial y}(\varepsilon\varepsilon'\frac{\partial\phi}{\partial x}) \tag{10.153}$$

In the above equation, $\varepsilon = \bar{\varepsilon}\,[1+\gamma_1\cos(4\theta)]$. Also, T represents the temperature, ϕ is the phase field, S is the supercooling temperature, $\bar{\varepsilon}$ is a mean value of ε which indicates interface thickness, γ_1 is the strength of anisotropy, and θ is the angle between the normal of the interface and the positive direction of x axis. Also, α and m are the dimensionless parameters related to the local real material properties such as interfacial energy, kinetic coefficient, specific heat, latent heat, thermal conductivity and the melting temperature. The detailed procedure to derive Equation 10.153 is given in Wang *et al.* [44].

10.7.4.2 Discontinuous Formulation

Equation 10.148 is basically a heat conduction equation, for which the discontinuous finite element solution was discussed in detail in Chapter 4. The matrix equation is the same as Equations 4.54–4.55. The only difference here is that the energy release due to solidification (i.e., the term associated with $g(\phi)$ in Equation 10.149 needs to be included as a source term, which is trivial once the evolution of phase field parameter is known.

The phase field equation, however, represents a rather complex nonlinear equation. To develop a discontinuous finite element formulation, the equation is split into the following first order differential equations:

$$\mathbf{R} = \boldsymbol{\Lambda} \cdot \nabla \phi \tag{10.154}$$

$$\tau(\mathbf{n}) \frac{\partial \phi}{\partial t} = Q(\phi, T) + \nabla \cdot \mathbf{R} \tag{10.155}$$

where $Q(\phi, T) = [\phi - \lambda T(1 - \phi^2)](1 - \phi^2)$.

The computational domain is now discretized into a tessellation of triangular finite elements in a 2-D geometry, or of tetrahedral elements in a 3-D geometry. Other shapes of elements are also possible. Following the procedures detailed in Section 4.5, the above two equations are multiplied by a pair of test functions $(\mathbf{w}, v)$ and integrated over element j,

$$\int_{\Omega_j} \mathbf{R} \cdot \mathbf{w} \, dV = \int_{\Omega_j} (\boldsymbol{\Lambda} \cdot \nabla \phi) \cdot \mathbf{w} \, dV \tag{10.156a}$$

$$\int_{\Omega_j} \tau(\mathbf{n}) \frac{\partial \phi}{\partial t} v \, dV = \int_{\Omega_j} (\nabla \cdot \mathbf{R} + Q(\phi, T)) v \, dV \tag{10.156b}$$

We now integrate by parts and replace fluxes at the element boundaries with the qualified numerical fluxes to obtain the following integral representation of the phase field model equations:

$$\int_{\Omega_j} \mathbf{R} \cdot \mathbf{w} \, dV = -\int_{\Omega_j} \phi \nabla \cdot (\boldsymbol{\Lambda} \cdot \mathbf{w}) dV + \int_{\partial \Omega_j} \hat{\phi} \mathbf{w} \cdot \boldsymbol{\Lambda} \cdot \mathbf{n}_j dS \tag{10.157a}$$

$$\int_{\Omega_j} \tau(\mathbf{n}) v \frac{\partial T}{\partial t} dV + \int_{\Omega_j} \mathbf{R} \cdot \nabla v \, dV = \int_{\Omega_j} Q(\phi, T) v \, dV + \int_{\partial \Omega_j} \hat{\mathbf{R}} \cdot \mathbf{n}_j v \, dS \tag{10.157b}$$

where $\mathbf{n}_j$ is the outward normal unit vector to $\partial\Omega_j$, the boundary of element j.

Approximating the unknowns using the interpolation functions, selecting appropriate numerical fluxes and applying the Galerkin procedure, we obtain a set of ordinary differential equations for the discretized values for the phase field parameter. These equations are summarized together with the equations for temperature,

$$\mathbf{U}_R = L_R(\underline{\phi}) \tag{10.158a}$$

$$\mathbf{M}_\phi \frac{d\underline{\phi}}{dt} = L_\phi(\underline{\phi}) \tag{10.158b}$$

$$\mathbf{U}_q = L_q(\underline{\mathbf{T}}) \tag{10.158c}$$

$$\mathbf{M}_T \frac{d\underline{\mathbf{T}}}{dt} = L_T(\underline{\mathbf{T}}) \tag{10.158d}$$

where L is the operator, $\underline{\phi}$ is the vector for nodal values of the phase field parameter ϕ, and $\mathbf{M}$ is the mass matrix. The subscripts refer to the relevant variables. The two unknown vectors $\mathbf{U}_q$ and $\mathbf{U}_R$ are defined as follows:

$$\mathbf{U}_q = \begin{pmatrix} \underline{\mathbf{q}}_x \\ \underline{\mathbf{q}}_y \\ \underline{\mathbf{q}}_z \end{pmatrix}; \ \mathbf{U}_R = \begin{pmatrix} \underline{\mathbf{R}}_x \\ \underline{\mathbf{R}}_y \\ \underline{\mathbf{R}}_z \end{pmatrix} \tag{10.159}$$

Equations 10.158a–d can be solved using an explicit time integration scheme. The computational procedure is given below:

(1) Given the initial and boundary conditions, $\mathbf{U}_R$ is obtained by solving Equation 10.158a;
(2) Using $\mathbf{U}_R$, Equation 10.158b is solved to obtain $\partial\underline{\phi}/\partial t$;
(3) Equation 10.158c is used to obtain $\mathbf{U}_q$;
(4) With $\mathbf{U}_R$ and $\partial\underline{\phi}/\partial t$, $\partial\underline{\mathbf{T}}/\partial t$ is obtained by solving Equation 10.158d;
(5) Advancing $\underline{\mathbf{T}}$ and $\underline{\phi}$ by $\underline{\mathbf{T}} = \underline{\mathbf{T}} + (\Delta t)\partial\underline{\mathbf{T}}/\partial t$ and $\underline{\phi} = \underline{\phi} + (\Delta t)\partial\underline{\phi}/\partial t$;
(6) Repeat steps (1) to (5) until convergence is achieved.

In performing time integration, the time step needs to be controlled to ensure numerical stability.

10.7.4.3 Numerical Examples
The above discontinuous finite element algorithm for coupled heat transfer and phase field distribution has been implemented to predict microstructure evolution

in solidification systems [45]. Numerical examples are presented here to illustrate certain features of the modeled microstructures. The examples include the 1-D, 2-D and 3-D calculations, with the 1-D data primarily used to test the accuracy of the phase field model.

Moving Boundary in 1-D Solidification. A 1-D solidification problem is useful for the purpose of checking the accuracy of the code and determining appropriate mesh sizes and other information used for adequate numerical simulations. The analytic solution to a 1-D problem is known [46]. A finite difference solution of the phase field model for 1-D solidification is also available in Fabbri and Voller [47]. For this problem, the initial temperature $T_0 = 0.015$, and the temperature at the cold end is $T_{cold} = -0.085$. To start the process, the temperature distribution and interface location are calculated by the analytic solution after a short period of solidification, $t_0 = 0.1846$ [47].

Figure 10.18 plots the temperature distribution at $t = 0.8$, from both the analytic solution and the numerical results obtained from the discontinuous Galerkin method described above. The CP and KP are two different versions of phase field models reported in the literature [4]. The mesh size is 0.004 and the time step is set to be 10^{-6}. From the figure, we can see that the KP model and the CP model render very similar results and both match well with the analytic solution [47].

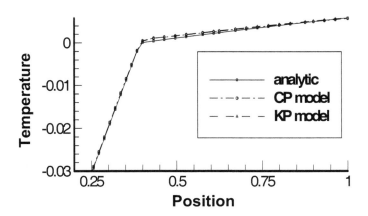

Figure 10.18. Comparison of 1-D solidification between analytic solution and finite element formulation

Microstructure During 2-D Solidification. Figure 10.19 illustrates the evolution of dendritic structures formed during solidification, the typical structure found in ice freezing from water. In this case, the growth starts from a circle of solid nucleus, with a radius of 0.1, located at the center of the domain. Growth gradually occurs as atoms are frozen from the adjacent liquid onto the solid from all the directions. The strength of anisotropy is 0.04, and six modes were used in this simulation. From these figures, it is clear that with an increase in latent heat, the solid crystal evolves from a hexagon shape to a snowflake shape

These results demonstrate that the discontinuous formulation of the phase field model is capable of modeling very complex moving boundary problems. In this sequence of figures, we see that the dendrite grows very fast in certain crystallographic orientations, of which ε has the maximum value and the fastest growth rate. Some small side dendritic tips grow from the main root branch. Finger-shaped dendrite growth is obtained with tip splitting. However, the needle-shaped crystal, which is demonstrated in Wheeler *et al.* [48], is not found in this calculation. The shape of dendritic tips can be determined by many parameters, such as the assumed interface thickness $\bar{\varepsilon}$, the strength of anisotropy γ, the supercooling parameter S, as well as the calculated domain and its spatial resolution.

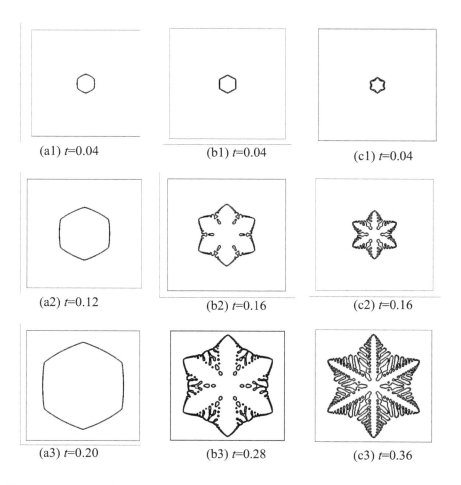

(a1) t=0.04   (b1) t=0.04   (c1) t=0.04

(a2) t=0.12   (b2) t=0.16   (c2) t=0.16

(a3) t=0.20   (b3) t=0.28   (c3) t=0.36

Figure 10.19. Dendrite growth under adiabatic conditions for various latent heat: $\Delta t = 5.0 \times 10^{-5}$, (a1)–(a3) $K = 0.8$; (b1)–(b3) $K = 1.2$; and (c1)–(c3) $K = 1.6$. K is non-dimensionalized latent heat

3-D Simulations. The discontinuous finite element phase field model is also applied to simulate a 3-D dendritic growth during solidification. As one might expect, these simulations are extremely computationally intensive because of very fine grids and the time resolutions required to obtain these fine features of microstructures. For this type of problem, the continuous finite element method would become rather inefficient, largely because of the huge global matrix formed. The discontinuous finite element method, however, does not require the assembly of a global matrix and it is thus less demanding for in-core memory. One of these structures obtained using the discontinuous finite element method is presented in Figure 10.20. The simulations used 200^3 linear elements.

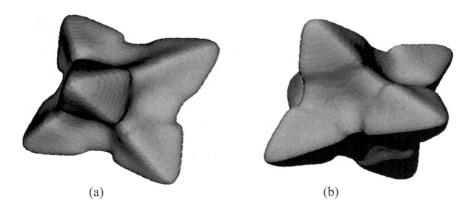

(a)   (b)

Figure 10.20. A 3-D view of microstructure of a single dendrite during solidification predicted by the phase field model: (a) viewed from $45°$ and (b) viewed from $135°$

10.7.5 Flow and Orientation Effects on Microstructure Evolution

Solidification involves complex local phenomena, which can be affected by various parameters of both liquid and solid during solid–liquid transitions and simulated using the phase field model. For instance, the formation of nuclei from the liquid phase can be simulated by introducing the noise term $\eta(\mathbf{r}, t)$ in Equation 10.138 [49]. Other effects such as fluid motion and crystal orientations on solidification microstructure formation can also be included in the phase field model. The discontinuous finite element method, being local in nature, presents a very powerful numerical tool for the phase field analysis of these problems. We present below the discontinuous finite element calculations of fluid flow and crystal orientation effects on the microstructure evolution during solidification.

10.7.5.1 Flow Effects on Microstructure Evolution
Under certain conditions during solidification, microstructure features such as dendrites branch into the surrounding liquid and are known to be affected by the liquid convection. The basic equations governing fluid flow, heat transfer and

phase field evolution during a dendritic solidification process were given in Section 10.7.2. These governing equations, that is, Equations 10.142–10.144 and (10.149), are solved using the discontinuous finite element method to study the effect of the fluid flow in the liquid pool on the microstructure formation.

Assuming that the liquid is incompressible, an algorithm for the solution may be described as follows. For the fluid flow and thermal equations, Equations 10.142–10.144 are solved using the discontinuous finite element methods presented in Chapter 6. The calculation of the phase field model, i.e., Equation 10.148, is the same as described in the previous section.

One of these calculations is given in Figure 10.21 for 2-D dendritic solidification for the non-dimensional times of 15, 66, and 96, with and without the fluid flow in the liquid pool considered. The shape of the dendrite is revealed by the phase field parameter. The left column shows the evolution of the dendrite with the liquid pool assumed to be quiescent at the three different times. The right column shows the velocity field around the dendrite and its effect on the morphology of the dendrite at the same times. Unlike the case of growth without flow, the temperature contours are not symmetric in all four directions, causing the dendrite to grow accordingly. The flow compresses the thermal boundary layer near the tip of the arm growing in the upstream direction while expanding it on the downstream side. The thermal boundary layer thickness near the tip of the dendrite arm perpendicular to the flow is not affected much by the flow. As a result of the smaller thermal boundary layer thickness near the tip of the upstream arm, and therefore higher temperature gradient, its growth rate is increased. The growth rate of the downstream arm, on the other hand, is reduced because of the lower temperature gradient there. The perpendicular arm is shifted slightly toward the flow direction with no significant effect on its growth rate. The higher temperature gradient on the upstream side also promotes the growth of side branches while on the downstream side the lower temperature gradient provides a more homogeneous temperature that inhibits the growth of side branches.

10.7.5.2 Microstructure Evolution During Polycystalline Solidification

The phase field model presented above assumes that a single grain grows into the liquid with and without being affected by local fluid flows. In many practical systems, grains with different orientations are nucleated and grow in a competitive environment. These grains will eventually meet and interact with each other. The evolution of a crystalline phase needs to include the physical effects of crystalline orientation, or of misorientation at grain boundaries. Developing numerical models to simulate these phenomena has received considerable attention recently and a comprehensive review on the subject and various models developed for this purpose is given by Granasy et al. [49].

The basic idea is to include extra terms in the free energy that are associated with the grain orientation effects. Various models have been proposed [48–50] and here we present the model for a pure material given by Warren et al. [50],

$$F = \int \left\{ \frac{1}{2} \xi^2 \Gamma^2 (\nabla \phi, \theta - \psi) + sg(\phi)|\nabla\theta| + \frac{1}{2}\varepsilon^2 h(\phi)|\nabla\theta|^2 + f(\phi,T) \right\} dV$$

(10.160)

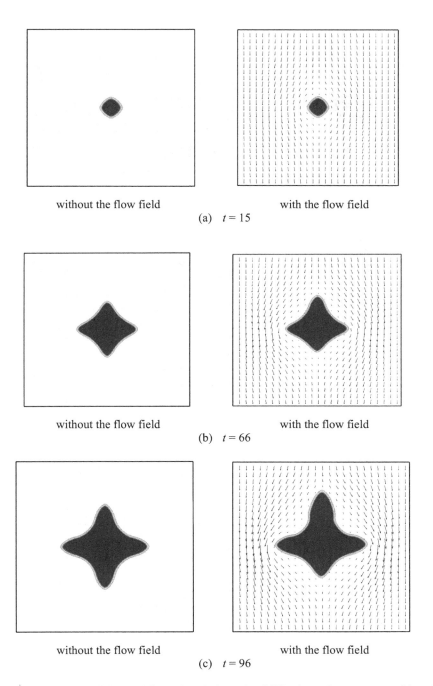

without the flow field   with the flow field

(a) $t = 15$

without the flow field   with the flow field

(b) $t = 66$

without the flow field   with the flow field

(c) $t = 96$

Figure 10.21. Phase field modeling of evolution of solidification microstructure with and without fluid flow

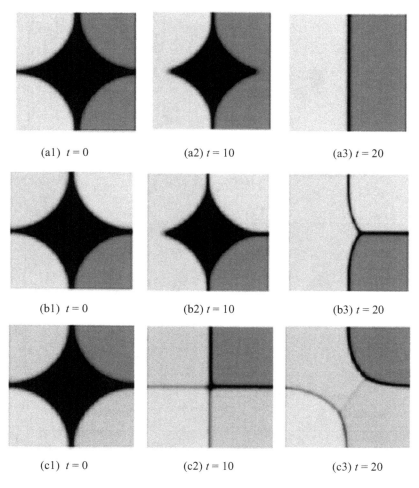

(a1) $t = 0$ (a2) $t = 10$ (a3) $t = 20$

(b1) $t = 0$ (b2) $t = 10$ (b3) $t = 20$

(c1) $t = 0$ (c2) $t = 10$ (c3) $t = 20$

Figure 10.22. Simulation of the impingement of four particles growing from the corners

where the homogeneous free energy density $f(\phi, T)$ is a double-well potential that has its local minima at the ϕ values corresponding to the solid ($\phi = 1$) and liquid ($\phi = 0$) phases, ξ is the gradient-energy coefficient related to the thickness of the interface, T is the temperature and Γ is the gradient penalty function [50].

The two terms (second and third under integral) have been added, which represent the energy cost of grain boundaries. Here, the parameter θ represents the local orientation measured with respect to a fixed axis of the crystal lattice. The values of θ span $-\pi/N < \theta \leq \pi/N$, where N is the rotational symmetry of the underlying crystal lattice. Also, ψ is defined as the direction of a normal to ϕ, i.e., $\tan(\psi) = (\partial\phi/\partial y)/(\partial\phi/\partial x)$, s and ε are the coupling constants, and g and h are specified as a function of ϕ^2. The monotonic nature of g and h is required if the effects of crystalline orientation are to be reduced or eliminated in the liquid phase.

By the same variational procedure given above, the following dimensionless governing equations for the evolution of phase parameter ϕ and crystal orientation θ are derived,

$$\tau_\phi \frac{\partial \phi}{\partial t} = \xi^2 \nabla^2 \phi + \phi(1 - \phi)m(\phi) - 2s\phi|\nabla \theta| - \varepsilon^2 \phi|\nabla \theta|^2 \tag{10.161}$$

$$P(\varepsilon|\nabla \theta|)\tau_\theta \phi^2 \frac{\partial \theta}{\partial t} = \nabla \cdot \left\{ \phi^2 \left[\frac{s}{|\nabla \theta|} + \varepsilon^2 \right] \nabla \theta \right\} \tag{10.162}$$

where $m(\phi) = \phi - 0.5 + \sigma\phi (1-\phi)$. The defintions and paremeters for calculations are given in [50].

Equations 10.161, 10.162 and 10.147 describe the effect of the crystallographic orientation of the solid grains, once formed during solidification, on the growth of the crystals. This set of equations may be solved using the discontinuous finite element method. For the purpose of demonstrating the effects of orientation only, the fluid flow is neglected. The impingement and coarsening of the grains as a change in orientation is displayed in Figures 10.22 and 10.23.

In Figure 10.22, three simulations of the impingement of four particles are plotted. In Figure 10.22a, the orientations of the two left grains are the same and differ by $\pi/4$ from the right gains. The right and left grains coalesce with each other, respectively. A grain boundary is formed along the vertical centerline. In Figure 10.22b, the orientations of the two left grains are the same but the two right grains have different orientations. It shows a dihedral angle after impingement. In Figure 10.22c, orientations of the four particles are all different. This leads to an unstable quadrijunction.

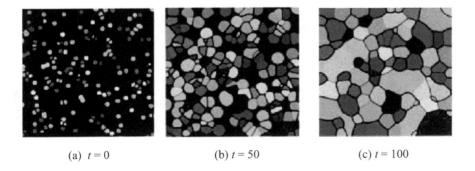

(a) $t = 0$ (b) $t = 50$ (c) $t = 100$

Figure 10.23. Simulation of the impingement and coarsening with many simultaneously introduced nuclei into an undercooled melt. Crystals with different orientations are obtained

Figure 10.23 shows an isothermal simulation of the impingement and coarsening with many simultaneously introduced nuclei into an undercooled melt. This is a case of multi-crystalline solidification, where grain boundary formation and grain coarsening are obvious.

Exercises

1. A local coordinate system erected at a surface point is given by (ξ, η, n) with their unit vectors defined as $\hat{\xi} = \xi_x \hat{i} + \xi_y \hat{j} + \xi_z \hat{k}$, $\hat{\eta} = \eta_x \hat{i} + \xi_y \hat{j} + \xi_z \hat{k}$ and $\mathbf{n} = \hat{n} = \eta_x \hat{i} + \xi_y \hat{j} + \xi_z \hat{k}$. Show that a transformation can be constructed such that

$$\begin{pmatrix} \dfrac{\partial}{\partial x} \\[2mm] \dfrac{\partial}{\partial y} \\[2mm] \dfrac{\partial}{\partial z} \end{pmatrix} = \begin{pmatrix} \xi_x & \eta_x & n_x \\ \xi_y & \eta_y & n_y \\ \xi_z & \eta_z & n_z \end{pmatrix} \begin{pmatrix} \dfrac{\partial}{\partial \xi} \\[2mm] \dfrac{\partial}{\partial \eta} \\[2mm] \dfrac{\partial}{\partial n} \end{pmatrix}$$

With the above transformation, further show that

$$\nabla = \frac{\partial}{\partial \xi}\hat{\xi} + \frac{\partial}{\partial \eta}\hat{\eta} + \hat{n}(\cdot\nabla) = \nabla_s + \mathbf{n}(\mathbf{n}\cdot\nabla)$$

where

$$\mathbf{n}(\mathbf{n}\cdot\nabla) = \mathbf{n}\frac{\partial}{\partial n} = \left(n_x i + n_y j + n_z k \right)\frac{\partial}{\partial n}$$

$$\nabla_s = \frac{\partial}{\partial \xi}\hat{\xi} + \frac{\partial}{\partial \eta}\hat{\eta}$$

With the additional relation,

$$0 = \frac{d(1)}{dn} = \frac{d(\mathbf{n}\cdot\mathbf{n})}{dn} = \mathbf{n}\cdot\frac{d\mathbf{n}}{dn} \text{ or } \frac{d\mathbf{n}}{dn} = \mathbf{n}\cdot\nabla\mathbf{n} = 0$$

show that

$$\nabla_s \cdot\mathbf{n} = \nabla\cdot\mathbf{n}|_S - \mathbf{n}(\mathbf{n}\cdot\nabla\mathbf{n}) = \nabla\cdot\mathbf{n}|_S$$

where S means evaluating the integral at the surface and ∇_s is the surface vector differential operator.

2. Consider a curved surface defined in a spherical coordinate system,

$$F = F(x, y, z) = F(R, \theta, \varphi) = R - r(\theta, \varphi) = 0$$

Show that the spatial gradient of F is given by

$$\nabla F = \hat{i}_r \frac{\partial F}{\partial R} + \hat{i}_\theta \frac{1}{R} \frac{\partial F}{\partial \theta} + \hat{i}_\varphi \frac{1}{R\sin\theta} \frac{\partial F}{\partial \varphi} = \hat{i}_r - \frac{\hat{i}_\theta}{R} \frac{\partial r}{\partial \theta} - \frac{\hat{i}_\varphi}{R\sin\theta} \frac{\partial r}{\partial \varphi}$$

where "^" denotes the unit vector.
Show that the tangential vectors in the polar and azimuthal directions for this surface are given by

$$t_\theta = \frac{r_\theta \hat{i}_r + r\hat{i}_\theta}{\sqrt{r^2 + r_\theta^2}}; \qquad t_\varphi = \frac{r_\varphi^2 \hat{i}_r + r\sin\theta\, \hat{i}_\varphi}{\sqrt{r^2 \sin^2\theta + r_\varphi^2}}$$

from which the normal vector can also be calculated,

$$\mathbf{n} = \mathbf{t}_\theta \times \mathbf{t}_\varphi$$

Show that this gives the same expresion as calculated by taking the spatial gradient of $\mathbf{n}$ as follows:

$$\mathbf{n} = \left.\frac{\nabla F}{|\nabla F|}\right|_{F=0} = \frac{1}{\sqrt{1 + \left(\frac{1}{r}\frac{\partial r}{\partial \theta}\right)^2 + \left(\frac{1}{r\sin\theta}\frac{\partial r}{\partial \varphi}\right)^2}} \left(\hat{i}_r - \frac{\hat{i}_\theta}{r}\frac{\partial r}{\partial \theta} - \frac{\hat{i}_\varphi}{r\sin\theta}\frac{\partial r}{\partial \varphi}\right)$$

Show that the curvature for the 3-D surface can be calculated by taking either the spatial gradient evaluated at $F = 0$ or the surface gradient, that is,

$$\frac{1}{R_1} + \frac{1}{R_2} = \left.\nabla \cdot \mathbf{n}\right|_{F=0} \qquad \text{or} \qquad \frac{1}{R_1} + \frac{1}{R_2} = \nabla_s \cdot \mathbf{n}$$

From the above curvature result, show that if the surface is axially symmetric, that is, independent of φ, then

$$\frac{1}{R_1} + \frac{1}{R_2} = \frac{1}{r^2 \sin\theta} \left(\frac{(2r^2 + r_\theta^2)\sin\theta}{\sqrt{r^2 + r_\theta^2}} - \frac{d}{d\theta}\left(\frac{r r_\theta \sin\theta}{\sqrt{r^2 + r_\theta^2}} \right) \right)$$

3. Consider a 3-D surface expressed in the spherical coordinates. Assume that the surface has a 3-D deformation given by the following expression,

$$r = a\left(1 + \left(\xi_{em} + R_{em}\right)Y_e^m(\Omega)\right)$$

where Y is the spherical harmonics and Ω is the solid angle.

Show that with the definition of F,

$$F = r - a\left(1 + (\xi_{em} + R_{em})Y_e^m(\Omega)\right) = 0$$

the surface normal $\mathbf{n}$ and curvature H for this deformed surface are calculated, to the first order in $(\xi_{em} + R_{em})$, by the following expressions:

$$\mathbf{n} = \frac{\hat{i}_r - \hat{i}_\theta \dfrac{a(\xi_{em}+R_{em})}{a\left(1+(\xi_{em}+R_{em})Y_e^m(\Omega)\right)} \dfrac{\partial Y_e^m}{\partial\theta} - \hat{i}_\varphi \dfrac{a(\xi_{em}+R_{em})}{a\left(1+(\xi_{em}+R_{em})Y_e^m(\Omega)\right)\sin\theta} \dfrac{\partial Y_e^m(\Omega)}{\partial\theta}}{\sqrt{1+\left(\frac{1}{r}\frac{\partial F}{\partial\theta}\right)^2 + \left(\frac{1}{Y\sin\theta}\frac{\partial F}{\partial\varphi}\right)^2}}$$

$$\cong \hat{i}_r - \hat{i}_\theta(\xi_{em}+R_{em})\frac{\partial Y_e^m}{\partial\theta} - \hat{i}_\varphi \frac{(\xi_{em}+R_{em})}{\sin\theta}\frac{\partial Y_e^m}{\partial\varphi}$$

and

$$2H \cong \frac{2}{r} + \frac{(\xi_{em}+R_{em})}{a}\left[\frac{1}{\sin\theta}\frac{\partial}{\partial\theta}\left(\sin\theta\frac{\partial}{\partial\theta}\right)+\frac{1}{\sin^2\theta}\frac{\partial^2}{\partial\varphi^2}\right]Y_e^m(\Omega)$$

$$\cong \frac{2}{a} + \frac{(\xi_{em}+R_{em})}{a}Y_e^m(\Omega)(e(e+1)-2)$$

4. Consider a solidification change problem that involves the motion of both solid and liquids, with $\mathbf{u}_s$ and $\mathbf{u}_l$ denoting the solid and liquid velocities, respectively. Show that for this case, the velocities of the liquid and the solid relative to the interface are given by the following expressions:

$$\mathbf{u}_s' = \mathbf{u}_s - \mathbf{u}_i ; \quad \mathbf{u}_l' = \mathbf{u}_l - \mathbf{u}_i$$

where $\mathbf{u}_i$ is the interface velocity. Show that the mass balance and energy balance across the interface are expressed by the following expression:

$$(\rho_s \mathbf{u}_s')\cdot\mathbf{n}_s + (\rho_l \mathbf{u}_l')\cdot\mathbf{n}_l = 0$$

and

$$\left(\rho_s \underline{H}_s \mathbf{u}_s' + \mathbf{q}_s\right)\cdot\mathbf{n}_s + \left(\rho_l \underline{H}_l \mathbf{u}_l' + \mathbf{q}_l\right)\cdot\mathbf{n}_l = 0$$

where $\underline{H}$ is the enthalpy and $\mathbf{n}_s$ points from the solid to the liquid. Let $L = \underline{H}_l - \underline{H}_s$, where L is the latent heat per unit mass, that is, the heat related when liquid converts into solid. Show further that the energy balance across the interface can be written as

$$(\mathbf{q}_l - \mathbf{q}_s)\cdot\mathbf{n} = \rho_s L \mathbf{u}_s'\cdot\mathbf{n} = -\rho_l L \mathbf{u}_l'\cdot\mathbf{n}$$

Show that the above equation reduces to Equation 10.44 if $\mathbf{u}_s = \mathbf{u}_l = 0$.

5. Integrate Equation 10.89 across a pillbox across a moving interface between two fluids and show that the resulting equation reduces to Equation 10.41, that is, the stress balance condition at the interface, as the size of the pillbox reduces to zero.

6. Starting with the Clausius–Clapeyron relation,

$$\frac{dp_{flat}}{dT_i} = \frac{\Delta s}{\Delta v} dT_i$$

show that for constant properties, the integration of the above relation yields the following relation for a material undergoing solidification:

$$\frac{(p_l - p_s)}{\rho_l^{-1} - \rho_s^{-1}} = L\left(\frac{T_i}{T_m} - 1\right) + \Delta C_p\left(T_i \ell n\left(\frac{T_i}{T_m}\right) + T_m - T_i\right) + \frac{2H\gamma}{\rho_s}$$

where $\Delta v = \rho_l^{-1} - \rho_s^{-1}$, T_m is the melting temperature, $\Delta s = s_l - s_s$, H the curvature, and T_i is the interface temperature.

Further show that if $\Delta C_p = 0$ and $\rho = \rho_l = \rho_s$, the above equation reduces to the classical Gibbs–Thompson relation,

$$T_i = T_m - \frac{2HT_m\gamma}{\rho L} = T_m - 2H\Gamma_G$$

For the above case, show that the interfacial energy balance yields the following relation:

$$k_s \mathbf{n} \cdot \nabla T_s - k_l \mathbf{n} \cdot \nabla T_l = \rho\big(L - \Delta C_p(T_m - T_i)\big)u_i \cdot \mathbf{n}$$

Clearly, for $\Delta C_p = 0$, we have the same relation as Equation 10.44.

7. A solid at the solidification (or melting) temperature T_m is confined to a half-space $x > 0$. At time $t = 0$, the temperature of the boundary surface at $x = 0$ is raised to T_0, which is higher than T_m and maintained at the temperature for times $t > 0$. As a result, melting starts at the surface $x = 0$ and the solid–liquid interface moves in the positive x direction. Write down the governing equations and boundary conditions for this problem and obtain the analytical solution for the problem using the method of separation of variables.

8. Develop a discontinuous finite element code to calculate the temperature distribution and phase model, and compare the numerical solution and analytic solution derived in Exercise 7.

9. Develop a discontinuous finite element code and incorporate the volume of fluid algorithm for free surface calculations. Compare the results with and without the global balance adjustment.

10. Modify the code developed for Problem 8 and use the level set method to track the surface. Apply the code to solve a boiling problem and compare the results with those reported in [38].

11. Develop a discontinuous finite element code for a 2-D simulation of solidification using the phase field model given in Section 10.7.4.

References

[1] Floryan JM, Rasmussen H. Numerical Methods For Viscous Flow with Moving Boundaries. Appl. Mech. Rev. 1989; 42: 323–40.

[2] Scardovelli R, Zaleski S. Direct Numerical Simulation of Free-Surface and Interfacial Flow. Annu. Rev. Fluid Mech. 1999; 31: 567–603.

[3] Sethian JA, Smereka P. Level Set Methods for Fluid Interfaces. Annu. Rev. Fluid Mech. 1999; 31: 567–603.

[4] Anderson DM, McFadden GB. Diffuse-Interface and Phase-Field Modeling. Phys. Fluids 1997; 9: 1870–1879.

[5] Weatherburn CE. Differential Geometry of Three Dimensions. London: Cambridge University Press, 1972.

[6] Aris R. Vectors, Tensors and the Basic Equations of Fluid Mechanics. New York: Dover Publications, 1962.

[7] Scriven LE. Dynamics of a Fluid Interface. Chem. Eng. Sci.1960; 12: 98–108.

[8] Saito H, Scriven LE. Study of Coating Flow by the Finite Element Method. J. Comput. Phys.1981; 42: 53–76.

[9] Popinet S, Zaleski S. A Front-Tracking Algorithm for Accurate Representation of Surface Tension. Int. J. Numer. Meth. Fluids 1999; 30: 775–793.

[10] Batchelor GB. An Introduction to Fluid Fynamics. London: Cambridge University Press, 1967.

[11] Flemmings MC. Solidification Processing. New York: McGraw-Hill, 1974.

[12] Donea J, Giuliani S, Halleux JP. An Arbitrary Lagrangian–Eulerian Finite Element Method for Transient Dynamic Fluid-Structure Interactions. Comput. Meth. Appl. Mech. Eng. 1982; 33: 689–723.

[13] Duarte F,Gormaz R, Natesan S. Arbitrary Lagrangian–Eulerian Method for Navier–Stokes Equations with Moving Boundaries. Comput. Meth. Appl. Mech. Eng. 2004; 193: 4819–4836.

[14] Bathe KJ, Zhang H. Finite Element Analysis of Fluid Flows Fully Coupled with Structural Interactions. Int. J. Numer. Meth. Fluids 2004; 60: 213–232.

[15] Peterson RC, Jimack PK, Kelmason MA. The Solution of Two-Dimensional Free-Surface Problems Using Automatic Mesh Generation. Int. J. Numer. Meth. Fluids 1999; 31: 937–960.

[16] Englemann MS, Sani RL. Finite Element Simulation of Incompressible Flows with a Free/Moving Surface. In Numerical Methods in Laminar and Turbulent Flows, London: Pineridge Press, 1984.

[17] Orr FM, Scriven LE. Rim Flows: Numerical Simulation of Steady, Viscous Free-Surface Flows with Surface Tension. J. Fluid Mech. 1978; 84: 145–165.

[18] Li K, Li BQ. A 3-D Model for g-Jitter Driven Flows and Solidification in Magnetic Fields. J. Thermophys. Heat Transf. 2003; 17: 498–508.

[19] Brown RA, Ettoney HM. Finite Element Methods for Steady Solidification Problems. J. Comput. Phys. 1983; 49: 118–150.

[20] Li K, Li BQ. Numerical analyses of g-jitter induced convection and solidification in magnetic fields. J. Thermophys. Heat Transf. 2003; 17: 199–209.

[21] Hirt CW, Nichols BD. Volume of Fluid (VOF) Method for the Dynamics of Free Boundaries. J. Comput. Phys. 1981; 39: 201–216.

[22] DeBar R. Fundamentals of the KRAKEN Code. Tech. Rep. UCIR-760. Lawrence Livermore Nat. Laboratories: 1974.

[23] Parker BJ, Youngs DL. 1992. Two and Three Dimensional Eulerian Simulation and Fluid Flow with Material Interfaces. Tech. Rep. Atomic Weapons Establishment, UK.

[24] Rider WJ, Kothey DB. Reconstructing Volume Tracking. J. Comput. Phy. 1998; 141: 112–152.

[25] Scardovelli R, Zaleski S. Interface Reconstruction with Least-Square Method and Split Eulerian–Lagrangian Advection. Int. J. Numer. Meth. Fluids 2003; 41:251–274.

[26] Ashgriz N, Barbat T, Wang G. A computational Lagrangian–Eulerian advection remap for free surface flows. Int. J. Numer. Meth. Fluids 2004; 44:1–32

[27] Shahbazi K, Paraschivoiu M, Mostaghimi J. Second order accurate volume tracking based on remapping for triangular meshes. J. Comput. Phy. 2003; 188:100–122.

[28] Huo Y, Ai X, Li BQ. Computation and Visualizaion of 3-D Marangoni and Magnetically-Driven Flows in Droplets. ASME Winter Meeting, Washington DC, 2003; Paper #: IMECE2003–42822.

[29] Zhang C. Data Visualization and Mesh Generation for Finite Element Simulation of Fluid Flow and Heat Transfer. MS Thesis. Pullman: Washington State University, 2001.

[30] Preparata FP, Shamos MI. Computational Geometry: an Introduction. New York: Springer, 1985.

[31] Harlow FH, Welch JE, Numerical Calculation of Time-Dependent Viscous Incompressible Flow of Fluid with Free Surface. Phys. Fluids 1965; 8: 2182–2189.

[32] Chen S, Johnson DB, Raad PE, Fadda D. The Surface Marker and Micro Cell Method. Int. J. Numer. Meth. in Fluids 1997; 25: 749–778.

[33] Iafrati A, Mascio AD, Campana EF. A Level Set Technique Applied to Unsteady Free Surface Flows. Int. J. Numer. Meth. Fluids 2001; 35: 281–297.

[34] Hu C, Shu C-W. A Discontinuous Galerkin Finite Element Method for Hamilton–Jacobi Equations. SIAM J. Sci. Comput. 1999; 21(2): 666–690.

[35] Sussman M, Smereka P, Osher S. A Level Set Approach for Computing Solutions to Incompressible Two-Phase Flow. J. Comput. Phys. 1994; 114: 146–159.

[36] Voller VR, Brent AD, Prakash C. The Modelling of Heat, Mass and Solute Transport in Solidification Systems. Int. J. Heat Mass Transf. 1989; 32(9): 1719–1731.

[37] Welch SWJ, Wilson J. A Volume of Fluid Based Method for Fluid Flows with Phase Change. J. Comput. Phy. 2000; 160: 662–682.

[38] Son G, Dhir AK. Numerical Simulation of Film Boiling Near Critical Point with a Level Set Method. ASME J. Heat Transf. 1998; 120: 183–192.

[39] Dalhuijsen AL, Segal A. Comparison of Finite Element Techniques for Solidification Problems. Int. J. Numer. Meth. Eng. 1986; 23: 1807–`829.

[40] Goldenfeld N. Lecture on Phase Transition and the Renormalization Group. Reading MA: Addison-Wesley, 1992.

[41] Isihara A. Statistical Physics. New York, Academic Press, 1971.

[42] Kubo R, Toda M, Hashitsume N. Statistical Physics II, Nonequilibrium Statistical Mechanics. Berlin: Spring-Verlag, 1991.

[43] Ala-Nissila T, Majaniemi S, Elder K. Phase-Field Modeling of Dynamical Interface Phenomena in Fluids. Lect. Notes Phys. 2004; 640: 357–388.

[44] Wang SL, Sekera RF, Wheeler AA, Murray BT, Coriell SR, Braun RJ, McFadden GB. Thermodynamically Consistent Phase-Field Models for Solidification. Physica D: Nonlinear Phenomena. 1993; 69: 189–200.

[45] Shu Y, Ai X, Li BQ. Phase Field Modeling of Solidification Processes. IMECE (ASME International Mechanical Engineering Congress and Exposition) 2004; Paper #: IMECE2004–59565.

[46] Özisik MN. Heat Conduction. New York: Wiley, 1980.

[47] Fabbri M, Voller VR. The Phase-Field Method in the Sharp-Interface Limit: A Comparison Between Model Potentials. J. Comput. Phys. 1997; 130: 256–265.

[48] Wheeler AA, Murray BT, Schaefer RJ. Computation of Dendrite Using a Phase-Field. Model. Physica D: Nonlinear Phenomena. 1993; 66: 243–262.

[49] Granasy L, Pusztai T, Warren JA. Modeling Polycrystalline Solidification Using Phase Field Theory. J. Phys. Condens. Matter. 2004; 16: R1205–R1235.

[50]   Steinbach I, Pezzolla F, Nestler B, Seeßelberg M, Prieler R, Schmitz GJ, L. Rezende JL. A Phase Field Concept for Multiphase Systems. Physica D: Nonlinear Phenomena. 1996; 94(3): 135–147.
[51]   Warren JA, Kobayashi R, Lobkovsky AE, Craig W. Extending Phase Field Models of Solidification to Polycrystalline Materials. Acta Mat. 2003; 51: 6035–6058.

11

Micro and Nanoscale Fluid Flow and Heat Transfer

Recently, there has been much interest in understanding micro/nanoscale fluid flow and heat transfer phenomena. This interest is largely driven by rapid advances in micro devices for microelectronic, microeletromechanical and biomedical applications. Two recent monographs have been devoted to the topic of fluid flow in micro and nano channels and structures, and that of microscale heat transfer in rapid thermal laser processing of thin films, respectively [1, 2].

Studies show that as the time and length scales are reduced, some of the assumptions used or implied in the macroscopic description of the fluid flow and heat transfer phenomena based on the continuum theorem may become invalid and modifications are needed to improve the mathematical description. For the cases of gas flow in microchannels, for example, the near-wall analysis suggests that the no-clip condition, which has been taken for granted in macroscopic fluid mechanics, is no longer valid. The collision of the molecules with the walls results in a slip of the molecules along the solid wall. There also have been attempts to directly apply the Boltzmann transport equation to describe the fluid flow phenomena at microscales.

One of the widely observed phenomena associated with the microscale thermal transport is the phase-lag behavior, which occurs in thin films under irradiation by a pulsing laser. When the pulse duration and the length scale are reduced to a level comparable with the mean free path of the phonons, the temperature gradient and the heat flux no longer travel at the same speed, making invalid the equal-speed assumption implied in the classical Fourier law for heat conduction. An appropriate description of the thermal phenomena at microscales thus requires the modification of the existing heat transfer equations.

Research indicates that a mathematical description of microscale heat transfer phenomena may be very similar to that for macro phenomena with relevant modifications for certain specific applications and that the description can be vastly different for other conditions.

In this chapter, we focus on two classes of mathematical description of fluid flow and heat transfer problems at micro and nanoscales for which the discontinuous finite element methods are particularly suited. The first class is based on the modification of the existing macroscopic theory for microscopic description, whereas the second class is concerned with a direct description of microscopic phenomena on the basis of the Boltzmann transport theorem. Relevant background information on the fundamentals for these two approaches is provided. The numerical solution of the relevant governing equations arising from these two classes of mathematical description by the discontinuous finite element method is discussed, along with numerical examples for microscopic heat transfer in laser annealing of thin films and for the lattice Boltzmann solution of Taylor vertex flows and flow over a cylinder.

11.1 Microscale Heat Conduction

The classical heat conduction equation and its discontinuous formulation were presented in Chapter 4. An important assumption embedded in the equation is that the temperature gradient and heat flux propagate at the same speed in the media. As the scales, either spatial or temporal or both, become smaller, the local level phonon–phonon interaction, the electron–phonon interaction, and the phonon–photon interaction may become significant and must be specifically taken into consideration. These interactions can be especially important when the spatial scales are on the order of the mean-free path of phonons, and the time scales are on the order of relaxation times characterizing these microscale interactions. For metals, the relaxation time is around the order of a picosecond, whereas for dielectric crystals and insulators, the relaxation time is on the order of nanoseconds to picoseconds. These fast transient effects may dramatically change the heat transfer phenomena at a local level and it is known that the classical heat conduction models are inadequate to describe the microscale interactions [2].

Different models have been proposed to characterize these interactions by adding additional terms to the classical heat conduction equation. For metals, the electron–phonon interaction is dominant and is modeled by the electron-gas model. By this model, the electron gas and metal lattice are heated up in the medium by a mechanism that involves the excitation of electron gas, and heating of the metal lattice through electron–phonon interaction in short time increments. For semiconductors, insulators and dielectric materials, however, the reference to free electrons is not applicable, because the electrons are bound more strongly to the lattices. The dominant local effect of these materials is phonon scattering and collision, and the contribution from electron gas is often neglected. The popular model describing these phonon–phonon interactions is the phase-lag model. Phonon interaction may also be described by the phonon radiative transfer model, which assumes that the phonons propagate in the medium following an equation similar to that of radiative heat transfer.

11.1.1 Two-temperature Equations

The photon–electron interaction is a dominant model for heat transport in metals. In the free electron model, the electrons are modeled as a gas and frequently collide with the lattices [3]. The heating mechanism involves two steps: the excitation of the electron gas, and the heating of the lattice through phonon–electron collisions in short time increments. The two-temperature model for electron–phonon interaction involves two sub-models, with one for the temperature of the electron gas and the other for the temperature of the lattice. Mathematically, these processes are described by the following two equations [4]:

$$C_e \frac{\partial T_e}{\partial t} = \nabla \cdot k \nabla T_e - G(T_e - T_l) \tag{11.1}$$

$$C_l \frac{\partial T_l}{\partial t} = G(T_e - T_l) \tag{11.2}$$

where the subscripts e and l refer to the electron gas and metal lattice, respectively. Also, C is the volumetric specific heat, and G is the electron–phonon coupling factor determined by

$$G = \frac{\pi^4 (nvk_b)^2}{18k} \tag{11.3}$$

where n is the number density of the electron gas, k_b is the Boltzmann constant, and v is the speed of sound. The values of G for various metallic materials are given in [2].

The above two equations can be combined to produce a single temperature equation either in terms of the electron gas temperature T_e or the lattice temperature T_l. In terms of the lattice temperature, the following single temperature is obtained:

$$\nabla^2 T_l + \left(\frac{C_e}{G}\right) \frac{\partial \nabla^2 T_l}{\partial t} = \left(\frac{C_e + C_l}{k}\right) \frac{\partial T_l}{\partial t} + \left(\frac{C_e C_l}{Gk}\right) \frac{\partial^2 T_l}{\partial t^2} \tag{11.4}$$

The electron gas temperature is then calculated by $T_e = T_l + (C_l/G)\partial T_l/\partial t$.

From the computational point of view, the two-temperature model may be more efficient, and indeed it is particularly suitable for treatment by the discontinuous finite element formulation [5, 6].

11.1.2 Phonon Scattering Equation

In dielectrics, electrons are bound and not able to move freely in the structure. The thermal transport is dominated by phonon collision and scattering. The popular model describing the heat transfer by the phonon–phonon interaction is the phase lag model, which consists of two equations,

$$C \frac{\partial T}{\partial t} + \nabla \cdot \mathbf{q} = Q(\mathbf{r}, t) \tag{11.5}$$

$$\frac{\partial \mathbf{q}}{\partial t} + \frac{c^2 C}{3} \nabla T + \frac{1}{\tau_R} \mathbf{q} = \frac{\tau_N c^2}{5} \left[\nabla^2 \mathbf{q} + 2\nabla(\nabla \cdot \mathbf{q}) \right] \tag{11.6}$$

where τ_R is the relaxation time for the momentum-nonconserving process, τ_N is the relaxation time for normal momentum-conserving process, C is the volumetric specific heat and c is the speed of sound (or phonon) in the solid. The first equation is the thermal balance equation for solids; Q is the external heating source. The second equation is a constitutive relation linking the heat flux $\mathbf{q}$ to the temperature gradient ∇T, and the equation is derived from the generalized phonon thermal conductivity relation. The two relaxation times represent the microscopic effects on the heat transfer in solids.

The above equations may be combined by eliminating the heat flux, and with Q set to 0, the resulting equation is written in terms of temperature,

$$\nabla^2 T + \frac{9\tau_N}{5} \frac{\partial \nabla^2 T}{\partial t} = \frac{3}{\tau_R c^2} \frac{\partial T}{\partial t} + \frac{3}{c^2} \frac{\partial^2 T}{\partial t^2} \tag{11.7}$$

which has the same form as the equation for the lattice temperature in the two-temperature model. Comparison of Equation 11.4 to the preceding equation shows that the two equations have the same form, though the microscopic mechanisms are different. Equation 11.7 sometimes is referred to as the thermal equation. It is seen that with both τ_R and τ_N set to zero, but $\tau_R c^2$ being finite, we easily recover the classical heat conduction equation from Equation 11.7.

Equations 11.5 and 11.6 are related to the dual-phase-lag model, which treats the temperature gradient and heat flux in a cause–effect relation. Mathematically, the model assumes the following form:

$$\mathbf{q}(\mathbf{r}, t + \tau_q) = -k\nabla T(\mathbf{r}, t + \tau_T) \tag{11.8}$$

The first-order expansion of the above equation in both τ_q and τ_T gives the following equation:

$$\mathbf{q}(\mathbf{r}, t) + \tau_q \frac{\partial \mathbf{q}(\mathbf{r}, t)}{\partial t} = -k\nabla T(\mathbf{r}, t) - k\tau_T \frac{\partial \nabla T(\mathbf{r}, t)}{\partial t} \tag{11.9}$$

Equations 11.5 and 11.9 may be combined to obtain a single equation for temperature distribution,

$$\nabla^2 T + \tau_T \frac{\partial \nabla^2 T}{\partial t} + \frac{1}{k} \left[Q + \tau_q \frac{\partial Q}{\partial t} \right] = \frac{1}{\alpha} \frac{\partial T}{\partial t} + \frac{\tau_q}{\alpha} \frac{\partial^2 T}{\partial t^2} \tag{11.10}$$

From Equations 11.9 and 11.10, the roles of τ_q and τ_T can be identified. Basically, τ_q and τ_T are two characteristic times or phase lags that relegate the behavior of heat flux and temperature at the microscale. They represent the properties of materials, reflecting their internal structure responses to the applied conditions. If $\tau_q < \tau_T$, the heat flux travels faster and is the cause for the temperature gradient, which is the effect, and travels slower. If $\tau_q > \tau_T$, however, the temperature gradient precedes the heat flow, or the temperature gradient is the cause, and heat flow is the effect. If $\tau = \tau_q = \tau_T$, then the heat flow and temperature gradient travel at the same speed and we recover the classical heat conduction equation. To see that, we set $\tau = \tau_q = \tau_T$ in Equation 11.10 to obtain

$$\left[\nabla^2 T - \frac{1}{\alpha} \frac{\partial T}{\partial t} + \frac{Q}{k} \right] + \tau \frac{\partial}{\partial t} \left[\nabla^2 T - \frac{1}{\alpha} \frac{\partial T}{\partial t} + \frac{Q}{k} \right] = 0 \qquad (11.11)$$

where $\alpha = k/C$ is the thermal conductivity. The above equation is satisfied by the following solution:

$$\frac{1}{\alpha} \frac{\partial T}{\partial t} = \nabla^2 T + \frac{Q}{k} \qquad (11.12)$$

which is nothing but the classical heat conduction equation. Mathematically, the equation is parabolic in nature, of which the discontinuous finite element solution was discussed in Chapter 4.

If $\tau_T = 0$ and $Q = 0$, Equation 11.10 is simplified as the thermal wave equation hypothesized by Morse and Feshback [2],

$$\nabla^2 T = \frac{1}{\alpha} \frac{\partial T}{\partial t} + \frac{\tau_q}{\alpha} \frac{\partial^2 T}{\partial t^2} \qquad (11.13)$$

The equation is hyperbolic in nature and its solution displays the wave behavior. Here the temperature disturbance propagates as a wave in a medium, and the thermal diffusivity appears as a damping effect in heat propagation. This is in contrast with the diffusive nature associated with the classical Fourier heat conduction equation. This is expected, in that when a cause–effect relation does not exist, or the temperature gradient and heat flux travel at the same speed, diffusion becomes the dominant mode for spreading the temperature distribution.

Tzou [2] further shows that the relaxations are related to the two-temperature model constants and the phonon scattering constants,

$$\frac{\tau_T}{2\tau_q} = \begin{cases} \dfrac{9}{5} \dfrac{\tau_N}{2\tau_R} & \text{(phonon scattering)} \\[2ex] \dfrac{1}{2}\left[1 + \dfrac{C_l}{C_g} \right] & \text{(phonon} - \text{electron intereaction)} \end{cases} \qquad (11.14)$$

In this way, the two phase-lag model includes both the two-temperature model and the thermal wave model.

11.1.3 Phonon Radiative Transfer Equation

This model is obtained by solving the Boltzmann equation for an acoustically thin medium, in which the phonon mean free path is smaller than, or comparable to, the thickness of the film. The starting point is the 1-D Boltzmann equation with the relaxation approximation,

$$\frac{\partial f}{\partial t} + v_x \frac{\partial f}{\partial x} = -\frac{1}{\tau} f + \frac{1}{\tau} f^{eq} \tag{11.15}$$

where f is the distribution function for phonons with a vibrating frequency ω, v_x is the 1-D phonon velocity, and t is the relaxation time. Also, superscript eq represents the equilibrium state.

The phonon intensity I in heat transport is obtained by summing up the distribution function over the three phonon polarizations p,

$$I(\omega, x, t, \Omega) = \sum_p v(\Omega) f(\omega, x, t) h\omega D(\omega) \tag{11.16}$$

where v is the speed of phonons in the direction defined by the solid angle Ω, h is the Planck constant and $D(\omega)$ is the density of states per unit volume in the frequency domain of lattice vibrations. The projection of the velocity to the x axis is $v_x = v(\Omega)\cos(\theta) = v(\Omega)\mu$. Note that use of f^{eq} in Equation will give I^{eq}.

Substituting Equation 11.16 into Equation 11.15 and carrying out the algebra, we have the phonon radiative transfer equation,

$$\frac{1}{v}\frac{\partial I}{\partial t} + \mu \frac{\partial I}{\partial x} + \frac{1}{v\tau} I = \frac{1}{2v\tau}\int_{-1}^{+1} I \, d\mu \tag{11.17}$$

which was first derived by Majumdar [7]. The above equation can be solved using the methods discussed in Chapter 9. The heat flux and internal energy at any point in space may be calculated by the following expressions (see also Chapter 9):

$$q = \int_{4\pi}\int_0^{\omega_D} I\mu \, d\omega d\Omega \,; \; E = \int_{4\pi}\int_0^{\omega_D} I v^{-1} \, d\omega d\Omega \tag{11.18}$$

where $d\Omega = \sin\theta d\theta d\varphi$ is the differential solid angle and ω_D is the Debye cutoff phonon frequency. The two terms are related by the 1-D energy balance equation,

$$\frac{\partial E}{\partial t} + \frac{\partial q}{\partial x} = 0 = 2\pi \int_{-1}^{+1}\int_0^{\omega_D} \frac{I^{eq} - I}{v\tau} d\omega d\mu \tag{11.19}$$

From the above equation, I^{eq} can be expressed as an integral of I because $I^{eq} = I^{eq}(\omega, T(x))$ only. This relation has been used in deriving Equation 11.17.

11.2 Discontinuous Finite Element Formulation

The discontinuous finite element formulation for the electron–lattice model and the radiative transfer model are very similar to that discussed in Chapters 5 and 9, and thus will not be elaborated upon here. We mention, however, that in the electron-gas temperature model the extra term can be simply treated as a source term. The radiative transfer equation describing the phonon interaction has a transient term and thus an additional time marching scheme is required to obtain the solution. Time integrators discussed in previous chapters (see Section 2.2.2, for example) can be used for this purpose. Such a scheme should be readily incorporated into the algorithms described in Chapter 9. It is worth noting that Equation 11.15 may be solved numerically using the lattice Boltzmann approach as discussed in Section 11.5.

Let us turn our attention to the two-phase lag model and consider the discontinuous finite element formulation for the numerical solution of the temperature distribution described by Equation 11.10. We study a general case of a pulsing laser heating process, where the length scale is comparable to the skin depth of the electromagnetic waves of the laser beam. To facilitate the discontinuous finite element formulation, Equation 11.10 is split into two first order partial differential equations,

$$Q - \nabla \cdot \mathbf{q} = C_p \rho \frac{\partial T}{\partial t} \tag{11.20}$$

$$\mathbf{q} + \tau_q \frac{\partial \mathbf{q}}{\partial t} = -k \left[\nabla T + \tau_T \frac{\partial}{\partial t} (\nabla T) \right] \tag{11.21}$$

For a thin film of thickness comparable to the skin depth of laser-induced electromagnetic waves, the general 3-D Gaussian laser pulses may be modeled as an internal energy source near the irradiating spot, namely,

$$Q = \frac{1-R}{\delta} \frac{I_0}{t_p \sqrt{\pi}} \exp \left(-\frac{z}{\delta} - \frac{x^2 + y^2}{r^2} - \frac{t^2}{t_p^2} \right) \tag{11.22}$$

where R is the surface reflectivity, t_p is the characteristic duration time of the laser pulse, I_0 is the laser intensity, δ is the penetration depth of heating energy, and r is the radius of the laser beam.

Because the physical dimensions of the media are small (in microns) and time duration is short (ps), a non–dimensionalized equation is easier to work with for this type of problem. By introducing the following dimensionless parameters:

$$\Delta = \frac{\delta}{2\sqrt{\alpha \tau_q}} \; ; \; \Delta_r = \frac{r}{2\sqrt{\alpha \tau_q}} \; ; \; (X, Y, Z) = \frac{(x, y, z)}{2\sqrt{\alpha \tau_q}}$$

$$\xi = \frac{t}{2\tau_q} \; ; \; \xi_p = \frac{t_p}{2\tau_q} \; ; \; \xi_T = \frac{\tau_T}{2\tau_q}$$

$$\theta = \frac{k\sqrt{\tau_q}\sqrt{\pi}[T(x,t) - T_0]}{(1-R)I_0\sqrt{a}} \; ; \; \mathbf{w} = \frac{\tau_q\sqrt{\pi}}{(1-R)I_0}\mathbf{q} \; ; \; \psi = \frac{2\tau_q\sqrt{\pi a \tau_q}}{(1-R)I_0}Q$$

where T_0 is the reference temperature. Equations 11.20 and 11.21 can be rewritten in the non–dimensionalized form of

$$\psi - \nabla \cdot \mathbf{w} = \frac{\partial \theta}{\partial \xi} \tag{11.23}$$

$$2\mathbf{w} + \frac{\partial \mathbf{w}}{\partial \xi} = -\nabla\theta - \xi_T\frac{\partial}{\partial \xi}(\nabla\theta) \tag{11.24}$$

with ψ being the dimensionless heating source expressed as

$$\psi(X,Y,Z,\xi) = \frac{1}{2\Delta\xi_p}\exp[-\frac{Z}{\Delta} - \frac{X^2 + Y^2}{\Delta_r^{\,2}} - (\frac{\xi}{\xi_p})^2] \tag{11.25}$$

and $\nabla = (\partial/\partial X, \partial/\partial Y, \partial/\partial Z)$.

The solution of the above equations requires boundary and initial conditions. For the calculations presented below, the following boundary and initial conditions are used, which in terms of dimensionless parameters, become

$$\mathbf{w} \cdot \mathbf{n} = w_x n_x + w_y n_y + w_z n_z = 0 \;, \quad X,Y,Z \in \partial\Omega \tag{11.26a}$$

$$\theta(X,Y,Z,\xi = 0) = 0; \quad \frac{\partial\theta}{\partial\xi}(X,Y,Z,\xi = 0) = 0;$$

$$w(X,Y,Z,\xi = 0) = 0\,, \quad\quad\quad X,Y,Z \in \partial\Omega \tag{11.26b}$$

where $\mathbf{n}$ is the outward normal vector at the domain boundary $\partial\Omega$. Note that the above boundary conditions on the temperatures approximate the situation where the heat transfer coefficient is very large at the two boundaries. Other boundary conditions such as periodic or reflection conditions may also be applied.

To develop an integral formulation that is suitable for discontinuous finite element solutions, the computational domain is first discretized into a tessellation of finite elements, that is, triangles in 2-D geometries and tetrahedrons in 3-D geometries. Then, the governing equations (Equations 11.23 and 11.24) are multiplied by a pair of test functions $(v, \mathbf{v})$ and integration over any of the elements Ω_j with the result,

$$\int_{\Omega_j} v\left(\frac{\partial\theta}{\partial\xi} + \nabla\cdot\mathbf{w} - \psi\right)dV = 0 \tag{11.27}$$

$$\int_{\Omega_j} \mathbf{v}\cdot\left(\frac{\partial\mathbf{w}}{\partial\xi} + 2\mathbf{w} + \nabla\theta + \xi_T\frac{\partial}{\partial\xi}(\nabla\theta)\right)dV = 0 \tag{11.28}$$

where $\mathbf{v} = (v_x, v_y, v_z)$. Integration by parts of the flux term gives

$$\int_{\Omega_j}\left(v\frac{\partial\theta}{\partial\xi} - \nabla v\cdot\mathbf{w} - v\psi\right)d\Omega + \int_{\partial\Omega_j} v\hat{\mathbf{w}}\cdot\mathbf{n}\,d\gamma = 0 \tag{11.29}$$

$$\int_{\Omega_j}\left(\mathbf{v}\cdot\frac{\partial\mathbf{w}}{\partial\xi} + 2\mathbf{v}\cdot\mathbf{w} - \theta\nabla\cdot\mathbf{v} - \xi_T\nabla\cdot\mathbf{v}\frac{\partial\theta}{\partial\xi}\right)d\Omega$$

$$+ \int_{\partial\Omega_j}\left(\hat{\theta} + \xi_T\hat{h}\left(\theta,\frac{\partial\theta}{\partial\xi}\right)\right)\mathbf{v}\cdot\mathbf{n}\,d\gamma = 0 \tag{11.30}$$

where use has been made of the numerical fluxes ($\hat{\theta}, \hat{h}(\theta, \partial\theta/\partial\xi), \hat{\mathbf{w}}$) to replace the function fluxes at the element boundaries.

By Galerkin's approach, the shape function is taken in the same manner as the test function. Thus, the variables are interpolated over an element as follows:

$$\theta(X,Y,Z,\xi) = \sum_{j=1}^{N_e}\phi_j(X,Y,Z)\theta_j(\xi) = \Phi^T\underline{\theta} \tag{11.31a}$$

$$\frac{\partial\theta}{\partial\xi}(X,Y,Z,\xi) = \sum_{j=1}^{N_e}\phi_j(X,Y,Z)\left(\frac{\partial\theta}{\partial\xi}(\xi)\right)_j = \Phi^T\underline{\dot{\theta}} \tag{11.31b}$$

$$\mathbf{w}(X,Y,Z,\xi) = \sum_{j=1}^{N_e}\phi_j(X,Y,Z)\mathbf{w}_j(\xi) = \Phi^T\underline{\mathbf{w}} \tag{11.31c}$$

where ϕ_j is the shape function, N_e is the number of nodes per element and the dot on the top of a variable denotes the time derivative.

As discussed in Chapters 4 and 5, a variety of numerical fluxes can be chosen for a discontinuous finite element solution. The fluxes in the above equations can be calculated by the following expressions:

$$\hat{\mathbf{w}}\cdot\mathbf{n} = \alpha(\mathbf{w}^+\cdot\mathbf{n}) + (1-\alpha)(\mathbf{w}^-\cdot\mathbf{n}) \tag{11.32}$$

$$\left(\hat{\theta}+\xi_T\hat{h}\right)n_i = \alpha\left[\theta^+ + \xi_T\left(\frac{\partial\theta}{\partial\xi}\right)^+\right]n_i + (1-\alpha)\left[\theta^- + \xi_T\left(\frac{\partial\theta}{\partial\xi}\right)^-\right]n_i$$

$$(11.33)$$

where the indices – and + denote the back and front sides of the vector, $\mathbf{n}$ is the outward normal, $\hat{i}$ is the unit vector and $n_i = \mathbf{n}\cdot\hat{i}$. Note that in the above definition, $\alpha = 1$ represents the use of the one-side upwinding scheme. Taking α to be ½, one has the central flux approximations, which were first used by Bassi and Rebay [8] and will be used for the results calculated below as well. The central flux approximations provide the simplest and most efficient expressions for numerical flux calculations and their mathematical properties have been studied recently [9]. Other numerical flux expressions summarized in Table 4.2 may also be used for the calculations.

Following the discontinuous finite element procedures outlined in Chapters 4 and 5 and carrying out the calculations at the elemental level, one has the following matrix equation for the element:

$$\mathbf{M\dot{U}} + \mathbf{KU} = \mathbf{F} \tag{11.34}$$

where the unknown vectors, force vectors and matrices take the following forms:

$$\mathbf{U} = \begin{Bmatrix} \boldsymbol{\theta} \\ \underline{\mathbf{w}}_x \\ \underline{\mathbf{w}}_y \\ \underline{\mathbf{w}}_z \end{Bmatrix} ;\ \dot{\mathbf{U}} = \begin{Bmatrix} \dot{\boldsymbol{\theta}} \\ \dot{\underline{\mathbf{w}}}_x \\ \dot{\underline{\mathbf{w}}}_y \\ \dot{\underline{\mathbf{w}}}_z \end{Bmatrix} ;\ \mathbf{F} = \begin{Bmatrix} \mathbf{F}_\theta \\ \mathbf{F}_{wx} \\ \mathbf{F}_{wy} \\ \mathbf{F}_{wz} \end{Bmatrix} ;$$

$$\mathbf{M} = \begin{bmatrix} \mathbf{M}_\theta & 0 & 0 & 0 \\ \mathbf{M}_x & \mathbf{M}_w & 0 & 0 \\ \mathbf{M}_y & 0 & \mathbf{M}_w & 0 \\ \mathbf{M}_z & 0 & 0 & \mathbf{M}_w \end{bmatrix} ;\ \mathbf{K} = \begin{bmatrix} 0 & \mathbf{K}_{\theta x} & \mathbf{K}_{\theta y} & \mathbf{K}_{\theta z} \\ \mathbf{K}_{wx} & \mathbf{K}_w & 0 & 0 \\ \mathbf{K}_{wy} & 0 & \mathbf{K}_w & 0 \\ \mathbf{K}_{wz} & 0 & 0 & \mathbf{K}_w \end{bmatrix} \tag{11.35}$$

with $\dot{\theta} = \partial\theta(\xi)/\partial\xi$, and $\dot{\mathbf{w}} = \partial\mathbf{w}(\xi)/\partial\xi$. The matrix elements are calculated using the following expressions:

$$\mathbf{M}_\theta = \int_{\Omega_j} \Phi\Phi^T d\Omega$$

$$\mathbf{M}_w = \int_{\Omega_j} \Phi\Phi^T d\Omega$$

$$\mathbf{K}_w = 2\int_{\Omega_j} \Phi\Phi^T dV$$

$$\mathbf{M}_k = -\xi_T \int_{\Omega_j} (\hat{e}_k \cdot \nabla\Phi)\Phi^T \, dV + \alpha\xi_T \sum_{i=1}^{NS} \int_{\partial\Omega_{j,i}} \Phi^T \Phi \mathbf{n}_{j,i} \cdot \hat{e}_k \, dS$$

$$\mathbf{K}_{\theta k} = \int_{\Omega_j} (\hat{e}_k \cdot \nabla\Phi)\Phi^T \, dV + \sum_{i=1}^{NS} \left(\int_{\partial\Omega_{j,i}} \alpha\Phi\Phi^T (\mathbf{n}_{j,i} \cdot \hat{e}_k) \, dS \right)$$

$$\mathbf{K}_{wk} = -\int_{\Omega_j} (\hat{e}_k \cdot \nabla\Phi)\Phi^T \, dV + \sum_{i=1}^{NS} \left(\int_{\partial\Omega_{j,i}} \alpha\Phi\Phi^T (\mathbf{n}_{j,i} \cdot \hat{e}_k) \, dS \right)$$

$$\mathbf{F}_\theta = \int_{\Omega_j} \Phi\psi \, dV - \sum_{i=1}^{NS} \left(\int_{\partial\Omega_{j,i}} (1-\alpha)\Phi\Phi^T (\mathbf{n}_{j,i} \cdot \hat{e}_k) \, dS \right) \mathbf{w}_{k,(NB,i)}$$

$$\mathbf{F}_{wx} = -\sum_{i=1}^{NS} \left[\int_{\partial\Omega_{j,i}} (1-\alpha)\Phi\Phi^T (\mathbf{n}_{j,i} \cdot \hat{e}_k) \, dS \right] \left(\theta + \xi_T \frac{\partial\theta}{\partial\xi} \right)_{(NB,i)}$$

$$k = x, y, z; \ \hat{e}_x = \hat{i}, \ \hat{e}_y = \hat{j}, \ \hat{e}_z = \hat{k}$$

where NS is the number of sides of element j.

The above algorithm has been applied to study the heat transfer phenomena in pulsing laser heating of thin films. The case studies include 1-D, 2-D and 3-D problems. These results are given in Figures 11.1–11.3.

Figure 11.1 compares the analytic solution and the numerical results obtained using the discontinuous Galerkin finite element method as described previously. This is a 1-D problem for which the analytic solution is obtained using the method of Green's function [6]. Numerical calculations used 200 linear elements and the time step is chosen as 0.001 to satisfy the stability criterion. As seen from the figure, excellent agreement exists between the analytic and the numerical solutions. At $\xi = 0.1$, a thermal wave peak has just formed and the wave behavior of the thermal signal propagation is apparent in Figure 11.1, where a sequence of peaks indicates the propagation of the thermal wave initiated by the pulsing laser. At $\xi = 2.4$, the thermal wave is reflected at the insulated end ($X = 2$) and propagates in the negative X direction. As time goes by, the temperature distribution along the bar becomes more and more even. Eventually the wave behavior disappears and a uniform temperature profile is attained.

It is known that the dual phase parameter τ_T provides the relaxation mechanism for the temperature gradient, and hence it is termed the phase lag of the temperature gradient. On the other hand, τ_q describes the relaxation mechanism for the heat flux q, and therefore is termed as the phase lag of the heat flux. Physically, these two parameters characterize the thermal wave behavior in the medium. If $\tau_T < \tau_q$, then the temperature gradient is ahead of the heat flux, which means that the temperature gradient is the cause and the heat flux is the result. This is the case where τ_T (or ξ_T) = 0, which is shown in Figure 11.1(a). If $\tau_T > \tau_q$ (i.e., $\xi_T > 0.5$), then the heat flux precedes the temperature gradient, which means that the heat flux is the cause and the temperature gradient is the effect. Thus, with an increase in τ_T, the temperature gradient across the film is relaxed, or more specifically, delayed. One such case is calculated using the discontinuous finite element model and the results are shown in Figure 11.1(b), along with the analytic solutions.

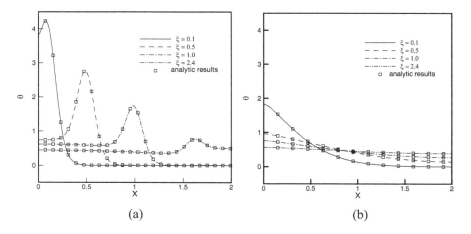

(a)   (b)

Figure 11.1. Computed results of a thermal wave in a 1-D thin film: (a) comparison of the propagation of temperature distributions in the 1-D problem obtained analytically using the method of Green's function, and numerically by the discontinuous Galerkin finite element method ($\xi_T = 0.0$); and (b) transient development of temperature distribution with dual phase lags along the 1-D domain that is irradiated by a pulsing laser beam ($\xi_T = 1.0$). Other conditions used for computations: $\Delta = 0.05$ and $\xi_p = 0.0$

Figure 11.2 plots the temperature propagation in a two-dimensional restricted converging–diverging channel. The domain can be divided into four regions: the uniform inlet region, converging region, diverging region and uniform outlet region, as shown in Figure 11.2(a). A steady pulse laser beam with a width of 0.2 and a characteristic duration of 0.1 irradiates a portion of the inlet of the channel with the heating source taken in the form of

$$\psi(X,Y,\xi) = \frac{1}{2\Delta\xi_p}\exp\left(-\frac{(1+X)}{\Delta} - \frac{Y^2}{\Delta_r^2} - (\frac{\xi}{\xi_p})^2\right) \qquad (11.36)$$

The calculations used an unstructured mesh shown in Figure 11.2(b), which is obtained using the front advancing automatic meshing generation scheme [6]. The mesh consists of a total of 7610 linear triangular elements. Figure 11.2(c) depicts the time snap shots of temperature distributions in the plate with ($\xi_T = 0.0$). At $\xi = 1.0$, the thermal waves arrives at the two boundary walls ($Y = \pm 0.5$) in the uniform inlet region, and the wave front starts to enter the compression region ($X = -0.5$).

At $\xi = 1.0$, the thermal wave front reaches the diverging region while the wave reflections occur at the walls in the converging region. At $\xi = 1.5$, the wave front is leaving the diverging region, and the whole thermal wave field evolves into a quite complex structure. Unlike the wave in the converging region, no strong wave reflection is observed at the boundary walls in this region because of the diffusive effect of the diverging region. At $\xi = 2.0$, the wave arrives at the right side wall ($X = 1.0$) and continues to evolve.

In Figure 11.3, a simple cubic geometry is considered to illustrate the thermal wave propagation in a 3-D domain. The laser beam arrangement is schematically shown in Figure 11.3(a). The computational domain is a $2 \times 2 \times 0.5$ cubic box, which is insulated from the ambient along the boundaries. A steady pulse laser beam with a radius of 0.1 and a characteristic duration of 0.1 irradiates a portion of the center of the top surface of the box with the heating source taken in the following form:

$$\psi(X,Y,Z,\xi) = \frac{1}{2\Delta\xi_p} \exp\left(-\frac{(0.5-Z)}{\Delta} - \frac{X^2+Y^2}{\Delta_r^2} - (\frac{\xi}{\xi_p})^2 \right) \tag{11.37}$$

The unstructured mesh for the discontinuous finite element computations has a total of 96000 linear tetrahedral elements. Figure 11.3(b–d) depicts the time evolution of the temperature distribution on the $Z = 0.5$ plane with $\xi_T = 0.0$. Once again, the wave-like characteristic of the propagation of the thermal wave in the system is observed.

11.3 Micro and Nano Fluid Flow and Heat Transfer

Study of fluid flows in microscale systems has been documented in a recent monograph by Karniadakis *et al.* [1], where fundamental equations and their application to micro systems have been discussed. For incompressible flows in microscale systems, the Navier–Stokes equations formulated for continuum media are also applicable under normal conditions, and the numerical techniques discussed in the previous chapters may be directly applied to perform flow simulations. One important point in this regard is that in order to numerically simulate these flows, the equations should be non-dimensionalized, using appropriate time and length scales, to prevent floating point problems from occurring and to ensure accuracy. For flows involving gases, however, the compressible effects may become important and the commonly used no-slip boundary conditions need to be modified to allow for the slip of fluid molecules along the walls.

Theoretically, the Navier–Stokes equations are the first order approximation of the Chapman–Enskog solution to the Boltzmann equation, and are accurate only up to $O(Kn)$ [10]. The Knudsen number (Kn) characterizes the regimes from free molecular to continuum flows and is defined as the ratio of the mean free path of the molecules over the characteristic dimension,

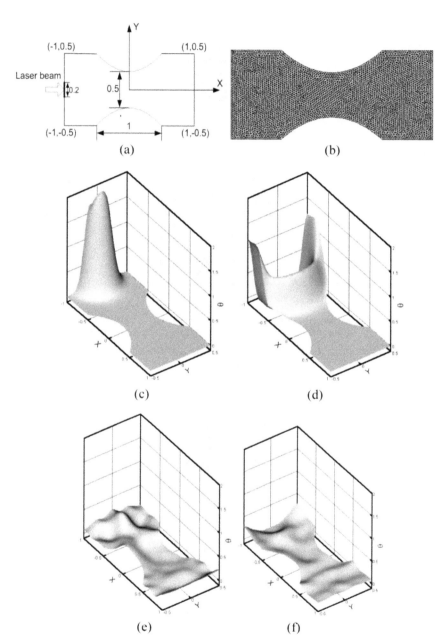

gure 11.2. Computed results of thermal wave propagation in a converging–diverging channel: (a) geometry and pulse laser heating arrangement for the 2-D converging–diverging channel problem, (b) the unstructured triangular mesh for the discontinuous finite element computations, and (c–f): evolution of temperature distributions at various instances: (b, wave-like) $\xi_T = 0.0$, (c) $\xi = 0.1$, (d) $\xi = 0.5$, (e) $\xi = 1.0$, and (f) $\xi = 1.5$. Parameters used for calculations: $\Delta = 1.0$ and $\xi_p = 1.0$

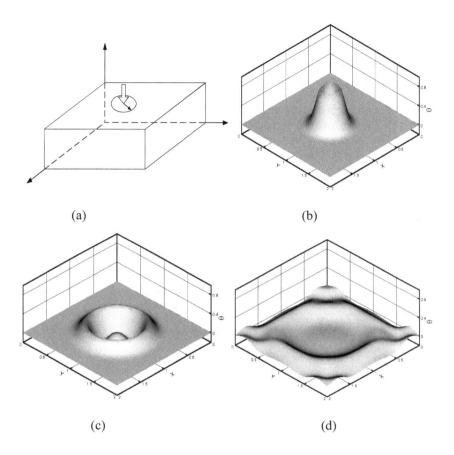

Figure 11.3. Evolution of temperature distributions of the 3D problem on the $z = 0.5$ plane at various instances (wave-like $\xi_T = 0.0$): (a) problem definition, (b) $\xi = 0.1$, (c) $\xi = 0.5$, and (d) $\xi = 1.5$. Parameters used for calculations: $\Delta = 1.0$ and $\xi_p = 1.0$

$$Kn = \frac{\lambda}{H} = \frac{k_b T}{\sqrt{2\pi d^2 pH}} \qquad (11.38)$$

where λ is the mean free path of the molecules, d is the diameter of the molecules, H is the characteristic length, k_b is the Boltzmann constant, T is the temperature, and p is the thermodynamic pressure. The Kn number may also be defined in terms of the non-dimensionalized numbers used in continuum fluid flow studies,

$$Kn = \sqrt{\frac{\pi\gamma}{2}} \frac{Ma}{Re} \qquad (11.39)$$

where γ is the specific heat ratio, $Ma = u/(\gamma RT)^{1/2}$ is the Mach number, R is the specific gas constant, $Re = \rho uH/\mu$ is the Reynolds number, and u is the flow velocity. The local Kn number is a measure of degree of rarefaction of gases in flows in microscale or nanoscale channels. Different regimes of fluid flow with Kn as the indicator are schematically sketched in Figure 11.4. The continuum description of fluid flow motion is applicable within the range of $Kn \to 0$ to $Kn = 0.1$. The no-slip boundary conditions at the walls, however, must be relaxed for the flows with $Kn = 0.001$ to 0.1. As the Kn number increases, the rarefaction effects become more pronounced and eventually the continuum assumption breaks down. The transition flow occurs when the characteristic dimension becomes comparable to the fluid mean path. In the range of $Kn = 1$ to $Kn \to \infty$, the solution of free molecule flows is required, which needs to take into account the individual molecule behaviors. Thus, the streaming velocity at the wall is comprised of flows of incident molecules and the scattered molecules by the wall.

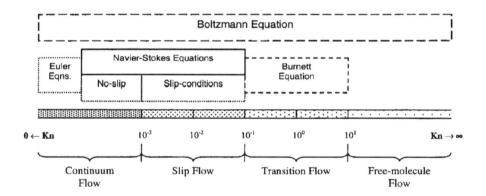

Figure 11.4. Classification of flows from free molecular flows to continuum [11]

In Chapman–Enskog's perturbation theory, the system is considered as a first order perturbation from the equilibrium Maxwellian distribution, and the distribution function f is expressed in a power series,

$$f = f^{(0)} + Kn f^{(1)} + Kn^2 f^{(2)} + \cdots \tag{11.40}$$

where the small perturbation parameter is taken to be Kn [11]. Consistent with the Navier–Stokes equations, which is accurate to $O(Kn)$, the boundary conditions to the same order accuracy are needed for the velocity. This leads to the following slip condition of gas velocity at the wall:

$$u_s = \frac{2-\sigma}{\sigma} HKn \frac{\partial u_s}{\partial n} = \frac{2-\sigma}{\sigma} \lambda \frac{\partial u_s}{\partial n} \tag{11.41}$$

where subscript s denotes the stream direction parallel to the wall, and n is the normal of the wall. Of course, in the direction perpendicular to the wall, the velocity is zero.

This is consistent with Maxwell's derivation for dilute, monatomic gases [12], which also includes the temperature effects and wall velocity u_w,

$$u_s - u_w = \frac{2-\sigma}{\sigma} \lambda \frac{\partial u_s}{\partial n} + \frac{3}{4} \frac{\mu}{\rho T_{gas}} \frac{\partial T}{\partial s} \tag{11.42}$$

The corresponding temperature jump relation at the wall was derived by von Smoluchowski [11] as

$$T - T_w = \frac{2-\sigma_T}{\sigma_T} \frac{2\gamma}{1+\gamma} \frac{\lambda}{\mathrm{Pr}} \frac{\partial T}{\partial s} \tag{11.43}$$

Here, the parameter σ measures the reflection of molecules diffusively from the walls. At $\sigma = 0$, the molecules reflect specularly, indicating a reversal of their normal velocity due to normal momentum transfer to the wall. At $\sigma = 1$, on the other hand, the molecules reflect diffusively when reflected from the wall with zero tangential velocity. Thus, the values of σ and σ_T depend on local characteristics near the wall including surface roughness, fluid temperature and pressure.

The above first order slip conditions are found to be applicable to the Navier–Stokes equations when the Kn number is in the range of $0.001 - 0.1$. For $Kn > 0.1$, further corrections may be needed. Karniadakis *et al.* [1] suggest the following correction in place of the no-slip condition:

$$u_s - u_w = \frac{2-\sigma}{\sigma} \frac{\lambda}{H-b\lambda} \frac{\partial u_s}{\partial n} + \frac{3}{4} \frac{\mu}{\rho T_{gas}} \frac{\partial T}{\partial s} \tag{11.44}$$

where b is an empirical constant. Other higher order corrections to the boundary conditions may be obtained from the direct solution of the Boltzmann equation [13, 14].

For these flows, it is not difficult to devise the discontinuous finite element solution. In fact, the algorithms presented in the previous chapters can be directly applied, with allowance made for the slip conditions in the stream directions.

11.4 The Boltzmann Transport Equation and Numerical Solution

One of the celebrated fundamental equations in statistical mechanics is the Boltzmann transport equation, which characterizes the kinetics and dynamics of the distribution of microscale particles, such as electrons, phonons, photons, molecules, etc. As suggested in Figure 11.4, the Boltzmann equation, which is an integral-differential equation, is applicable over the entire spectrum of Kn. We discuss the basics of the equation and its numerical solution in this section.

11.4.1 The Boltzmann Integral-Differential Equation

The Boltzmann integral-differential equation (or the Boltzmann transport equation) describes neutral and charged particle transport phenomena and expresses the global non-equilibrium distribution in terms of the local equilibrium distributions. The equation enables application of the properties of equilibrium systems to the study of a non-equilibrium system. For a system with a non-uniform particle density and temperature, in each place there is a local range where the thermal velocities are given by an equilibrium distribution function. The distribution is temperature dependent and varies from place to place. Whenever a particle is scattered, or collides with the medium, its thermal velocity immediately after the collision will be that of the equilibrium distribution at the collision point. The equation is derived based on the balance of randomly moving particles within a medium with a temperature gradient and takes the following form [15, 16]:

$$
\frac{\partial f}{\partial t} + \mathbf{v} \cdot \nabla f + \frac{\mathbf{F}}{m} \cdot \nabla_{\mathbf{v}} f
$$
$$
= \int_{\mathbf{v}_1} \int_{\mathbf{v}'} \int_{\mathbf{v}_1'} (f' f_1' - f f_1) u \sigma_c(\mathbf{v}, \mathbf{v}_1 \to \mathbf{v}', \mathbf{v}_1') d\mathbf{v}_1' d\mathbf{v}' d\mathbf{v}_1 \quad (11.45)
$$

where $f(\mathbf{r}, \mathbf{v}, t)$ is the distribution function and its physical meaning is interpreted as the particle distribution at $\mathbf{r}$, $\mathbf{v}$ and t, m is the mass of the molecule, $\mathbf{F}$ is the external force, and $\mathbf{v}$ is the velocity of individual molecules. The integration is carried out over the entire space of the velocity. Note that this equation involves 7 independent variables (3 spatial variables, x, y, z; 3 velocity variables, v_x, v_y, v_z; and time t). Also, $\sigma_c(\mathbf{v}, \mathbf{v}_1, \mathbf{v}', \mathbf{v}_1')$ is the differential scattering cross-section for the collision of two particles. These two particles have velocities $(\mathbf{v}, \mathbf{v}_1)$ before collision and $(\mathbf{v}', \mathbf{v}_1')$ after collision. In the equation, $\mu = |\mathbf{v} - \mathbf{v}_1| = |\mathbf{v}' - \mathbf{v}_1'|$ is the relative speed and $f' = f(\mathbf{r}, \mathbf{v}', t)$, etc.

The above equation may also be written in a simplified form,

$$
\frac{\partial f}{\partial t} + \mathbf{v} \cdot \nabla f + \frac{\mathbf{F}}{m} \cdot \nabla_{\mathbf{v}} f = \int_{\mathbf{v}_1} \int_{\Omega} (f' f_1' - f f_1) \mu \sigma_c(\mu, \Omega) d\Omega d\mathbf{v}_1 \quad (11.46)
$$

where $d\Omega = \sin\theta d\theta d\varphi$ is the differential solid angle subtended at the center of mass. In writing Equation 11.46, we have also used the following definitions:

$$
\nabla = \hat{i}\frac{\partial}{\partial x} + \hat{j}\frac{\partial}{\partial y} + +\hat{k}\frac{\partial}{\partial z}; \quad \nabla_{\mathbf{v}} = \hat{i}\frac{\partial}{\partial v_x} + \hat{j}\frac{\partial}{\partial v_y} + +\hat{k}\frac{\partial}{\partial v_z} \quad (11.47)
$$

The distribution function f is defined such that $f(\mathbf{r}, \mathbf{v}, t)dxdxdydv_xdv_ydv_z = $ probability of finding a particle phase space volume $dxdxdydv_xdv_ydv_z$ centered at $\mathbf{r}$, $\mathbf{v}$ and t. The distribution function f has a unit of $m^{-3}(ms^{-1})^{-3}$ and it satisfies the following conservation relation:

$$\int_{\mathbf{v}} \int_{\mathbf{r}'} f(\mathbf{r}, \mathbf{v}, t) \, d\mathbf{r} d\mathbf{v}_1 = \int_{\mathbf{r}'} n(\mathbf{r}, t) d\mathbf{r} = N \tag{11.48}$$

where n is the density of particles and N is the total number of the particles in the system.

The Boltzmann integral-differential equation may also be written in a simplified form involving differential operators only,

$$\frac{df}{dt} = \left(\frac{df}{dt}\right)_{coll} \tag{11.49}$$

where the left hand side represents the dynamics of the particle and the right hand side is the collision term,

$$\frac{df}{dt} = \frac{\partial f}{\partial t} + \mathbf{v} \cdot \nabla f + \frac{\mathbf{F}}{m} \cdot \nabla_{\mathbf{v}} f \tag{11.50}$$

$$\left(\frac{df}{dt}\right)_{coll} = \int_{\mathbf{v}_1} \int_{\Omega} (f' f_1' - f f_1) \mu \sigma_c(\mu, \Omega) \, d\Omega \, d\mathbf{v}_1 \tag{11.51}$$

This form is convenient for the derivation of the lattice Boltzmann equation, which has been used for simulation of various microscale flows (see Section 11.6 below).

As the Boltzmann equation describes the microscopic phenomena, its use for the study of microscale thermal and fluid flow phenomena is obvious. It is known that a macroscopic description of transport phenomena is essentially an assemble average of microscopic phenomena. In fact, the continuity, the Navier–Stokes and the energy balance equations can be directly derived from the Boltzmann equation. This allows us to establish the direct link between the microscopic and macroscopic descriptions of physical phenomena in thermal fluids systems. We outline basic procedures by which these macroscopic equations are derived from the microscopic Boltzmann equation.

Let $\psi(\mathbf{r}, \mathbf{v}, t)$ be a generic variable for thermal fluids such as the flow velocity, momentum and kinetic energy. Microscopically, this variable is transported along with molecules. We obtain the transport equations for these quantities by integrating them with the Boltzmann equation over the entire velocity space, viz.,

$$\int_{\mathbf{v}} \psi \frac{\partial f}{\partial t} d\mathbf{v} + \int_{\mathbf{v}} \psi \mathbf{v} \cdot \nabla f \, d\mathbf{v} + \int_{\mathbf{v}} \psi \frac{\mathbf{F}}{m} \cdot \nabla_{\mathbf{v}} f \, d\mathbf{v}$$

$$= \int_{\mathbf{v}} \psi \int_{\mathbf{v}_1} \int_{\Omega} (f' f_1' - f f_1) u \sigma(u, \Omega) \, d\Omega \, d\mathbf{v}_1 d\mathbf{v} \tag{11.52}$$

In the derivations given below, we will use the averages that are defined as follows:

$$\bar{\psi} = \frac{1}{n} \int_v \psi f \, d\mathbf{v} \; ; \qquad\qquad \overline{\frac{\partial \psi}{\partial t}} = \frac{1}{n} \int_v \frac{\partial \psi}{\partial t} f \, d\mathbf{v} \qquad (11.53)$$

where the integration is carried out over the entire molecular velocity space. With the above definitions and noticing that $\mathbf{F}$ is a function of $\mathbf{r}$ only, but not of $\mathbf{v}$, and making use of the symmetry condition of the scattering cross-section, we have the following transport or transfer equation:

$$\frac{\partial (n\bar{\psi})}{\partial t} + \nabla \cdot (n\overline{\mathbf{v}\psi}) - n\overline{\frac{\partial \psi}{\partial t}} - n\overline{\mathbf{v} \cdot \nabla \psi} - n\frac{\mathbf{F}}{m} \cdot \overline{\nabla_v \psi}$$

$$= \frac{1}{4} \int_v \int_{v_1} \int_\Omega [\psi' + \psi_1' - \psi - \psi_1](f' f_1' - f f_1) \mu \sigma(\mu, \Omega) \, d\Omega \, d\mathbf{v}_1 \, d\mathbf{v}$$

$$(11.54)$$

where n is the number of particles per unit volume.

It is known that before and after a collision of molecules, the mass, momentum and energy are conserved. These conservation properties are written as

$$\psi' + \psi_1' - \psi - \psi_1 = 0 \qquad (11.55)$$

This will make the collision term vanish, which basically means that the collision does not create or destroy conservation properties at a fixed location but only shift them in the velocity space. For example, a collision process conserves the momentum. The specific conservation properties of interest to thermal fluids, that is, mass (m), momentum ($m\mathbf{u}$) and energy (mu^2), are a function of velocity only,

$$\frac{\partial \psi}{\partial t} = \nabla \psi = 0 \qquad (11.56)$$

or

$$n\overline{\frac{\partial \psi}{\partial t}} - n\overline{\mathbf{v} \cdot \nabla \psi} = 0 \qquad (11.57)$$

Combining Equations 11.54–11.57, we have the governing equation for the conservation properties,

$$\frac{\partial (n\bar{\psi})}{\partial t} + \nabla \cdot (n\overline{\mathbf{v}\psi}) - n\frac{\mathbf{F}}{m} \cdot \overline{\nabla_v \psi} = 0 \qquad (11.58)$$

from which the equations for the conservation of mass, momentum and energy are derived.

Letting $\psi(\mathbf{r}, \mathbf{v}, t) = m$ and substituting into the above equation, we have the continuity equation,

$$\frac{\partial \rho}{\partial t} + \nabla \cdot (\rho \mathbf{u}) = 0 \tag{11.59}$$

where $\rho = nm$ and $\mathbf{u} = \overline{\mathbf{v}}$. Note that $\mathbf{u}$ is the macroscopic velocity used in the Navier–Stokes equations, which is just an assemble average of molecular velocity.

To obtain the momentum equation, we let $\psi(\mathbf{r}, \mathbf{v}, t) = m\mathbf{v}$, and note that $\partial v_i / \partial v_j = \delta_{ij}$, whence we have the macroscopic momentum equation for $\mathbf{u}$,

$$\frac{\partial (\rho \mathbf{u})}{\partial t} + \nabla \cdot (\rho \, \overline{\mathbf{vv}}) - \rho \frac{\mathbf{F}}{m} = 0 \tag{11.60}$$

The second term can be further simplified such that

$$\overline{\rho v_i v_j} = \rho \overline{v}_i \overline{v}_j + \overline{\rho (v_i - u_i)(v_j - u_j)}, \quad \text{with } u_j = \overline{v}_j \tag{11.61}$$

where by definition the second term is calculated by

$$\overline{\rho (v_i - u_i)(v_j - u_j)} = m \int_{\mathbf{v}} (v_i - u_i)(v_j - u_j) f \, d\mathbf{v} \tag{11.62}$$

Here $\mathbf{w} = \mathbf{v} - \mathbf{u}$ is the velocity of the particle relative to the local macroscopic flow velocity $\mathbf{u}$ and $\overline{\mathbf{w}} = \mathbf{0}$. Further, the fluctuating velocity correlation term (i.e., the second term in Equation 11.56) may be decomposed into two components,

$$p = \tfrac{1}{3} \overline{\rho (v_i - u_i)^2}; \quad \sigma_{ij} = p \delta_{ij} - \overline{\rho (v_i - u_i)(v_j - u_j)} \tag{11.63}$$

With $\mathbf{u} = \overline{\mathbf{v}}$ substituted, we thus have the momentum balance equation,

$$\frac{\partial (\rho \mathbf{u})}{\partial t} + \nabla \cdot (\rho \mathbf{uu}) = -\nabla p + \nabla \cdot \boldsymbol{\sigma} + \rho \mathbf{f}_{tot} \tag{11.64}$$

where $\mathbf{f}_{tot} = \mathbf{F}/m$ is the external force per unit mass including the contribution from gravity. This is the Navier–Stokes equations for momentum transport. We see here that the shear stress is basically the transport of the fluctuating velocity momentum and the viscosity goes to oppose the shear motion and inter-molecular penetration.

If the above equation is further combined with Equation 11.59, we have the following well known expression for the macroscopic momentum balance,

$$\rho \frac{\partial \mathbf{u}}{\partial t} + \rho \mathbf{u} \cdot \nabla \mathbf{u} = -\nabla p + \nabla \cdot \boldsymbol{\sigma} + \rho \mathbf{f} + \rho g \tag{11.65}$$

where $\mathbf{f}_{tot} = \mathbf{f} + \mathbf{g}$ has been subsituted and f is the external force excluding the gravitational force.

Letting $\psi(\mathbf{r}, \mathbf{v}, t) = m\mathbf{v}^2$, and going through the same procedure above, one has the energy balance equation,

$$\rho\frac{\partial E}{\partial t} + \rho\mathbf{u}\cdot\nabla E = -p\nabla\cdot\mathbf{u} - \nabla\cdot\mathbf{q} + \boldsymbol{\sigma}:\nabla\mathbf{u} \tag{11.66}$$

where the last term on the right side represents the viscous dissipation, E is the internal energy per unit mass and $\mathbf{q}$ is the heat flux representing energy flow per unit area per unit time,

$$E = \frac{1}{2n}\int_{\mathbf{v}} |\mathbf{v} - \mathbf{u}|^2 f\, d\mathbf{v} \tag{11.67}$$

$$\mathbf{q} = \frac{m}{2}\int_{\mathbf{v}} f(\mathbf{v} - \mathbf{u})\,|\mathbf{v} - \mathbf{u}|^2 d\mathbf{v} \tag{11.68}$$

In deriving the energy balance equation, we have used the continuity equation and the momentum equation to eliminate the term associated with $\mathbf{F}$. In addition, the following statistic-averaging relations have also been applied,

$$\overline{v_i^2} = \overline{(v_i - u_i)^2} + \overline{u_i^2} \tag{11.69}$$

$$\overline{v_i^2 v_j} = \overline{(v_i - u_i)^2 (v_j - u_j)}$$
$$+ \overline{2u_i(v_i - u_i)(v_j - u_j)} + \overline{u_j(v_i - u_i)^2} + \overline{u_i^2 u_j} \tag{11.70}$$

Equations 11.59, 11.65 and 11.66 constitute the macroscopic description of hydrodynamics of a fluid in motion, which is derived from the statistical average of the Botlzmann transport equation. It is clear that the macroscopic variables, that is, mass (ρ), velocity ($\mathbf{u}$), heat flux ($\mathbf{q}$), stress (σ_{ij}), pressure (p) and the internal energy (E), are linked to the microscopic variables through averaging processes. We summarize these macroscopic and microscopic relations below,

$$\rho = m\int_{\mathbf{v}} f\, d\mathbf{v} \tag{11.71}$$

$$\mathbf{u} = \frac{1}{n}\int_{\mathbf{v}} f\mathbf{v}\, d\mathbf{v} \tag{11.72}$$

$$\mathbf{q} = \frac{m}{2}\int_{\mathbf{v}} f(\mathbf{v} - \mathbf{u})\,|\mathbf{v} - \mathbf{u}|^2 d\mathbf{v} \tag{11.73}$$

$$E = \frac{1}{2n} \int_{\mathbf{v}} f \mid \mathbf{v} - \mathbf{u} \mid^2 d\mathbf{v} \tag{11.74}$$

$$p = \tfrac{1}{3} \rho \overline{(v_i - u_i)^2} \tag{11.75}$$

$$\sigma_{ij} = p\delta_{ij} - \rho \overline{(v_i - u_i)(v_j - u_j)} \tag{11.76}$$

The Boltzmann integral-differential equation is also useful for the calculation of transport properties such as viscosity, thermal conductivity and diffusivity for fluids. The mathematical theory developed for this purpose has been discussed in detail by Chapman and Cowling [10]. The use of the fluctuating theory for calculating the transport properties is also discussed by Isihara [15] and Kubo *et al.* [16]. Boundary conditions for the solution of the Botlzmann equations in confined regions are discussed by Cercignani [17] and Harris [18]. They can also be used to provide higher order slip boundary conditions for gas flows in micro/nano structures.

11.4.2 Numerical Solution of the Boltzmann Transport Equation

The Boltzmann transport equation (or the Boltzmann integral-differential equation) is a very complex mathematical expression. It has seven independent variables and it would be a formidable task to solve the Boltzmann transport equation directly. Two classes of computational techniques are devised to solve the transport equation [19–22].

The first class involves the deterministic methods, and the transport equation is discretized using a variety of methods and then solved directly or iteratively. Different types of discretization give rise to different deterministic methods. Methods in this category includes discrete ordinates, spherical harmonics, collision probabilities, nodal methods, spectral Galerkin methods, and others. By these methods, a discretization of the velocity space is made first, which transforms the equation into a system of linear, hyperbolic partial differential equations. The discontinuous Galerkin finite element method or other numerical methods are then used to discretize the physical space and solve the resulting system of differential equations.

The second class of techniques includes the Monte Carlo methods. By these methods, a stochastic model is constructed in which the expected value of a certain random variable is equivalent to the value of a physical quantity to be determined. The expected value is estimated by the average of many independent samples representing the random variable. Random numbers, following the distributions of the variable to be estimated, are used to construct these independent samples. There are two different ways to construct a stochastic model for Monte Carlo calculations. In the first case the physical process is stochastic and the Monte Carlo calculation involves a computational simulation of the real physical process. In the other case, a stochastic model is constructed artificially, such as the solution of deterministic equations by a Monte Carlo technique.

The direct solution of the Boltzmann equation has been used to study the flow, heat transfer and temperature distribution in a binary mixture of rarefied gases. To obtain the high order accuracy of the velocity distribution function, the complicated nonlinear collision integrals are computed by the deterministic numerical kernel method. The overall quantities (the heat flow in the mixture, etc.) and the profiles of the macroscopic quantities (the molecular number densities of the individual components, the temperature of the total mixture, etc.) are obtained once the distribution function is known. The use of the Boltzmann equation for the study of microscopic transport phenomena has been reported for a wide range of the *Kn* number [19–21]. In a recent monograph, Aristov [22] has also documented various numerical schemes for the direct solution of the Boltzmann equation and their applications.

Perhaps the major difficulty with the direct solution of the equation is the precise treatment of the differential scattering cross-section associated with the molecular collision process. Fortunately, for a majority of thermal and fluids applications, approximations can be made to simplify the equation [23]. In the two sections below, we discuss the Bhatnagar–Gross–Krook (BGK) approximation and the discontinuous finite element solution of the Boltzmann–BGK equation.

11.5 The Boltzmann–BGK Equation and Numerical Solution

In this section, we discuss the continuous Boltzmann–BGK equation and the numerical solution of the equation using the discontinuous finite element method.

11.5.1 The Boltzmann–BGK Equation

As discussed above, the Boltzmann equation describes the evolution of the distribution function f of a fluid. The fluid density, momentum and energy can all be found from the distribution function by considering the appropriate integral. In theory this appears straightforward; however in practice it can be difficult because of the complicated form of the collision term on the right hand side of Equation 11.45. A large amount of the detail of the two-body interaction, which is contained in the Boltzmann collision operator, is unlikely to significantly influence the values of the macroscopic quantities [18]. Thus, the integral can be replaced by a simplified collision operator that retains only the qualitative and average properties of the actual collision operator. Any replacement collision function, however, must satisfy the conservation properties, such as the conservation of mass, momentum and energy, imposed on the fluid system [23].

A widely used approximate collision operator for thermal fluids applications is the BGK assumption by which the molecular collision operation is approximated by a time differencing using a single relaxation time [23],

$$\left(\frac{df}{dt}\right)_{coll} = \frac{f - f^{(0)}}{\tau} \tag{11.77}$$

where τ is the relaxation time characterizing the molecular collision, and $f^{(0)}$ is the Maxwell–Boltzmann equilibrium distribution function,

$$f^{(0)}(\mathbf{r},\mathbf{v},t) = \frac{n}{(2\pi\, k_b T\,/\,m)^{d/2}} \exp\left(-\frac{(\mathbf{v}-\mathbf{u})^2}{2k_b T\,/\,m}\right) \tag{11.78}$$

in which k_b is the Boltzmann constant, d is the dimension of the space, m is the mass of the particle, and n, $\mathbf{u}$ $(\mathbf{u} = \overline{\mathbf{v}})$ and T are the macroscopic number density, velocity and temperature, respectively. It is noted that strictly speaking, the Maxwell–Boltzmann equilibrium distribution function used in statistical mechanics is Equation 11.78 with $\mathbf{u} = \mathbf{0}$ [15–18]; Equation 11.78 was used in the BGK approximation to the collision operator [23]. It is noteworthy also that other equilibrium distribution functions, such as the Bose–Einstein distribution, may be used for other particle systems.

Once the distribution function is known, the macroscopic properties can be calculated using the following definitions:

$$n = \int_{\mathbf{v}} f\,d\mathbf{v}\;; \qquad n\mathbf{u} = \int_{\mathbf{v}} f\mathbf{v}\,d\mathbf{v}\;; \qquad n E = \int_{\mathbf{v}} f(\mathbf{v}-\mathbf{u})^2\,d\mathbf{v} \tag{11.79}$$

The derivative $\nabla_{\mathbf{v}} f$ cannot be calculated directly because the dependence of the distribution function on the microscopic velocity is unknown. Following the procedures stated in [24, 25], the derivative may be approximated by

$$\nabla_{\mathbf{v}} f \approx \nabla_{\mathbf{v}} f^{(0)} \tag{11.80}$$

Then, defining an equilibrium distribution function as

$$f^{eq} = f^{(0)} - \tau(\mathbf{F}/m)\cdot\nabla_{\mathbf{v}} f = (1 - \tau(\mathbf{F}/m)\cdot\nabla_{\mathbf{v}})f^{(0)}$$

$$= (1 + \tau(\mathbf{F}/m)\cdot\nabla_{\mathbf{u}})f^{(0)} \tag{11.81}$$

we have the well known Boltzmann–BGK transport equation,

$$\frac{\partial f}{\partial t} + \mathbf{v}\cdot\nabla f = \frac{f - f^{eq}}{\tau} \tag{11.82}$$

We note here that if the external force is absent (i.e., $\mathbf{F} = 0$), $f^{eq} = f^{(0)}$.

11.5.2 Discontinuous Finite Element Formulation

For the purpose of numerical solution, Equation 11.82 may be more conveniently written with $\mathbf{v}$ normalized as a unit vector. This can be done by introducing a velocity scale v_0, and thus we have the following equation:

$$\frac{\partial f}{v_0 \partial t} + \hat{\mathbf{v}} \cdot \nabla f = \frac{f - f^{eq}}{v_0 \tau} \tag{11.83}$$

with $\hat{\mathbf{v}}$ being the unit velocity vector, i.e., $\hat{\mathbf{v}} = \mathbf{v}/v_0$.

The above equation is very similar in form to the radiative transfer equation for radiation intensity I, whose solution was discussed in Chapter 9, the only difference being the presence of a time derivative term here. Consequently, the discontinuous finite element procedure detailed in Chapter 9 for radiative transfer calculations can be directly applied here, with a straightforward modification for the transient term.

Integrating the above equation with respect to a weighting function over element j yields the following integral representation:

$$\int_{\Delta\Omega_i} \int_{\Omega_j} \left(\frac{\partial f}{v_0 \partial t} + \hat{\mathbf{v}} \cdot \nabla f - \frac{f}{v_0 \tau} \right) dV d\Omega(\hat{\mathbf{v}}) = \int_{\Delta\Omega_i} \int_{\Omega_j} \left(-\frac{f^{eq}}{v_0 \tau} \right) dV d\Omega(\hat{\mathbf{v}}) \tag{11.84}$$

where $\Delta\Omega$ is the control angle for the velocity space. Following the procedure given in Chapter 9, we can easily derive the following discretized formulation for element j,

$$\mathbf{M} \frac{\partial \underline{\mathbf{f}}_{(j)}}{\partial t} + \mathbf{K}\underline{\mathbf{f}}_{(j)} = \underline{\mathbf{F}}_{(j)} \tag{11.85}$$

where $\underline{\mathbf{f}}_{(j)} = [f_i^{(1)}, f_i^{(2)}, \dots, f_i^{(N_e)}]_{(j)}^T$, N_e being the number of nodes per element, $\mathbf{M}$ is the mass matrix, and $\mathbf{K}$ and $\mathbf{F}$ are caluclated in the same fashion as discussed in Chapter 9. The solution procedure involves an iterative process by which the numerical solution is obtained element-by-element, once an explicit time integration scheme is applied.

11.6 The Lattice Boltzmann Equation and Numerical Solution

The Boltzmann–BGK equation may also be integrated in certain discrete directions of velocity, similar in the way the discrete ordinates are applied in the solution of radiative transfer equation (see also Chapter 9). In the literature, the equation is discretized based on the quadrature rule that leads to the lattice Boltzmann equation, which has been widely used in simulating a variety of fluid flows and heat transfer at microscales [26, 27]. In this section, we discuss the derivation of the lattice Boltzmann equation from the Boltzmann–BGK equation, the boundary conditions characterizing the interactions between gas molecules and the solid walls, the discontinuous finite element formulation, and the numerical procedures for the solution of the lattice equation. Numerical examples using the discontinuous formulation are given for Taylor vortex flows and flow pass over a cylinder.

11.6.1 Derivation of the Lattice Boltzmann Equation

The lattice Boltzmann equation can be derived in two ways [26]. One of them originates from the lattice gas automata model and the other from the Boltzmann transport equation. An important observation in the lattice gas automata simulations is that the fluid motion can be calculated by assuming that the molecules in the system are massless and the molecules move at the same speed but with different directions. From the lattice gas dynamics point of view, the density of molecules change as a result of collision and thus an accurate representation of the collision operator is important. The other approach to derive the lattice Boltzmann equation is by the direct integration of the Boltzmann transport equation. This approach is considered more rigorous and is discussed below.

The starting point is the Boltzmann–BGK equation (i.e., Equation 11.82). For the purpose of selecting appropriate discrete directions for integration over the velocity space, the equation is non-dimensionalized by using the reference time t_s, number density n_0, temperature T and mass m_0. In this case, the reference velocity can be chosen as $c_s = (k_b T/m_0)^{1/2}$, which is the speed of sound for an ideal gas consisting of molecules with mass m_0 at temperature T. Non–dimensionalized, the Boltzmann–BGK equation takes the following form:

$$\frac{\partial f^*}{\partial t^*} + \mathbf{v}^* \cdot \nabla^* f^* = -\frac{f^* - f^{*(eq)}}{\tau^*} \tag{11.86}$$

where the non-dimensionalized equilibrium distribution function with $\mathbf{F} = \mathbf{0}$ is given by the Maxwellian distribution,

$$f^{*(eq)} = f^{*(0)} = n^* \left(\frac{1}{2\pi}\right)^{d/2} \exp\left[-\frac{(\mathbf{v}^* - \mathbf{u}^*)^2}{2}\right] \tag{11.87}$$

In the above equations, d is the spatial dimension, and the superscript * denotes the dimensionless quantity, $f^* = f/f_0$, $\mathbf{u}^* = \mathbf{u}/c_s$, $n^* = n/n_0$, $t^* = t/t_s$, and $\mathbf{v}^* = \mathbf{v}/c_s$. Also, the references of length l_0, acceleration a_0 and distribution function f_0 are $c_s t_s$, c_s/t_s, and n_0/c_s^d, respectively.

In order to solve Equation 11.86 numerically, a discretized velocity and spatial space needs to be chosen. For this purpose, the distribution function $f^*(\mathbf{x}^*, \mathbf{v}^*, t^*)$ is expanded as a power series in $\mathbf{v}^*$. At a low Mach number, a Hermite polynomial is generally used because of its symmetric property [28–34]. The Hermite polynomial of order n is defined as [28]

$$H^{(n)} = \frac{(-1)^n}{\varpi} \nabla^n \varpi \tag{11.88}$$

where the weighting function is defined as

$$\varpi(\mathbf{v}^*) = \frac{1}{(2\pi)^{d/2}} \exp\left(-\frac{1}{2}\mathbf{v}^* \cdot \mathbf{v}^*\right) \tag{11.89}$$

In the above equations, $H^{(n)}$ is the nth-order tensor and a polynomial of order n. The differential operator is defined in [28] and some of the typical operations are given as follows:

$$\nabla = \nabla_i = \frac{\partial}{\partial v_i^*}; \qquad\qquad \nabla^2 = \nabla_i \nabla_j = \frac{\partial^2}{\partial v_i^* \partial v_j^*};$$

$$\nabla^3 = \nabla_i \nabla_j \nabla_k = \frac{\partial^3}{\partial v_i^* \partial v_j^* \partial v_k^*}; \text{ etc.} \tag{11.90}$$

The first few polynomials are given below [28]:

$$\begin{cases} H^{(0)} = 1 \\ H_i^{(1)} = v_i^* \\ H_{ij}^{(2)} = v_i^* v_j^* - \delta_{ij} \\ H_{ijk}^{(2)} = v_i^* v_j^* v_k^* - (v_i^* \delta_{jk} + v_j^* \delta_{ik} + v_k^* \delta_{ij}) \end{cases} \tag{11.91}$$

where the index i, j, k refers to the component of the velocity vector v_i^*.

With the Hermite polynomials defined above, the particle distribution function can be expressed as

$$f^*(\mathbf{x}^*, \mathbf{v}^*, t^*) = \omega(\mathbf{v}^*) \sum_{n=0}^{\infty} \frac{1}{n!} a_i^{(n)}(\mathbf{x}^*, t^*) H_i^{(n)}(\mathbf{v}^*) \tag{11.92}$$

Here the subscript i is an abbreviation for the multiple indices $\{i_1, i_2, \ldots, i_n\}$. Written in full, we should have the following expression:

$$f^* = \omega(\mathbf{v}^*) \left\{ a^{(0)} H^{(0)} + \sum_{i=0}^{N} a_i^{(1)} H_i^{(1)} + \sum_{i,j=0}^{N} a_{ij}^{(2)} H_{ij}^{(2)} + \sum_{i,j,k=0}^{N} a_{ijk}^{(3)} H_{ijk}^{(3)} + \cdots \right\} \tag{11.93}$$

If the velocity $\mathbf{v}$ has 3 components, then $N = 3$. In Equation 11.92, $a_i^{(n)}$ is the Hermite polynomial coefficient, which is calculated by

$$a_i^{(n)} = \int f^* H_i^{(n)}(\mathbf{v}^*) d\mathbf{v}^* \tag{11.94}$$

For the Maxwell–Boltzmann equilibrium distribution function (Equation 11.87), the first few Hermite coefficients are calculated with the result,

$$a^{(0)} = n^*; \quad a_i^{(1)} = n^* u_i^*; \quad a_{ij}^{(2)} = n^* u_i^* u_j^* \tag{11.95}$$

Since the Hermite expansion has the feature that a velocity moment of a given order is solely determined by the Hermite coefficients up to that order [29], the summation in Equation 11.92 can be truncated to several lower-order terms,

$$f^*(\mathbf{x}^*, \mathbf{v}^*, t^*) = \omega(\mathbf{v}^*) \sum_{n=0}^{M} \frac{1}{n!} a_i^{(n)}(\mathbf{x}^*, t^*) H_i^{(n)}(\mathbf{v}^*) \tag{11.96}$$

For a momentum calculation, M equals 2; for an energy calculation, M equals 3.

We may now discretize the velocity space at each discretized space point $\mathbf{x}$ using quadratures. Let $\mathbf{v}_i^*$ and w_i ($i = 0, 1, ..., N_L$) be the nodes and weights of a numerical quadrature (see also Chapter 3). If $p(\mathbf{v}^*)$ is a polynomial with a degree not greater than $2N_L+1$, we have

$$\int \varpi(\mathbf{v}^*) p(\mathbf{v}^*) d\mathbf{v}^* = \sum_{i=1}^{N_L} w_i p(\mathbf{v}^*) \tag{11.97}$$

where N_L is the order of integration. Thus, the Hermite polynomial coefficient can be calculated using the numerical quadrature,

$$a_i^{(n)} = \int \varpi(\mathbf{v}^*) \frac{f^*}{\varpi(\mathbf{v}^*)} H_i^n(\mathbf{v}^*) d\mathbf{v}^* = \sum_{i=1}^{N_L} \frac{w_i f_i^* H_i^n(\mathbf{v}_i^*)}{\varpi(\mathbf{v}_i)} \tag{11.98}$$

For the same reason, we use numerical quadrature for macroscopic property calculations,

$$n^* = \sum_{i=1}^{N_L} \frac{w_i f_i^*}{\varpi(\mathbf{v}_i^*)}; \quad n^* \mathbf{u}^* = \sum_{i=1}^{N_L} \frac{w_i f_i^* \mathbf{v}_i^*}{\varpi(\mathbf{v}_i^*)};$$

$$n^* E^* = \frac{1}{2} \sum_{i=1}^{N_L} \frac{w_i f_i^* (\mathbf{v}_i^* - \mathbf{u}^*)^2}{\varpi(\xi_i)} \tag{11.99}$$

where $E^* = E/E_0$ with $E_0 = c_s^2$. Equation 11.100 may be rewritten using the relation $\rho^* = m^* n^*$, with $m^* = m/m_0$, whence we have the following expressions for macroscopic variables:

$$\rho^* = \sum_{i=1}^{N_L} \frac{w_i m^* f_i^*}{\varpi(\mathbf{v}_i^*)}; \quad \rho^* \mathbf{u}^* = \sum_{i=1}^{N_L} \frac{w_i m^* f_i^* \mathbf{v}_i^*}{\varpi(\mathbf{v}_i^*)};$$

$$\rho^* E^* = \frac{1}{2} \sum_{i=1}^{N_L} \frac{w_i m^* f_i^* (\mathbf{v}_i^* - \mathbf{u}^*)^2}{\varpi(\boldsymbol{\xi}_i)} \tag{11.100}$$

By neglecting the high order terms, Equation 11.87 can be expanded as a power series up to the second order in $\mathbf{u}^*$ [30–32],

$$f^{*(0)} = n^* \left(\frac{1}{2\pi}\right)^{d/2} \exp\left[-\frac{\mathbf{v}^* \cdot \mathbf{v}^*}{2}\right] \exp(\mathbf{v}^* \cdot \mathbf{u}^*) \exp\left[-\frac{\mathbf{u}^* \cdot \mathbf{u}^*}{2}\right]$$

$$= n^* \varpi(\mathbf{v}^*) \left[1 + \mathbf{v}^* \cdot \mathbf{u}^* + \frac{(\mathbf{v}^* \cdot \mathbf{u}^*)^2}{2} - \frac{\mathbf{u}^* \cdot \mathbf{u}^*}{2}\right] \tag{11.101}$$

By defining the following new variables,

$$f_i = w_i m^* f_i^* / \varpi(\mathbf{v}_i^*) \tag{11.102}$$

$$\mathbf{v}_i^* = \frac{\mathbf{v}_i}{c_s} = \frac{\mathbf{v}_i}{c} \frac{c}{c_s} = \frac{\mathbf{v}_i}{c} \sqrt{3} = \mathbf{e}_i \sqrt{3} \tag{11.103}$$

$$\mathbf{u}^* = \frac{\mathbf{u}}{c_s} = \frac{\mathbf{u}}{c} \frac{c}{c_s} = \frac{\mathbf{u}}{c} \sqrt{3} = \mathbf{u}^{**} \sqrt{3} \tag{11.104}$$

we obtain the lattice Boltzmann equation in the form of,

$$\frac{\partial f_i^{**}}{\partial t^{**}} + \mathbf{e}_i \cdot \nabla^{**} f_i^{**} = -\frac{f_i^{**} - f_i^{**(eq)}}{\tau^{**}} \tag{11.105}$$

where superscript ** means $c = (3)^{1/2} c_s$ is used as the velocity scale instead of c_s, and the equilibrium distribution function is given by

$$f_i^{**(eq)} = \rho^{**} w_i \left(1 + 3\mathbf{e}_i \cdot \mathbf{u}^{**} + \frac{9(\mathbf{e}_i \cdot \mathbf{u}^{**})^2}{2} - \frac{3\mathbf{u}^{**} \cdot \mathbf{u}^{**}}{2}\right) \tag{11.106}$$

Dropping out the superscript as often is done in the fluid mechanics literature, we have the lattice Boltzmann equation written in the following familiar form [26]:

$$\frac{\partial f_i}{\partial t} + \mathbf{e}_i \cdot \nabla f_i = -\frac{f_i - f_i^{(eq)}}{\tau} \tag{11.107}$$

and other related expressions are given as

$$f_i^{(eq)} = \rho w_i \left[1 + 3\mathbf{e}_i \cdot \mathbf{u} + \frac{9(\mathbf{e}_i \cdot \mathbf{u})^2}{2} - \frac{3\mathbf{u} \cdot \mathbf{u}}{2} \right] \tag{11.108}$$

$$\rho = \sum_{i=0}^{N_L} f_i; \quad \rho\mathbf{u} = \sum_{i=0}^{N_L} \mathbf{e}_i f_i; \quad \rho E = \frac{1}{2}\sum_{i=0}^{N_L} f_i (\mathbf{e}_i - \mathbf{u})^2 \tag{11.109}$$

where $\mathbf{F} = 0$ has been assumed.

At this point, we may recall Equation 11.15, which was used in the derivation of the phonon radiative transfer equation in Section 11.1.3. Comparison of Equations 11.107 and 11.15 suggests that the BGK assumption was implied in Equation 11.5. For phonon scattering applications, however, the Bose-Einstein distribution function may be used instead [2].

Turning to Equation 11.107, the numerical quadrature provides guidance on selecting the direction i in the lattice Boltzmann equation. Various quadrature schemes can be used. One popular scheme is the 9-bit lattice, which is shown in Figure 11.5(a). By this scheme 9 directions are selected in the phase space for 2-D problems, viz.,

$$\mathbf{e}_0 = (0,0), \qquad i = 0, \tag{11.110a}$$

$$\mathbf{e}_i = (\cos(\pi(i-1)/2), \sin(\pi(i-1)/2)), \qquad i = 1,2,3,4 \tag{11.110b}$$

$$\mathbf{e}_i = \sqrt{2}(\cos(\pi(i-4.5)/2), \sin(\pi(i-4.5)/2)), \qquad i = 5,6,7,8 \tag{11.110c}$$

with $w_i = 4/9$, for $i = 0$; $1/9$ for $i = 1,...,4$; $1/36$ for $i = 5,...,8$. The 9-bit lattice can be easily extended to 3-D calculations. Application of the same numerical quadrature results in a 27-bit lattice for 3-D problems,

$$\mathbf{e}_0 = (0,0,0), \qquad i = 0 \tag{11.111a}$$

$$\mathbf{e}_i = (\pm 1,0,0),\ (0,\pm 1,0),\ (0,0,\pm 1), \qquad i = 1,2,...,6 \tag{11.111b}$$

$$\mathbf{e}_i = (\pm 1,\pm 1,0),\ (\pm 1,0,\pm 1),\ (0,\pm 1,\pm 1), \qquad i = 7,8,...,18 \tag{11.111c}$$

$$\mathbf{e}_i = (\pm 1,\pm 1,\pm 1), \qquad i = 19,20,...,26 \tag{11.111d}$$

with $w_i = 8/27$ for $i = 0$; $2/27$ for $i = 1, 2, ..., 6$; $1/54$ for $i = 7, 8, ..., 18$; $1/216$ for $i = 19, 20, ..., 26$. Construction of this 27-bit lattice for a cube can be a straightforward superposition of the 2-D 9-bit lattice in three dimensions. For 3-D problems, a 15-bit scheme may also be employed [35],

$$\mathbf{e}_0 = (0,0,0), \qquad i = 0 \tag{11.112a}$$

$$\mathbf{e}_i = (\pm 1, 0, 0), \ (0, \pm 1, 0), \ (0, 0, \pm 1), \qquad i = 1,2,...,6 \tag{11.112b}$$

$$\mathbf{e}_i = (\pm 1, \pm 1, \pm 1), \qquad i = 7,8,...,14 \tag{11.112c}$$

with $w_i = 2/9$ for $i = 0$; $1/9$ for $i = 1, ..., 6$; $1/72$ for $i = 7, ..., 14$. The 15-bit lattice for 3-D calculations is illustrated in Figure 11.5(b).

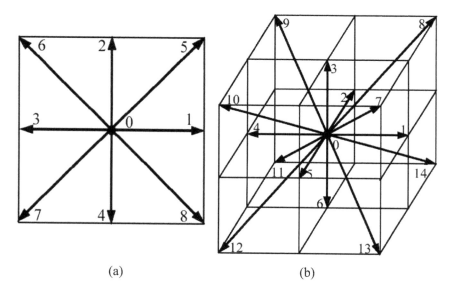

(a)   (b)

Figure 11.5. Velocity lattices for lattice Boltzmann calculations: (a) a 9-bit lattice for 2-D computations and (b) a 15-bit lattice for 3-D computations [35]

The lattice Boltzmann equation can also be used to derive the macroscopic Navier–Stokes equations [26]. The procedure involves the small parameter expansion of the distribution function f similar to that used by Chapman–Enskog's approach [10] to the solution of the Boltzmann integral-differential equation and summing up in all discrete directions i. The detailed process is given in [26, 30], where it is shown that the relaxation time τ is related to the kinematic viscosity as follows:

$$v = \frac{2\tau - 1}{6} \tag{11.113}$$

It is important to note that the quantities in the above equation are dimensionless.

11.6.2 Boundary Conditions

In the lattice Boltzmann simulation, boundary conditions on the fluid velocity are usually imposed on the particle distribution function. Chen and Doolen [23] discuss this point in their review. Typically, the following boundary conditions are applied.

11.6.2.1 Bounce Back Boundary Conditions (No Slip)
The no-slip velocity condition on a motionless wall is modeled by a particle distribution function bounce-back scheme. Bounce-back means that, when a particle distribution streams to a wall node, the particle distribution scatters back to the node it came from. For the 9 velocity 2-D lattice as shown in Figure 11.5(a), for instance, the idea of bounce-back can be illustrated in Figure 11.6(a). In this diagram, the physical boundary is assumed to lie midway between the closest lattice points in the flows and the closest boundary point (i.e., a point that lies inside the solid surface). This assumption is motivated by the analysis of Ziegler [36] who showed that, if the rigid boundary was located midway between the nearest lattice sites, the bounce-back scheme would produce second order accuracy.

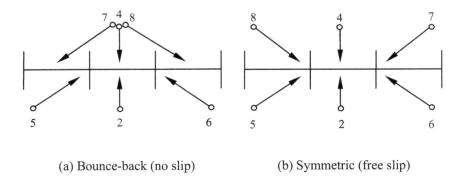

(a) Bounce-back (no slip)   (b) Symmetric (free slip)

Figure 11.6. Schematic illustration of the bounce-back (a) and symmetric (b) boundary conditions for lattice Botlzmann simulations

11.6.2.2 Symmetric Boundary Conditions (Free Slip)
On a free stress surface, one can use a symmetric boundary condition ("free slip") on a particle distribution function. This condition states that a particle distribution function is equal to that on the opposite side of the symmetric surface with equal and opposite normal components of velocity. The free slip boundary condition is shown in Figure 11.5(b).

11.6.2.3 Inflow and Outflow Boundary Conditions (No Gradient)
The additional two boundary conditions are inflow and outflow conditions, which are illustrated in Figure 11.7(a) and 11.7(b), respectively. These arrangemens result in no gradient of a particle distribution function in the inflow or the outflow.

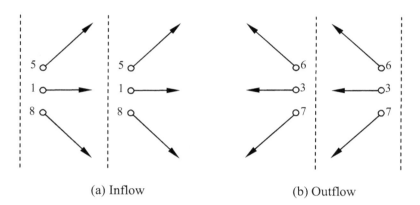

(a) Inflow   (b) Outflow

Figure 11.7. Inflow (a) and outflow (b) boundary conditions used for lattice Boltzmann simulations

11.6.2.4 Force Field Conditions
Another way to treat a boundary is to impose an artificial force field to the fluid. This method was proposed by Goldstein *et al.* [37]. The main idea is to add an artificial body force to the Navier–Stokes equation and choose an appropriate value so that the points inside the solid objects move with the correct velocity. This method was to impose boundary conditions on curved and moving surfaces in a lattice Boltzmann simulation of a turbulent stirred tank, and good agreement with experimental results was obtained for the mean flow and turbulent statistics [38, 39]. The advantage of this technique is that it can provide a relatively simple way of handling complex geometries and moving objects such as the impeller blades in the stirred tank.

11.6.2.5 Moving Wall Conditions
In the case of moving solid wall or moving wall in the shear flow, besides the artificial force field, there is another approach to impose the boundary condition. The particle distribution function can be set as a bounce-back plus an extra term in order that the velocity on the solid wall is the same as the real value. This approach was used to simulate the solid–fluid suspension systems [40, 41].

11.6.3 Discontinuous Finite Element Formulation

To develop a discontinuous finite element formulation for the lattice Boltzmann equation (Equation 11.107), we integrate the equation with respect to the weighting function ϕ over element j,

$$\int_{\Omega_j} \phi \left(\frac{\partial f_i}{\partial t} + \mathbf{e}_i \cdot \nabla f_i \right) dV = \int_{\Omega_j} \phi \left(-\frac{1}{\tau} f_i + \frac{1}{\tau} f_i^{(eq)} \right) dV \tag{11.114}$$

The above equation is integrated once and the fluxes at the element boundaries are replaced by numerical fluxes. This procedure yields the final integral representation for the lattice Boltzmann equation,

$$\int_{\Omega_j} \phi \frac{\partial f_i}{\partial t} \, dV - \mathbf{e}_i \cdot \int_{\Omega_j} f_i \nabla \phi \, dV + \int_{\partial \Omega_j} \hat{f}_i \mathbf{e}_i \cdot \mathbf{n} \phi \, d\Gamma$$

$$+ \frac{1}{\tau} \int_{\Omega_j} \phi f_i dV = \int_{\Omega_j} \phi f_i^{(eq)} dV \tag{11.115}$$

With an appropriate choice of numerical fluxes at the element boundaries and interpolation basis functions, the above equation can be readily integrated numerically. Following the procedures developed in Chapter 5, one reduces the original partial differential equation to a system of ordinary differential equations,

$$\frac{d\mathbf{U}_{i,(j)}}{dt} = L(\mathbf{U}_{i,(j)}) \tag{11.116}$$

where L is the operator and $\mathbf{U}_{i,(j)} = [f_i^{(1)}, f_i^{(2)}, \ldots, f_i^{(2)}, \ldots, f_i^{(Ne)}]_{(j)}^T$. As usual, subscript (j) refers to the jth element and superscript (k) on f to the kth node local to element j. The equations can be integrated using the Runge–Kutta time marching scheme to obtain a numerical solution.

The discontinuous formulation presented above is applied to simulate a two dimensional Taylor vortex flow in a rectangular domain (30×120) with periodic boundary conditions. The initial velocity field is as follows:

$$u_x(x, y, 0) = -\cos(k_1 x) \sin(k_2 y) \tag{11.117a}$$

$$u_y(x, y, 0) = (k_1 / k_2) \sin(k_1 x) \cos(k_2 y) \tag{11.117b}$$

where $k_1 = k_2 = \pi/16$.

For this problem, the analytic solution is also available, which takes the following form:

$$u_x(x, y, t) = -\exp(-vt(k_1^2 + k_2^2)) \cos(k_1 x) \sin(k_2 y) \tag{11.118a}$$

$$u_y(x, y, t) = (k_1 / k_2) \exp(-vt(k_1^2 + k_2^2)) \sin(k_1 x) \cos(k_2 y) \tag{11.118b}$$

which can be used to compare with numerical simulations. The simulation uses a structured mesh, with linear triangular elements used to approximate the local

distribution of the partition function f. The Euler forward time integration is used
and the time step size is 0.01 to satisfy the *CFL* condition.

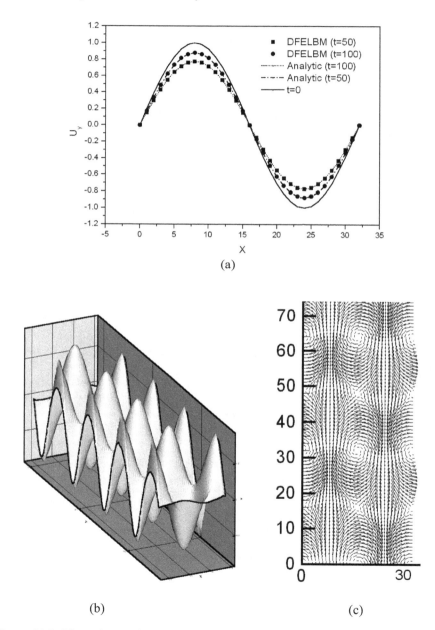

(a)

(b)   (c)

Figure 11.8. Discontinuous finite element solution of the lattice Boltzmann equation: (a)
comparison with analytic solution, (b) 3-D view of the y component of the velocity field at t
$= 100$ and (c) vortex structure in the flow field at $t = 100$

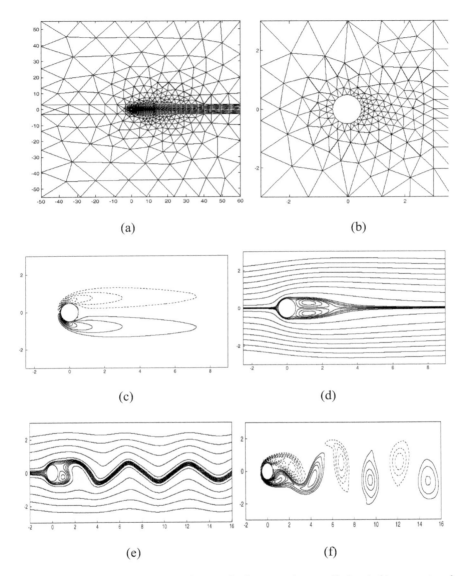

Figure 11.9. Simulation of steady and unsteady flows passing a cylinder: (a, b) unstructured triangular mesh, (c,d) stream lines and vorticity distribution in a steady flow and (e,f) stream lines and vorticity distribution in an unsteady flow [42]

The computed results for the Taylor vortex flow are given in Figure 11.8. As is seen, the computed results compare well with the analytical solution, validating the discontinuous finite element method presented. The 3-D view of the instantaneous flow distribution and the vector flow field at $t = 100$ are also plotted. Strong vortex flow is observed.

The discontinuous finite element procedure has recently been applied by Shi *et al.* [42], who used a spectral basis function for local interpolation. The numerical flux is approximated by Roe's flux model. They applied the discontinuous finite element model to study the flow passing a circular cylinder. Figure 11.9, taken from their work, shows the unstructured meshes (a,b) and computed results include both steady state flows passing a cylinder (c,d) and vortex shedding (e,f).

Exercises

1. Solve Equation 11.15 using the discontinuous finite element method.
2. Consider that a 1-D domain is heated by a pulsing laser heating source irradiating at $X = 0$ along the X direction, with the heating source given by

$$\psi(X,\xi) = \frac{1}{2\Delta\xi_p}\exp[-\frac{X}{\Delta}-(\frac{\xi}{\xi_p})^2]$$

Show that the analytic solution for the temperature distribution takes the following form:

$$\theta(X,\xi) = \int_x^X \int_{\xi_0}^{\xi} G(X,\xi\mid\tilde{X},\tilde{\xi})u(\tilde{X},\tilde{\xi})d\tilde{\xi}d\tilde{\eta}$$

Here, the Green function $G(X,\xi\mid\tilde{X},\tilde{\xi})$ is in the form of

$$G(X,\xi\mid\tilde{X},\tilde{\xi}) = \frac{1}{X_L}e^{\tilde{\xi}-\xi}\sinh(\xi-\tilde{\xi})$$

$$+\begin{cases} \dfrac{2}{X_L}\displaystyle\sum_{m=1}^{N}e^{-(1+\xi_T\lambda_m^2/2)(\xi-\tilde{\xi})}\cos(\lambda_m X)\cos(\lambda_m\tilde{X})\dfrac{\sinh[(\xi-\tilde{\xi})\sqrt{\beta}]}{\sqrt{\beta}} \\ \text{if } \beta > 0 \\[2em] \dfrac{2}{X_L}\displaystyle\sum_{m=N+1}^{\infty}e^{-(1+\xi_T\lambda_m^2/2)(\xi-\tilde{\xi})}\cos(\lambda_m X)\cos(\lambda_m\tilde{X})\dfrac{\sinh[(\xi-\tilde{\xi})\sqrt{-\beta}]}{\sqrt{-\beta}} \\ \text{if } \beta < 0 \end{cases}$$

and

$$u(\tilde{X},\tilde{\xi}) = \frac{1}{\Delta\xi_p^2}\left(\xi_p-\frac{\tilde{\xi}}{\xi_p}\right)\exp\left(-\frac{\tilde{X}}{\Delta}-\left(\frac{\tilde{\xi}}{\xi_p}\right)^2\right)$$

where X_L is the dimensionless length of the solid, $\beta = \lambda_m^2 - (1 + \xi_T \lambda_m/2)^2$, $\lambda_m = m\pi/2$, and N is the number of the terms when β is changing from positive to negative.

3. Develop a discontinuous finite element formulation for a 1-D laser heating problem and compare the results with the analytic solution shown above.

4. The correlation function is defined by

$$\langle X(t)X(t+\tau)\rangle = \lim_{s\to\infty} \frac{1}{s}\int_{-s/2}^{s/2} X(t)X(t+\tau)\,dt$$

By the Fourier transformation, show that the above equation can also be written as

$$\langle X(t)X(t+\tau)\rangle = \frac{1}{2\pi}\int_0^\infty G(\omega)\cos(\omega\tau)\,d\omega$$

$$G(\omega) = \lim_{s\to\infty} \frac{2}{s}\left|\int_{-s/2}^{s/2} X(t)e^{-j\omega t}\,dt\right|^2$$

and further show that

$$\int_0^\infty \langle X(t)X(t+\tau)\rangle\,d\tau = \frac{1}{4}G(0^+)$$

5. The Langevin equation is given by the following ordinary differential equation:

$$\frac{dv}{dt} + \zeta v = X(t)$$

Using the equations derived in Problem 4, derive the relation,

$$\langle v^2(t)\rangle = \frac{1}{2\pi}\int_0^\infty \frac{G(\omega)}{\zeta^2 + \omega^2}\,d\omega$$

Furthermore, show that if $G(\omega) = $ const and $<v^2(t)> = k_b T/m$, then

$$\zeta = \frac{m}{k_b T}\int_0^\infty \langle X(t)X(t+\tau)\rangle\,d\tau$$

which is the well known Green–Kubo relation for calculating transport coefficients.

6.   From the definition of the heat flux,

$$\mathbf{q} = \frac{1}{2}m\int_{\mathbf{v}} (\mathbf{v} - \mathbf{u})|\,\mathbf{v} - \mathbf{u}\,|^2\, f^{(0)}\, g\, d\mathbf{v}$$

where

$$g = -\frac{1}{n}\left(\frac{2k_b T}{m}\right)^{1/2} \mathbf{A}\cdot\nabla(\ell nT)$$

derive the following relation:

$$\mathbf{q} = -k\nabla T$$

where k is the thermal conductivity, and is calculated by

$$k = \frac{2k_b^2 T}{3m}\int_{\mathbf{v}} \mathbf{A}\cdot\mathbf{w}_1 f^{(0)}(\mathbf{w}_1^2 - \tfrac{5}{2})d\mathbf{v}$$

and

$$\mathbf{w}_1 = \left(\frac{m}{2k_b T}\right)^{1/2}(\mathbf{v} - \mathbf{u})$$

7.   Derive the Navier–Stokes equations, the equation of continuity, and Equation 11.113 starting with the lattice Boltzmann equation.
8.   Develop a discontinuous finite element code for the solution of the lattice Boltzmann equation for a 2-D channel flow.

References

[1]   Karniadakis G, Beskok A, Narayan A. Microflows and Nanoflows: Fundamentals and Simulation. New York: Spinger-Verlag, 2002.
[2]   Tzou DY. Macro-to-Microscale Heat Transfer. Washington D.C.: Taylor & Francis, 1997.
[3]   Reitz JR, Milford FJ, Christy RW. Foundations of Electromagnetic Theory. Reading MA: Addison-Wesley, 1992.
[4]   Qiu TQ, Tien CL. Short-Pulse Laser Heating on Metals. Int. J. Heat Mass Transf. 1992, 35: 2799–2808.
[5]   Xu B, Li BQ. Finite Element Solution of Non-Fourier Thermal Wave Problems. Numer. Heat Transf. Part B. 2003; 44: 45–60.
[6]   Ai X, Li BQ. A Discontinuous Finite Element Method for Hyperbolic Thermal Wave Problems. J. Eng. Comput. 2004; 21(6): 577–597.

[7]   Majumdar, A. Microscale Heat Conduction in Dielectric Films. ASME J. Heat Transf. 1993; 117: 7–16.
[8]   Bassi F, Rebay S. A High-Order Accurate Discontinuous Finite Element Method for the Numerical Solution of the Compressible Navier–Stokes Equations. J. Comput. Phys. 1997; 131: 267–279.
[9]   Arnold DN, Brezzi F, Cockburn B, Marini LD. Unified Analysis of Discontinuous Galerkin Methods for Elliptic Problems. SIAM J. Numer. Anal. 2002; 39(5): 1749–1779.
[10] Chapman S, Cowling TG. The Mathematical Theory of Non-Uniform Gases. London: Cambridge University Press, 1970.
[11] Roy S, Raju R, Chuang HF, Cruden BA, Meyyappan M. Modeling of Gas Flow Through Microchannels and Nanopores. J. App. Phys. 2003; 93(8): 4870–4879.
[12] Maxwell JC. Philos. Trans. R. Soc. London. 1879; 170: 231–235.
[13] Dadzie SK, Méolans JG. Anisotropic Scattering Kernel: Generalized and Modified Maxwell Boundary Conditions. J. Math. Phys. 2004; 45(5): 1804 – 1809.
[14] Sakiyama Y, Takagi S, Matsumoto Y. Multiscale Analysis of Nonequilibrium Rarefied Gas Flows with the Application to Silicon Thin Film Process Employing Supersonic Jet. Phys. Fluids 2004; 16(5): 1620 1626.
[15] Isihara A. Statistical Physics. New York: Academic Press, 1971.
[16] Kubo R, Toda M, Hashitsume N. Statistical Physics II: Nonequilibrium Statistical Mechanics. Berlin: Springer-Verlag, 1991.
[17] Cercignani C. Theory and Application of the Boltzmann Equation. Scotland: Scottish Academic Press, 1975.
[18] Harris S. An Introduction to the Theory of the Boltzmann Equation. Austin: Holt, Rinehart and Winston, 1970.
[19] Siewert CE. Viscous-Slip, Thermal-Slip, and Temperature-Jump Coefficients as Defined by the Linearized Boltzmann Equation and the Cercignani–Lampis Boundary Condition. Phys. Fluids 2003; 15(6): 1696–1702.
[20] Kosuge S, Aoki K, Takata S. Heat Transfer in a Gas Mixture Between Two Parallel Plates: Finite-Difference Analysis of the Boltzmann Equation. AIP Conf. Proc. 2001; 585(1): 289–296.
[21] Christlieb AJ, Hitchon WNG, Sun Q, Boyd ID. Application of the Transition Probability Matrix Method to High Knudsen Number Flow Past a Micro-Plate. AIP Conf. Proc. 2003; 663(1): 768–773.
[22] Aristov VV. Methods of Direct Solving the Boltzmann Equation and Study of Nonequilibrium Flows. The Netherlands: Kluwer Academic Publishers, 2001.
[23] Bhatnagar PL, Gross EP, Krook M. A Model for Collision Processes in Gases I: Small Amplitude Processes in Charged and Neutral One-Component Systems. Phys. Rev. 1954; 94 (3): 511–525.
[24] Shan X, Chen H. Lattice Boltzmann Model for Simulating Flow with Multiple Phases and Components. Phys. Rev. E. 1993; 47: 1815–1819.
[25] Martys NS, Shan X, Chen H. Evaluation of the External Force Term in the Discrete Boltzmann Equations. Phys. Rev. E. 1998; 58: 6855–6857.
[26] Chen S, Doolen GD. Lattice Boltzmann Method for Fluid Flows. Annu. Rev. Fluid Mech. 1998; 30: 329–364.
[27] Chen S, Chen HD, Martinez D, Mattheus W. Lattice Boltzmann Model for Simulation of Magnetohydrodynamics. Phys. Rev. Lett. 1991; 67: 3776–3779.
[28] Grad, H.(1949) Note on N-Dimensional Hermite Polynomials. Commun. Pure Appl. Math. 1949; 2: 331–336.
[29] Shan, X. and He, X. Discretization of the Velocity Space in the Solution of the Boltzmann Equation. Phy. Rev. Letters. 1998; 80: 65–68.

[30] Chen S, Wang Z, Shan X, Doolen GD. Lattice Boltzmann Computational Fluid Dynamics in Three Dimensions. J. Stat. Phys.1992; 68: 379–400.

[31] Luo LS. Unified Theory of the Lattice Boltzmann Models for Nonideal Gases. Phys. Re. Lett. 1998; 81(8): 1618–1621.

[32] He X, Luo L. A Priori Derivation of the Lattice Boltzmann Equation. Phy. Rev. E. 1997; 55: R6333–R6336.

[33] He X, Luo L. Theory of the Lattice Boltzmann Method: from the Boltzmann Equation to the Lattice Boltzmann Equation. Phy. Rev. E. 1997; 55: R6333–R6336.

[34] Cao N, Chen S, Jin S, Martinez D. Physical Symmetry and Lattice Symmetry in the Lattice Boltzmann Method. Phys. Rev. E. 1996; 55: R21–R24.

[35] Qian D. Bubble Motion, Deformation and Breakup in Stirred Tanks. Ph.D. Thesis. New York: Clarkson University, 2003.

[36] Ziegler DP. Boundary Conditions for Lattice Boltzmann Simulations. J. Stat.Phys. 1993; 71: 1171–77.

[37] Goldstein D, Handler R, Sirovich L. Modeling a No-slip Flow Boundary with an External Force Field. J. Comput. Phys. 1993; 105: 354–366.

[38] Eggels JGM, Somers JA. Numerical Simulation of Free Convective Flow Using the Lattice-Boltzmann Scheme. Int. J. Heat Fluid Flow, 1995; 16: 357–364.

[39] Derksen JJ, Van den Akker HEA. Large Eddy Simulations on the Flow Driven by a Rushton Turbine. AIChE J. 1999; 45: 209–221.

[40] Ladd AJC. Numerical Simulation of Particulate Suspensions via a Discretized Boltzmann Equation, Part 1: Theoretical Foundation. J. Fluid Mech. 1994; 271: 285–309.

[41] Ladd AJC. Numerical Simulation of Particulate Suspensions via a Discretized Boltzmann Equation, Part 2: Numerical Results. J. Fluid Mech. 1994; 271: 311–339.

[42] Shi X, Lin J, Yu Z. Discontinuous Galerkin Spectral Element Lattice Boltzmann Method on Triangular Element. Int. J. Numer. Meth. Fluids 2003; 42: 1249 –1261.

Fluid Flow and Heat Transfer in Electromagnetic Fields

In many thermal and fluids systems, external fields such as the electromagnetic fields are imposed to achieve a certain desired performance through the interaction of the fluids, or fluid motion with the imposed fields. Practical systems include induction and microwave heating, electromagnetic stirring, magnetic control of turbulent flows and thermal fluctuations, plasma spaying and micro actuation in fluidic devices, etc. Strictly speaking, the general, rigorous mathematical description of the electromagnetic field in a moving medium and its mutual coupling with the thermal field and fluid motions in the medium should be made within the framework of Einstein's relativistic theory. For most engineering applications, however, the speed of the motion is much smaller than that of the light, and thus the commonly known magnetohydrodynamic theory provides an adequate theoretical basis for the description of the electromagnetic, thermal and fluid flow fields and their interactions. The magnetohydrodynamic equations are a nonlinear set consisting of the Maxwell equations, the Navier–Stokes equations and the thermal and species transport equations, with the coupling between the fields made through constitutive relations and sources/sinks.

In order to accurately interpret the behavior of an electromagnetically assisted thermal fluid system for both fundamental understanding and process design, information on the distribution of the thermal, fluid flow and electromagnetic fields is required. This in turn requires the solution of the magnetohydrodynamic governing equations. Thus far, we have discussed extensively the application of the discontinuous finite element methods to the solution of the thermal and fluid flow equations for various systems. These methods, as will be shown below, may be applied with straightforward modifications in the source/sink terms, for the analysis of thermal and fluid flow systems under the influence of an electromagnetic field. The solution of the Maxwell equations is a subject of computational electromagnetics and many different techniques, both analytical and numerical, have been developed. Recently, the discontinuous finite element formulation has also been extended to solve the complete set of the Maxwell equations. One important aspect of the Maxwell equations, which is different from the thermal fluids equations, is that the former will in general transcend the region of a conducting fluid and, ideally, extend to all of space. Consequently, the solution of these equations needs to discretize the entire space, even though the interest of the solution is primarily in a confined region for thermal fluids applications. This is in contrast with the solution of the thermal and fluid flow equations, which is often obtained only in a region of interest.

This chapter is concerned with the discontinuous finite element solution of these electromagnetically induced thermal and fluid flow problems. It starts with a brief discussion of the magnetohydrodynamic theory of the electromagnetic field and its interaction with the thermal and fluid flow fields. The use of the discontinuous finite element method is then presented for the solution of the Maxwell equations. The coupled solution of electromagnetic and thermal fluids systems is also discussed. Numerical simulation of electromagnetically driven fluid flow and thermal systems is illustrated through examples taken from the applications in the areas of microwave heating, electrokinetically driven flows in microchannels and electrically induced free surface deformation.

12.1 Maxwell Equations and Boundary Conditions

In this section, we briefly discuss the Maxwell equations and the boundary conditions that are relevant to electromagnetically induced thermal and fluid flow applications. An in-depth discussion of the electromagnetic theory and its applications has been documented in well known textbooks [1–3].

12.1.1 Maxwell Equations

The electromagnetic field distributions in continuum media are governed by the Maxwell equations, which represent one of the most elegant and concise ways to describe the fundamentals of electricity and magnetism. The Maxwell equations are four vector equations summarizing the basic laws governing the electromagnetic field behavior in a medium [1–3],

$$\nabla \times \mathbf{E} = -\frac{\partial \mathbf{B}}{\partial t} \tag{12.1}$$

$$\nabla \times \mathbf{H} = \frac{\partial \mathbf{D}}{\partial t} + \mathbf{J} \tag{12.2}$$

$$\nabla \cdot \mathbf{B} = 0 \tag{12.3}$$

$$\nabla \cdot \mathbf{D} = \rho_e \tag{12.4}$$

where $\mathbf{E}$ is the electric field, $\mathbf{D}$ is the electric displacement field, $\mathbf{B}$ is the magnetic induction (magnetic flux density) field, and $\mathbf{H}$ is the magnetic field. Also, ρ_e is "free" (including both induced and impressed) electric charges, which do not include bounded charges such as induced dipoles in dielectrics. The current density consists of both the impressed and induced contributions: $\mathbf{J} = \mathbf{J}_e + \mathbf{J}_i$, $\mathbf{J}_i$ being impressed and $\mathbf{J}_e$ induced. The relations between $\mathbf{E}$ and $\mathbf{D}$, and between $\mathbf{B}$ and $\mathbf{H}$, are specified by the constitutive equations,

$$\mathbf{D} = \varepsilon_0 \mathbf{E} + \varepsilon_0 \mathbf{P} = \varepsilon_0 \left(1 + \chi_e\right)\mathbf{E} = \varepsilon\,\mathbf{E} \qquad (12.5)$$

$$\mathbf{B} = \mu_0 \mathbf{H} + \mu_0 \mathbf{M} = \mu_0 \left(1 + \chi_m\right)\mathbf{H} = \mu\,\mathbf{H} \qquad (12.6)$$

where ε_0 is the permittivity of free space, $\mathbf{P}$ is the polarization, μ_0 is the permeability of free space, and $\mathbf{M}$ is the magnetization. In a linear, isotropic medium, which we consider in this chapter, ε and μ are constants. In general $\mathbf{H}$ (or $\mathbf{D}$) is not a unique function of $\mathbf{B}$ (or $\mathbf{E}$), but depends upon the earlier time evolution (hysteresis). Also, ε_0 and μ_0 are related through the following relation:

$$\mu_0 \varepsilon_0 = c^{-2} \qquad (12.7)$$

where c is the speed of light.

Just as the mass is conserved in a fluid flow system, so are the electric charges, which cannot be created or destroyed. The continuity equation for charge conservation has the form of

$$\frac{\partial \rho_e}{\partial t} + \nabla \cdot \mathbf{J} = 0 \qquad (12.8)$$

which follows from the Maxwell equations. Specifically, Equation 12.8 is a combination of Equations 12.2 and 12.4.

12.1.2 Boundary Conditions

As for other boundary value problems, boundary conditions need to be specified to obtain the electromagnetic field distribution in a domain. The boundary conditions that the electric and magnetic fields must satisfy can be deduced by a standard procedure, which involves creating a pillbox-shaped differential volume at the interface between two media, integrating the Maxwell equations over the volume and taking the relevant limits. We give below the boundary constraints for the electric and magnetic fields but omit the detailed procedures to derive these conditions, which one can find in standard textbooks [1, 2].

12.1.2.1 Interface Boundary Condition
At the interface between two media, the tangential electric field must be continuous and the normal component of the electric displacement field suffers a jump. These two conditions are mathematically stated as

$$\mathbf{n} \times \left(\mathbf{E}_1 - \mathbf{E}_2\right) = 0 \qquad (12.9)$$

$$\mathbf{n} \cdot (\mathbf{D}_1 - \mathbf{D}_2) = \sigma_s \qquad (12.10)$$

where σ_s is the free surface charge density. Also, at the interface between two media, the tangential magnetic field experiences a jump, and the normal component of the magnetic induction field is continuous,

$$\mathbf{n} \times (\mathbf{H}_1 - \mathbf{H}_2) = \mathbf{J}_s \qquad (12.11)$$

$$\mathbf{n} \cdot (\mathbf{B}_1 - \mathbf{B}_2) = 0 \qquad (12.12)$$

where $\mathbf{J}_s$ is the tangential surface currents along the interface. Here the normal is outward from medium 2 to medium 1. We note that only two of the above four boundary conditions are independent: one from Equations 12.9 and 12.10 and the other from Equations 12.11 and 12.12.

Sometimes, when the current density is used as a variable, the boundary condition needs to be prescribed for it. The same pillbox procedure gives the surface charge conservation at the interface in the following form [4, 5]:

$$\mathbf{n} \cdot (\mathbf{J}_1 - \mathbf{J}_2) + \nabla_s \cdot \mathbf{J}_s + \frac{\partial \sigma_s}{\partial t} = 0 \qquad (12.13)$$

where ∇_s is the surface derivative operator. Here, σ_s includes both external and induced surface charges, but does not include the polarization surface charge on dielectric surfaces.

The above boundary conditions are general. For special cases, these conditions take simpler forms. We consider several special and yet commonly encountered situations below.

12.1.2.2 Perfect Conducting Surface
If medium 2 is assumed to be a perfect conductor, then $\mathbf{E}_2$ and $\mathbf{B}_2$ are zero and we have, $\mathbf{E} = \mathbf{E}_1$ and $\mathbf{B} = \mathbf{B}_1$,

$$\mathbf{n} \times \mathbf{E} = 0 \text{ and } \mathbf{n} \cdot \mathbf{B} = 0 \qquad (12.14a)$$

where $\mathbf{n}$ is outnormal pointing away from the conducting medium. Note that in this approximation, the conducting boundary can support a surface current and surface charge,

$$\mathbf{n} \cdot \mathbf{D} = \sigma_s \text{ and } \mathbf{n} \times \mathbf{H} = \mathbf{J}_s \qquad (12.14b)$$

Here the physics is such that $\mathbf{n} \cdot \mathbf{D}_1 = \sigma_{es} + \mathbf{n} \cdot \mathbf{D}_2 = \sigma_{es} + \sigma_{Ds}$, where σ_{es} are "free" surface charges and $\sigma_{Ds} = \mathbf{n} \cdot \mathbf{D}_2$ induced charges. Thus strictly speaking, $\mathbf{n} \cdot \mathbf{D}_2$ is not "truly" zero but its effect is simulated by σ_{Ds}. These conditions are often used to model a metal surface of a waveguide or cavity, such as a microwave oven for thermal processing or hyperthermia.

12.1.2.3 Impedance Condition

The impedance condition is often used in computational electromagnetics. By definition, impedance Z ($Z = R + j\omega L + 1/(j\omega C)$) is the input impedance for an R-L-C circuit and we also have $1/Z$ as the *admittance* for an R-L-C circuit. But in general for an electric circuit, $V = ZI$, V being the voltage, and I the current. In this sense, the impedance acts as a resistance.

For a distributed electromagnetic field, the impedance is calculated by the following expression:

$$Z = \frac{|\mathbf{n} \times \mathbf{E}|}{|\mathbf{H}|} = \frac{|\mathbf{E}|}{|\mathbf{n} \times \mathbf{H}|} \quad (\text{e.g., } Z = \frac{E_x}{H_y} \text{ for a plane wave}) \tag{12.15}$$

Note that the analogy is such that $E \Leftrightarrow V$ and $H \Leftrightarrow I$.

If medium 2 is an imperfect conductor (for example, a conductor coated with a thin layer of dielectric on the surface), then the following impedance condition applies at the interface:

$$\mathbf{E} - (\mathbf{n} \cdot \mathbf{E})\mathbf{n} = \eta Z_0 \mathbf{n} \times \mathbf{H} \quad \text{or} \quad \mathbf{n} \times \mathbf{E} = \eta Z_0 [\mathbf{H} - (\mathbf{n} \cdot \mathbf{H})\mathbf{n}] \tag{12.16}$$

Note that $\mathbf{E} - (\mathbf{n} \cdot \mathbf{E})\mathbf{n} = \mathbf{n} \times (\mathbf{n} \times \mathbf{H})$. For the 2-D case, we have, with $\mathbf{E} = \hat{z}E_z$ and $\mathbf{H} = \hat{z}H_z$,

$$\frac{\partial E_z}{\partial n} = jk\frac{\mu_{r1}}{\eta}E_z \quad \text{or} \quad \frac{\partial H_z}{\partial n} = jk\varepsilon_{r1}\eta H_z \tag{12.17}$$

where $\eta = (\mu_{r1}/\varepsilon_{r2})^{1/2}$ is the normalized intrinsic impedance of medium 2 and $Z_0 = (\mu_0/\varepsilon_0)^{1/2}$ is the intrinsic impedance of free space. By definition, $1/\eta$ is the intrinsic admittance of medium 2.

12.1.2.4 Sommerfield Radiation Condition

If all sources are immersed in free space, then the electric and magnetic fields are required to satisfy the Sommerfield radiation condition,

$$\lim_{r \to \infty} r \left[\nabla \times \begin{pmatrix} \mathbf{E} \\ \mathbf{H} \end{pmatrix} + jk\hat{r} \times \begin{pmatrix} \mathbf{E} \\ \mathbf{H} \end{pmatrix} \right] = \begin{pmatrix} 0 \\ 0 \end{pmatrix} \tag{12.18a}$$

where $r = \sqrt{x^2 + y^2 + z^2}$. For a 2-D problem, $r = \sqrt{x^2 + y^2}$ and the above condition simplifies to

$$\lim_{r \to \infty} r \left[\frac{\partial}{\partial r} \begin{pmatrix} E_z \\ H_z \end{pmatrix} + jk \begin{pmatrix} E_z \\ H_z \end{pmatrix} \right] = \begin{pmatrix} 0 \\ 0 \end{pmatrix} \tag{12.18b}$$

12.1.2.5 Symmetry Boundary Condition
The symmetry boundary condition for the electromagnetic fields is applied as follows:

For calculating **E**,

$$\mathbf{H} \times \mathbf{n} = 0 \quad \text{(perfect magnetic conductor)} \tag{12.19a}$$

$$\nabla \times \mathbf{E} \times \mathbf{n} = 0 \quad \text{(perfect magnetic conductor)} \tag{12.19b}$$

For calculating **H**,

$$\mathbf{E} \times \mathbf{n} = 0 \quad \text{(perfect electrical conductor)} \tag{12.20a}$$

$$\nabla \times \mathbf{H} \times \mathbf{n} = 0 \quad \text{(perfect electrical conductor)} \tag{12.20b}$$

12.2 Maxwell Stresses and Energy Sources

The Maxwell stresses and energy flux are important quantities that are directly responsible for fluid motion and thermal balance. They appear either as a source term in the momentum and thermal balance equations or as additional terms in the boundary conditions.

The Maxwell stresses represent the interaction of the electric and magnetic fields and are of the following general form [2]:

$$T = \mathbf{DE} + \mathbf{BH} - 0.5\,\mathbf{I}\left[\mathbf{E}\cdot\mathbf{E}\left(\varepsilon - \left(\frac{\partial\varepsilon}{\partial\rho}\right)_T \rho\right) + \mathbf{H}\cdot\mathbf{H}\left(\mu - \left(\frac{\partial\mu}{\partial\rho}\right)_T \rho\right)\right] \tag{12.21}$$

In the Maxwell stresses, the first two terms are directly responsible for bulk flows and the last two terms cause the interface shape to change because they add to the pressure if ε and μ are constant. In the case that these properties are a strong function of the density, the last terms will also contribute to the flow motions, which are often referred to as the electrorestrictive and magnetorestrictive stresses. For simplicity, we take both $\partial\varepsilon/\partial\rho$ and $\partial\mu/\partial\rho$ to be zero.

When an electromagnetic field is present, the momentum balance for the fluid flow needs to include the Maxwell stress tensor,

$$\rho\frac{D\mathbf{u}}{Dt} = -\nabla p + \nabla\cdot\boldsymbol{\sigma} + \nabla\cdot(\mathbf{DE} + \mathbf{BH}) \tag{12.22}$$

and the thermal balance equation needs to incorporate the Joule heating effects,

$$\rho C\frac{D\mathbf{u}}{Dt} = -\nabla\cdot\mathbf{q} + \mathbf{J}\cdot\mathbf{E} \tag{12.23}$$

where $\mathbf{q}$ is the heat flux and the last term results from the self-interaction of the electric field, namely the Joule heating source.

The mathematical description of electromagnetically induced thermal and fluid flow problems consists of the Maxwell equations 12.1–12.4 and Equations 12.21–12.23. These equations constitute the basis for the mathematical description of magnetohydrodyanmic phenomena.

12.3 Discontinuous Formulation of the Maxwell Equations

The use of the discontinuous finite element method for the solution of the Maxwell equations has recently received much attention. Some useful algorithms have been proposed and are discussed in this section. Their use for practical process design of electromagnetically assisted thermal and fluids systems, however, has yet to be tested. More information is also needed to assess the numerical performance of the discontinuous schemes in comparison with other established methods such as the finite element method, the boundary element method, the finite difference time domain method and the method of moments.

12.3.1 Solution in Time Domain

The Maxwell equations are first order vector partial differential equations. As a result, the discontinuous finite element method may be applied directly to solve these equations. To develop a discontinuous formulation, the Maxwell equations are re-written in the following conservation form [6, 7]:

$$\frac{\partial \mathbf{Q}}{\partial t} + \nabla \cdot \mathbf{F} = \mathbf{S} \qquad (12.24a)$$

where the variables are defined as follows:

$$\mathbf{Q} = \begin{pmatrix} \mathbf{B} \\ \mathbf{D} \end{pmatrix}; \ \mathbf{F}_i = \begin{pmatrix} \mathbf{e}_i \times \mathbf{E} \\ -\mathbf{e}_i \times \mathbf{H} \end{pmatrix}; \ \mathbf{S} = \begin{pmatrix} (\mu - \mu_{ref})\partial \mathbf{H}^i / \partial t \\ (\varepsilon - \varepsilon_{ref})\partial \mathbf{E}^i / \partial t \end{pmatrix} \qquad (12.24b)$$

In the above equations, the superscript i denotes the incident fields, and $\mathbf{e}_i$ is the unit vector in the ith Cartesian coordinate direction.

Following the general procedure for a discontinuous formulation of boundary value problems, the above equation is integrated over element j with respect to testing functions. With unknowns approximated using the polynomial basis function, followed by elemental calculations, one has the following equation:

$$\frac{d\mathbf{Q}}{dt} = L(\mathbf{F}, \mathbf{S}) \qquad (12.25)$$

where L is the discretized matrix operator.

The above matrix equation can then be integrated in time using the Runge–Kutta integrator, once the numerical fluxes are determined. Kopriva *et al.* [6] tested the above algorithm and used the fluxes by solving a Riemann problem. Their results compare well with the analytic solution for a 2-D scattering problem. Hesthaven and Warburton recently studied the stability, convergence and accuracy of the method [7]. The construction of a locally divergence-free function space for the discontinuous solution of the Maxwell equations has also been proposed, which uses the Lax–Friedrichs fluxes [8, 9].

12.3.2 Solution in Frequency Domain

For applications in electromagnetic wave propagation, the vector wave form of the Maxwell equations is may also be written in frequency domain. With $E(x, t)$ (or $H(x, t)) = E(x)\exp(j\omega t)$ (or $H(x)\exp(j\omega t)$), the frequency-based vector wave form for a charge-free medium can be obtained by combining the original Maxwell equations,

$$\nabla \times (\mu_c^{-1} \nabla \times E) - k_0^2 \varepsilon_c E = -j\omega\mu_0 J \qquad (12.26)$$

$$\nabla \cdot E = 0 \qquad (12.27)$$

where $E = E(x)$ and, for convenience, this holds true for frequency-domain based method from here on unless indicated otherwise. In solving the above equations, the $\nabla \cdot E = 0$ condition may pose a problem. Perugia *et al.* [10] proposed a discontinuous finite element formulation with the constraint $\nabla \cdot E = 0$ enforced by an internal penalty approach, which is often used in the calculations of incompressible fluid flows. Houston *et al.* [11] recently presented a non-stabilized discontinuous finite element formulation for the solution of the Maxwell equations in frequency domain.

12.3.3 Solution in Other Forms

For many thermal and fluid flow applications, the Maxwell equations may be reduced to the other forms that can be solved using the techniques discussed in Chapters 4 and 5. This is illustrated through numerical examples discussed below.

12.4 Electroosmotic Flows

Over the past decade, considerable attention has been received in the research area of electroosmotic flows in micro- and nano-channels, which are essential components for microfluidics or on-chip laboratories for biochemical applications. As the ratio of the volume over the surface area becomes small in these fluid systems, the surface and interfacial phenomena become increasingly important. In this section, we discuss the basic principle of this type of electrically driven flow and the discontinuous finite element algorithms for the flow simulation.

12.4.1 Governing Equations

Electroosmotic flows, which were discovered many decades ago, are driven by the interfacial electric forces near the interface between the electrolyte and a solid wall [12]. The polarization of charges near the solid surface results in a double layer or Debye layer at the channel walls, where electrochemical reactions at the wall–liquid interface cause a surplus of ions in the liquid near the wall surface. These ions are closely adsorbed near the wall surface and balance the negative charges on the wall so that the bulk of the liquid remains electroneutral. When an external electric field is applied along the channel, however, a shear force gradient is produced in the double layer, which causes a motion of bulk fluid or electroosmotic flow, thereby pumping the bulk electrolyte in the direction of the electric field. As a result of the motion, a drag will be produced at the wall, which as usual opposes the fluid motion. This flow is illustrated in Figure 12.1.

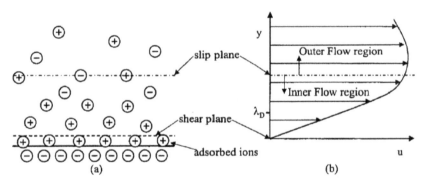

Figure 12.1. Illustration of the Debye double layer near the solid wall and flow regions separated by the slip plane [11]

The thickness of the double layer or the Debye layer is typically on the order of 1–10 nm. If the channel size is larger than the Debye thickness, then the description of the flow in the Debye layer may be decoupled from the bulk flow. Santiago [13] has recently analyzed the inner motion near a charged surface. According to his analysis, the electroosmotic flow velocity of the bulk fluid is proportional to the magnitude of the electric field E applied parallel to the wall, with a constant proportionality termed the electroosmotic mobility μ_e. For these cases, the flow in the bulk channel can be modeled by a slip condition, where the velocity parallel to the walls u_s is given by

$$u_s - u_{wall} = \mu_e E \tag{12.28}$$

with subscript s referring to the direction along the surface of the wall, $\mu_e = - \varepsilon \zeta / \mu$, and ζ the ζ-potential [13]. This condition is similar to the Maxwell correction to the gas flows in small channels. The slip condition, however, becomes invalid when the size of the channel is comparable to the Debye layer. Experiments suggest that the slip model breaks down in a channel of width below 100 nm [14, 15].

Studies further show that near the wall surface, the ion density follows a Maxwellian distribution [12],

$$\frac{c_i}{c_{\infty,i}} = \exp\left(-\frac{z_i e (y - y_w) \psi}{k_b T} \right) \tag{12.29}$$

where c_i is the concentration of species i, z_i is the valence of species, e is the electric charge, ψ is the electric potential in the double layer, and k_b is the Boltzmann constant.

For this type of problem, electrostatics is applicable and thus two of the Maxwell equations (Equations 12.1 and 12.4) can be combined, giving rise to the Poisson equation. Substituting Equation 12.29 into the Poisson equation, we have the following Poisson–Boltzmann equation for ψ:

$$\nabla^2 \psi = -\frac{e}{\varepsilon} \sum z_i c_{\infty,i} \exp\left(-\frac{z_i e (y - y_w) \psi}{k_b T} \right) \tag{12.30}$$

where the right hand side represents the contribution from the ionic charges in the liquid solution $(-\rho_e / \varepsilon)$, and the sum is taken over all ionic species. If we apply $e^{-x} = 1 - x + O(x^2)$ and note that $\Sigma c_{\infty,i} z_i = 0$, then the above equation simplifies to

$$\nabla^2 \psi = \frac{\psi e^2}{\varepsilon k_b T} \sum z_i^2 c_{\infty,i} = \frac{\psi}{\lambda_D^2} \tag{12.31}$$

where λ_D is the Debye length defined as

$$\lambda_D = \sqrt{\frac{\varepsilon k_b T}{e^2 \sum z_i^2 c_{\infty,i}}} \tag{12.32}$$

We note that the Debye length increases with temperature and that the charge density in the system is given by

$$\rho_e = -\varepsilon \frac{\psi}{\lambda_D^2} \tag{12.33}$$

Equation 12.31, coupled with the potential equation for the applied electric field distribution and the momentum balance equation, completes the mathematical description of electroosmotic flows in micro or nanochannels. We summarize these equations in non-dimensionalized form below for such flow in a microchannel,

$$\nabla^2 \psi = (kh)^2 \psi \tag{12.34}$$

$$\nabla^2 \phi = 0 \tag{12.35}$$

$$\nabla \cdot \mathbf{u} = 0 \tag{12.36}$$

$$\mathrm{Re}\left[\frac{\partial \mathbf{u}}{\partial t} + (\mathbf{u} \cdot \nabla)\mathbf{u}\right] = -\nabla p + \nabla^2 \mathbf{u} + (k^2 hH)\,\psi \nabla \phi \tag{12.37}$$

where the length is scaled by the channel width h, the potential ϕ by the value of the applied potential (φ), the potential ψ by the zeta potential ζ, the velocity u by $\varphi\varepsilon\zeta/\mu H$, where H is the distance between reservoirs 1 and 2 (or two electrodes), and the pressure by $\varphi\varepsilon\zeta/hH$. In the above equation, Re is the flow Reynolds number defined as

$$\mathrm{Re} = \rho\left(\frac{\varphi\varepsilon\zeta}{\mu H}\right)\left(\frac{h}{\mu}\right) \tag{12.38}$$

It is noted that Equations 12.34 and 12.35 represent two contributions to the total electric field in the system, $\mathbf{E} = -\nabla\psi - \nabla\phi$.

12.4.2 Discontinuous Finite Element Formulation

To implement the discontinuous finite element method, the governing equations (Equations 12.34–12.37) are rewritten as a set of first order differential equations,

$$\mathbf{s} = \nabla\psi \tag{12.39}$$

$$\nabla \cdot \mathbf{s} = (kh)^2 \psi \tag{12.40}$$

$$\mathbf{q} = \nabla\phi \tag{12.41}$$

$$\nabla \cdot \mathbf{q} = 0 \tag{12.42}$$

$$\nabla \cdot \mathbf{u} = 0 \tag{12.43}$$

$$\underline{\tau} = \nabla\mathbf{u} \tag{12.44}$$

$$Re\left[\mathbf{u}_t + (\mathbf{u} \cdot \nabla)\mathbf{u}\right] = -\nabla p + \nabla \cdot \underline{\tau} + \mathbf{f} \tag{12.45}$$

$$\mathbf{f} = (k^2 hH)\psi\, \mathbf{q} \tag{12.46}$$

Multiplying Equations 12.39–46 by smooth test functions v, $\underline{\sigma}$, $\mathbf{v}$, $\mathbf{r}$, r, $\mathbf{w}$ and w respectively, and integrating by parts over an arbitrary element Ω_j, followed by replacing the fluxes at the element boundary by numerical fluxes, we have the following integral representations:

$$\int_{\Omega_j} s_h \cdot \mathbf{r}\, dV = -\int_{\Omega_j} \psi_h \nabla \cdot \mathbf{r}\, dV + \int_{\partial\Omega_j} \hat{\psi}_h \mathbf{n}_j \cdot \mathbf{r}\, dS \tag{12.47}$$

$$-\int_{\Omega_j} s_h \cdot \nabla r\, dV + \int_{\partial\Omega_j} \hat{s}_h \cdot \mathbf{n}_j r\, dS = (kh)^2 \int_{\Omega_j} r\psi_h\, dV \tag{12.48}$$

$$\int_{\Omega_j} \mathbf{q}_h \cdot \mathbf{w}\, dV = -\int_{\Omega_j} \phi_h \nabla \cdot \mathbf{w}\, dV + \int_{\partial\Omega_j} \hat{\phi}_h \mathbf{n}_j \cdot \mathbf{w}\, dS \tag{12.49}$$

$$\int_{\Omega_j} \mathbf{q}_h \cdot \nabla w\, dV = \int_{\partial\Omega_j} \hat{\mathbf{q}}_h \cdot \mathbf{n}_j w\, dS \tag{12.50}$$

$$-\int_{\Omega_j} \mathbf{u}_h \cdot \nabla v\, dV + \int_{\partial\Omega_j} v\hat{\mathbf{u}}_h^p \cdot \mathbf{n}_j\, dS = 0 \tag{12.51}$$

$$\int_{\Omega_j} \underline{\tau}_h : \underline{\sigma}\, dV = -\int_{\Omega_j} \mathbf{u}_h \cdot \nabla \cdot \underline{\sigma}\, dV + \int_{\partial\Omega_j} \hat{\mathbf{u}}_h^\sigma \cdot \underline{\sigma} \cdot \mathbf{n}_j\, dS \tag{12.52}$$

$$Re \int_{\Omega_j} (\mathbf{u}_h)_t \cdot \mathbf{v}\, dV - Re \int_{\Omega_j} \mathbf{u}_h \cdot \nabla(\mathbf{v} \otimes \mathbf{u}_h)\, dV$$

$$+ Re \int_{\partial\Omega_j} \mathbf{u}_h \cdot \mathbf{n}_j \hat{\mathbf{u}}_h^c \cdot \mathbf{v}\, dS - \int_{\Omega_j} p_h \nabla \cdot \mathbf{v}\, dV + \int_{\partial\Omega_j} \hat{p}_h \mathbf{v} \cdot \mathbf{n}_j\, dS$$

$$+ \int_{\Omega_j} \underline{\tau}_h : \nabla \mathbf{v}\, dV - \int_{\partial\Omega_j} \hat{\underline{\tau}}_h : (\mathbf{v} \otimes \mathbf{n}_j)\, dS = \int_{\Omega_j} \mathbf{f} \cdot \mathbf{v}\, dV \tag{12.53}$$

As discussed in Chapter 5, the numerical fluxes need to be selected based on the diffusion and convection mechanisms. With appropriate fluxes selected, and the unknowns approximated using the local interpolation functions, the matrix equations can be obtained for the above integral equations,

$$\mathbf{K}\underline{\mathbf{U}}_{(j)} = \mathbf{F} \tag{12.54}$$

where $\underline{\mathbf{U}}_{(j)}$ is the vector containing the unknowns local to element j, ψ, $\mathbf{q}$, ϕ, $\mathbf{u}$, $\underline{\tau}$, and p. The solution procedure is iterative and involves an element-by-element sweep over the entire computational domain as detailed in Chapters 4 and 6.

The above algorithm has been applied to simulate electroosmotic flows in a cross-channel configuration. An unstructured triangular mesh is used, as shown in Figure 12.2a. The conditions used for the calculations are such that an insulated condition is imposed at all walls for ϕ, and $\phi=1$ at the left entrance, $\phi=0$ at the right outlet, and $\phi=0.5$ at the top and bottom outlets. For the potential field ψ due to the surface charge, the insulation condition is imposed at all entrances, and $\psi = -1$ at the walls. For the flow field, $\mathbf{u} = (1,0)$ at the entrance, and the outflow conditions are applied at all other outlets. The computed results are plotted in Figure 12.2.

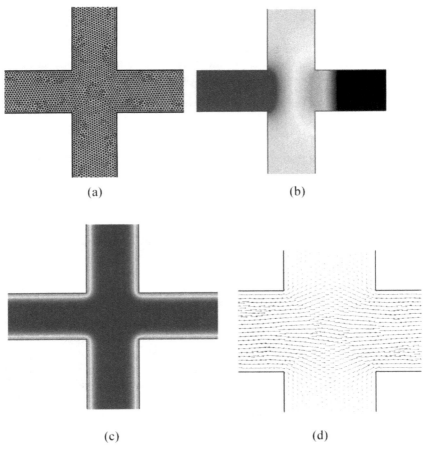

(a)   (b)

(c)   (d)

Figure 12.2. Computed results for electroosmotic flows: (a) unstructured mesh used for computations, (b) the potential distribution for the external electric field, (c) the electric potential distribtution due to charge distribution and (d) the velocity field

12.5 Microwave Heating

Microwave heating has been widely used in the food and materials processing industries, and in hyperthermia treatment of cancer patients. The essential idea of microwave heating is that the resonance excitation of the dipoles in dielectric materials generates a heating source. The governing equations for the electromagnetic and thermal field distributions in general consist of the frequency-based vector wave equation derived from the Maxwell equations and the energy equation [16],

$$\nabla \times \frac{1}{\mu_c} \nabla \times \mathbf{E} - k_0^2 \varepsilon_c \mathbf{E} = -j\omega\mu_0 \mathbf{J} \tag{12.55}$$

$$\rho C \frac{\partial T}{\partial t} = -\nabla \cdot (\kappa \nabla T) + \mathbf{J} \cdot \mathbf{E}^* \tag{12.56}$$

where superscript * denotes the complex conjugate. The thermal radiation boundary conditions are prescribed for the temperature solution. The condition for the electromagnetic field on the microwave cavity wall is such that the tangential electric field is zero,

$$\mathbf{E}_t = \mathbf{n} \times \mathbf{E} = 0 \tag{12.57}$$

where subscript t refers to the tangential component. Also, for the port with wave incident,

$$\mathbf{n} \times (\nabla \times \mathbf{E}) + \gamma \mathbf{n} \times (\mathbf{n} \times \mathbf{E}) = \mathbf{U}^{inc} \tag{12.58}$$

where $\mathbf{U}^{inc}$ is due to the incident wave and for the port with electromagnetic wave being transmitted,

$$\mathbf{n} \times (\nabla \times \mathbf{E}) + \gamma \mathbf{n} \times (\mathbf{n} \times \mathbf{E}) = 0 \tag{12.59}$$

where $\mu_c (=\mu/\mu_0)$ is the relative magnetic permeability, $\varepsilon_c (= \varepsilon' - j\sigma_e/(\varepsilon_0\omega))$ results from a combination of the induced current $(\sigma \mathbf{E})$ and the displacement current $(j\omega\varepsilon_0(\varepsilon'-j\varepsilon'')\mathbf{E})$, with $\sigma_e = \varepsilon_0\omega\varepsilon''+\sigma$, and k_0 the system parameter, $k_0^2 = \omega^2\mu_0\varepsilon_0$.

We consider a hybrid continuous edge finite element and discontinuous finite element method for the solution of the electromagnetic and thermal field distribution in a microwave heating system.

The edge finite element formulation starts with the three-dimensional wave equation in frequency domain, or Equation 12.55. After multiplying the equation by a vector weighting function $\mathbf{W}$ and integrating over the microwave cavity (or computational domain) and making use of the vector Green's theorem identity and the boundary conditions, we have the integral representation of the vector wave equation [17–19],

$$\iiint_{\Omega} \left(\frac{1}{\mu_r} \nabla \times \mathbf{E} \cdot \nabla \times \mathbf{W} - k^2 \varepsilon_c \mathbf{W} \cdot \mathbf{E} \right) dV = -\iiint_{\Omega} j\omega\mu \mathbf{W} \mathbf{J}_i \, dV$$

$$-\oiint_{\partial\Omega} \left[\gamma_e (\mathbf{n} \times \mathbf{E}) \cdot (\mathbf{n} \times \mathbf{W}) + \mathbf{W} \cdot \mathbf{U}^{inc} \right] dS \qquad (12.60)$$

Following the standard finite element procedure and using the edge elements to approximate the unknown field, we have the final global matrix,

$$\mathbf{K}\underline{\mathbf{E}} = \mathbf{F} \qquad (12.61)$$

where $\underline{\mathbf{E}}$ is the unknown vector of nodal values of the electric field over the entire computational domain. Note that Equation 12.61 results from conventional edge finite element formulation and cross-element continuity is strongly enforced. This is in contrast with the discontinuous finite element formulation, which weakly enforces the across-element continuity. The solution of Equation 12.61 is obtained using the sparse matrix solver coupled with the matrix rearrangement using the minimum degree dissection. The detailed solution procedure is given in a recent paper [20].

The temperature calculations are obtained using the discontinuous finite element formulation stated in Chapter 4. In this case, the Joule heating is treated as a heating source for the balance equation and easily incorporated once the electric field distribution is known. The computational procedure for this type of problem can be either iterative or hierarchical. If the electrical conductivity is a function of temperature, then the solution requires an iterative procedure between the temperature and electromagnetic calculations over each time step. If the electrical conductivity can be taken as a constant, then a hierarchical coupling is possible. In this case, the electric field needs to be calculated once, and the Joule heating source is input into the thermal balance equation for the temperature distribution calculations. A set of these calculations is given in Figure 12.3.

12.6 Electrically Deformed Free Surfaces

Here we consider a problem involving a deformation of droplet surfaces by an electric force. The application of this phenomenon has been in electrospray and electrostatic levitation. Figure 12.4 illustrates the system to be analyzed. An electrically conducting droplet is charged positively and placed in an electrostatic field generated by a pair of electrodes far apart. The field is positioned upward such that the Coulomb force resulting from the interaction of the charges and the applied electric field will be able to levitate the droplet in air by counterbalancing the downward gravitational force. Aside from supporting the weight of the droplet, the electric field will also produce a normal stress. This electric stress is distributed non-uniformly along the surface, causing the shape of the droplet to deform. The equilibrium surface of the droplet is determined by the balance of normal stresses acting on the surface, which include the Maxwell stress, the hydrostatic stress and surface tension effects. We note also that the tangential stress for the droplet under an isothermal condition is constant and thus does not induce an internal flow in the droplet. A thermally induced flow inside the droplet may also occur, the study of which has been presented elsewhere for electrostatic levitation applications [21]. Here we are concerned about the numerical calculation of the equilibrium shape of the droplet under both normal and microgravity conditions.

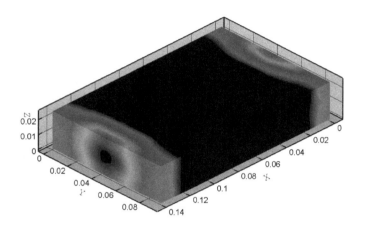

(a)   Calculated temperature profile

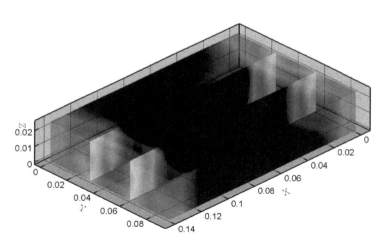

(b) Cut-view of calculated temperature distribution

Figure 12.3. Numerical results from a hybrid continuous edge finite element and discontinuous node element model for a microwave heating system used for food processing: (a) 3-D view of temperature distribution, and (b) cut-view of the temperature distribution in a food package subjected to microwave irradiation

For a general case of a droplet with a dielectric constant ε placed in an electric field, the Maxwell equations can be simplified in terms of a scalar potential or electric potential Φ,

$$\nabla^2\Phi = -\frac{\rho_e}{\varepsilon} \quad \in \Omega_1 \bigcup \Omega_2 \tag{12.62}$$

$$\Phi_1 = \Phi_2 \quad \in \Omega_1 \bigcap \Omega_2 \tag{12.63}$$

$$\varepsilon\frac{\partial\Phi_1}{\partial n} = \varepsilon_0\frac{\partial\Phi_2}{\partial n} \quad \in \Omega_1 \bigcap \Omega_2 \tag{12.64}$$

where ρ_e is the electrical charge and ε the electric permittivity. Here the domain of the droplet is denoted by Ω_1 and the free space outside the droplet by Ω_2. When Equation 12.62 applies to the free space, $\rho_e = 0$. If the droplet is electrically conducting, for example, a semiconductor or metal melt droplet, the whole droplet is at the same potential and all the charges are distributed on the surface of the droplet. Consequently, the above equations are simplified as

$$\nabla^2\Phi = 0 \quad \in \Omega_2 \tag{12.65}$$

$$\Phi = \Phi_{inside} = \Phi_0 \quad \in \Omega_1 \bigcap \Omega_2 \tag{12.66}$$

$$\varepsilon_0\frac{\partial\Phi}{\partial n} = -\sigma_e \quad \in \Omega_1 \bigcap \Omega_2 \tag{12.67}$$

$$\oiint_{\partial\Omega_2} \sigma_e ds = -\oiint_{\partial\Omega_2} \varepsilon_0\mathbf{n}\cdot\nabla\Phi ds = Q \quad \in \Omega_1 \bigcap \Omega_2 \tag{12.68}$$

$$\Phi = -Er\cos\theta \quad r \to \infty \tag{12.69}$$

where Φ_0 is a constant and is determined by Equation 12.69. In the above equations, Φ is the electric potential outside the droplet, σ_e is the surface charge, ε_0 the electric permittivity of the free space, E the applied electric field, and Q the total charge impressed on the droplet. Equation 12.68 represents the conservation of electric charges and the electric field generated by the electrodes is described by Equation 12.69.

The surface deformation of an electrically levitated droplet is determined by the balance of the electrostatic pressure, hydrostatic pressure and surface tension along the surface. This balance equation is given by

$$\gamma\nabla_s \cdot\mathbf{n} = \rho gz + P_0 - 0.5\varepsilon_0(\mathbf{n}\cdot\nabla\Phi)^2 \tag{12.70}$$

Equation 12.70 is basically the same as the normal stress balance equation discussed in Section 10.3, with modification made to take into account the electric effect.

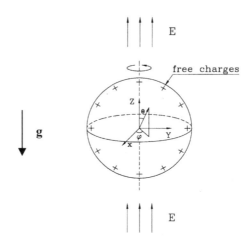

Figure 12.4. Schematic representation of an electrically conducting droplet in an applied electric field. The electric field generated by a pair of electrodes placed far apart points upward to counterbalance the effect of gravity

   To compute the free surface shapes, the electric pressure must be resolved first. This requires the solution of the electric field, which in turn is affected by the shape deformation. Thus, an iterative procedure is needed to solve the field distribution and the force balance equation simultaneously. To calculate the electric field defined by Equation 12.65 for an electrically conducting droplet, the entire free space outside the droplet must be considered. We consider the use of the boundary integral method for the electric field calculations, which is then coupled with the weighted residuals method for the determination of the equilibrium shape of the droplet.

   To develop a boundary integral formulation of the electric potential in Ω_2 that involves a boundary at infinity, it is perhaps more constructive to consider the closed computational domain as shown in Figure 12.5, and then to let the outer boundary approach to infinity. To facilitate the treatment of the boundry condition at infinity, we conisder a simple transformation, $\Phi_1 = \Phi + E_0 r \cos\theta$ such that $\Phi_1 = 0$ at infinity, a condition we need to take advantage of a boundary formulation. From the theory of electrostatics, the boundary integral formulation of the electric potential Φ may be obtained for any point r_i in Ω_2 using Green's theorem [2],

$$\frac{1}{2\pi}C(\mathbf{r}_i)\Phi_1(\mathbf{r}_i) + \oint_{\partial\Omega_2} \Phi_1(\mathbf{n}\cdot\nabla G)rd\Gamma + \oint_{\partial\Omega} \Phi_1(\mathbf{n}\cdot\nabla G)rd\Gamma = \cdot$$

$$\oint_{\partial\Omega_2} G(\mathbf{n}\cdot\nabla\Phi_1)rd\Gamma + \oint_{\overline{\partial\Omega}} G(\mathbf{n}\cdot\nabla\Phi_1)rd\Gamma + \int_{\Omega_2} \mu JGrd\Omega \qquad (12.71)$$

where $C(\mathbf{r}_i)$ is a geometric constant. The Green function G for an axisymmetric vector potential and its normal derivative can be derived from the consideration of a single current loop in free space [22],

$$G(\mathbf{r}_i,\mathbf{r}) = \frac{4}{\sqrt{(r_i+r)^2 + (z-z_i)^2}} K(\kappa) \qquad (12.72)$$

$$\frac{\partial G}{\partial n} = \frac{4}{\sqrt{(r+r_i)^2 + (z-z_i)^2}}$$
$$\times \left\{ \frac{n_r}{2r}[E(\kappa) - K(\kappa)] - \frac{n_r(r-r_i)+n_z(z-z_i)}{(r-r_i)^2 + (z-z_i)^2} E(k) \right\} \qquad (12.73)$$

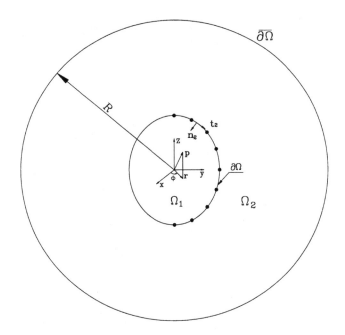

Figure 12.5. Boundary element discretization. Ω_2 is the exterior region modeled by boundary elements, $\partial\Omega_2$ the interface, $\overline{\partial\Omega}$ the boundary at infinity, and • the node points along the interface, which are used for both the electric field and shape deformation calculations

where $K(\kappa)$ and $E(\kappa)$ are the elliptical integrals of the first and second kinds, respectively, and κ the geometric parameter defined as

$$\kappa^2 = \frac{4r_i r}{(r_i + r)^2 + (z_i - z)^2} \tag{12.74}$$

The two integrals involving $\overline{\partial\Omega}$ represent the contribution from the boundary at $R \to \infty$. With the following asymptotic behavior of G and Φ_1,

$$\Phi_1(r_i, R) \approx O(R^{-2}) , \quad \frac{\partial\Phi_1}{\partial n}(r_i, R) \approx O(R^{-3}) \text{ as } R \to \infty \tag{12.75}$$

$$G(r_i, R) \approx O(R^{-2}) , \quad \frac{\partial G}{\partial n}(r_i, R) \approx O(R^{-3}) \text{ as } R \to \infty \tag{12.76}$$

and also $d\Gamma = R(\theta) d\theta$, it is straightforward to show that the two integrals each approach to zero as $R \to \infty$,

$$\oint_{\partial\Omega} \Phi_1 (\mathbf{n} \cdot \nabla G) r d\Gamma \to 0 \text{ and } \oint_{\partial\Omega} G(\mathbf{n} \cdot \nabla\Phi_1) r d\Gamma \to 0 \text{ as } R \to \infty \tag{12.77}$$

With this relation, Equation 12.71 can be simplified to involve integrals along the boundary of the droplet only. This way, the condition at $R \to \infty$ is directly incorporated into the boundary integral formulation.

We now apply the inverse transformation, $\Phi = \Phi_1 - Er\cos\theta$ and obtain the final boundary integral equation for the electric potential Φ,

$$C_i\Phi(\mathbf{r}_i) + \int_{\partial\Omega_2} \Phi\frac{\partial G}{\partial n} d\Gamma = \int_{\partial\Omega_2} \left[\frac{\partial\Phi}{\partial n}G + En_zG - Ez\frac{\partial G}{\partial n}\right] d\Gamma - C_iEz_i \tag{12.78}$$

Following the boundary element discretization [23, 24] and recognizing that the potential on the surface is a constant, one has the final matrix form for the unknowns on the surface of the droplet,

$$\mathbf{H}\{\Phi_0\} = -\mathbf{G}\left\{\frac{\partial\Phi}{\partial n}\right\} + E\mathbf{G}\left\{\frac{\partial z}{\partial n}\right\} - \mathbf{H}E\{z\} \tag{12.79}$$

This is then solved along with Equation 12.68.

For the purpose of droplet shape calculations, the normal stress balance equation (i.e., Equation 12.70) may be more conveniently written in spherical coordinate system,

$$\frac{1}{r^2 \sin \theta} \left[\frac{\left(2r^2 + r_\theta{}^2\right)\sin \theta}{\sqrt{r^2 + r_\theta{}^2}} - \frac{d}{d\theta}\left(\frac{rr_\theta \sin \theta}{\sqrt{r^2 + r_\theta{}^2}} \right) \right] = -K - Br \cos \theta - \frac{a}{\gamma} P_m$$

(12.80)

where a is the radius of the undeformed sphere, r the non-dimensionalized radial coordinate, $K = a\, P_o/\gamma$, $B = \rho g a^2/\gamma$ and $P_m = -\varepsilon_0 (\mathbf{n} \cdot \nabla \Phi)^2/2$ for electrostatically levitated droplets. The Weighted Residuals method may be applied to solve the above equation once the potential field distributions are known. To do that, the surface of the droplet is discretized and defined by r_i, the distance between the surface node and the center of the droplet, as shown in Figure 12.5. The solution of Equation 12.80 by the Weighted Residual method is constructed by integrating the equation with respect to a weight function ψ_i along the droplet surface,

$$\int_{\partial \Omega_1} \left\{ \frac{1}{r^2 \sin \theta} \left[\frac{\left(2r^2 + r_\theta{}^2\right)\sin \theta}{\sqrt{r^2 + r_\theta{}^2}} \right. \right.$$

$$\left. \left. - \frac{d}{d\theta}\left(\frac{rr_\theta \sin \theta}{\sqrt{r^2 + r_\theta{}^2}} \right) \right] + K + Br \cos \theta + \frac{a}{\gamma} P_m \right\} \psi_i ds = 0 \qquad (12.81)$$

where $r_\theta = dr/d\theta$. Integrating by parts, one reduces the order of derivatives by one, that is, the second order derivative is reduced to the first order derivative. This allows Equation 12.81 to be written as

$$\int_0^\pi \left[\frac{rr_\theta \dfrac{d\psi_i}{d\theta} + \psi_i \left(2r^2 + r_\theta^2\right)}{\sqrt{r^2 + r_\theta^2}} + r^2 \psi_i \left(K + Br \cos \theta + \frac{a}{\gamma} P_m \right) \right] \sin \theta \, d\theta = 0$$

(12.82)

The variables r and r_θ are interpolated by

$$r = \sum_{i=1}^{N_e} \psi_i r_i \, ; \qquad\qquad r_\theta = \sum_{i=1}^{N_e} r_i \frac{d\psi_i}{d\xi} \frac{d\xi}{d\theta} \qquad (12.83)$$

where N_e is the number of nodes per element and ψ_i is the shape function.

The constraints of the volume conservation and the center of the mass of the levitated sphere are needed to determine the shape and position of the droplet. The aspect of imposing constraints for free surface problems was discussed in Chapter 10. In dimensionless form, the two constraints are expressed as

$$\int_0^\pi r^3 \sin \theta \, d\theta = 2 \qquad (12.84a)$$

$$\frac{3}{8}\int_0^\pi r^4 \cos\theta \sin\theta d\theta = z_c \tag{12.84b}$$

where z_c is the center of mass. The free surface may now be discretized into N elements and Equations 12.82–12.84 are integrated numerically. Both continuous and discontinuous elements can be applied. If a continuous element is applied, the results may be arranged in the following matrix form:

$$\begin{bmatrix} a_{1,1} & a_{1,2} & & & & & b_1 \\ a_{2,1} & a_{2,2} & a_{2,3} & & & & b_2 \\ & \cdot & \cdot & \cdot & & & \\ & & \cdot & \cdot & \cdot & & \\ & & a_{n-1,n-2} & a_{n-1,n-1} & a_{n-1,n} & b_{n-1} \\ & & & a_{n,n-1} & a_{n,n} & b_n \\ b_1 & b_2 & \cdot & \cdot & b_{n-1} & b_n & 0 \\ c_1 & c_2 & \cdot & \cdot & c_{n-1} & c_n & 0 & -1 \end{bmatrix} \begin{Bmatrix} r_1 \\ r_2 \\ \cdot \\ \cdot \\ r_{n-1} \\ r_n \\ K \\ z_c \end{Bmatrix} = \begin{Bmatrix} F_1 \\ F_2 \\ \cdot \\ \cdot \\ F_{n-1} \\ F_n \\ 2 \\ 0 \end{Bmatrix} \tag{12.85}$$

where the coefficients a_{ij}, b_j, c_j and F_j are calculated by

$$a_{i,j} = \int_{-1}^1 \left\{ \frac{\psi_i\left(2r\psi_j\left(\frac{d\theta}{d\xi}\right)^2 + \frac{dr}{d\xi}\frac{d\psi_j}{d\xi}\right) + \psi_j\frac{d\psi_i}{d\xi}\frac{dr}{d\xi}}{\sqrt{\left(\frac{dr}{d\xi}\right)^2 + \left(r\frac{d\theta}{d\xi}\right)^2}} + r\psi_i\psi_j\frac{aP_m}{\gamma}\left|\frac{d\theta}{d\xi}\right| \right\} \sin\theta d\xi \tag{12.86}$$

$$b_i = \int_{-1}^1 r^2\psi_i \sin\theta\left|\frac{d\theta}{d\xi}\right| d\xi \tag{12.87}$$

$$c_j = \frac{3}{8}\int_{-1}^1 r^3\psi_j \cos\theta \sin\theta\left|\frac{d\theta}{d\xi}\right| d\xi \tag{12.88}$$

$$F_i = -\int_{-1}^1 Br^3\psi_i \cos\theta \sin\theta\left|\frac{d\theta}{d\xi}\right| d\xi \tag{12.89}$$

Both discontinuous and continuous element approximations have been used for calculations. The accuracy for both approximations is the same, as expected. It is noted that even for continuous element approximations, the discontinuous elements provide a natural choice for the fluxes at the sharp corners, where a discontinuity in flux occurs. While other types of approximation may also be made to treat the discontinuity (for example, the finite difference approximation may be used just for the elements having their notes at the corners), numerical experience suggests that discontinuous elements perform the best for this type of problem [24].

The free surface deformation of a copper droplet levitated in a uniform electric field against gravity is shown in Figure 12.6. The droplet is charged positively and the total charge is 1.56×10^{-9} (C) which is within the limit of $Q_c = 1.06 \times 10^{-6}$ (C) for droplet rupture [25]. The droplet deformation is such that the lower part of the surface is flatter and the droplet points upward in the direction opposite to that of gravity. For the system shown in Figure 12.4, the electrostatic force (the electric charge times the electric field) acting on the lower portion of the droplet is smaller and in the negative z direction, while the force on the upper portion is bigger and in the positive z direction. This, combined with gravity and the surface tension effect, gives the final shape as shown in Figure 12.6.

As a contrast, Figure 12.6 also plots the surface deformation of a copper droplet in microgravity. This case represents a somewhat idealized situation in that there is no free charge impressed on the droplet against gravity. The deformation of the droplet in this case is caused by the interaction of the induced charges on the surface of the droplet with the applied electric field. As the induced surface charges are distributed symmetrically, the electrostatic forces are equal in magnitude and acting in the opposite directions on the lower and upper portions of the droplet, causing the droplet to deform symmetrically.

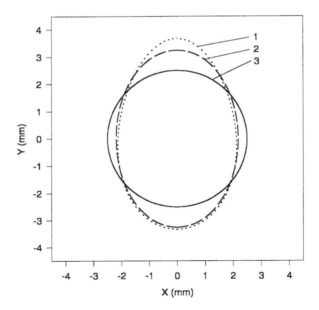

Figure 12.6. Comparison of free surface profiles of a Cu droplet in normal and microgravity: (1) $E_0 = 3.3 \times 10^6$ V/m and $Q = 1.56 \times 10^{-9}$ C (Coulombs) (normal gravity), (2) $E_0 = 3.3 \times 10^6$ V/m and $Q = 0$ C (microgravity gravity), and (3) un-deformed liquid sphere

12.7 Compressible Flows in Magnetic Fields

As a last numerical example, we consider a full 3-D calculation of compressible flows under the influence of magnetic fields. The discontinuous finite element algorithm for this type of calculation has been developed by Warburton and Karniadakis [26]. A locally divergence-free discontinuous finite element formulation was also presented in a recent paper by Li and Shu [27]. We consider below some of the essential ideas of the algorithm given by Warburton and Karniadakis and details of the algorithm development can be found in their original paper.

The equations for compressible magnetohydrodynamics (MHD) may be written in conservative form using fluxes,

$$\frac{\partial \mathbf{U}}{\partial t} = -\frac{\partial \mathbf{F}_x^{Ideal}}{\partial x} - \frac{\partial \mathbf{F}_y^{Ideal}}{\partial y} - \frac{\partial \mathbf{F}_z^{Ideal}}{\partial z} + \frac{\partial \mathbf{F}_x^{Visc}}{\partial x} + \frac{\partial \mathbf{F}_y^{Visc}}{\partial y} + \frac{\partial \mathbf{F}_z^{Visc}}{\partial z} \qquad (12.90)$$

$$\nabla \cdot \mathbf{B} = 0 \qquad (12.91)$$

$$\mathbf{U} = (\rho, \rho u, \ \rho v, \rho w, B_x, B_y, B_z, E) \qquad (12.92)$$

where the flux function is split into the inviscid and viscous fluxes, an approach similar to that discussed in Chapter 5 for general convection–diffusion problems.

The presence of the constraint $\nabla \cdot \mathbf{B} = 0$ implies that the equations do not have a strictly hyperbolic character and thus require special treatment. Two ideas have been implemented. One is based on the introduction of a stream function for the magnetic field, similar to that used in incompressible flow calculations discussed in Chapter 6. The other is to reformulate the Jacobian matrix to include the divergent mode, which is suitable for 3-D simulations. This term is then included as a source contribution.

The inviscid fluxes and their derivatives in the interior of the elements are evaluated and the correction terms (jumps) for the discontinuities in the flux between any two adjacent elements are added. A one-dimensional Riemann solver is used to model a numerical flux at the element interface. At a domain boundary the specified conditions are used. The exterior boundary is treated as the boundary of a "ghost" element, so that the same Riemann solver can be used at all element boundaries.

The viscous terms are treated in two steps. First, the spatial derivatives of the primitive variables are calculated using the discontinuous Galerkin approach. This process is then repeated for each of the viscous fluxes using these derivatives. The Dirichlet boundary conditions for the momentum and energy characteristic variables are imposed weakly or explicitly after the fluxes have been evaluated and then the result is projected using the orthogonal basis.

The main algorithm was developed by Karniadakis and Warburton for the discontinuous finite element solution of Equations 12.90–12.92, which involves 12 major steps:

- Step 1. Read in initial conditions $\mathbf{U}(\mathbf{x}, 0)$ and evaluate all fields at all element quadrature points. Set $n = 0$.
- Step 2. Calculate the fluxes $\hat{\mathbf{F}}_n$ at the Gauss quadrature points Q^I on the element interfaces. At domain boundaries use the prescribed boundary conditions for the exterior values of the fields. Interpolate the fluxes $\hat{\mathbf{F}}_n$ to the quadrature points Q^E. Scale the fluxes with the edge Jacobian divided by the volume Jacobian.
- Step 3. Calculate the inviscid flux terms $\mathbf{F}_x^{Ideal}$, $\mathbf{F}_y^{Ideal}$, $\mathbf{F}_z^{Ideal}$ at the element quadrature points.
- Step 4. For each component of the state vector $\mathbf{U}_k = (\mathbf{U}(\mathbf{x}, t^n))_k$ calculate $(\partial \mathbf{F}_x^{Ideal} / \partial x + \partial \mathbf{F}_y^{Ideal} / \partial y + \partial \mathbf{F}_z^{Ideal} / \partial z)$.
- Step 5. Form $\hat{\mathbf{F}}_n - \mathbf{F}_I$ (where F_I is the flux at interior edge side) and add it to the divergence of the inviscid fluxes calculated in Step 4.
- Step 6. Calculate the spatial derivatives of the primitive fields.
- Step 7. Use the derivatives of the primitive fields to construct the viscous flux terms $\mathbf{F}_x^{Visc}$, $\mathbf{F}_y^{Visc}$, and $\mathbf{F}_z^{Visc}$.
- Step 8. Take the divergence of the viscous flux terms and subtract the results of Step 5.
- Step 9. Take the inner product of the result from Step 8 with the orthogonal basis. Evaluate the resulting polynomials at the quadrature points and place it in $\mathbf{Uf}(\mathbf{x}, t^{n-q})$.
- Step 10. Update the vector of the unknowns, that is, $\mathbf{U}(\mathbf{x}, t^{n+1}) = \mathbf{U}(\mathbf{x}, t^n)$ $+ \Delta t \sum_q \beta_q \mathbf{Uf}(\mathbf{x}, t^{n-q})$ using an Adams–Bashforth integration scheme.
- Step 11. Increase n by one. If $t^n <$ the termination time return to Step 2.
- Step 12. Output final values of the state vector $\mathbf{U}(\mathbf{x}, t^{end})$.

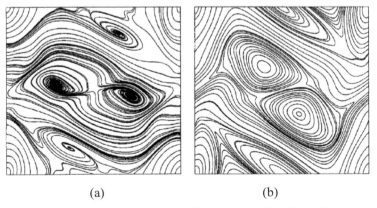

(a)   (b)

Figure 12.7. Computed results for compressible magnetically driven flows in a 2-D geometry: (a) flow stream lines and (b) magnetic stream lines [24]

Their algorithm has been tested for a 2-D compressible magnetohydrodynamic problem and the results shown in Figure 12.7.

Exercises

1. Consider a conducting medium (1) in air (2). The medium is charged with surface charge density being σ_s. Create a Gaussian surface as appears in the figure below. Show that

$$E_s = \frac{\sigma_s}{\varepsilon_2} \quad \text{(points in the outnormal direction)}$$

and that the normal stress due to the charge and E_s is given by

$$F_s \mathbf{n} = \sigma_s E_s \mathbf{n} = \frac{\sigma_s^2}{2\varepsilon_0} \mathbf{n}$$

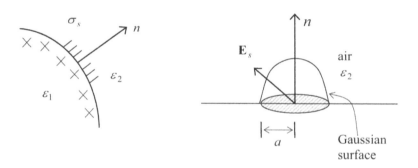

This force tends to pull the interface towards the air, and to tear the medium one apart. Further show that the normal stress balance along the interface is given by

$$-P_0 + \rho g \psi + Pa - \frac{\sigma_s^2}{2\varepsilon_0} = 2H\gamma$$

with H being the mean curvature.
2. Develop a discontinuous finite element code for the simulation of electroosmotic flows in a microchannel following the integral formulation given in Section 12.5.
3. Derive the boundary conditions (Equations 12.9–12.12) from the Maxwell equations.
4. Assuming that the electromagnetic field oscillates time-harmonically with a frequency of ω, derive Equation 12.26 from the Maxwell equations.
5. Starting with a differential volume, and applying the Newton's second law, derive Equation 12.22.

6. Following the discontinuous finite element solution procedures, obtain the detailed matrix expression for the operator L in Equation 12.25.

References

[1]   Reitz JR, Milford FJ, Christy RW. Foundations of Electromagnetic Theory. Reading MA: Addison-Wesley, 1992.
[2]   Jackson JD. Classical Electrodynamics, New York: John Wiley, 1975.
[3]   Landau LD. Electrodynamics of Continuous Media. Oxford: Pergamon Press, 1960.
[4]   Collin RE. Field Theory of Guided Waves. New York: IEEE Press, 1991.
[5]   Van Bladel J. Electromagnetic Fields. New York: McGraw Hill, 1964.
[6]   Kopriva DA, Woodruff SL, Hussaini MY. Discontinuous Spectral Element Approximation of Maxwell Equations. In: Cockburn B, Karniadakis GE, Shu CW, editors. Discontinuous Galerkin Method: Theory, Computation and Applications. Lecture Notes in Computational Science and Engineering, 11. New York: Springer-Verlag, 2000.
[7]   Hesthaven JS, Warburton T. Nodal High-Order Methods on Unstructured Grids, Time-Domain Solution of Maxwell's Equations. J. Comput. Phys. 2002; 181: 186–221.
[8]   Chen MH, Cockburn B, Reitich F. High-order RKDG Methods for Computational Electromagnetics. J. Sci. Comp. 2005; 22(1): 205–226.
[9]   Cockburn B, Fand L, Shu, CW. Locally Divergence-Free Discontinuous Galerkin Methods for the Maxwell Equations. J. Comput. Phys. 2004; 194: 588–610.
[10] Perugia I, Schötzau D, Monk P. Stabilized Interior Penalty Methods for the Time-Harmonic Maxwell Equations. Comp. Meth. Appl. Mech. Eng. 2002; 191: 4675–4697.
[11] Houston P, Perugia I, Schötzau D. Mixed Discontinuous Galerkin Approximation of the Maxwell Operator: Non-Stabilized Formulation. J. Sci. Comp. 2005; 22(1): 315–346.
[12] O'Bockis JM, Reddy AKN. Modern Electrochemistry. New York: Plenum Publishing Company, 1977.
[13] Santiago JG. Electroosmotic Flows in Microchannels with Finite Inertial and Pressure Forces. Anal. Chem. 2001; 73: 2353–2365.
[14] Yan F. Numerical Simulations of High Knudsen Number Gas Flows and Microchannel Electrokinetic Liquid Flows. Ph.D. Thesis. Drexel University, 2003.
[15] Olesen LH. Computational Fluid Dynamics in Microfluidic Systems. MS. Thesis, Philadelphia: Technical University of Denmark, 2003.
[16] Metaxas AC. Foundations of Electroheat: A Unified Approach. New York: John Wiley & Sons, 1996.
[17] Volakis JL, Chatterjee A, Kempel LC. Finite Element Method for Electromagnetics. New York: IEEE Press, 1998.
[18] Jin J. The Finite Element Method in Electromagnetics. New York: Wiley, 1993.
[19] Dibben DC, Metaxas AC. Finite Element Time Domain Analysis of Multimode Applicators Using Edge Elements. Journal of Microwave Power and Electromagnetic Energy 1994; 29: 242–251.
[20] Huo Y, Li BQ, Akarapu R. A finite Element-Boundary Integral Method for 3-D Electromagnetic Heating Analysis. In: ASME National Heat Transf. Conf., Charlotte, NC, 2004, Paper #: HT-FED2004-56392.
[21] Song SP, Li BQ. Free Surface Profiles and Thermal Convection in Electrostatically Levitated Droplets. Int. J. Heat Mass Transf. 2000; 43(19): 3589–3606.

[22] Smith WR. Static and Dynamic Electricity. New York: McGraw-Hill, 1968.
[23] Li BQ, Evans JW. Boundary Element Solution of Heat Convection-Diffusion Problems. J. Comput. Phys. 1991; 93(2): 255–272.
[24] Song SP, Li BQ. A Coupled Boundary/Finite Element Method for the Computation of Shapes Of Magnetically and Electrostatically Levitated Droplets. Int. J. Numer. Meth. Eng. 1999; 44(8): 1055–1077.
[25] Adornato PM, Brown, RA. Shape and Stability of Electrostatically Levitated Drops. Proc. R. Lond. A. 1983; 389: 101–117.
[26] Warburton TC, Karniadakis GE. A Discontinuous Galerkin Method for the Viscous MHD Equations. J. Comput. Phys. 1999; 152: 608–641.
[27] Li F, Shu CW. Locally Divergence-Free Discontinuous Galerkin Methods for MHD Equations. J. Sci. Comp. 2005; 22(1): 413–442.

Index

Printed in the USA
CPSIA information can be obtained
at www.ICGtesting.com
LVHW011318201223
766701LV00084B/69

9 781852 339883